T0177769

The Life and Work of James Bradley

The Life and Work of James Bradley

The New Foundations of 18th Century Astronomy

JOHN FISHER

Great Clarendon Street, Oxford, OX2 6DP,
United Kingdom

Oxford University Press is a department of the University of Oxford.
It furthers the University's objective of excellence in research, scholarship,
and education by publishing worldwide. Oxford is a registered trade mark of
Oxford University Press in the UK and in certain other countries

Published in the United States of America by Oxford University Press
198 Madison Avenue, New York, NY 10016, United States of America

British Library Cataloguing in Publication Data

Data available

Library of Congress Control Number: 2023934090

ISBN 9780198884200

DOI: 10.1093/oso/9780198884200.001.0001

Printed and bound by
CPI Group (UK) Ltd, Croydon, CR0 4YY

For Deborah

Melius sero quam numquam.
(Better late than never.)

Foreword

If asked about famous eighteenth-century astronomers, most members of the British general public would be able to identify Edmond Halley, the second Astronomer Royal, who in 1705 showed that the eponymous comet of 1682 had returned at roughly seventy-six-year intervals and would continue to do so. Many would mention Isaac Newton, inventor of the first working reflecting telescope and author of the 1687 *Principia Mathematica*, whose contents paved the way for the dramatic flourishing of astronomy in the following decades. At the other end of the eighteenth century, the scientific achievements of the one-time musician William Herschel remain astonishing; among other intellectual feats he discovered Uranus in 1781 (and later two of its moons, as well as two of Saturn's), identified thousands of nebulae, and discovered infrared radiation while pioneering the measurement of stellar spectra. Some might have heard of Nevil Maskelyne, the fifth Astronomer Royal, but only because he is rather unfairly cast as the anti-hero to the great horologist John Harrison in Dava Sobel's 1995 book *Longitude*. Still others will know the names of Charles Mason and Jeremiah Dixon, whose surveying and astronomical work in North America from 1763 to 1767 was the basis for what came to be called the Mason–Dixon line, the boundary between the northern 'free' states and the southern slave-owning states following the Missouri Compromise of 1820.

As John Fisher shows in his excellent book, it is extraordinary that the name and accomplishments of James Bradley (1692–1762), the third Astronomer Royal, are not better known. Bradley was the outstanding positional astronomer of his time and was recognized as such for many decades after his death. His reputation was sealed by the publication of his paper on a 'new discovered motion of the fixed stars' at the start of 1729; aside from providing overwhelming evidence in favour of the Copernican system, the discovery of the apparent annual motion of stars (which he later termed 'aberration') also provided the basis for producing an accurate calculation of the speed of light. In the 1730s, Bradley's expertise was crucial in corroborating the pro-Newtonian findings of various scientific expeditions regarding the oblate spheroidal shape of the earth. However, his reputation for the unrivalled exactitude of his work was sealed by his announcement in 1748 of the existence of 'nutation', a oscillation in the cyclical 'wobble' of the earth that gives rise to the precession of the equinoxes. In 1818, Friedrich Bessel published *Fundamenta Astronomiae*, his great work that reduced the positions of 3,222 stars (to a 1755 base) from data produced by Bradley and his assistants at Greenwich between 1750 and 1762. In the title of Bessel's work, Bradley was hailed as 'that incomparable man'.

Born in Sherborne, Gloucestershire, Bradley attended the Grammar School at Northleach and went up to Balliol College Oxford in 1711. Two years after the award of an MA in 1717, he became vicar of Bridstow, but by then he had already fallen in love with astronomy, and his tenure as a cleric was short lived. Bradley's first astronomical observations were made in 1712, and from 1715 he worked closely with his maternal uncle, the astronomer James Pound, at Pound's observatory in Wanstead in Essex. His precocious talent was recognized by his election as a Fellow of the Royal Society in 1718, and in the following years he engaged in an intense study of the satellites of Jupiter, passing on his observations to Halley for use as part of his efforts to find a method that would allow the determination of longitude at sea.

Two inanimate objects stand out as constant companions in Bradley's endeavours, without which his discoveries would have been impossible. The first was the orange giant γ *Draconis* (Eltanin), the star chosen by Robert Hooke in the late 1660s in his attempt to prove the motion of the Earth by determining its annual parallax. Assuming that stars were close enough to the Earth for parallax to be detected, the truth of the Copernican system might be shown by a shift of position of a star against the background of other stars when observations were taken at diametrically opposed positions in the Earth's orbit (i.e. six months apart). γ *Draconis* was the star of choice for London-based astronomers because it was a second-magnitude star that passed very close to the zenith. Since the observation of such stars were least affected by atmospheric refraction, their movements could be tracked more accurately than others. γ *Draconis* would be the most important celestial object for Bradley in the programme of observations that underpinned his accounts of aberration and nutation.

The second was the 12.5 foot zenith telescope or sector made for Bradley by the great instrument-maker George Graham. The sector enabled astronomers to make highly accurate measurement of stars passing near the zenith, and Samuel Molyneux, a close associate of Pound and secretary to the Prince of Wales (George II from 1727), had commissioned a 24.5 foot sector from Graham in 1725 to examine the motion of γ *Draconis* at his home at Kew. Bradley's smaller instrument, which was set up at Elizabeth Pound's house in

August 1727, had a much greater freedom of movement either side of the zenith than the sector at Kew (more than 6° each way), allowing many more stars to be taken into account.

The support of Halley and Pound led to Bradley's election as Savilian Professor of Astronomy at Oxford in 1721, but he spent most of his time during the 1720s at Wanstead and Kew. *y Draconis* was the focus of Molyneux and Pound's work at Kew in the early 1720s, which was designed in part to test Hooke's claims about stellar parallax. Following the death of Pound in 1724, Bradley continued to work on the project with Molyneux and Graham at Kew, and he detected anomalous southwards movements of *y Draconis* almost immediately following the installation of the sector in November 1725. Molyneux initially believed that they confirmed Hooke's conclusions about stellar parallax, which implied that *y Draconis* and other stars were much closer to the solar system than Newton had allowed in his theory of the heavens. As Molyneux rather unwisely told Newton himself, the findings suggested that the universe was much more densely crowded with stars and more unstable than Newton had claimed in the *Principia*. Although Newton apparently told Molyneux that there was no argument against empirical evidence, he had shown to his satisfaction in the *Principia* that the closest stars to the solar system were too far away for any parallax to be observed.

As Fisher points out, a substantial part of Bradley's great gifts as an astronomer was the fact that he was comfortable working to a level of precision of one arc second or even less, when many contemporaries such as Halley dismissed such precision as irrelevant or impossible to achieve. In the case of *y Draconis*, Bradley detected 'entirely unexpected' changes in its position of one arc second every three days, and having rejected Molyneux's idea that this movement was evidence of stellar parallax, he considered that it might be due to nutation. However, by comparing it with another star (35 Camelopardalis), he rejected this and other hypotheses, convinced nevertheless that the motions of the two stars conformed to some regular pattern. It was as a result of this that he decided to commission the second instrument from Graham for use at Wanstead. Over two decades, Bradley performed painstaking observations that combined a commitment to precision with a concern that his instruments should be capable of providing the most accurate data possible. As Fisher shows with great clarity, the annual apparent motion of the general law of aberration emerged as a result of detailed observations of a number of stars, and only later did Bradley concoct his explanation for it.

Bradley's other great discovery, that of nutation, stemmed from his initial observations of *y Draconis* at Kew, but ultimately was determined by a research programme at Wanstead that lasted over two decades. Early on he had rejected nutation as an explanation of the 'unexpected' movement of γ *Draconis* and other stars, but his attention to detail made him aware that aberration did not explain all of the anomalous motions he witnessed. In this case, stars had changed their declinations a few arc seconds *less* than the 50.4 arc seconds of annual movement caused by the precession of equinoxes. Again, Bradley checked to see if his instrument was responsible for deviations from expected values but having found that it was not, he set out to ascertain whether these new motions conformed to a periodic law. Fisher's fine account of this research shows that Bradley's belief that the anomalous motions were due to a lunar-influenced nutation had hardened by the early 1730s, and he was now convinced that the nutation was linked to the period of revolution of the motions of the nodes of the lunar orbit (lunar nodes are those points where the orbiting moon crosses the plane of the earth's equator). This being 18.6 years, Bradley decided to perform observations for slightly longer than this, in order to demonstrate that the phenomenon conformed to a general law. He announced the 'discovery' of nutation in a letter addressed to the Earl of Macclesfield and read to the Royal Society in January 1748. He was awarded the Copley Medal, the Society's highest award, a few months later.

Bradley's researches into the phenomena of aberration and nutation ought to be better known in the history of science, since they exemplify the close entanglements between state-of-the-art observation and the development of astronomical theory. Although one of his discoveries concerned an 'apparent' motion of the stars, and the other a 'real' motion of the earth, Bradley showed that they could both be subsumed under general laws. Each of these findings provided valuable information about the real world, confirmed Newtonian theory, and could be used to correct observations. The 'discoveries' were thus not momentary 'aha' moments that occurred to a prepared mind, but the results of fastidiously performed observational programmes designed to prove a hunch that over time became a conviction.

In the 1730s, Bradley was one of the most eminent members of the Royal Society. In the early 1730s, he persuaded Graham to supply a zenith sector to the impending French expedition to Lapland, which was to measure the length of a degree in as northerly a latitude as was possible. The venture, led by the French natural philosopher Pierre-Louis Moreau de Maupertuis, was designed to test the validity of Newton's demonstration that the earth was an oblate spheroid, in contrast with the generally held French view that it was a prolate spheroid. Immediately following the conclusion of the expedition in 1737, Maupertuis asked for Bradley's help in applying his theory of aberration to correct the observational data collected from the French expedition to Lapland. Bradley was delighted that the Lapland data confirmed Newton's account but disappointed that the French team had not reversed the sector to test its accuracy in extreme conditions. Here he demonstrated his characteristic concern with the reliability of instrumentation and with the need to remove every possible error from its use. Once installed as Astronomer Royal in the following decade, he would insist that every observation taken by himself or his assistants would be accompanied by barometric and thermometric data, accompanied by corrections for aberration and nutation, and it was this information that would be used to such good effect by Bessel.

By the end of the 1730s, it was clear that the octogenarian Halley was ailing, and that he was no longer able to perform the duties of Astronomer Royal at Greenwich. Bradley was the obvious successor when Halley died in 1742 and, having gained substantial experience from stocking the Earl of Macclesfield's excellent observatory at Shirburne Castle, he was in a prime position to assess the state of the equipment and buildings at Greenwich. He immediately set out to repair old instruments and stock the Observatory with new ones, ordering items from craftsmen, such as the clockmaker John Shelton, the telescope-maker James Short, and John Bird, who made a new transit instrument for the Observatory in 1750. Bradley paid for these items with a grant of £1,000 that the king allowed him in 1749, and at the same time he designed the 'New Observatory' that was erected in 1749/50. For £300 Bird built a new brass mural quadrant to replace the iron version that had damaged the supporting wall and was now useless, while Bradley procured £45 from the total in order to bring the Wanstead sector to the Observatory. One of the most significant instruments ever deployed in the history of science, it now sits on the west wall of the Transit Room at the Old Royal Greenwich Observatory.

Beyond Greenwich and Wanstead, Bradley spent at least three months every year at Oxford, where – in addition to his position as Savilian Chair – he had been appointed lecturer in experimental philosophy in 1729. He gave three courses of twenty lectures each year until 1749, and then two per year until 1760. The lectures offered theoretical and practical demonstrations of the doctrines found in Newton's *Principia* and *Opticks*, though there was little astronomy in them. Because more than forty per cent of students in any year signed up to hear him, Bradley made a tidy sum from these lectures (c. £400 p.a.) which supplemented his salary as Savilian Professor (£150 p.a.) and the pension of £250 p.a. he was awarded by George II after 1752. A surviving student notebook from a course given in 1754 shows that Bradley closely followed Newton and his public defenders in appealing to the authority of rigorously conducted observations and experiments, and in inveighing against hypotheses and systems. Although hundreds of students attended the course over three decades, the content of Bradley's lectures was held by a number of reactionary scholars to be inimical to true religion.

Concerning Bradley the private man, there remains a conspicuous lack of direct evidence. In 1832, Stephen Rigaud, Savilian Professor of Astronomy and editor of Bradley's *Works and Correspondence*, appended a final chapter on Bradley's family to his account of the latter's scientific achievements. He started with Bradley's marriage in 1744 at the age of 51 to Susannah Peach, a union that produced a daughter in the following year. His wife died in 1757, and Rigaud noted that he had afterwards remained on good terms with her family, though he added that Bradley suffered greatly from abdominal pains and 'the last two years of his life were spent under a melancholy depression of spirits'. Other than that, Rigaud recorded that in his prime Bradley was capable of tremendous periods of concentration, and he was 'humane, benevolent and kind'. Fisher points out, as Rigaud could not, that Pound's wife Elizabeth accompanied Bradley to the Savilian Chair's lodgings at New College in 1732 and lived with him until soon before she died in 1740. Bradley was prevented by law from marrying his uncle's widow, but waited until she died before courting and marrying Susannah. It is likely that concerns about the family's reputation lay behind their refusal to release Bradley's personal papers, whose whereabouts – if they still exist – remain unknown.

For over three decades, John Fisher has been the leading expert on Bradley's life and work, and his biography is definitive, superseding Rigaud's comprehensive introductory biography to the edition of Bradley's *Works and Correspondence*. Fisher emphasizes throughout Bradley's cautious attitude to theories and his meticulous working habits, notably in regard to his moral commitment to precision and his constant checking of the quality and reliability of his instruments. Artisans as well as elite patrons are centre stage in Fisher's story, and he devotes as much attention to Bradley's role in training assistants as he does to those figures of eminence who made Bradley's success possible in Georgian Britain. Bradley's reputation suffered in the nineteenth century and he was virtually unknown in the twentieth. Tedious, painstaking work of the sort Bradley practised is not as riveting as the discovery of planets or moons, but his work on aberration and nutation nevertheless stands as a brilliant exemplar of astronomical theory and practice. Thanks to Fisher, Bradley is restored to his rightful place as a great scientist – and an astronomer's astronomer.

Robert Illife

Preface

This book is the result of a long-standing labour of love, extended over several decades of research and reflection. Much of the research for this work pre-dates the digital revolution that is transforming historical research and necessitated extensive travel and investment. For much of my working life I was employed as a public librarian, only able to access Bradley's archives two or three days a month, sometimes relinquishing annual leave, to enable me to undertake the extensive research that forms the basis of this work.

Without the friendship, support, and encouragement of many different people I am sure this book would never have been written. I left school at the age of fourteen bereft of qualifications or attainments and was offered a life of few if any prospects. Measles at the age of two left me with a permanent hearing deficit which in part explains why I finished school early. My father Joseph instilled me with a love of natural history and astronomy, teaching me about the constellations before he died, from the injuries he sustained during the Second World War, when I was just five years of age. My mother Irene introduced me to a love of the arts, particularly of music, she being a trained lieder singer, a graduate of the Birmingham School of Music (now the Birmingham Conservatoire). I was a quiet, shy, studious boy, who unfortunately attracted bullies. Subsequently I failed at school. A friend has suggested that school failed me, but during the immediate post-war years with a shortage of schoolteachers, with class sizes often approaching fifty, I cannot blame my teachers, most of whom were dedicated to their work and are still remembered fondly.

At the absurdly early age of eleven, the 'sheep' were supposedly separated from the 'goats', a minority of boys being sent to the local Bishop Vesey Grammar School in Sutton Coldfield whilst we of 'the great unwashed' were sent to fend for ourselves at Riland-Bedford Secondary Modern Boys School, where we were prepared for the Birmingham metal bashing industries. True to my calling, I was for many years employed as a machinist in various factories in Birmingham. At home I was surrounded by music, books, reproductions, and original works of art. My scientific interests, including astronomy and natural history, led to my spending much time in Sutton Park an area of great natural beauty, now designated as a site of special scientific interest. Sutton Park contains about 2,400 acres of diverse habitats that allowed me to study a great variety of fauna and flora. I became something of an autodidact, my continuing education being based at the local public library where the Borough Librarian Mr Sykes was so supportive. The fact that later in my life I became a public librarian is in no small part a result of the respect I had for him.

In 1971, at the age of twenty-eight, I was one of the very first intake of students by the Open University. My diverse interests were continued as an undergraduate. During that first year 25,000 students were enrolled, 5,000 of whom were offered two foundation courses. Only four students opted for the combination I enrolled for: humanities and science. This pattern was continued throughout my undergraduate studies, reading history and philosophy and the life and earth sciences, capping my studies in ecology. I also studied the history of science and the history of mathematics, having taken the shorter version of the mathematics foundation course. I met several influential tutors, but one was outstanding: Dr John Mason, an American scholar lecturing in history at Birmingham University. He firmly instilled many of the skills I used later in my research and writings. It was his prompting that gave me the much-needed confidence to gain a distinction in a course entitled 'War and Society'. I worked full time as a clerical assistant, and after coming fifth out of almost 40,000 candidates in the Civil Service Examination became a tax officer with the Board of Inland Revenue. After graduating from the Open University with first-class honours, I decided to become a librarian.

I began work at Birmingham Central Library, and after enrolling in the post-graduate course in librarianship at the City of Birmingham Polytechnic (now Birmingham City University) I qualified as a chartered librarian. It was while I was at library school that I met Deborah Collinson who became my life partner. I acquired a post at Haringey Public Libraries in north London before gaining an MA in Librarianship and Arts Administration at City University. My involvement with the life and work of James Bradley first began shortly after I married. My mother-in-law Joyce Hall mentioned that the third Astronomer Royal had been born and raised in the small Cotswold village of Sherborne where Deborah and her sister Frances had also been raised. Joyce was the village schoolteacher who, when Deborah and I married in 1982, was educating the grandchildren of the children she first taught when she arrived in Sherborne in the 1950s. She had a keen appreciation of the life of the village. I knew a few things about James Bradley. I had been an amateur astronomer since childhood, but my knowledge of Bradley was sketchy at best.

After moving into our house in Forest Gate in east London, we were surprised to discover that Bradley lived in Wanstead only a mile or so away. There he lived from 1711 to 1732, retaining a connection with what was then a small prosperous township until 1747. It was in Wanstead that Bradley made the two discoveries associated with his name, the aberration of light and the nutation of the Earth's axis. The serendipitous double coincidence connecting Deborah with Bradley prompted me to seek Bradley's biography. I was disappointed to discover that the only account of Bradley's life and work based substantially on archival materials was published in 1832. Stephen Peter Rigaud's *Memoir* was attached to Bradley's *Miscellaneous Works and Correspondence*. This important collection, compiled by Rigaud, has formed the basis of most of the secondary and tertiary accounts of Bradley's work ever since.

In 1985, I sought advice from the Department for the History and Philosophy of Science at Cambridge. To my surprise, I was invited to meet Dr Jim Bennett, then the Curator of the Whipple Museum. I told him of my intention to attempt to write an account of the life and work of James Bradley, arguing that it was long overdue. He was supportive, but also cautionary, for he was aware of the comparative paucity of archival material upon which to work. This was surely a prime reason why historians had largely avoided research in the life and works of the third Astronomer Royal. My meeting with Jim Bennett had two important consequences. The first was to join the British Society for the History of Science. This was important. It gave me real insight into a discipline I found interesting, and recognizing I needed guidance, it put me in touch with some of the leading historians of science in the country. Secondly, Jim put me in touch with Mari Williams who had written a doctoral thesis on the history of attempts to measure annual parallax from Hooke to Bessel. She proved to be both approachable and helpful. She loaned me her valuable thesis, allowing me to copy it.

At this time I had also just completed my MA degree at City University, during which I was offered the option to study English Law for two years at the feet of a Parliamentary draughtsman Professor Charles Arnold-Baker, with tutorials at his chambers in the Inner Temple. I was fortunate to meet such an outstanding teacher. My studies of case law and the ways that English Law, particularly Common Law, worked later gave me important insights into the origins and development of English empirical science during the seventeenth and eighteenth centuries. These important connections are touched upon in Chapter 6 of this work, where I discuss Bradley's attitudes to natural philosophy and the establishment of the laws of nature.

I traced a copy of *The Miscellaneous Works and Correspondence of James Bradley* to the Royal Society Library. Working as a senior librarian at Haringey Public Libraries, I was only able to study the work two or three days a month. This I did for several weeks, when this was noticed by the then Librarian of the

Royal Society, Norman Robinson, who kindly allowed me to borrow the work to enable me to study it at my home. I had this copied and returned the work, so thankful for his generous intervention. For almost forty years, this has been a constant source of reference to me. This research increased my awareness that I needed a solid grounding in the history of science. I enrolled at the London Centre for the History and Philosophy of Science, Technology and Medicine gaining an MSc with distinction in the History and Philosophy of Science at Imperial College London, also being awarded the Diploma of Imperial College. Two tutors developed my thinking. First was Professor Vivian Nutton whose expertise about Galen and ancient medicine made its mark. My undergraduate studies in the life sciences enabled me to capitalize on his teaching and to appreciate his deep learning about his subject, which ever since became an inspiration to me. Second was Professor Piyo Rattansi whose knowledge about Isaac Newton guided me during my studies of the scientific revolution and the work of James Bradley. My master's dissertation on Bradley's discovery of the aberration of light was highly regarded.

The composition of this book is not just the work of this author, for without the skills and knowledge acquired from so many others it could not have been written. It was when I began my doctoral studies on the work of James Bradley that I met with the person who not only supervised them but strengthened my resolve and encouraged me to bring my studies to an initial fruition. He proved to be an outstanding teacher, who both encouraged me to develop my own ideas whilst offering guidance and support, at a time when there was serious illness in my family. Professor Rob Iliffe, is now head of the Faculty of History at Oxford University, and has remained a steadfast supporter of my studies of the work of James Bradley and his contemporaries. My viva voce examination was conducted by Professor Piyo Rattansi of University College London and Professor Simon Schaffer of Cambridge University.

Although my thesis passed muster, I realized that two things were still required before I could attempt to write a satisfactory account of the life and work of Bradley. Firstly, it required much more research, particularly concerning Bradley's early work, his lectures in experimental philosophy, and his Greenwich observations, as well as the work undertaken by Bessel. Secondly, I needed time to reflect, time to connect and contextualize Bradley's work. I retired only three years after gaining my doctorate. I was given much of the time denied me when I was working for my living, but now my health and mobility began to deteriorate. This in itself gave me valuable insights into the decline that comes with old age and growing infirmity, particularly in my understanding of the decline of Bradley's abilities in the final active decade of his life. Without the support of my younger wife Deborah, my labours would have been much more truncated.

Yet I must thank so many more people who have supported my studies over the past four decades beginning with the late and much missed Eric Dodson, who was not only a character actor who played roles so different from his real self, but was also a harpsichord restorer, and serious amateur astronomer who lived in Sherborne, not far from my mother-in-law Joyce Hall and her husband Brian. Eric Dodson, Joyce Hall, and I arranged for a plaque, celebrating James Bradley's connections with the village, to be placed in St Mary Magdalene's Church. Unfortunately the plaque gives the year of Bradley's birth as 1693. Bruce Vickery has since revealed that Bradley was baptised on 3 October 1692 as recorded in recently digitized diocesan records. Two other late friends must also be mentioned and thanked. Deborah's uncle, David Simonson, the one-time classics master at the King's School at Gloucester, very kindly translated several chapters of Eustachio Manfredi's *De annuis inerrantium stellarum aberrationibus*. I stressed to David that just because Manfredi remained geocentrically inclined, that this did not in any way detract from his abilities as a fine observational astronomer. I also formed a growing friendship with Dr Eric Aiton. His modest demeanour disguised a highly perceptive scholar. He advised me extensively on Cartesian natural philosophy, and particularly on Descartes' theory of vortices. We discovered that we shared a common love of music. He played violin in a string quartet.

A valued and long-standing friendship is with Dr Allan Chapman of Wadham College and the Faculty of History at Oxford. I first met Allan at a day conference at the Royal Society when we shared lunch and briefly discussed our mutual interests in the work of James Bradley. I still appreciate his support and correspondence which has always been encouraging. Working as I have done, largely in isolation, such collegial support is so essential to my mental health and continuing research. The entirety of my work on James Bradley was undertaken whilst a member of the North East London Astronomical Society (NELAS) based in Wanstead, about 200 yards from where Bradley made his momentous discoveries. I have given many talks to the society, not just about the work of Bradley, but on many current topics in astronomy, astrophysics, and planetary science, as well as talks to other astronomical societies, not least to the Loughton Astronomical Society. I really must thank all of my friends and colleagues in NELAS for their unfailing support.

Few outside the scholarly community have any real understanding of the amount of work that goes into a project of this nature. This work has been written and rewritten a dozen times at least. My wife Deborah is witness to this and a valued partner in all of my work. Her consistent support and encourage-ment, along with her criticisms and contributions have always been valued. Her good humour and good-natured intrusions, as well as her work in the preparation for the publication of this book took such a weight off my shoul-ders. Three American scholars and friends must be thanked. I first met all three at the combined Manchester Conference of the British, American, and Cana-dian societies in the history of science in 1988. Professor David Hill, then of Augustana College in Illinois and now living in Nevada, Missouri has long encouraged me to exploit my knowledge and insights. As a scholar of Galileo's life and work I am sure of his interest in the person who vindicated the Ital-ian mathematician and philosopher. My Baltimore-based friends Ed Morman, once of Johns Hopkins University and his very special partner Julie Solomon, have long remained supportive of my efforts. We also met at the Manchester Conference when Ed surprised me about his knowledge of English music and shared my love of the music of Ralph Vaughan Williams. Their friendships began just two months after the passing of my mother and meant so much to me, when I felt my loss at its greatest

Life is so often full of coincidences and serendipities. The external examiner of my MSc dissertation Dr John Henry of Edinburgh University, undertook his initial teacher training at Sutton Coldfield College of Further Education, the very place where I studied for my 'O' Levels in Mathematics, English, History, and Geography in the year before I began my studies with the Open University. As far as I am aware our time there did not coincide. John Henry's evaluation of my dissertation on Bradley's discovery of the aberration of light at Impe-rial College was both generous and encouraging. How extraordinary that our paths should cross in the way they did.

The life of Bradley too is redolent with paths crossing and re-crossing. It is a simple story of someone of modest origins, fulfilling the promise of their early years and abilities. But he lived in an age when so many lives were transformed by fundamental changes in society. It witnessed the burgeoning expansion of the London scientific instrument trade, itself the result of the growth of trade generally and the changing patterns of social life in England and Scotland. Within this complex nexus bounded by ambition, knowledge, skills, and the opportunities opened by patronage and mutual interest, Bradley was able to carve out a career that was resplendent in its achievements and yet so modest in its claims.

Dr John Fisher FRAS
4th January 2023

Acknowledgements

Most of my research was undertaken at the Bodleian Library at Oxford and at Cambridge University Library. At Oxford I consulted James Bradley's miscellaneous papers and correspondence, together with Roger Heber's memorandum of Bradley's lectures in experimental philosophy. My time at Cambridge was chiefly to consult Bradley's Greenwich registers and other papers connected to his work at the Royal Observatory. In all the many years I have been visiting the Bodleian Library, whether at Duke Humphrey's Library, over the Divinity School, or at the Mackerras Reading Room at the Weston Library, I have known nothing but immense courtesy and support at all times. In my many visits to the Department of Manuscripts at Cambridge the help I received was on occasion anticipatory, with materials being provided ahead of my realization that I might need them. For this I must specifically thank the then Curator of the Royal Observatory Archives Adam Perkins, whom I found so helpful as I began to feel my way into my subject. In more recent times my studies met with the support of Dr Emma Saunders.

In London I used several libraries. These included the Royal Society Library where the help I was offered by the former Librarian Norman Robinson was exemplary. I thank the staff for their immense help in providing various documents, including many editions of the *Philosophical Transactions*. At a time when the cost of repeated travel to Oxford and Cambridge became a real burden to me, I thank the Royal Society for a grant that was set against these necessary expenses. I also used the Library of the Royal Astronomical Society, of which I have been a Fellow since 2007. Here I gained access to a copy of Bessel's *Fundamenta Astronomiæ* that I regularly referred to until I was able to procure a copy of my own. When I began my doctoral studies under the direction of Professor Rob Iliffe I had access to Imperial College Library and to the Science Museum Library where the staff became familiar with my repeated demands on their time. Since losing access to these two libraries, I have been a member of the London Library and its various reader services for which I remain grateful. Having trained as a chartered librarian, working with Birmingham City Libraries, Haringey Public Libraries, and the Library of the City Literary Institute in Holborn, I was only too thankful for the services I received at all of these libraries, including those where I was employed.

I must thank several people who have been especially helpful during my research. Dr Jim Bennett advised and cautioned me at the outset of my work on the life and work of James Bradley. More than this, at my earliest meetings of the British Society for the History of Science he kindly introduced me to members of the society. He was instrumental in introducing me to Dr Mari Williams whose doctoral thesis was to become a valuable resource in my own research. It was at a couple of day conferences at the Royal Society that I met Dr Allan Chapman of Wadham College Oxford, who has been of immense help and support in our meetings and correspondence. I came to value his collegiality, generosity, and friendship over many years. I also wish to thank Dr John Henry of Edinburgh University who gave me sound advice during the early stages of my work on Bradley. He gave me encouragement. I was after all a middle-aged man with much of the outlook of a recent graduate.

Dr John Mason, my history tutor when I was an undergraduate of the Open University, instilled some of the fundamental skills of historical research that were so essential to my project. Some of the guidance he gave me has remained with me to this day. I also thank the late Professor Charles Arnold-Baker of City University for introducing me to the intricacies of English Case Law and English Common Law that later allowed me to interpret some important characteristics of English empirical science during the seventeenth and eighteenth centuries. The inculcation of the knowledge and skills required to begin my work with confidence, began during my studies at the London Centre for the History of Science, Technology and Medicine. My debt to Professor Piyo Rattansi, who was my tutor during my studies of the scientific revolution, is inestimable. Due to my advanced years, I was unable to acquire a British Academy bursary even after gaining my MSc with distinction. This was the result of a change in government policy.

My doctoral studies at Imperial College were supervised by a remarkable and dedicated scholar whose knowledge sometimes astounded me. Professor Rob Iliffe, now head of the Faculty of History at Oxford University, allowed me to develop my ideas about Bradley whilst offering so much sound advice. My doctoral thesis was not so well developed as my master's dissertation. Some of my arguments were, in my later opinion, rather forced.

In the absence of a bursary I was employed as a graduate teaching assistant which so increased my confidence, giving lectures, leading seminars, and helping to mark undergraduate examinations that I applied to become an Associate Lecturer with the Open University, the very institution that gave me my earliest opportunities. At Imperial College I received sound advice and support from Professor David Edgerton, when teaching during his course on twentieth-century European history and from Professor Andrew Warwick when presenting his course on the history of science. The encouragement I received from Professor Simon Schaffer when our paths crossed many times at conferences, often in convivial pubs over a favourite tipple or two, was greatly appreciated. These meetings reminded me of the coffee house culture that was often associated with formal meetings of the Royal Society. Professors Schaffer and Rattansi's examination of my doctoral thesis was, however, fiercely searching when I was only too well aware of some of its shortcomings. Yet they were both assertive in stating that this was merely a stage in my own intellectual development. Almost two decades later I am able to attest to the truth of their sound advice. I was to become aware of the rather forced nature of many of my arguments. Critical voices will certainly argue that the same shortcomings can still be laid at my charge. It is inordinately difficult to deal with the life of a man where over half of the 'script' is missing. Too often I have relied on inferences and circumstantial evidences, though to be fair I am the first to admit this.

While working on my doctoral thesis as well as being employed as an Associate Lecturer with the Open University, I was also employed part-time at two notable libraries, the Lindley Library at the headquarters of the Royal Horticultural Society at Vincent Square, Westminster, the largest horticultural library in the world, where I helped to organize the archives of the society. The then Lindley Librarian Dr Brent Elliott was a convivial colleague who was very supportive of my latter-day researches, for which I was so grateful when this period coincided with serious illness in my family. I was also fortunate enough to be employed by the Library of the Royal College of Surgeons of England at Lincoln's Inn Fields where my skills as a cataloguer were matched by my knowledge of the history of the biological sciences.

I made visits to Gloucestershire Records Office at Gloucester where I spent days going through parish records, not only of Sherborne, but of various nearby villages and hamlets associated with the Bradley family in and around the township of Northleach, where entries in the Sherborne Parish Records for 1690 are immediately followed by those for 1706 in the same neat hand. I discovered that St Mary Magdalene's Church at Sherborne was closed for renovations during this period. The entries were all so neatly written that I was quite sure it was a copy in fair hand. Only recently, following the digitization of diocesan records, Bruce Vickery located the entry for James Bradley and kindly informed me through the offices of Ian Ridpath. He as the editor of the *Antiquarian Astronomer*, the house journal for the Society for the History of Astronomy, informed me of this when I was preparing a paper on Bradley's discovery of the aberration of light published in June 2022. Bradley's baptism was located on 3 October 1692, several months earlier than Rigaud's estimate of 23 March 1693. I also made a visit to Southampton Records Office to seek Samuel Molyneux's papers. Southampton was where Molyneux's widow Elizabeth eloped with Nathaniel St Andre immediately following Samuel's death. I am grateful for the help I received from the staff of Gloucestershire and Southampton Records Offices.

More recently, I am thankful for the exemplary help and support I have received from James Dawson, Librarian of the Society for the History of Astronomy. He accepted earlier drafts of this work for safe-keeping. James played an important role in putting me in touch with editors at Oxford University Press who were ready to accept my submission concerning this work. In the preparations to this work, I am also grateful to friends who have provided photos of Bradley's memoria. Margaret and Graham Garnett kindly provided photos of Bradley's tomb in the churchyard of Holy Trinity Church, Minchinhampton and two more inside the nave of the church. In addition, I thank Pat Elliott for her photo of the memorial stone placed by James Pound's grave by members of the Royal Astronomical Society, in what had originally been the nave of the Church of St Bridget or St Bride in Wanstead. The new church of St Mary the Virgin, consecrated in 1791, was opened immediately to the north of the site of the old demolished church. I thank Barbara Elliott for her photo of the green plaque placed on the wall of the Corner House in Grove Park Wanstead, close to the site of Elizabeth Pound's town house. This commemorated Bradley's discovery of the motion of the Earth.

I thank the National Maritime Museum, the National Portrait Gallery, the Royal Society, the Science Museum, and the Science Photo Library for permissions to publish important illustrations and for the help given by members of staff in obtaining these. I thank Ian Ridpath, editor of the *Antiquarian Astronomer* for permission to use the tables included in my paper on *The discovery of the aberration of light by James Bradley* in June 2022. His patience and support in the preparation of this paper was gratefully accepted. I also thank Simon Schaffer, the then editor of the *British Journal for the History of Science* who published a very different account of Bradley's discovery of aberration as *Conjectures and Reputations: The composition and reception of James Bradley's paper on the aberration of light with some reference to a third unpublished version. British Journal for the History of Science* 43 (1) 19–48 March 2010. My own diagram, showing the differences between annual aberration and annual parallax, included with this paper has been edited and adapted for use in this work. For permission to use this diagram I thank Cambridge University Press.

I thank various members of the North East London Astronomical Society, past and present, for their continuing support of my many interests. I thank Bernard Beeston for his friendship and support and Steve Karpel for accepting an early draft of this work for safe keeping. Jack Martin whose spectrographic work stands four square in the Bradleian tradition, has always supported my research and I thank the late Andrew Lawrence who sadly passed away before I was able to publish this work. Now that I am only partially mobile, I thank Ken and Shirley Salmon for their many kindnesses. I value the convivial and long-standing support of the membership of this society.

Throughout the entire length of my work on James Bradley my debt to my wife Deborah has been without equal. Her patience, her criticisms, her contributions, her good humour, and her work, most particularly during the publication of this book, as well as her knowledge and skills, have been such that I am very sure that without her this work would never have seen the light of day. This book is dedicated to her.

I recognize that this is an imperfect work, for scholarship is the severest of taskmasters, but I hope that for all its possible and clearly recognized short-comings, it can be accepted as a hoped for first move in a new conversation about the work of James Bradley and his immediate contemporaries. I believe it helps to fill a gap in the historiography of eighteenth century astronomy between the work of Flamsteed and his generation and Herschel and his con-temporaries. The work of Bradley forms the keystone that connects the work of his predecessors with those who were able to capitalize on his contributions to the astronomical sciences. May I thank my editors at Oxford University Press, Dan Taber and Hayley Miller, Vicky Harley my patient copyeditor as well as Rajeswari Azayecochi and his colleagues, who put my mind at rest many times during the publication of this work.

The astronomical work practised by Bradley came before the introduction of compound achromatic lenses and the important innovations in the preparation of blanks, when the available glasses used were often less than perfect. The astronomical interests associated with the work of James Bradley are no longer redolent with the concerns of contemporary astronomy, as is the work for instance of William Herschel. Yet the age of positional astronomy has been revivified with the launch of the European Space Agency *Gaia* satellite that is locating the positions of untold billions of stars as well as their trajectories. James Bradley and his great predecessors, including Tycho Brahe, Robert Hooke, Johannes Hevelius, Gian Domenico Cassini, Jean Picard, Ole Rõmer, Jacques Cassini. John Flamsteed, James Pound, and Edmond Halley all stand in a tradition now being fulfilled by *Gaia*.

For all those many people over a period of almost forty years, that I may have consulted and have now forgotten, I can only offer my sincerest apologies and my thanks.

Dr John Fisher FRAS

Contents

List of Illustrations

Thomas Hudson's portrait of James Bradley, painted c.1742–1747, after he became the third Astronomer Royal.

Introduction

Contexts and Connections

Celebrity is obscurity biding its time.

Carrie Fisher.

James Bradley, once the most celebrated astronomer in Europe, has become an obscure figure, mainly relegated to the footnotes of scientific history. The reasons for this obscurity are both simple and complex. Simple, in that for all but a handful of figures in any generation that is the fate of almost all of humankind. Complex, because the fate of his finest work fell into the hands of philistines who denied access to it until it became to be regarded as obsolete by the very people who should have most protected it. This book is not merely an attempt to open up a new conversation about the work of James Bradley (1692–1762), the third Astronomer Royal of England, but also about the development of what might be termed modern science, of which Bradley was a leading exponent. This book is not just about the theory and practice of science in the eighteenth century, or as it was then termed natural philosophy, but the various contexts within which this science was practised. It will examine the role of patronage, together with the diminishing influence of the court in England. This will be explored by a comparison of the careers of John Flamsteed, the first Astronomer Royal, and James Bradley, the third person to hold that august office.

Both men suffered through the obsolete arrangement of the Royal Observatory at Greenwich, designed by an astronomer of a previous generation Christopher Wren, and built using bricks from the demolished Greenwich Castle. The design, especially of the now famous Octagon Room, which may have been the work of Robert Hooke, in what is now called Flamsteed House, was ill-equipped for the new astronomy. The positions of the stars were no longer determined primarily by triangulation, but by right ascension and declination, the celestial equivalents of longitude and latitude. The architecture of the newly built Royal Observatory was ill-suited to the new way of determining stellar positions. It drove the first three Astronomers Royal into the open

The Life and Work of James Bradley. John Fisher, Oxford University Press. © John Fisher (2023).
DOI: 10.1093/oso/9780198884200.003.0001

air or into temporary outhouses. This was changed when Bradley reformed the institution by commissioning new instruments and constructing the New Observatory, the building through which the prime meridian passes. If Flamsteed, a man who suffered greatly with arthritis of his joints, was forced into the open air he was even more handicapped by a lack of patronage. Perceiving of himself as a court-appointed astronomer like his hero Tycho Brahe, the great Danish astronomer, he seemed oblivious to the changing structures of English society where the influence of the court was diminished. With the increasing secularization of society, new forms of patronage were growing at the expense of the crown. Because of this Flamsteed found himself bereft of support during his bitter dispute with Isaac Newton. Add to this Flamsteed's character, described as morose by Jim Bennett, and it is easy to perceive why his career was comparatively stymied. Even though Flamsteed was an all but brilliant observational astronomer, his achievements were undermined by various factors. These not only came from his shortcomings as an astronomer but also from his personal and social failings that left him short of friends. This was particularly so in the Royal Society from which he was ejected in a rather ignoble fashion.

The climax of the conflict between Flamsteed and Newton, two men who evidently detested each other, came with the 'enforced' publication of Flamsteed's observations made at Greenwich. Flamsteed was compelled to release all of his observations under a Royal Warrant so that refusal could be interpreted as treason. This was a tragic situation for an astronomer who believed he had the protection of the crown. During the reign of Queen Anne, Newton could do little wrong, having proved his value as the Master of the Mint in helping to save the English currency. The interplay of politics, statecraft, and the work of the Royal Observatory is one of the important backdrops to the dramas that unfolded at Greenwich. As will be revealed, Bradley proved to be extremely astute in his dealings both with the crown and with the ruling administrations. His skills as an astronomer were matched by his dealings with those who were in power. Flamsteed, for all of his observational abilities was unable to deal with the shifting centres of power and patronage.

I dwell on the work of John Flamsteed as a contrast to the career of James Bradley. Flamsteed died on the last day of 1719 (OS) whilst Bradley's astronomical career began some time during 1711 when he took up residence at his maternal uncle's home, the parsonage at Wanstead in Essex. In the same year he matriculated as a commoner at Balliol College at Oxford. The first Astronomer Royal was dismissed by many members of the Royal Society, particularly by the Curator of Experiments Robert Hooke who had expectations of filling the post himself rather than seeing it being acquired by an unknown provincial. As an autodidact who taught himself Latin and mathematics, Flamsteed saw himself as an outsider, never feeling comfortable within the coffee house culture that followed formal meetings of the Royal Society. In contrast, James Bradley was soon valued by several influential members of the society, including Edmond Halley and Newton, who were both mightily impressed with the young astronomer's abilities.

The publication, against Flamsteed's will, of the observations of the Royal Observatory, raised important questions over the ownership of observations made there. In Flamsteed's case, the overwhelming number of observations at Greenwich were made with instruments that had been commissioned out of his own resources. On a meagre income of £100 per annum, he was supposed to provide his own instruments, and train and pay his own assistants. Flamsteed did not have the financial resources expected of a gentleman, so he took on students to make ends meet. As a result, Flamsteed's observations could be seen as his own property. The problem of ownership was not resolved. At his death, his widow Margaret removed most of the instruments her husband had used at Greenwich, leaving his successor Edmond Halley with an almost empty observatory. Halley was provided with £500 to purchase the instruments required to make astronomical observations at the Royal Observatory. He commissioned an eight-foot quadrant and a transit instrument from George Graham, beginning his observations of the Moon in the pursuit of an adequate lunar theory. This was to enable mariners to determine their longitude at sea. The motions of the Moon, however, proved to be far more complex than most astronomers at first recognized.

When Bradley was appointed as the third Astronomer Royal in 1742, the issue of ownership of the observations made at the Royal Observatory remained unresolved. When Bradley died in 1762, fully eighty-seven years after the foundation of the Royal Observatory, his executor took issue over the ownership of Bradley's observations at Greenwich. The lawsuit between the Board of Longitude and the executor of Bradley's will lasted fourteen years until 1776, during which time access to Bradley's legacy was denied to the entire astronomical community. The finest positional observations of the first three-quarters of the entire eighteenth century fell into the hands of his executor Samuel Peach, a man who only conceived of their monetary value. On the death of Peach, the registers were surrendered to Lord North as the Chancellor of Oxford University. He gave instructions for the registers to be published as soon as possible. Unfortunately, Bradley's successor as the Savilian Professor of Astronomy at Oxford, Thomas Hornsby, seems to have regarded the task as irksome and of little worth. He relinquished the editorship, publishing only the first six years of observations from 1750 to 1755, twenty-two years after being given the task. This work was completed by Abram Robertson, when the observations were published in two volumes in 1805. It was Friedrich Wilhelm Bessel who revealed the riches contained in the observations made by Bradley and his three assistants, which he described as 'incomparable'. Bessel, having been provided with the two volumes by Heinrich Olbers, spent eleven years reducing the observations to epoch 1 January 1755, publishing a catalogue of 3,222 stars of unprecedented reliability in the *Fundamenta Astronomiæ* in 1818. To this day, it remains the earliest stellar catalogue used by astronomers for purposes of comparison, such as the proper motion of any of these 3,222 stars down to the eighth magnitude. Bradley's registers remained the property of Oxford University until April 1861, when they were surrendered to the Royal Observatory to be placed alongside the registers of all the other Astronomers Royal. This was a direct consequence of the lack of a legal determination of ownership dating from 1675.

The observations made by Bradley and his assistants had never been dealt with urgently by the Syndics of the Clarendon Press and least of all by Hornsby, the first editor. Hornsby possessed an excellent reputation at Oxford, largely because of his lectures in experimental philosophy. He rarely, if ever, observed the night sky at the newly instituted Radcliffe Observatory, preferring to concentrate on daylight observations. This may have coloured his perceptions of the importance of Bradley's *stellar* observations. This may be conjectural, but one thing is certain: he placed little or no value to Bradley's surviving papers. Bradley's first and only biographer Stephen Peter Rigaud was vexed by the negligent manner in which Hornsby treated Bradley's archive, which was often interfiled carelessly with his own. Rigaud clearly expressed his anger at the actions of his predecessor as Savilian Professor of Astronomy. To this day, various papers remain out of reach, including Bradley's record of his students from 1729 to 1745. His record from 1746 to 1760 is such a useful source, so we must assume that his earlier records have been misplaced or destroyed.

The world that Flamsteed inhabited was profoundly different from that in which Bradley lived and worked. The transformation in the social status of skilled artisans in Great Britain, particularly in the City of London, is evident. The insults that Robert Hooke directed towards his clockmaker Thomas Tompion during the 1680s were accepted by the craftsman because his patron gave him status in polite society. In an unstable and violent society, such recognition was vital in the maintenance of his workshop and the work he provided for his journeymen and apprentices. By the time that James Bradley was commissioning his zenith sector from George Graham in 1727, the latter was regarded as a gentleman in his own right. In 1722, Graham was elected to the Council of the Royal Society and also became Master of the Clockmakers Company. During this year he revealed his scientific credentials by discovering the diurnal variation of the compass using an instrument of his own design and construction. Bradley and Graham appear to have met during 1723 when they worked together observing the notable comet of that year. Bradley was fortunate to have a man like Graham as a close friend and confidant. Graham's position in polite society was also reflective of some radical transformations of English and British society. These changes also helped Bradley, the third son of a country steward, to become the Savilian Professor of Astronomy at Oxford and the third Astronomer Royal of England at Greenwich. The rise of new ideas, political and social as well as scientific, together with the changing processes of socialization enabled men of modest origins to rise to positions of trust and respect.

Some of the imputed causes of this upward social shift of many artisans may in fact be better characterized as effects. Integral to this was the patent system that protected the innovations of craftsmen for twenty-five years. The English system was probably the first modern patent system in Europe. It gave protection to innovators in ways denied to those on much of the continent, and particularly in France where the social status of skilled workers remained little better than that of common labourers. France was not short of fine craftsmen producing many ingenious automatons – toys for the aristocracy. Yet French artisans did not enjoy the same status or protections as those enjoyed by English craftsmen. French natural philosophers sometimes turned to English artisans. For example, Pierre Louis Maupertuis commissioned George Graham to design and construct a zenith sector for the geodesic expedition to Lapland in 1735–36.

The rise of James Bradley into prominence was made possible by the life, work, and actions of his maternal uncle. The Rev. James Pound entered into the service of the English East India Company during 1699. The original intention was that Pound be appointed the company chaplain for the Honourable Company at Fort St George. The fort had been built in 1639 near the Indian port of Madras (now known as Chennai). Pound travelled accompanied by his manservant Moses Wilkins. Before Pound left England, he was engaged by John Flamsteed to make observations on his behalf. Flamsteed was to provide Pound with a portable quadrant and a set of tables for the satellites of Jupiter. Pound received the quadrant, although following several voyages throughout the Indies, and East Asia, he was often separated from this useful instrument. But for the period of seven years, he never received the promised tables. Whether Flamsteed sent them is a moot point.

Pound was never far from the threat of sudden death. The mortality rates for Europeans in the Indies could only be described as horrific. As the chaplain of the community serving the Company on the small island of Pulo Condore near the mouth of the Mekong delta, he buried thirty-seven men, six women, and a child, all in a period of about eighteen months. On 3 March 1705 Pound survived a massacre of most of the European inhabitants – English, Dutch, and Swedish – by a detachment of locally employed mercenaries who had been dismissed by the Dutch for their unreliability. These Macassars, believing they had lost face, destroyed the Company's presence on the island. Pound was one of only fifteen who sailed in a small sloop for about ten days. The survivors included eleven Europeans and four black slaves or servants.

Pound returned to England in July 1706, and one year later he was appointed as the Rector of Wanstead in Essex by Sir Richard Child, the third son of the former governor of the East India Company. Sir Richard was the recipient of an income of £10,000 per annum, the possible equivalent now of about £2.5 million. He used this to support the arts, commissioning the new Wanstead House, the earliest Palladian mansion in England. The architect, Colen Campbell, the author of *The British Vitruvius*, was given a free hand to introduce this innovation in English architecture. It was completed in 1722 and it was probably during the celebrations for the opening of this great house that James Pound and Samuel Molyneux began their discussions about a revival of Hooke's famous attempt to observe annual parallax with a zenith sector.

At the time that Wanstead House was being built, from 1715 to 1722, the growing wealth of the middling classes in England partly rode on the back of the profits of the tripartite Atlantic slave trade. The South Sea Bubble of 1720 exposed the growing greed of many in middling society, greed that often led to financial ruin. The wealth of the new moneyed classes in the guise of the Whigs rivalled that of the established landed classes, who were often associated with the interests of the Tories. The latter party tended to support traditional values. The Whigs seized their opportunities after the accession of the Hanoverian dynasty. During the early years of the new Hanoverian line, the Tories became politically suspect because of lingering loyalties to the previous dynasty. Many Tories, justly or unjustly, were associated with the Jacobite cause, as indeed many were during the 1715 Jacobite rising. The Tories remained out of office until after the 1745–46 attempt by the Jacobites to seize power had failed. This was at a time when Great Britain was in conflict with France during the War of the Austrian Succession up to 1748.

I mention this political and social background because it took intelligence, diplomacy, and political nous to negotiate the fracture lines that could be met by anyone who sought any form of support or patronage. England was a divided society at the time of Bradley's rise to prominence. Certainly, we know that Bradley had close friends who had connections with the Tories, including Halley and Newton; their scientific credentials placed them largely beyond suspicion, even though Newton's religious beliefs could have been useful to his enemies. Many Tories still thrived within society, but they could always come under some measure of suspicion. Much enmity persisted between Tories and Whigs. The conduct of the non-juror Thomas Hearne at Oxford must be considered as evidence of Bradley's social and political skills. From opposing Bradley's election as the Savilian Professor of Astronomy in 1721, a reflection of his opposition to a man supported by a Whig powerbase in London, Hearne became Bradley's only supporter when he applied to become the Keeper of the Ashmolean Museum in 1731.

Bradley's more important supporters and patrons were firm Whigs. These included friends and associates of his uncle James Pound, notably Thomas Parker, later Lord Chancellor and the first Earl of Macclesfield, who acted as the Regent of Great Britain during the brief interregnum between the death of Queen Anne and the arrival of George I in London from Hanover in 1714. Also of note was the Rt Revd Dr Benjamin Hoadly, who acquired some of the leading bishoprics in the land and was regarded by many as the most controversial churchman in England. Another important patron was Samuel Molyneux, the Prince of Wales' private secretary, who acquired for Bradley a half living in Pembrokeshire in the gift of the prince and who became Bradley's senior partner in the repeat of Hooke's attempt to observe annual parallax. All of Pound's and Bradley's patrons were Fellows of the Royal Society which throughout much of the eighteenth century acquired many of the characteristics of a gentleman's club.

Bradley completed his observations of the nutation of the Earth's axis in September 1747 when Great Britain and France were in conflict during the War of the Austrian Succession which ended in 1748. France's leading positional astronomer Nicolas-Louis de Lacaille remained entirely ignorant of the completion of Bradley's programme of observations leading to the discovery of the nutation of the Earth's axis. Bradley's reputation in France was without rival. Even though the two countries were at war again from 1756 to 1763, during the conflict better known in Great Britain and America as the 'Seven Years War', the Royal Academy of Sciences bestowed the full status of Academician on Bradley in 1761, the year before his death. The expedition to the southern hemisphere commissioned by Bradley, as the Astronomer Royal and as a member of the Board of Longitude, to observe the rare event of the transit of Venus due on 6 June 1761, was placed in jeopardy by the actions of a French frigate which attacked the British frigate carrying the astronomers in the very year Bradley was honoured in Paris. The British vessel suffered extensive damage and loss of life, sufficient to force a return to port to make necessary repairs. As well as being recognized in France, by the end of his life Bradley was honoured by the Imperial Academy of Sciences at St Petersburg, the Prussian Academy in Berlin, and by the Institute of Bologna.

This book has eleven chapters covering the work of James Bradley, his predecessors, and his successors. Because of the actions of his executor Samuel Peach and Bradley's successor at Oxford Thomas Hornsby, there is a truncated archive almost completely devoid of all personal correspondence and other private papers, as well as some of his scientific archive. Even his surviving working papers were affected by the careless, even negligent, actions of the man entrusted to publish Bradley's observations made at Greenwich. Bradley's first biographer Stephen Peter Rigaud was greatly vexed by this neglect, because he had to go through all the papers of Bradley's successor at Oxford in order to extract those specific to Bradley. Hornsby, it appears, enjoyed a good reputation at Oxford but, as already touched upon, failed to comprehend the significance of the 60,000 plus immaculate stellar observations made by Bradley and his assistants.

The first chapter of the book, The King's Observator, is about the work of Bradley's astronomical predecessors, with an emphasis on the working career of the first Astronomer Royal John Flamsteed. The chapter reveals the many difficulties he had dealing with the Royal Society in an environment of comparative neglect. His contemporaries in France were directly employed as servants of the state and were given all possible support. From 1675 Flamsteed, with his paltry income of £100 per annum for the administration of an institution that was considered important enough to warrant a Royal Commission, was throughout his career dependant on his own resources as was then expected of a 'gentleman'. In England the activities that we now call science were the province of independent gentlemen acting according to their own interests and within their financial capacity. In Flamsteed's case this was little until his father Stephen died in 1688.

The second chapter, May It Please Your Honours, covers the life and career of Bradley's maternal uncle James Pound, both before and after he became the Rector of Wanstead in Essex. Without Pound, it is wholly inconceivable that the third son of an obscure country steward would have become one of the leading astronomers of the eighteenth century. Not only was Pound a prosperous figure enjoying the patronage of the one-time Regent of Great Britain, later the Lord Chancellor, but he was also, in the breakdown of all possible intercourse between Newton and Flamsteed, the person to whom both Newton and Halley turned to in the pursuit of their own interests. The title of the chapter is taken from a letter he addressed to the managers of the English East India Company that gave a complete account of the massacre of the company fort and station on the island of Pulo Condore. Pound's survival in such a dangerous environment may yet have affected him, for his sudden and unexpected demise at the age of fifty-five left his nephew bereft and in a state of financial dependency on his aunt.

The third chapter, An Ingenious Young Man, covers the early years of Bradley's career as a young, highly promising astronomer from 1711 to 1725. No young man at this time could have enjoyed the instruction of two more outstanding figures than his uncle James Pound, who instructed him in the tacit skills of observation, and his friend Edmond Halley, who passed on much of his knowledge of Newton's *Principia*. The latter was responsible for enticing Newton to compose the *Principia*, widely accepted as one of the most important works of science ever written. Bradley may have been given outstanding tuition but it was his native intelligence that allowed him to take advantage. This chapter concludes with Bradley's first published paper in the *Philosophical Transactions*. It reveals his ingenuity when he used disparate observations of the innermost of the Jovian satellites in Wanstead, London, New York, and Lisbon to determine the longitudes of New York and Lisbon. It was always entirely impractical to expect mariners to take such sightings on the moving deck of a ship at sea. Even so, Bradley's observations of all four of the Jovian satellites over several years led to his discovery of the gravitational resonances between them. His knowledge of the motions of these satellites and the necessity of applying an equation of time due to the finite or progressive motion of light became a contributory factor towards the hypothesis of the aberration of light, later in the decade.

The fourth chapter, A New Discovered Motion, focuses on James Bradley's involvement with the attempt led by Samuel Molyneux to repeat Robert Hooke's celebrated attempt to measure the annual parallax of a star in 1669. The evidence supplied by James Pound's Accounts Book reveals repeated visits by Pound to Molyneux's residences in Richmond and Kew. We can be sure that the demise of Pound interfered with the planning of the attempt to be made by Molyneux for he lacked the precise technical knowledge required to proceed with measuring the annual parallax. Molyneux turned to George Graham in order to study Hooke's original attempt and to design an instrument that would observe the passage of a star through a meridian close to the zenith. He invited Bradley to participate in place of his uncle. Molyneux required the support of these highly skilled partners to make accurate observations and to validate the attempt. Graham introduced an important innovation that increased the accuracy of this instrument over that used by Hooke. Hooke simply suspended his instrument and then estimated how far or near the passage of the object star was from the zenith. Graham's instrument was designed so that the passage of the star passed through the centre of the field of vision as it might through a transit instrument, the direction of the tube being adjusted by an accurate screw micrometer. The attempt to observe annual parallax at Kew failed. It did so for two reasons. Initially the angles to be observed to determine annual parallax were much smaller than most astronomers at this time surmised, the stars being at inconceivable distances from the Earth. In addition the motion of the object and control stars revealed a motion that could not be reduced to the theory of annual parallax. Some accounts which suggest that Bradley discovered the aberration of light or annual aberration in December 1725 at Kew are simply incorrect.

The fifth chapter, And Yet It Moves, discusses Bradley's observations of a persistent anomaly. Bradley needed to study many more stars in order to clarify whether this 'anomalous' motion could also be observed in them. He sought to determine a general law that described this newly observed motion. Unfortunately the instrument commissioned by Molyneux was only capable of observing stars that passed very close to the zenith at the latitude of London and was insufficient to observe the large number of stars required. Bradley commissioned his friend George Graham to construct a zenith sector capable of observing stars up to several degrees from the zenith ideally taking in the first magnitude star *Capella*. This instrument was suspended at Bradley's aunt's modest townhouse in Wanstead in August 1727. Over the next year he observed the motion of about seventy stars which included an intensive study of the motion of eleven bright stars throughout the year including daylight hours. By April 1728, Bradley had discovered this motion in all of the stars which led to the identification of the law of 'the new discovered motion' in every star he was observing. By October 1728 he believed he had recognized the cause of this motion. It was a product of the motion of the Earth in its orbit around the Sun relative to the finite velocity of light. This was the motion now called annual aberration. It was the first hard empirical evidence of the motion of the Earth in its orbit as a planet around the Sun. In his discovery of the aberration of light, Bradley revealed his skills as an observer and satisfied his expectations that the motion of γ Draconis would prove to be the result of a 'law of nature'. The new sector at Wanstead was specifically designed to discover or uncover this natural law.

The Laws of Nature, the sixth chapter, is an examination of Bradley's lectures on experimental philosophy at Oxford, lectures he presented from 1729 to 1760. They reveal a mind firm in its grasp of contemporary physics. His understanding of the contents of Newton's *Principia* and *Opticks*, together with his well-rehearsed comprehension of the work of Robert Boyle, Robert Hooke, and many others, is confidently displayed. He reveals his knowledge of Aristotelian natural philosophy and Cartesian physics which, when he began his lectures, were rivals to Newtonian physics in Paris and other key centres in Europe. To Bradley both the teachings of the Peripatetic School and the claims of the Cartesian forms of natural philosophy were worthless because they were both examples of systems based on first principles. He considered this was arguing backwards. For Bradley the laws of nature were to be discovered by an open-ended investigation based on observation and experiment leading to the fundamental principles we call natural laws.

Although Bradley left various papers concerning his lectures, as well as a notebook detailing his course, posterity is fortunate in having a notebook written by one of his students Roger Heber dating from 1754. Although his student occasionally appears to misunderstand some aspects of his presentations, his record confirms Bradley's original intentions. Bradley presented his course three times a year until the demands of his post as the Astronomer Royal enforced a change to two presentations. The lectures in experimental philosophy began at a time when Bradley remained dependent on his aunt's financial support. As he was giving an extra-curricular course, he was able to charge fees which became his main source of income. His lectures were popular. This is indicated by the attendance of forty per cent of the student body at Oxford. However, there was a small yet significant number of a conservative persuasion who were concerned about Bradley's teaching and also the entire development of the new knowledge and its claims. There were fears that these lectures were an encouragement to atheism and godlessness. That Bradley was an ordained priest of the Church of England mattered little to such critics, and many saw the new philosophy as a threat to Christian society. Such was the fear to many, that it united High Church members with those of an evangelical stamp such as Wesley. Even so, Bradley was awarded a doctorate by the University and later he received an address in his honour at the Sheldonian Theatre.

The seventh chapter, On the Figure of the Earth, sets Bradley's work against the background of the scientific and philosophical conflicts then taking place in Paris. It was a conflict that transcended all rational and civil discussion. At issue was a conflict over the shape of the Earth. It exposed an argument between the 'Cartesians' and the 'Newtonians' which elicited profound patriotic and emotional responses. The conflict, however, was based on a fundamental dispute over the validation of knowledge claims. The Cartesian refusal to accept the arguments of the *Principia* as a description of 'reality' was based on a rejection of the Renaissance belief in the occult, specifically in Newton's apparent belief in 'action at a distance'. Newton never assented to such a doctrine. In Paris the *Principia* was generally admired as a work of geometry. Cartesian natural philosophy still upheld the epistemological division between physics and mathematics initiated by the Peripatetic schools. Subsequently, the claims that the *Principia* was a mathematical analysis of 'reality' was rejected out of hand by the supporters of the Cartesian school.

Against this background, Bradley and Graham undertook various experiments on pendulums. An experiment organized by George Graham with the cooperation of Bradley compared the vibrations of a pendulum in London with those in Jamaica. This experiment was undertaken to test Newton's calculations on the figure of the Earth, to determine whether or not the Earth was shaped as an oblate spheroid. Clocks with pendulums lose time as they approach the equator due to an associated measurable weakening of gravitation as the equatorial 'bulge' was approached. The shape of the major planets Jupiter and Saturn also displayed clearly that they were oblate. The Cartesians who held sway in Paris saw gravity as the effect of the pressure of the material circulating in vortices of fine matter. The universe of Newton's *Principia* is one of matter in a void; the cosmos of the Cartesians was a plenum, it is all matter.

The title of the eighth chapter, The Triumph of Themistocles, is a quotation from Bradley's first biographer. It referred to the persistent actions of the Athenian general who built their navy sufficient to withstand the might of the Persians who he was certain would return. Likewise Rigaud saw Bradley's work on the nutation of the Earth's axis as being equally determined and as successful. Bradley was certain that within the residuals of his observations of the aberration of light, there was a much smaller phenomenon hiding in plain sight. He first checked to determine whether this small deviation from the theory of aberration could be attributed to the effects of annual parallax, which he was certain could not be greater than about one second of arc. Bradley sought to determine the causes of the observed deviation.

Bradley began to revive the possibility that it might be an annual nutation which had been rejected as an explanation for the motion identified as the aberration of light. After observing the phenomenon with the zenith sector at Wanstead for two or three years, he rejected the annual nutation for a second time. When he moved from Wanstead to Oxford due to his growing workload at the University, he was certain he was observing a nutation of the Earth's axis induced by the differential gravitational effects of the retrogression of the nodes of the lunar orbit around the Earth. Bradley's twenty-year study of the nutation of the Earth's axis was known throughout the astronomical community in Europe before his paper was read to the Royal Society on 7 and 14 January 1748. He was awarded the Royal Society's highest award of scientific achievement, the Copley Medal. Not only did Bradley achieve the discovery of a phenomenon almost infinitesimally small by eighteenth-century standards, he enabled astronomers to calculate the annual precession of the Earth's axis, a feat that had remained beyond the ability of astronomers for almost two millennia. Bradley received honours from every major scientific institution in Europe for this extraordinary achievement.

The title of the ninth chapter, If Such a Man Could have Enemies, is a quote from one of Bradley's supporters, expressed before he had been appointed the third Astronomer Royal. It was indicative of how well Bradley was thought of in polite society. It was during the period before his appointment that Sir Robert Walpole lost a division in Parliament over a minor issue. Bradley was appointed by Walpole in the period between this division and the prime minister's resignation from office. There was a fear that a non-astronomer would be offered the post and then employ someone qualified to fulfil its demands. Such procedures were common at this time. The annual income of the office, a mere £100 per annum, hardly attracted any attention and Bradley was soon appointed as the obvious candidate.

Bradley was in for a shock when he reached Greenwich. He knew that the great eight-foot quadrant constructed for Halley by Graham had been distorted by the weight of the instrument but he found that the entire Royal Observatory was in disarray. Due to the paralysis of his right hand, Halley had sought to resign from his post in favour of Bradley but was refused. As well as major problems with the great quadrant, Bradley was also taken aback by the state of the transit instrument. Requiring support by hand, it was completely useless as a precision instrument. The collimation marks inscribed on the wall of Greenwich Park could not be used because a tree had grown to block the view. With the help of his fourteen-year-old nephew John Bradley his assistant, James Bradley set about a complete renovation of the Observatory.

As well as repairing many instruments, he acquired new clocks, barometers, and thermometers, all out of his own pocket. Over the next decade the Royal Observatory was reformed until it became the finest in Europe. This included a new observatory designed to observe stars in right ascension and declination. Bradley commissioned high precision instruments and regulators. John Bird built an eight-foot quadrant completed and suspended in 1750. Bradley was now equipped to undertake a survey of stars down to the eighth magnitude visible from Greenwich.

The tenth and penultimate chapter, Observations Beyond Compare, covers the period of Bradley's greatest single achievement, the 60,000+ high precision observations made by him and his three assistants. Each was trained to observe to Bradley's exacting standards. The reform of the Royal Observatory was a product of the patronage and support Bradley enjoyed as well as his own abilities. The New Observatory building was constructed as it should have been in 1675 to allow astronomers to observe the stars in right ascension and declination sheltered from the worst of the elements. For three-quarters of a century astronomers had been forced to observe, either in the open air in all weathers, or in temporary wooden buildings. Such was Bradley's reputation that he was granted all the resources he required. George II took an interest in Bradley's work beginning with his association with Samuel Molyneux, his private secretary up to 1727. Bradley's international prestige in 1748 led to the king granting £1,000 to re-equip the observatory. Other sums were made available to construct the new observatory although the source of the funding is unclear.

Increased precision brought with it new problems associated with the reduction of observations. Bradley was able to reduce all of his Wanstead observations which were made close to the zenith without any need to take atmospheric refraction into account. The effect of observing stars closer to the horizon was a more difficult problem to resolve. Bradley developed an ad hoc solution based upon his own observations of atmospheric refraction but was not satisfied with it. In France, Lacaille wrestled with the same problem and developed a complex formula used over much of the Continent. The results derived from applications of Bradley's and Lacaille's formulae could produce differences of almost an entire minute of arc. Bradley did not reduce his observations. Instead, he recorded the bolometric pressure, the atmospheric temperature inside and outside of the observatory, the calibration of his clocks and instruments, and the seeing conditions for every single observation made. Together with calculations for aberration, nutation, and precession, Bradley expected his successors at Greenwich to reduce these observations when satisfactory solutions to these problems had been established.

Bradley left a national observatory that ran continuously. It was organized, disciplined, and precise. It ran like a ship of the Royal Navy. Francis Baily, who was Flamsteed's earliest biographer, denigrated Bradley for his failure to reduce his observations. John Flamsteed, his hero, had reduced all his observations. Yet all these reductions counted for nought after Bradley discovered the aberration of light, a phenomenon that altered the position of the stars according to the dates and times when they were observed. Bradley was more than grateful for Flamsteed's catalogue of 2,833 stars which largely contributed to his own of 3,222 stars. What Bradley left was 60,000+ observations that were beyond compare in 1762. However, his expectations that these observations would be reduced at the Royal Observatory were to be confounded by the actions of his executors.

The final chapter, *Fundamenta Astronomiæ*, resolves many of the problems faced by Bradley. Bessel's reductions of Bradley's observations were published in 1818, some fifty-six years after his death, a period longer than his working life from 1711 to 1761. The interminable amount of time it took to publish Bradley's Greenwich observations after the registers had been surrendered to Lord North in 1776 until they were published as two volumes in 1805 was unpardonable. That twenty-nine years were spent producing an unreduced edition of these observations in addition to the fourteen years when access to them had been denied by Bradley's executors in a law suit with the Board of Longitude had converted what were 'observations beyond compare' into a historical oddity, thought to be of little worth. Positional astronomy had advanced markedly since the period when Bradley was at the forefront of astrometric practice. At the age of nineteen in 1804, Bessel sent a paper to Olbers on the orbit of Halley's comet, using data from the observations made by Thomas Harriot in 1607. Olbers was the leading cometary observer of his generation and was so impressed with Bessel's work that he later offered him the two volumes of Bradley's observations believing he possessed the required skills needed to reduce them.

Bessel proved to be a mathematical genius. The problems besetting Bradley had become urgent as observations had become even more precise. A constant such as the aberration of light might lead one astronomer to record it as $20.234''$. Another, working to equally precise parameters might come to a value of $20.262''$. There was a proliferation of such results between different astronomers for an increasing number of constants including atmospheric refraction, using different equations in the reduction of the positions of the stars and other celestial objects. It became impossible to gain agreement between astronomers. Bessel openly challenged the notion that a constant such as the aberration of light possessed a *specific* value. Instead he assigned the value of $20.255'' + \Delta$ for the constant of aberration. With the proliferation of different values for an increasing number of constants, he appreciated that these values might be inaccurate and for the first time he assessed the amount by which they were inaccurate. Bessel assigned the parameters within which a constant must lie. In the *Fundamenta Astronomiæ* what really impresses is the way Bessel calculates errors at various stages of his work. By carrying the errors through all of his calculations, they determined that the final errors were more convincing than anything previously derived. They had been calculated as rigorously as the data had allowed.

All of these advances in the reduction of observations were developed by Bessel as he worked on his reduction of the observations made by Bradley and his three consecutive assistants from 1750 to 1762. Bradley left all of the data required to make as accurate a reduction as possible. Bessel's methodology as developed in the *Fundamenta* could be applied to any set of observations or any number of constants. This great work, ostensibly undertaken in order to reduce Bradley's great series of observations, now opened the way to new developments in the science of positional astronomy. The lengthy delays meant few grasped the enormous amount of reliable information contained in them. Until Bessel reduced them and revealed the many riches contained in the data, the observations remained a dead letter. Bessel produced a dependable catalogue of 3,222 stars based on Bradley's observations and included in the *Fundamenta Astronomiæ*.

The publication of the *Fundamenta Astronomiæ* in 1818 in Königsberg not only laid the foundations of astronomical and astrometrical reduction, it also revealed something of the extent of the achievement of James Bradley and his assistants, John Bradley, Charles Mason, and Charles Green. Such a wealth of dependable data had fallen into the hands of men like Samuel Peach whom Bradley entrusted to be his executor and Thomas Hornsby his successor in both of the posts he held at Oxford, denying access to Bradley's reduced observations from 1762 to 1818. Abram Robertson completed his task with greater urgency. Thanks to the foresight of Olbers and the hard work and genius of Bessel, this wealth of raw data proved to be as Bessel described it: 'incomparable'. Every last jot of possible information of each and every observation had been recorded allowing a complete reduction of the recorded data.

Fourteen years after the publication of the *Fundamenta Astronomiæ*, Bradley's earliest biographer Rigaud published the *Miscellaneous Works and Correspondence of the Rev. James Bradley DD, FRS*, at Oxford in 1832. Rigaud also wrote his *Memoirs of Bradley* which was included in the same volume with the *Miscellaneous Works.* The publication of Bessel's reduction of Bradley's Greenwich observations revived interest in the work of Bradley particularly by Rigaud, who possessed a remarkable collection of the works of science and mathematics from the seventeenth and eighteenth centuries. He also became the Savilian Professor of Astronomy after holding the Savilian chair in Geometry. Not only did Rigaud publish much of what remained of Bradley's archive, he also first spent effort tracing this archive, having been compromised by Hornsby's neglect and carelessness. Without the work of Bessel and Rigaud, even this truncated archive might have been lost. I am sure more of the Bradley archive is yet to be located.

1

The King's Observator

A Brief Introduction to the Development of Astronomical Observation using Telescopic Sights and Micrometers up to 1719 with an emphasis on English Practice, highlighting the Work of John Flamsteed, First Astronomer Royal of England.

> I crave the liberty to conceal my name, not to suppress it. I have composed the letters of it written in Latin in this sentence – *In Mathesi a sole fundes.*
>
> John Flamsteed to Henry Oldenburg 4 November 1669.

In 1608, Hans Lippershey a spectacle or eyeglass maker of Middelburg attempted to claim a patent[1] with the States General on an optical device that is now known as a telescope. An application was made on 2 October 1608 for a 'kijker' (looker) but was refused on the grounds that the device was already common knowledge and was being sold in street markets throughout the Netherlands. Nevertheless, the Dutch government granted a contract to Lippershey to make the device to copies of his own design. These early 'spy glasses' were being sold all over the Netherlands and into the Empire as amusing toys that magnified objects about three times. It is entirely impossible to discern the identity of the first person to turn one of these spy glasses to the heavens to observe the Moon, the planets, and the stars. It is generally agreed that the Italian natural philosopher and mathematician Galileo Galilei (1564–1642) was the first to use a telescope to observe the night sky and fully comprehend the philosophical consequences of what he was observing. A case has been made to suggest that the first person to observe the four large moons of Jupiter was Simon Marius, although his reputation was besmirched by his earlier association with Baldassarre Capra's conflict with Galileo.[2] Marius's claims to have observed the satellites before Galileo were challenged as was his integrity.

When Galileo first acquired news of the new optical device he devised his own from first principles. His first attempts magnified up to about ten times and later, his best, about thirty times. All these early devices used *divergent*

The Life and Work of James Bradley. John Fisher, Oxford University Press. © John Fisher (2023).
DOI: 10.1093/oso/9780198884200.003.0002

convex objective lenses and *convergent concave* eyepieces. The resultant image was upright but with a narrow field of view. This arrangement is sometimes referred to as 'Galilean' even though it had been in use before Galileo constructed his earliest devices. He used it to observe the mountains of the Moon and the satellites of Jupiter, both of which contradicted the central contentions of Aristotelian physics and cosmology. According to Aristotle all celestial bodies orbited the earth in perfect regular spheres.[3] All motions were centred on the earth. The observation of the planet Jupiter with its retinue of satellites clearly refuted this assertion. There is no surprise that Aristotelian philosophers argued that these 'moons' were illusions, artefacts of this new unsubstantiated device. Until there was a convincing theory that explained how the new instrument worked, many if not most Aristotelians resisted the reality of the observations recorded in Galileo's small book *Sidereus Nuncius* ('The Starry Messenger' or conversely 'The Message of the Stars') published in 1610 in Venice.

During the following year of 1611, Johannes Kepler (1571–1630) greatly improved the new instrument. As proposed by Kepler in his *Dioptrice* a *convergent convex* object glass and a *convergent convex* eyepiece greatly widened the field of view, but at the cost of inverting the image. For astronomers, this was no disadvantage once it was accepted that south was observed at the top of the field of view. Galileo's telescopes were quite sufficient for his requirements allowing him to observe the phases of Venus as well as permitting him to observe that the Milky Way was made up of untold numbers of stars invisible to the naked eye. Higher magnifications were possible with Kepler's arrangement. To overcome spherical and chromatic aberration, long-focus object glasses were used, leading to the use of aerial telescopes.

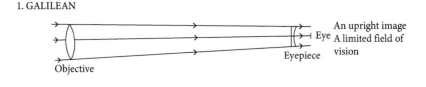

Telescopic Systems–comparing Galilean and Keplerian systems

The next major innovation in the astronomical or astrometrical use of the telescope was made possible by the use of the Keplerian arrangement to measure small angles. In the late 1630s, William Gascoigne (1612–1644) from Thorp-on-the-Hill near Leeds in Yorkshire was working with a Keplerian optical arrangement when he noticed that a thread from a spider's web was caught up in the combined optical focus of both lenses. This allowed it to be observed sharply within the field of view. He recognized that if a wire was placed at this common focus, it would aid in the alignment of the telescope. From this, Gascoigne developed a telescopic sight with a micrometric device situated at the focal point of a Keplerian telescope. He attached it to a quadrant modelled on an instrument that had been used by Tycho Brahe (1546–1601). Gascoigne's five-foot quadrant measured angles that were inconceivable to the great Danish astronomer or his assistant Longomontanus only forty years earlier. Gascoigne recognized that the introduction of two points whose separation could be adjusted by a screw would allow the measurement of the angular size of any image. By knowing the pitch of the screw and the focal length of the lens producing the image, he could calculate the angular size of the object to previously unattainable limits. In 1640, Gascoigne contacted a Lancastrian astronomer from Broughton near Salford, William Crabtree (1610–1644). It was after visiting Gascoigne that Crabtree wrote to his friend Jeremiah Horrocks (1618–1641), describing these remarkable innovations. Understanding the significance of the device, Crabtree wrote to Gascoigne saying, 'My friend Mr. Horrox professeth that little touch which I gave him hath ravished his mind quite from itself and left him in an Exstasie between Admiration and Amazement. I beseech you Sir, slack not your intentions for the Perfection of your begun Wonders.'[4]

Gascoigne received a commission in the army of Charles I and perished during the Battle of Marston Moor, an engagement of the English Civil War on 16 July 1644. William Crabtree, a substantial merchant resident in Manchester and a supporter of the Parliamentary cause in the Civil War was either unable or disinclined to correspond with Gascoigne after the opening conflicts between the Royalist and Parliamentary forces. Crabtree made a will on 19 July 1644 and was interred on 1 August 1644 which suggests he knew his life was ending. The greatest loss, however, was that of the precocious Jeremiah Horrocks. Coming from a Puritan family Horrocks was born in Toxteth Park near Liverpool and attended the University of Cambridge.[5] In 1632, he matriculated at Emmanuel College as a sizar. He was one of few followers of Copernican astronomy at the University at this time. He studied the work of Kepler and Tycho. He was, with Crabtree, one of only two people *known* to have observed the transit of Venus in 1639. Horrocks calculated that the transit of Venus would begin at about 3pm on 24 November 1639 OS. At his location in Much Hoole, he viewed an image of the Sun projected by his telescope onto a plane surface observing the small black circular image of Venus as it began to cross the Sun at 3.15pm. It was clearly visible even though it was a cloudy day. He continued observing the phenomenon until he lost sight of it because the Sun set below his horizon. His treatise *Venus in sole visa* (Venus seen on the Sun) was later published at his own expense by the Danzig-based astronomer Johannes Hevelius (1611–1687). It created a lot of excitement at the nascent Royal Society in 1662 over twenty years after the treatise was written. After returning to Toxteth Park during 1640, Horrocks died from unknown causes on 3 January 1641 aged just twenty-two. Crabtree wrote, 'What an incalculable loss!' Horrocks was the first to show that the Moon in accordance with Kepler's theories moved in an elliptical orbit as did comets. What is remarkable about Horrocks was his dedication to what H. C. Plummer called *scientific determinism* in spite of his pious upbringing and background. There was uncertainty at the time concerning the true model of the world. Criticisms were directed by many of the 'religious' against cosmological schemes that appeared to contradict Holy Writ. Though coming from a devout Puritan background and remaining a devout Christian gentleman, Horrocks supported the claims of scientific evidence. He wrote,

> It is wrong to hold the most noble Science of the Stars guilty of uncertainty on account of some people's uncertain observations. Through no fault of its own it suffers these complaints which arise from its uncertainty and error not of the celestial motions... I do not consider any imperfections in the motions of the stars have so far been detected, nor do I believe that they are ever to be found. Far be it from me to allow that God has created the heavenly bodies more imperfectly than man has observed them.[6]

With the premature deaths of these three remarkable astronomers, William Gascoigne, William Crabtree, and Jeremiah Horrocks, the last clearly a man of immense gifts, the science of astronomy lost much observational and theoretical development. With much of Europe in the throes of war and fratricide, with England, Scotland, and Ireland about to enter a period of bloody civil war[7] astronomical advances counted for little. The death of these three, especially Horrocks, ended a promising chapter.

The subject of eyepiece micrometers became a topic of lively interest during the early years of the Royal Society. In letters exchanged between

Robert Hooke (1635–1703) and Adrien Auzout (1622–1691), the French-man promised that he would explain a way to assign 'the Bigness of the Diameter of all the Planets'.[8] Michael Hunter explains that when Henry Old-enburg (1619–1677) reported Auzout's method of using a filar micrometer to a meeting of the Royal Society on 9 January 1667,

> Dr Wren and Mr Hooke, having related to the society several ways, which they had known long before, of taking the diameters of the planets to sec-onds, were desir'd briefly to describe them, so that it might be signified, to the Parisian philosophers, that it was a thing not at all new among the English.[9]

Auzout was a founder member of the Paris Observatory and briefly a member of the Royal Academy of Sciences from 1666 to 1668.[10] It is not known why he left, though there is suspicion of a dispute within the Academy. David Sellers writes that 'Upon reading Towneley's letter claiming priority for Gascoigne in relation to the micrometer, Auzout was more intrigued than offended. Henri Justel (1620–1693) one of Oldenburg's other French correspondents – reported that "Mr Auzout behaved very well" in face of the controversy'.[11]

As it appears that Auzout took such little offence at being corrected by his English contemporaries, it seems odd that he left Paris in what appeared to be gloomy circumstances. During the early 1660s, Auzout observed various comets and argued that they followed, as did Horrocks over twenty years earlier, elliptical or parabolic paths.

With the Restoration of the English Crown in 1660 and the founding of the Royal Society, organized scientific activity in London revived and the application of the filar micrometer was reinstated. In the Dutch Republic in 1656, Christiaan Huygens described a clock that used a pendulum which con-trolled the timepiece by the stability of the oscillations, greatly increasing its dependability and accuracy. This was a vitally important technical advance when applied to the determination of the position of celestial bodies in right ascension, opening the way to the 'tyranny of the meridian'.[12] Prior to the invention of the pendulum clock even the most accurate timepieces could lose or gain fifteen minutes or more a day, inhibiting the attempted location of celestial bodies by right ascension. By 1657, the clockmaker John Fromanteel (1638–c.1682) the son of Ahasuerus Fromanteel (1607–1693) who had Flem-ish connections was selling pendulum clocks in London based on Huygens' novelty with a manifold improvement in performance over previous clocks, accurate to within a few seconds a day. There are claims that the pendulum clock was first introduced to England by John Harris who had premises at

Holborn Bridge, or even Robert Hooke, who was certainly involved with the early introduction of pendulum clocks into England.[13]

There can be no doubt that Robert Hooke was an immensely talented mechanic. It is difficult to categorize his work in terms that fit the various internal boundaries within which modern natural science has developed. For long when compared with a figure like Newton, he was regarded as a gad-fly, someone who jumped from subject to subject without mastering any of them. This is a shallow analysis. He possessed an ever questing mind. Modern studies of Hooke and his work are drawing out the interconnections between so many of his interests, revealing the continuity of so much of his work. Hooke was the critically important patron of Thomas Tompion (1639–1713) giving him access to Royal patronage. Even so he appears to have abused Tompion in part because of his modest origins. He was the son of a black-smith who possibly worked with his father before he was apprenticed in 1664. Tompion appears to have been connected with Joseph Knibb (1640–1711) who may have put him into contact with Hooke. Knibb and Tompion were both highly regarded as clockmakers. Tompion was employed by Hooke to make a watch of his own design during his dispute with Christiaan Huygens. Hooke developed the balance spring about five years before the Dutch savant who published a paper in the *Journal de Scavans* in 1675. Hooke put forward his watch as a means of determining longitude. Any possibility of the found-ing of a national observatory would undoubtedly have seen him as a possible future director of such an institution. It was therefore a surprise not merely to Hooke but to many of the virtuosi at the Royal Society that a provincial with few apparent claims was appointed as the King's Astronomical Observator in 1675.

Ian Stewart has suggested that the Derbyshire-born Flamsteed cultivated the self-image of an unknown solitary, writing to the Royal Society very much in the well-recognized trope as a dedicated, hard-working astronomer-mathematician.[14] On 4 November 1669, he sent his calculations of an eclipse of the Sun, as well as five occultations of stars by the Moon to the society's secretary Henry Oldenburg, signing his communication with an anagram for 'Johannes Flamsteedius'. Oldenburg replied,

Though you did what you could to hide your name from us, yet your inge-nious and industrious labours for the advancement of Astronomy addressed to the noble President of the Royal Society,[15] and some others of that illus-trious body, did soon discover you to us, upon our solicitous inquiries after their worthy author.[16]

After congratulating the young man upon his skills (he was an autodidact who taught himself mathematics and Latin), Oldenburg signed his letter: 'Your very affectionate friend and real servant'. This revealed Oldenburg's high evaluation of Flamsteed's skills in astronomy.[17] Flamsteed made friends within the London-based cognoscenti and, over the four years up to 1675, his health improved. He came to the notice of the Surveyor of the Ordnance Sir Jonas Moore who presented him with Richard Towneley's micrometer, promising to furnish him with object glasses at a very reasonable price.[18] Later Flamsteed wrote that,

> A Frenchman that called himself Le Sieur de St. Pierre, having some small skill in astronomy and made an interest with a French lady, then in favour at Court,[19] proposed no less than the discovery of the Longitude, and had procured a commission from the King to the Lord Brouncker [et al]. ... Sir Jonas Moore carried me with him to one of their meetings, where I was chosen into their number.[20]

Describing St. Pierre's proposals Flamsteed then asserted,

> It was easy to perceive, from these demands, that the sieur understood not that the best lunar tables differed from the heavens; and that, therefore, his demands were not sufficient for determining the longitude of the place where such observations were, or should be, made, from that to which the lunar tables were fitted, which I represented immediately to the company.[21]

Maunder writes attesting that Flamsteed apprised St. Pierre that his proposal was neither original nor practicable. He stated that if the Frenchman had consulted his countryman Morin's writings he would have discovered that an almost identical proposal was placed before Cardinal Richelieu as early as 1634. It was rejected as impracticable given the then state of astronomical knowledge. Maunder adds, 'Possibly Flamsteed meant further to intimate that St. Pierre had simply stolen his method from Morin, hoping to trade it off upon the government of another country; in which case he would no doubt regard Flamsteed's letter as a warning that he had been found out'.[22]

Charles II reacted with vehement anger at the suggestion that the positions of the stars in the catalogue were in fact false, arguing that they should be observed, examined, and corrected for the use of seamen.[23] Maunder continued his narrative,

Thus, in his twenty-ninth year Flamsteed became the first Astronomer Royal. In many ways he was an ideal man for the post. In the twelve years which had passed since leaving school he had accomplished an amazing amount of work. Despite his ill-health and severe sufferings, and the circumstance – which may be inferred from many expressions in his autobiographies – that he assisted his father in his business, he had made himself master, perhaps more thoroughly than any of his contemporaries, of the entire work of a practical astronomer as it was then understood.[24]

The early years of Flamsteed's incumbency at the Royal Observatory at Greenwich were affected by the over-intrusive interventions of Robert Hooke who must have been shocked that such an office was given to an unknown provincial astronomer. Even before he crossed swords with Hooke, the 'King's Astronomical Observator' expressed his disapproval of the man. This disapprobation was shared, when in correspondence with Halley, he referred to the conflict between Hooke and Hevelius. Flamsteed was fully taken aback when Hooke challenged the eminent, respected astronomer in intemperate and disrespectful tones. Halley was impressed with the quality of Hevelius's instruments. After visiting Hevelius in 1679 Halley wrote to Flamsteed asserting that that both he and the German astronomer reached mutually agreeable readings both to an accuracy of ten arcseconds. Allan Chapman writes,

> much of the anger stemmed not so much from Hooke's scientific claims as from the manner in which he pursued them against the revered German astronomer who was thirty years his senior. His snide references to Hevelius, the sarcastic tone of his Cutlerian Lecture, and his other 'unmannerly railings', served only to anger Hooke's English colleagues and intensify the sense of persecution felt by the Curator of Experiments.[25]

Hevelius used plain sights, but what really characterized Hevelius as an 'ancient' astronomer was the multiplicity of instruments he used. At the close of the seventeenth century, a 'modern' astronomer used fewer instruments; by the middle of the eighteenth century, Bradley used only mural quadrants, transit instruments, a zenith sector, and regulators for the majority of his positional work.[26] A few other instruments were occasionally used, but the work of a positional astronomer relied on the few instruments itemized above. Flamsteed was amongst those astronomers who led the way to the use of a smaller range of instruments though initially he was thwarted by Hooke's intrusions. Undoubtedly, Hooke regarded Flamsteed as a provincial outsider unworthy of his position. Each man in turn supported quite different models of the role of an astronomer. Hooke was from an academic, Oxford collegial heuristic in which the interplay of ideas leading to the development of new theories was valued far more than approaches to astronomy practised by the likes of Flamsteed. The Derbyshire man valued the dedicated hard work displayed by men like Hevelius. His hero was Tycho whose achievements he extolled in the preface to the authorized edition of the *Historia Coelestis Britannica* of 1725.[27]

Flamsteed was substantially superior to Hooke in the disciplines required as a technically proficient positional astronomer. Any future advances were to be made with the use of telescopic sights. Jim Bennett reveals that,

> Flamsteed came to Greenwich already with a small quadrant fitted with such sights. He sought to place this instrument and more ambitious subsequent designs directly in the Gascoigne/Towneley tradition and then minimise any acknowledgement to, in his words, 'a certain member of the Royal Society who always showed off his inventions excessively'[28], namely Robert Hooke.[29]

Hooke's ill-judged assault on the standing of the highly respected Baltic astronomer Hevelius in 1673 lost him a lot of support. When the appointment was made of the King's Astronomical Observator in 1675 he was passed over. It was Hooke's character that had undermined his own expectation of acquiring the post that Flamsteed now filled. What Flamsteed now required to fulfil the tasks he set himself at Greenwich was accurate, reliable instruments, particularly an accurate, dependable quadrant. There had been several years of development in astronomical instrumentation including the development of systems of precision measurement and the accurate division of quadrants and other similar instruments. It seemed the least that Flamsteed must have looked forward to was the acquisition of accurately divided instruments, preferably a good-sized mural quadrant. But as Allan Chapman expresses it, what Flamsteed acquired was something far short of what he had every right to expect as the 'King's Observator'. He writes, 'Flamsteed's early years at Greenwich provide an interesting insight into what were considered important astronomical preliminaries, and how an ambitious research programme was mounted with a collection of instruments that were little more than untried novelties of a few years' standing.'[30]

Flamsteed inherited an empty building and the only instruments he possessed were those he had acquired by his own initiative or through his patron Sir Jonas Moore. These included clocks which had thirteen-foot pendulums beating 'double seconds', a Towneleian micrometer given him in 1670 along with what Allan Chapman describes as 'an indifferent quadrant' and several miscellaneous pieces which he brought from Derbyshire. From 1675 to 1689

Flamsteed undertook the task of 'rectifying the positions of the fixed stars' using as his main instrument a 7-foot iron sextant constructed by Thomas Tompion. It incorporated Hooke's screw graduation[31] which Flamsteed found completely unsatisfactory. In his correspondence with William Molyneux (1656–1698), he expressed his criticisms about Hooke's device although as Allan Chapman suggests much of this criticism may have been more than tinged with his detestation of Hooke. Jim Bennett writes,

> The micrometer screw movement of the telescopic sight along the limb was a feature Flamsteed regarded as a failure, because of the effects of wear, and he replaced this method of subdivision by the more standard diagonal division of Tycho and Hevelius. Here, in the case of a fault, he admitted the influence of Hooke, saying that he had persuaded Moore to adopt this unfortunate feature.[32] It had been executed by Thomas Tompion, whom Hooke had recommended for such work, in explaining his equatorial quadrant in the *Animadversions*.[33] Hooke's *Diary* records a number of discussions of instrument design with Flamsteed in 1674, though they were not always entirely congenial: 'I shewd Flamsteed my quadrant. He is a conceited cocks comb'.[34,35]

With Hooke's tasteless attack on the methods of a highly respected elder practitioner like Hevelius fresh in everyone's minds, he was for Flamsteed a man difficult to respect. Flamsteed believed that the Curator of Experiments had abused his influence at the Royal Society and more directly with his patron Sir Jonas Moore, to foist this device on him. Already, before Flamsteed became the first Astronomer Royal, it is plain to see that he regarded Hooke as a conceited, arrogant person he did not respect. Neither man much cared for the other. But Hooke was a figure who 'did not suffer fools gladly' or at least those he did not respect.

Margaret 'Espinasse captured a sense both of Tompion's sense of dependence on Hooke and Hooke's often expressed impatience, not always recognizing the difficulties a master craftsman had to face putting Hooke's often brilliant ideas into practice. She writes,

> Tompion's realization of his good fortune in Hooke's acquaintance is suggested by his perpetual haunting of him: 'Tompion here till 10 p.m.', 'Mr. Tompion here from 10 to 10. He brought clockwork to show'. The rapid, dynamic and indefatigable Hooke sometimes became impatient of an ordinary pace, and two or three times he rated Tompion for slowness ('Tompion a Slug') and very occasionally Tompion offended him personally: 'Dec. 31,

1675. Tompion a clownish churlish Dog. I have limited him to 3 day and will never come neer him more'. But as usual Hooke's child-like absence of rancour prevailed, and in two days he was back at Tompion's workshop.[36]

Tompion was the finest instrument maker in London at a time when the capital was becoming the leading centre in Europe for the manufacture of precision instruments. Not even he could make the gearing to the order of perfection that was required. The diversity of metals used made them open to differential wear. In a letter to Towneley, Flamsteed wrote that he was,

> much troubled with Mr Hooke who, not being troubled with the use of any instrument, will needs force his ill-contrived devices on us. Hee talks much of such things as none but those yt understand them not can esteem them possible or probable: yt an instrument of no more yn 18 inches radius should measure an angle to less than 6 seconds; that he has an instrument, or quadrant of 36-foot radius yt weighs not a pound and which he can put in his pocket; and several things of the same sort but larger far.[37]

Obviously, the instrument of thirty-six-foot radius that 'weighs not a pound' refers to the zenith sector Hooke used in his attempt to measure annual parallax. It is obvious that Flamsteed had lost all patience with a man he felt as a thorn in his side. Later, when in correspondence with William Molyneux, he adds to his low opinions of Hooke's intrusions the voice of experience. In response to Molyneux he wrote, 'no screw whatever, is so equally made as to have all wayes the same number of revolves in an inch'.[38]

Hooke provided Flamsteed with a ten-foot quadrant[39] that was unreliable and dangerous. The huge weight of the two alidades[40] was located on a single pivot which led to it acting similarly to a pair of scissors. He tore his hands on the instrument, complaining to Sir Jonas Moore that it had deprived his assistant Cuthbert of his fingers.

By 1678, Flamsteed had abandoned the instrument. For several years in lieu of the absence of a reliable quadrant he devised a method of using his sextant in the plane of the meridian in order to record zenith distances. By observing upper and lower transits of *Polaris* in 1677, it allowed him to obtain a moderately accurate estimate of the latitude for the Royal Observatory which after reduction came to 51°28′30″.[41] It was soon evident that the sextant was no more reliable than the quadrant when it was used at his meridian. He discovered that the plane of the limb moved from the perpendicular each time he moved the alidade. The result was that the observed position of a star could vary by as much as 30″ on consecutive observations. Without predictable consistency, an astronomical instrument is useless in a precision science such as positional astronomy.

To add to Flamsteed's woes, a month after his patron Sir Jonas Moore died in August 1679, a delegation was sent to the Royal Observatory when,

> Mr Hooke produced an order to remove the instruments of the Royal Society to Gresham College. They took away: the small quadrant of 5 inches radius with the screwed limb, another quadrant with two telescopes on it, a dividing plate, an instrument of 3 rulers, Mr Hooke's 3 foot quadrant.[42].

Allan Chapman suggests that none of these instruments were vital to Flamsteed's work but it seems that Hooke's actions were singularly vindictive. Given many of Hooke's actions it suggests that he really did resent that he had been overlooked and instead some complete nonentity from Derbyshire had been appointed as the King's Observator in his place. It never seems to have occurred to Hooke that his petulant attitude to all and sundry may have undermined any chance he had for this appointment at Greenwich. Yet one thing is certain: Flamsteed had to cope with inadequate instruments and a lack of resources. A gentleman was expected to provide for his own needs. Unfortunately, until his father died, Flamsteed lacked the personal resources sufficient for these needs and was forced to take on students to help make ends meet.

Given many of the difficulties experienced by Flamsteed in the acquisition of adequate resources during his stewardship of the Royal Observatory, combined with the almost complete lack of support in the provision of tried and tested instruments, we must question whether the government or the Royal Society had any comprehension of the purpose of the observatory at Greenwich. Some of the misunderstanding may have come from the contrasts between Flamsteed's model of the role of an observational astronomer in which industriousness and integrity were valued above the development of philosophical or cosmographical theories. In this sense, Flamsteed perceived the calling of an astronomer as a moral vocation as much as an intellectual occupation.

Did Charles II commission the building of the observatory because his cousin Louis XIV had equipped a handsome building to observe the stars? The Sun King built the Observatoire de Paris to glorify his name but other authorities possessed more prosaic objectives such as the determination of the longitude at sea in order to aid France's extensive maritime and naval interests. In consequence, no expense was spared. Much effort was expended on the production of an accurate map of the Kingdom of France. The initial driving force behind its construction was Jean-Baptiste Colbert (1619-1683).[43] Given the lack of investment for the Royal Observatory at Greenwich, I think it opportune to contrast its resources with those extended to the Paris Observatory. The observatory at Greenwich was constructed to Christopher Wren's design, though it seems Hooke may have designed the Octagon Room. In a distinctly English fashion, the observatory was built with cost borne in mind using bricks from the recently demolished Greenwich Castle.[44] Whereas the first observers[45] at the Paris Observatory, including such leading lights as Giovanni Domenico Cassini (1625-1712), were offered the best instruments available and given the resources needed, the King's Observator John Flamsteed was treated with neglect bordering on contempt. In a period of growing naval and imperial rivalry with France, there appears to have been little appreciation of this in ruling circles in London.[46] Cassini used the Paris Observatory along with Jean Picard (1620-1682), Ole Römer (1644-1710), and Christiaan Huygens (1629-1695). During the 1680s after the death of his patron, Flamsteed was largely bereft of support and being seen by some of the leading lights within the Royal Society as an 'outsider'. He also lacked effective support to give sufficient succour to the Royal Observatory.

Jim Bennett appraised the character of Flamsteed as morose. Long before he broke with Halley, there is strong evidence that the first Astronomer Royal was rather sullen. In a period when he was short of resources, his somewhat melancholic outlook did little to support his cause.[47] The situation changed for Flamsteed with the death of his father on 8 March 1688. The forty-two-year-old Flamsteed inherited his father Stephen's wealth. His father had been a successful maltster who also owned a lead mine in the Peak District of Derbyshire. John's education had been compromised by his father who had withdrawn him from school when he was fourteen. He was a sickly child suffering from arthritis in his joints. In September 1670, he went up to Jesus College, Cambridge but he never took up full residence. He was there for a couple of months in 1674. Ordained a deacon he was preparing to take up a living in Derbyshire before he was called back to London by his patron Sir Jonas Moore,

He arrived in London on 2 February 1675 staying with Moore at the Tower of London. Moore had been in contact with the Royal Society with an offer to build an observatory. This objective was overtaken by Charles II's appointment of a Royal Commission in December 1674 leading to the construction of the Royal Observatory in order to map the stars and determine the longitude from the position of the Moon. Flamsteed was taken by Silius Titus (1623-1704) who was a member of the Royal Commission to meet the King, and was admitted as an assistant to the commission, supplying observations and comments on the original proposals made by Sieur de St Pierre. On 4 March 1675, Flamsteed was appointed by royal warrant as 'the King's Astronomical Observator' with an allowance of just £100 per annum.[48] On the one hand, it seems that everything proceeded with urgency and yet on the other this important task was ill-financed. The allowance was completely inadequate for what seems to have been regarded as an important task of state sufficiently important for the King to call for a royal commission and appoint someone in the King's name.

Flamsteed worked under parlous circumstances for several years. It didn't help that being resident several miles from the City of London, he remained an 'outsider', usually divorced from the centres of scientific discourse. The years from 1675 to 1689 were years of lost opportunity. Allan Chapman writes,

> Not until 1689 did the Observatory possess an adequate meridian instrument, which Flamsteed provided from his own pocket. Following the death of his assistant, Stafford in 1688 Flamsteed engaged the 'very able mechanic and ... expert mathematician' Abraham Sharp[49] to take his place, and construct a new mural arc. It took 14 months to build and cost £120.[50] Like the unsuccessful instrument of 1681 upon which it was based the new arc was of 7-ft radius and covered 140° of sky, so that it would be possible to obtain the Observatory's latitude by direct observations of the Pole Star, instead of having to triangulate its position from the circumpolar stars with the sextant.[51]

After Flamsteed laid the foundation stone in August 1675, it seems evident that those in governing circles paid little attention to the provision of instruments for the Royal Observatory in order to make the astronomical observations required to fulfil its function. Though nominally under the authority of the Royal Ordnance, the demise of Sir Jonas Moore in 1679 left Flamsteed largely bereft of support. Flamsteed was further handicapped by his hesitancy to approach those in positions of authority after the death of his patron. The resources available to Cassini and his fellow academicians in Paris compared with those made available to Flamsteed at Greenwich appear almost insulting. After a decade and a half of neglect and intrusive interference by Hooke, he took possession of Sharp's new mural arc which featured degree graduations of great quality. In contrast to later generations of instrument makers, Sharp divided the meridian instrument whilst it was in situ on the meridian wall. Later craftsmen divided such instruments flat on a workshop floor or on a supporting work bench. There are accounts about how the instrument was divided, such as by Allan Chapman in *Dividing the Circle*,[52] although he finds some parts of the process unclear. He writes,

> the Flamsteed mural arc was itself the last major observatory instrument to use the 'diagonal' method of subdividing degrees. There is no written evidence to suggest this line of development from Hevelius through Hooke to Flamsteed, for they were not much inclined to paying mutual compliments, though when one examines their descriptions and engravings, a distinct thread emerges.[53]

As well as being a more than competent artisan, Abraham Sharp (1653–1742) was also an excellent astronomer. In this, he prefigures the multifaceted skills that were later displayed by George Graham. His practice of observing the positions of zenith stars with the alidade of the meridian instrument preliminary to making the divisions enabled the alidade to be used as a later check on the collimation of the instrument, in a similar way that Bradley was later to achieve with the zenith sector erected at Elizabeth Pound's town house in Wanstead from August 1727. By 1688, Flamsteed appreciated the advantage of having two independent means of observing any angle so one acted as a control over the other. Chapman states that it wasn't until Bradley took over the Royal Observatory that an independent instrument, a transit or much more accurately his zenith sector, was used to check the collimation of the other instruments in the observatory. As I have already intimated, Nevil Maskelyne later verified that the sector was accurate to within a half-second of arc.

From 1689, Flamsteed was able make observations with more certainty and less inconvenience. He was able to locate and measure declinations with ease, as well as the latitude of the Pole Star by direct observation. With the aid of a trustworthy clock, Flamsteed was also able to observe the differences in the times of transits of any two objects giving the differences in their right ascensions. Using these instruments, he was able to locate right ascensions and declinations together in a single operation. Tompion's sextant fell into disuse. Flamsteed suffered greatly from arthritis so the new instrument possessed another advantage being housed within a small wooden building so he avoided having to make observations exposed to the elements. He discovered that the roof slit reduced glare, enabling him to observe dim stars with the naked eye.[54] Flamsteed was able derive accurate solar coordinates from which he was able to obtain more accurate details of the tropical year.

Flamsteed's experiences as the first Astronomer Royal were climaxed by his dispute with the Board of Visitors of the Royal Observatory, a body instituted by the Royal Society under the direction of Isaac Newton as a means of gaining control over Flamsteed's work at Greenwich. It was the culmination of the growing antipathy that had developed between Flamsteed and Newton. The Astronomer Royal was comparatively isolated downriver at Greenwich and was largely ignored by many of the virtuosi based at the Royal Society in London. An attentive reader will ask why Flamsteed did not try to improve his situation by approaching those in authority long before this confrontation with the President of the Royal Society. Flamsteed perceived himself as an astronomer in a trope modelled on predecessors such as Tycho or Regiomontanus, as a court-based *mathematicus*. He had been appointed by Royal Warrant and other than the patronage he received from the Surveyor-General of the Ordnance did not seek or receive patronage from any other source.

The growth of influence of the Royal Society together with other wider sources of patronage did not fit Flamsteed's self-image. He was an outsider in several ways not merely in being an unknown from the provinces or some-one located outside London several miles downriver. He was an outsider in the sense that he saw himself as someone apart from many of the London-based cognoscenti. His difficulties with the Royal Society began well before his conflict with Newton. With Sir Jonas Moore's passing only four years after he gained the post as the King's Observator, he lost much of his support within the Royal Society. His subsequent lack of confidence and his defensiveness served him ill. His taciturn character hardened into mistrust and suspicion. When he finally acquired the financial resources he required after the death of his father, he commissioned the instruments he needed from his own assets. Flamsteed's fractured relationship with the Royal Society was constructed in part at least on the self-image he maintained. He held a low opinion of many of its leading members and this explains much about his relations before and after January 1709.[55] At this juncture, the society's council ruled that all members whose dues were in arrears should pay for a year or be struck off the lists. It provided the mechanism required for Flamsteed's expulsion from the Royal Society on 9 November 1709.

During June 1709, William Derham corresponding with Flamsteed made clear his support. Derham was a member of a small group within the society who were opposed to the dominance of Newton. On 14 December 1710, Flamsteed learned of the appointment of the President and Council Members of the Royal Society as Visitors with powers to supervise the Royal Observatory following a Royal Warrant dated 12 December 1710. Queen Anne was easily influenced by a man she believed had saved the English currency from forgers and clippers. Her late escort Prince George of Denmark had admired Newton's scientific achievements. Flamsteed was appalled by the notion of having to answer to Visitors he regarded as a tool created to demean or humiliate him. He earnestly petitioned the Queen, seeking a revocation of the decision. Conceding that there would be some limitations placed on him he attempted to ensure that the Board of Visitors be made up of 'more qualified' figures, people experienced in observational astronomy. All such attempts failed and the Royal Society proceeded with the issue of orders for 'the annual handing over of observations'. The Royal Society interpreted the Royal Warrant as a command to give instructions about repairs to the observatory's instruments and as authorization to take over control of them, even though most of them were the personal property of the Astronomer Royal.

It would be easy to suppose neither Newton nor the Council had any understanding of Flamsteed's situation from 1675 to 1689 or that the instruments he was using were his own property, purchased out of his pocket. This charitable interpretation does not bear investigation. In March 1711, Dr John Arbuthnot (1667–1735) suggested that the Royal Society buy the instruments. Flamsteed said he would not sell them. Any doubts that there was a lack of appreciation that Flamsteed owned the instruments he used at the Royal Observatory must be dispelled at the outset. Seven months after his encounter with Arbuthnot the Astronomer Royal met with Newton and two Council members. Flamsteed wrote his account of this bruising encounter,

> I have had another contest with the PR.RS who had formed a plot to make my Instruments theirs and sent for me to a Committee where onely himselfe and two Physicians Dr Slone and another as little skilful as himself [Richard Mead] were present the pr[esident] ran himselfe into a great heat and very indecent passion I had resolved aforhand his kn—sh talk should not move me Shewed him that all the Instruments in the Observatory were my own ... this netled him ... and he sayd as good have no Observatory as no Instruments. I comp[lained] then of my Catalogue being printed by Raymer[56] without my knowledge and that I was Robd of the fruits of my labor at this he fired and cald me all the ill Names Puppy etc [that he] could think of[57]

The 'unpleasantness' that existed between Flamsteed and Newton had been years in gestation. Some writers have sought reasons for this breakdown between the Astronomer Royal and the President of the Royal Society. No matter how many sound notions have been manufactured to explain the episode, it remains that they simply loathed each other. To emphasize the exceptional way that Flamsteed was treated by the Council of the Royal Society, I quote Mordechai Feingold who wrote concerning his expulsion from the Society,

> Most remarkable about Flamsteed's ouster, however, was not simply the Astronomer Royal alone in 1709 suffered such indignity. Rather, it was the unwillingness on the part of the Society to extend to Flamsteed the exemption that in the previous two years had been granted to no fewer than 18 Fellows – especially in light of the fact that he had been the beneficiary of such an exemption for more than thirty years! As for sheer irony, it cannot be ignored that three of the six Council Members who resolved on Flamsteed's removal were among those exempted, while a fourth, Newton, had enjoyed such a condition until his election as President in 1703.[58]

Relations between Newton and Flamsteed deteriorated from when Newton discerned that Flamsteed had failed to supply him with data on the 'double comet' of 1680 while Flamsteed took exception to Newton's apparent offhanded manner. Flamsteed perceived that Newton regarded him as an official functionary. It was Flamsteed who first suggested that the two comets might be one in the face of Newton's initial scepticism. Flamsteed was openly sceptical about Newton's theoretical 'conjectures'. He could not accept that the motion of the planets could affect the motions of other planets, believing that the influence of the Sun was so great that the influence of the planets would be of no consequence. It is possible that Flamsteed may have conceived this force as sharing characteristics with magnetism. The influence of magnets he observed was severely constrained by distance, becoming virtually nonexistent more than a few yards away. When a handful of years later in 1687 Newton published the *Principia*, Flamsteed remained unimpressed with its theoretical conjectures. Flamsteed may have underestimated Newton, associating him in his mind with the empty claims of Newton's *bête noir* Robert Hooke. In this period during the early 1680s, Flamsteed fell out with Halley who had assisted him at Greenwich. Halley corrected several errors in a table of tides that Flamsteed had compiled and the tetchy Astronomer Royal took offence at his 'friend's' actions.

Apparently Flamsteed and Halley were respectful colleagues up to the early 1680s, though the King's Observator explained his objections to Halley citing his 'atheism'.[59] There is some evidence that Flamsteed began to mistrust Halley in the midst of the coffee house culture where so much scientific and social gossip found expression. Flamsteed's discomfiture in the midst of these surroundings was palpable. Adrian Johns writes, 'It is thus understandable that from his distant vantage-point of Greenwich the new Astronomer Royal could come to see the capital's coffee houses as powerful and incorrigible tribunals, standing ready to condemn him. His rivals were not so fearful.'[60]

Hooke's antipathy towards Flamsteed was probably openly expressed within the discussions at Garraway's coffee house following meetings at the Royal Society. Halley was much more at ease in this company, and it appears that Hooke sought to have Flamsteed replaced with his fellow Tory. Without doubt Flamsteed saw Halley as a personal and political rival and as someone he mistrusted. This mistrust dated to the early 1680s.

On 24 January 1684, at the Royal Society and in discussions later at Garraway's coffee house Wren, Hooke, and Halley discussed Halley's demonstration that Kepler's third law of planetary motion, the harmonic law, necessarily implied that the attraction of the Sun on the planets was the inverse of the square of the distance between them. Wren offered a prize of some books to the first of Hooke or Halley to show that the inverse square law led to elliptical orbits. Both Hooke and Halley were possibly aware that the problem could be solved by Newton. For Hooke an approach to Newton was quite unthinkable, for he would lose face after indulging in too many disputes with his imperious contemporary. It took a few months before Halley finally approached Newton. He claimed to have solved the problem. Newton was a hoarder and was unable to locate his solution to the problem when he proved that the path of a body in motion around the Sun assuming its gravity diminished according to the square of its distance would always be an ellipse. Alan Cook surmised that without Halley's approach to Newton and his part in the 'conception, development, printing and publication of the *Principia*', it would very likely have not been composed.[61] Halley had the work published from his own financial resources because the Royal Society's coffers were almost empty at the time. Halley's star within the Royal Society rose in inverse proportion to the setting of Flamsteed's. It is possible to discern that amongst the various reasons why Flamsteed remained singularly unimpressed with Newton's *Principia* was that the man he perceived as his bitter rival was behind its publication.

In the breakdown of respect between Flamsteed and Newton, it has to be admitted that Newton became more vindictive in his handling of the Astronomer Royal. During the reign of Queen Anne he began to use his courtly influence to push for a stellar catalogue to be printed under the auspices of the Royal Society. This was to be with or without Flamsteed's participation or permission. This dimension of Newton's influence at court is sometimes lost sight of.

Towards the end of the seventeenth century, England's currency had become so debased that foreign traders and markets commonly refused to accept English coinage. It threatened William III's plans in his conflict with Louis XIV. Newton, in his role as Warden and then Master of the Royal Mint, reformed the coinage and pursued the counterfeiters and clippers with the determination that was characteristic of all his work. If the most persistent and inventive forger William Chaloner had been more aware of the way Newton treated those who crossed swords with him he would have been more circumspect. Newton watched Chaloner hang at Tyburn in 1699. It was for the manner that Newton had ruthlessly pursued the counterfeiters and clippers that he was knighted, not for his scientific achievements. After the death of his nemesis Robert Hooke in 1703 and his subsequent accession to the Presidency of the Royal Society Newton became almost unchallengeable. I say 'almost' for if his enemies had ever become aware of Newton's unorthodox religious beliefs, particularly his 'anti-trinitarianism', he might have been undone.

Against a Royal Warrant, Flamsteed was forced to hand over his observations to the Royal Society in the clear knowledge that they would come under the control of his now bitter enemy Sir Isaac Newton who Flamsteed increasingly referred to as S I N. Flamsteed now delayed as much as possible. He raised objections to Newton's measurement of the size of the stars in the *Opticks*. Yet he was excluded from all discussions about the catalogue. In 1711, Newton was able to convince Queen Anne to take up her late husband's support for the publication of the catalogue.[62] Newton now made it clear to Flamsteed that any future subterfuges to delay would be regarded as disobeying the Queen's command, an act of treason. It was the solar eclipse of 4 July 1711 that brought everything to a head. The observations would prove invaluable to the advancement of Newton's lunar theory but Flamsteed adamantly refused to observe the phenomenon, an extraordinary step for any astronomer to make let alone the Astronomer Royal. Flamsteed was brought before the Council under the chair of Newton who ordered the immediate publication of Flamsteed's observations in spite of his objections that many of them (made with the sextant) were at fault and needed revising. Under the editorship of Halley, they were published as the *Historia Coelestis* in 1712 against Flamsteed's will and without his participation.

The influence of Newton at Court waned after the death of Queen Anne on 1 August 1714. With the accession of the German-speaking George I of the new Hanoverian line, Flamsteed's influence recovered for he was well established as a Whig partisan and therefore perceived as a reliable supporter of the new regime, while Newton's association with Anne and her Tory associates made him slightly suspect in the eyes of the new King. Added to this was Newton's long-standing feud with Gottfried Wilhelm Leibniz (1646–1716) whose patron was the Elector of Hanover now George I of England who demanded a reconciliation. After Leibniz's death in 1716, Newton continued to do all in his power to diminish the reputation of his opponent over the priority of the invention of the calculus. In 1726, with the publication of the third edition of the *Principia* Newton expunged every reference to Leibniz (just as in the first edition he did the same to Hooke). Even though most scholars agree that Newton was the first to develop the calculus (as his method of fluxions) he still felt he had to diminish the contributions made by the German mathematician. Yet it was Leibniz's clearer notation that won the day.

Although some have expressed a high opinion of Flamsteed and suggested that he was an approachable even a kindly man, there were others who were treated with suspicion and hostility. The mutual hostility of Newton, Hooke, and Flamsteed is well recorded but Edmond Halley was amongst those who raised Flamsteed's ire. Various reasons have been suggested for his turning on his protégé but they seem on reflection to be more in the way of symptoms rather than causes.

With the publication of Halley's *Catalogum stellarum australium sive supplementum catalogi Tychonici* in 1679, the King's Observator was soon accusing Halley of various wrongdoings. Flamsteed was already discomforted by his rival's observations made at St. Helena when Halley was a Captain in the Royal Navy. The catalogue became a weapon in the hands of Flamsteed's adversaries in the Royal Society when they asked why the King's Observator hadn't produced a similar catalogue of the northern stars. Feingold writes, 'Halley's rise in the Society's affairs witnessed a corresponding withdrawal by Flamsteed. Thus, despite the fact that he was a member of Council in 1685–86. Flamsteed failed to attend most meetings'.[63]

Flamsteed's disaffection from the Royal Society was gradual but noticeable, certainly by those who were inclined to dislike him or those who objected to his opposition to the political stance of the society.[64] Even so, he was elected to Council in 1692, 1694, and 1699. Feingold comments that the routine business of the Royal Society fell into the hands of men that Flamsteed perceived as his enemies, most notably Halley and Hans Sloane (1660–1753).[65]

Demands from within the Royal Society for Flamsteed to publish were stoutly resisted. He remained dissatisfied with his results, most notably those he obtained with the sextant. By the turn of the century, he sought to deflect his detractors seeking the support of various dignitaries including the Archbishop of Canterbury Thomas Tenison, and various members of the Royal Society he personally respected, including Sir Christopher Wren and Thomas Smith. He explained the delays. It was because he was not satisfied with his results, for to be kind to him Flamsteed was a perfectionist and wanted to delay publication until 'the world will be satisfied that I have not misspent my time'.[66] In spite of Smith's sympathy and a promise to defend the Astronomer Royal from further attack, his relations with the Royal Society continued to deteriorate. Smith suggested he write an 'apologetical' letter to the Society.[67] As Feingold observes, just before Newton took control of the Royal Society in 1703, Flamsteed had become a nominal member with barely a friend among the Fellows. Feingold writes that an entry in the Council's minutes for 16 February 1703 is revealing. It states, 'Sr Christopher Wren proposed that the Telescope given by Mr Huygens to ye Society should be set up in [St.] Pauls and astronomical observations made. The Council thanked him & desired him to take care of it'. It betrays the Royal Society's indifference to the Astronomer Royal. That Wren suggested making independent observations reveals why Newton gained supervisory control over the Royal Observatory and gained possession and control over Flamsteed's stellar catalogue. Flamsteed lacked the pragmatism to meet the society half way. For him the catalogue had to be perfect before releasing it for fear it would undermine his reputation. He failed to see that if he had courted friends where it mattered in the Royal Society, his reputation would never have come into question. His posthumous reputation was rescued by the publication, five years after his death in 1725, of the *Historia Coelestis Britannica,* his authorized catalogue, by his widow Margaret. It led to the publication in 1729 of the *Atlas Coelestis* which at this time was the largest stellar atlas published. It was the earliest comprehensive telescopic star atlas published in England. It contained twenty-six maps of the constellations visible from Greenwich with drawings in the Rococo style by James Thornhill. It was a handsome volume that was an immediate success, and it remained a standard reference for astronomers for a century.

My account of the work of James Bradley, the third Astronomer Royal, is a lesson in the careful acquisition of resources, the pragmatic acceptance of patronage and open friendship. It opened many doors to an astronomer who was as gifted as Flamsteed certainly appeared. The real tragedy of Flamsteed's life recognized after his death is that his determination to achieve perfection was undermined by a two fatal flaws. First was his fractious character which denied him the trust and friendship of many who would have supported him and second was his failure to investigate his observations of *Polaris* which may have led him to the discovery of aberration. If he had pursued the reasons why the pole star regularly wandered over the years he observed it, he may have latched on to the explanation. A directed programme of observation made by Bradley from 1725 to 1728, identifying the phenomenon of a new discovered motion, later called the aberration of light, could so easily have been Flamsteed's redemption. The greatest irony lay in the fact that Bradley's discovery made all prior positional observations, no matter how carefully or precisely procured, invalid, and inaccurate. So much of Flamsteed's work was in vain though his catalogue of 2,883 stars and his nomenclature was to prove a foundation for the later achievements not only of Bradley but many other astronomers too. For this, posterity owes the first Astronomer Royal immense gratitude.

Notes

1. The Venetian patent system giving protection for ten years dates from 1474. It led to a wide diffusion of local patents of the sort that was sought by Lippershey. The English patent system evolved from various trade practices into the earliest modern patent system. It formed the fundamental legal protection that laid the foundations of the commercial and industrial revolutions in England.

2. In 1607, Baldassarre Capra (1580–1626) published the tract *Usus et fabrica circini cuiusdam proportionis, …* which was very largely a translation into Latin of Galileo's *Le operazioni del compasso geometric et military* of 1606, accusing Galileo of plagiarism. Having already crossed swords with Capra in a dispute over Kepler's star (now designated Kepler's supernova) and having defended himself against an earlier claim of plagiarism, Galileo took his complaint to the rectors of his university where the testimony of Paolo Sarpi was crucial in having Capra's claim thrown out. Galileo believed that Simon Marius had been the real source of Capra's false claim. When it was suggested that Marius claimed priority over the observation of Jupiter's satellites, his integrity was called into question.

3. In accord with the cosmology devised by the geometrician Eudoxus, an acolyte of Plato at the Academy.

4. William Crabtree to William Gascoigne, 28 December 1640.

5. Jeremiah Horrocks must have been a member of the Church of England to allow him to matriculate at Emmanuel College, Cambridge.

6. H. C. Plummer, 'Horrocks and his Opera Posthuma', *Notes and Records of the Royal Society of London*, Vol. 3, 1940–1941, pp. 39–52.

7. The Thirty Years War ravaged Germany from 1618 to 1648 and the British Civil Wars lasted from 1642 to 1651.

8. *Philosophical Transactions,* Vol. 1, p. 63.

9. Michael Hunter, *Robert Hooke: Tercentennial Studies*, London, 2003.

10. Adrien Auzout (1622–1691) was elected a Fellow of the Royal Society in 1666. Little is known of Auzout after this period. He went to live in Italy and died in Rome.

11. Henri Justel to Henry Oldenburg, mid-June 1667.

12. This term was used during the transition from the location of celestial bodies by triangulation to their positioning by right ascension and declination, the celestial equivalents of longitude and latitude.

13. Herbert Cescinsky and Malcolm R. Webster, *English Domestic Clocks*, New York, 1968, p. 285.

14. Ian G. Stewart, *'Professor' John Flamsteed* in *Flamsteed's Stars: New Perspectives on the Life and Work of the First Astronomer Royal (1646–1719),* ed. Frances Willmoth et al., London, 1997. pp. 145–166.

15. From 1662 to 1677, this was the noble member Viscount Brouncker, a mathematician who would have appreciated Flamsteed's efforts.

16. Henry Oldenburg to John Flamsteed, 14 January 1670.

17. E. Walter Maunder, *The Royal Greenwich Observatory: A Glance at its History and Work,* London, 1900. Reprinted Cambridge University Press, 2013. p. 30.

18. Ibid. p. 31.

19. Louise de Kérouaille, (1649–1734) who became the Duchess of Portsmouth, was a mistress of Charles II.

20. E. Walter Maunder, *The Royal Greenwich Observatory: A Glance at its History and Work,* London, 1900. Reprinted Cambridge, 2013. p. 32.

21. Ibid. p. 33.

22. Ibid. p. 34.

23. Ibid. p. 34.

24. Ibid. pp. 34–37.

25. Allan Chapman, *Dividing the Circle: The Development of Critical Angular Measurement in Astronomy 1500–1850.* 2nd edn. Chichester, 1995. p. 39.

26. A regulator is an accurate clock.

27. John Flamsteed, *Preface to Historia Coelestis Britannica,* edited and introduced by Allan Chapman, based on a translation by Alison Dione Johnson, Greenwich, National Maritime Museum, Maritime Monographs and Reports, No. 52, 1982, vi + 222 pp.

28. Ibid. p. 118.

29. Jim Bennett, *Flamsteed's Career in Astronomy: Nobility, Morality and Public Utility.* In *Flamsteed's Stars. New Perspectives on the Life and Work of the First Astronomer Royal (1646–1719),* ed. Frances Willmoth et al., London, 1997, p. 24.

30. Allan Chapman, *Dividing the Circle: The Development of Critical Angular Measurement in Astronomy 1500–1850.* 2nd edn. Chichester, 1995, p. 49.

31. Ibid. p. 50.

32. Baily, Francis, An Account of the Rev. John Flamsteed, the First Astronomer Royal, compiled from his Manuscripts and other authentic Documents, never before published, London, 1835, p. 39.

33. Robert Hooke, *Lectiones Cutlerianæ,* London, 1679, p. 54.

34. H. W. Robinson and W. Adams, *The Diary of Robert Hooke 1672–1680,* London, 1935, p. 105.

35. Jim Bennett, *Flamsteed's Career in Astronomy: Nobility, Morality and Public Utility*. In *Flamsteed's Stars. New Perspectives on the Life and Work of the First Astronomer Royal (1646–1719)*, ed. Frances Willmoth et al., London, 1997, p. 26.

36. Margaret 'Espinasse, *Robert Hooke*, London, 1956, p. 136.

37. John Flamsteed to Richard Towneley, 3 July 1675, Royal Society Library, MS 243, fol. 8r. I thank Allan Chapman for the reference to this item, which I recently located in my Royal Society Notebook.

38. John Flamsteed to William Molyneux, 4 November 1686. MS. Southampton Records Office. D/M 1/1, fols, 99v & 100v. I located this source when I visited Southampton when seeking Samuel Molyneux's papers. I made a note of it dated October 1990, in a miscellaneous notebook, only to rediscover it as a footnote in Allan Chapman's *Dividing the Circle*. A case of two drinking from the same well. I was working on sources which I later used in my MSc dissertation.

39. Again an example of the impracticality of the instruments that Hooke hoisted on to Flamsteed. Add to this Hooke's bullying attitude towards a man he evidently held in contempt and it is all too easy to understand the causes of the Astronomer Royal's further resentments, additional to his personal dislike following the episode with Hevelius.

40. An alidade is a pointer used for measuring angles. In divided instruments at this period, they usually carried telescopic sights.

41. Allan Chapman, *Dividing the Circle: The Development of Critical Angular Measurement in Astronomy 1500–1850*. 2nd edn. Chichester, 1995, p. 54.

42. John Flamsteed, Cambridge University Library, *Royal Greenwich Observatory Papers*, RGO 1/188, 'Foul Observations Book', 26 September 1679.

43. One of Colbert's grandfathers was Scottish, a reflection of the Auld Alliance. This is a long standing memory from my own extensive reading.

44. The Royal Exchequer was short of money. This was reflected in the secret clauses of the Treaty of Dover whereby Charles II received moneys from Louis XIV with England becoming a client state of France in opposition to the Dutch Republic.

45. There was no director as such, though G. D. Cassini took the lead in the work of the observatory. The first officially appointed director was Cassini's grandson Cassini de Thury in 1771.

46. Given the secret clauses of the Treaty of Dover in 1670, it may well have been difficult for Charles II to make the Royal Observatory at Greenwich into an effective rival to the Paris Observatory. In no real sense was Charles in a position to attempt to rival the resources of his cousin Louis XIV of France.

47. Jim Bennett, *Flamsteed's Career in Astronomy: Nobility, Morality and Public Utility*. In *Flamsteed's Stars. New Perspectives on the Life and Work of the First Astronomer Royal (1646–1719)*, ed. Frances Willmoth et al., London, 1997.

48. With this annual income of £100 (the equivalent of £25,000 pa) Flamsteed was supposed to purchase instruments and pay for any assistants. As a gentleman he was expected to provide for his own needs, but in reality the King's Astronomical Observator needed to supplement his income by taking on students.

49. Abraham Sharp of Little Horton, Bradford, Yorkshire, 1653–1742.

50. John Flamsteed, *Historia Coelestis Britannica*, Vol III, p. 108.

51. Allan Chapman, *Dividing the Circle: The Development of Critical Angular Measurement in Astronomy 1500–1850.* 2nd edn. Chichester, 1995, p. 54

52. Ibid. pp. 55–56.

53. Ibid. p. 56.

54. Ibid. p. 57.

55. January 1708/09 OS.

56. 'Raymer' is what he calls Halley. In this he is referring to Tycho's rival who claimed Tycho's system as his own.

57. John Flamsteed, *The Correspondence of John Flamsteed, the First Astronomer Royal,* Vol. 3, Introduction, p. l.

58. Mordechai Feingold, 'Astronomy and Strife: John Flamsteed and the Royal Society' in *Flamsteed's Stars. New Perspectives on the Life and Work of the First Astronomer Royal (1646–1719),* ed. Frances Willmoth et al., London, 1997. p. 31.

59. There is no evidence that Halley was an atheist in the sense known in the twentieth century onwards. What some people objected to was the casual off-hand way he regarded many religious doctrines and practices. Some, including Flamsteed, found this objectionable.

60. Adrian Johns, *Flamsteed's Optics and the Identity of the Astronomical Observer* in *Flamsteed's Stars, New Perspectives on the Life and Work of the First Astronomer Royal,* ed. Frances Willmoth et al., London, 1997. p. 81.

61. Alan Cook, Edmond Halley and Newton's Principia, *Notes and Records of the Royal Society of London,* Vol. 45, No. 2, 1991, pp. 129–138.

62. Prince George of Denmark died 28 October 1708, aged fifty-five.

63. Mordechai Feingold, 'Astronomy and Strife: John Flamsteed and the Royal Society' in *Flamsteed's Stars. New Perspectives on the Life and Work of the First Astronomer Royal (1646–1719),* ed. Frances Willmoth et al., London, 1997, p. 46.

64. The Royal Society was associated in the minds of many Royalists and Tories with seditious sympathies which is why the society valued Royalists such as Christopher Wren amongst their number. The very first society meeting following a lecture given at Gresham College by Christopher Wren on 28 November 1660 indicates that this was an attempt to be seen supportive of the Restoration. In the mind of John Flamsteed a strong supporter of the Whigs, the Royal Society appeared to have too many Tory sympathies and even though the motto of the society *Nullius in verba* (take no one's word for it) was adopted, not only because of its pursuit of natural and useful knowledge by means of experiments and observations, but also because it forbade all discussion of matters political and religious.

65. Ibid. p. 46.

66. Ibid. p. 47.

67. Francis Baily, *An Account of the Revd. John Flamsteed, the First Astronomer-Royal* London, 1835, pp. 744–747.

2

May It Please Your Honours

The Critical Importance to James Bradley of the Life and Work of James Pound

> and finding we had Rice for the rest for 10 days at near half a pint each
> man diem after we had taken in Water and Ballast we heaved upon our
> Anchor the same 3rd Day a little after Sun Set, and resolved to try for
> Malacca.[1]
>
> James Pound.

In any account of the life and work of James Bradley, the critical importance of the life of his maternal uncle James Pound (1669–1724)[2] looms large. Without Pound, it is entirely inconceivable that Bradley would have become an astronomer, the discoverer of the gravitational resonances of the Jovian satellites, of the aberration of light, and the nutation of the Earth's axis, or the author of one of the most important series of stellar observations of the eighteenth century. This chapter is an appraisal of the life of a man even more obscure than his nephew. What is so important about Pound is that he was responsible for Bradley's education and for his introduction to some of the leading figures of the Royal Society. He was not only responsible for Bradley's introduction to astronomy but also for placing him in contact with influential patronage. The title of this chapter comes from a letter written to the managers of the East India Company[3] from the Dutch stronghold of Batavia in Java[4] informing them that their station in the South China Sea had been reduced to ashes in a treacherous uprising and massacre.

In his journal, Pound listed the names of those who perished on Pulo Condore before the massacre. In a period that couldn't have been longer than just under two years, this tiny outpost witnessed the deaths of thirty-one men, six women, and a child. Pound does not give the causes of their deaths, but they probably perished after contracting any of a dozen or more tropical diseases. Richard Griffith, his wife, and their child Florentia were all among the dead.[5] Pound states they all died from natural causes. Pound might have contracted any of these illnesses so it was singularly fortunate that he survived the years he travelled throughout eastern Asia from 1699 to 1706. During this period

The Life and Work of James Bradley. John Fisher, Oxford University Press. © John Fisher (2023).
DOI: 10.1093/oso/9780198884200.003.0003

Pound was a friend[6] of Flamsteed's, engaged by the Astronomer Royal to make astronomical observations on his behalf. Larry Stewart refers to various 'hitchhikers', naturalists, and astronomers, who were regarded as 'supercargo'. Stewart clearly perceives that ships, chronometers and compasses, charts, and telescopes were the means by which European empires and companies could manage the world.[7] That so many died on the island of Pulo Condore was a reflection of the risks imposed by life within a stockade in the midst of pestilential waters. Those in the service of the Company tended to live in close proximity to each other. Contagion met with little resistance in closely housed settlements. Stewart identifies European expansionism as an imposition on the societies they traded with, but it was only made possible by many quite ordinary[8] men and women prepared to take risks. Pound was aware that he was subjecting himself to these risks when in 1699 he entered into the service of the Honourable Company.

In his first letter from China to Flamsteed on 18 December 1700 (which was received in Greenwich on 4 July 1701) Pound writes informing him that he was just finding a convenient place for his instrument. He describes the lunar eclipse he observed at sea whilst the ship was becalmed near Emoy,

I had as good an observation of the Lunar Eclipse on Aug. 18 as could be expected at sea; at ye time of observation we were almost becalmed being as near as we could guess (by seeing ye Hills in ye Evening near Emoy) in the Meridian of Emoy, and as I inferred from the preceding Noon, in the Lat. of 24° 54'. Twas somewhat cloudy when the Moon arose, but it immediately cleared up, and with a glass about 3 foot long.[9] I observed the Beginning at 6h 52', Immersion at 7h 57', Emersion at 9 40, End at 10 45, therefore the middle 8. 48½' by my watch. At 8h 44½' by the Watch wth a Cross Staff (not being able at sea to use a better Instrument) I found Arcturus 17° 35' high. His distance from ye Pole 69° 12' therefore his Hour 5h 16' 24'' pm. His Right Ascension 14h 02' 11'' and the R Ascens. of the Sun 10h 31. 17. therefore the Hour of the night was 8h 47' 18'' So that the watch went 3' too slow.[10]

For a few years Pound regularly sent lengthy letters to Flamsteed detailing astronomical data and interesting pieces of information about life in and around the China seas. He received a medical degree and a licence to practice medicine in 1697 and was in Oxford up to that date. Pound took priestly orders before going to the Indies as a chaplain, employed by the Company. His regular letters to the Astronomer Royal between 1700 and 1705, reveal Pound as an experienced and competent astronomer. He complains about delays in

transporting the quadrant that was provided by Flamsteed from one place to
another, as he was relocated several times by the Company. In a letter sent by
Pound to Flamsteed from Chusan on 18 December 1700, he opens,

Sir

On the 18th October we arrived at the port of Chusan on the island of the
same name which is about 20 miles long and 10 broad. I believe it to be the
most easterly of any land belonging to China, the Junks go hence to Japan in
3 days, sometimes in less time, and to Limpo (which lies west) in 10 hours.
Since I have had my instruments afloat I have not had the opportunity of
making any observation.[11]

When Pound was united with the quadrant and other instruments, he made
accurate latitudinal observations in Chusan, an island under the governance
of the Emperor of China. These observations were forwarded to Flamsteed
dating from 31 March 1701 OS to 28 September 1701 OS. It would be inap-
propriate to recount all of the observations made by Pound but initially he
determined the latitude of Chusan.

I am unable to determine when Pound acquired his astronomical skills. He
made observations on Flamsteed's behalf, aside from all his duties as a servant of the East India Company. Flamsteed's tables of the motions of Jupiter's
satellites never arrived.[12] Pound was a busy person conducting the Company's
services as a chaplain with medical qualifications.[13] He made observations
both as a naturalist and an astronomer taking precise astronomical measurements. By calculating the latitude of Chusan, Pound was able to determine
its distance from the pole precisely and accurately, a solid achievement given
the instruments at his disposal. Over the next few months, he made repeated
observations, to determine with greater exactitude the latitude and the longitude of the port of Chusan. On 28 September 1701, Pound determined that
Chusan was 8hrs 02min 58sec ahead of Greenwich in right ascension. I have
already suggested that Flamsteed had a rather mercenary attitude to Pound's
work. Lesley Murdin is apparently quoting Flamsteed's communication to
Pound (unfortunately without references): 'I expect you should send me ye
measured distances as you take them for haveing contrived and formed the
instrument *I have a right to them*' (my italics). [14]

The Astronomer Royal later required a full programme of observations that
were made by Pound on the island of Pulo Condore. I am convinced Flamsteed
sought to rid himself of his dependence on the now-hated Halley's observations made at St Helena. I lean to the view that Pound was simply undertaking

astronomical observations on the Astronomer Royal's behalf, unaware of any possible ulterior motives.

My narration of Pound's experiences in the East Indies reveals the character of the man. On his return to England, he moved easily amongst the highest in the land.[15] His character must have greatly influenced Bradley. Much undoubtedly came from a shared genetic propensity, but Pound's ability to become all things to all men was a characteristic often evinced by Bradley. Pound's time in East Asia and the East Indies and his experiences at Pulo Condore together with the perilous voyage to Malacca must have made a marked impression upon his psyche. His account reveals a mind that was always in control of the situation, desperate though at times it must have been, subjected as the survivors were to extremes of terror and despair. It reveals a person genuinely disposed to think well of his fellow men. This characteristic surely affected his capacity to attract friends and patrons which later advanced Bradley's career. His experiences in the Indies must have advanced his independence of mind. No finer teacher could have been capable of introducing the young Bradley to the exacting science of positional astronomy. Pound appears to have been regarded as a person who was easy to work with. When all hopes of Newton obtaining observations from Flamsteed had gone, Pound became the man he and Halley turned to. As well as being a skilled astronomer, Pound was far less inclined to take issue over perceived slights than the first Astronomer Royal.

After leaving Chusan the first letter from Pound to Flamsteed is dated 14 March 1702 which was written in Batavia the Dutch stronghold in Java. This was the chief centre of Dutch power in the Indies and offered extensive facilities for the repair and maintenance of ships. Even so, outside the fort and the walled boundaries of the settlement, it was considered highly dangerous for Europeans to wander. He opens,

Revd Sr,
Being forced away from Chusan on the 2nd of Feb. We arrived here the 10th instant: nothing remarkable has occurred. In our passage hither but the Appearance of a Comet for about a week I doubt not but that you saw it however.

The rest of the letter gives an account of his onboard observations from 13 to 19 February 1702. On the 13th, Pound states that he observed the tail of the comet at sunset and he saw it again the following evening. It is his account of his qualitative observations on the 15th which become more detailed. Using a cross-staff, Pound was able to determine that he was located at 18°32′ N latitude, somewhere off the southern coast of China. He attempted to locate the position of the comet in the long winding constellation of Eridanus. During the following evenings, he gradually saw less of the comet until he finished his account on the 19th stating that the 'hazyness' of the weather and the rising of the Moon made further observations impossible.

The assertion that the party was 'forced away' from Chusan is explained by an ongoing controversy within the Imperial Court in China, affecting the influence of the Jesuits which created a growing political situation that the Honourable Company found impossible to countermand. Batavia was sought as a destination because the Dutch stronghold was regarded as a safe haven.[16] On 9 June 1702, Pound penned another letter to Flamsteed offering more detail about why they left Chusan. He opens,

> Sir!
>
> Yours dated July 11th 1701 I received by ye Macklesfield Frigt with ye Historia Cultus Sinansis[17] for wch I heartily thank you: in my opinion ye Jesuits have lost ground in ye Controversy, tho they have printed a Book at Pekin exhibiting the Testimonies of ye Emperour and many of ye chiefest Mandarines in favour of their Assertions: Having had the use of this Book for 3 days a little before we left Chusan I transcribed it. (being lent to me by 2 Jesuits who could not part wth ye original) and sent it to my Friend Mr Hodges Rector of St Swithins by the hand of Capt Philipps in ye Eaton Frigt to be delivered (if Mr Hodges saw fit) to his Grace of Canterbury: But I believe what they have said there (tho wrote in a handsome Style) will scarcely answer many things alleged by Maigrot and Charmot.[18]

The attempts by the Honourable Company to enter into negotiations with the Chinese authorities were thwarted by a growing controversy that threatened the ability of all Western powers to trade with China.

On 23 March 1693, Charles Maigrot, the apostolic vicar of Fujian of the Missions Etrangères de Paris (MEP), prohibited the terms 'Heaven' (tian) and 'Sovereign from Above' (shangdi) that had been offered by the emperor. Maigrot's 'mandate' was controversial because he attacked the protection offered by the Kangxi emperor and the 'Edict of Tolerance' in favour of the Christian faith. In 1700 at the request of the Jesuits, the Kangxi emperor declared that these rites were the civil heritage of the Chinese nation and did not involve religious beliefs. It appears that Pound and his party arrived in China in the midst of a complex and sensitive cultural, theological, and political crisis, hardly a propitious moment to arrive to set up a trading station.

Bishop Maigrot sent Nicolas Charmot MEP to Rome to ask Pope Innocent XII to re-examine and question these rites. The Catholic Church attacked the position upheld by the Jesuits without any understanding of the different value systems and the role of the 'three teachings' in Chinese life.[19]

At some time between July 1702 and January 1703, Pound was relocated to the small but strategically placed island settlement of Pulo Condore, close to the mouth of the Mekong Delta, part of the realm of the Kings of Cambodia. At some time during 1702, the Honourable Company founded a station on the island in a sheltered bay ideal for provisioning the Company's ships on the voyage between China, Japan, and Madras. Located close to the mouth of a small river from the heavily forested interior of the island, a wooden palisade was constructed. When Pound arrived on the island he took up residence in a small wooden dwelling about 400 yards from the fort. He shared this with another servant of the Company[20] as well as a slave employed as a domestic servant.

It was fortuitous that Pound was an astronomer needing to be located away from the noise and distractions of a small, densely populated settlement. With infections carried by ships plying the South China Seas, life inside the stockade would have exposed him to many hazardous diseases. Any such fort was a perfect breeding ground for many tropical infections. Batavia the most important centre for re-provisioning in the Indies was notorious for the incubation of such illnesses. The location of Pound's house was his greatest protection against disease, but its location also saved his life during the massacre in the settlement on the night of 3 March 1705.

In references to the massacre at Pulo Condore it has been referred to as a revolt by the 'natives'. Murdin writes, 'Pound was awakened in the early morning of 3 March by the noise and smoke of a fire and the screams of his companions being slaughtered by mutinous natives'.[21]

This is misleading in several ways. Before describing this 'mutiny' more attention needs to be paid to the other inhabitants of the island. The natives were an estimated 200 Cochin-Chinese[22] who mainly earned their living as fishermen. They lived on the northern coast on the opposite side of the island. The other inhabitants were people who were loyal to the Kings of Cambodia so they may be described as 'Cambodians' or 'Cambojans' as they were called by Pound in his writings. The inhabitants of the station were on good terms with this diverse community, several of whom seem to have performed various services for the Company. They numbered no more than thirty-eight individuals.[23] Just prior to the uprising, plans were finalized to send a delegation to the Cambodian Court. The station offered prosperity to those who lived in its proximity.

The perpetrators of the massacre came from a tribe who had been employed by the Dutch as mercenaries. They were not natives in the accepted sense. They may have come from Borneo. They were banned from the territories ruled by the Kings of Siam. In Pound's account, they are usually described as the 'Macassars'. The Company employed them to support the soldiers guarding the fort. As well as being trained and armed with European firearms, they also habitually carried swords and daggers. It appears that the Macassars were generally feared by the local native peoples, this being the reason that they came to the notice of the Dutch. There is evidential support from the governor of the settlement when he 'threatened to treat them as they were by the Dutch'. It appears that the governor intended to dispense with their services just as the Dutch had before.

The best account of the massacre, the only one giving much detail, is that given by Pound in his letter to the Court of Managers of the Honourable Company in London. He described the events of 3 March 1705 explaining the impossibility of using the Company's facilities on the island of Pulo Condore. In his account of 2 May 1705, Pound gives reasons why the managers received no mails from Condore 'this last year', stating that the Company's ships no longer dropped anchor there on their return from China, the brief exception being the *Montague* which raised anchor immediately.[24] He paints a picture of the Company's activities in the area. He wrote,

> In January last the King of Camboja sent a vessel well man'd (and built for this very purpose) to Condore, to bring over Mr Henry Greenhil the Purser of the Seaford who (by the ignorance of those who were to conduct him down the River with a Boat full of Provisions) missed the Ship and was left behind. The Capt. of the Vessel was a Mandarine of that Country, who brought a small present to our Governour from the King with a Letter to invite us to come and settle in his Country: part of the present being live cattle were left behind, the Boat being to[o] small to bring them.

Pound explained that the governor determined to send with the 'Camboja' vessel a small sloop belonging to the Company, to fetch provisions that were needed. Mr Abraham Chitty, a merchant was to go likewise, to procure other goods. This suggests that the station was barely self-sustaining, but it had more than sufficient funds to buy local provisions that were essential for the maintenance of the settlement. The sloop *Rose* was brought out from the river and brought to anchor in front of the fort on 2 March 1705, a fortuitous event for it was the means by which the survivors of the massacre made their escape. If

Pound had lived in the settlement or there had been no sloop at anchor in the harbour then he would have perished along with the other victims.

Pound presented a graphic account of 'the villainy' as he terms it, corroborating evidence as well as adding the testimony of his own eyes. At 1 o'clock on the Saturday morning of 3 March 1705, Macassar soldiers set fire to the large thatch-covered store house within the fort. Abraham Chitty, the merchant, had quarters close to the commotion and was woken. Out of his window, he saw Mr Ridges dragging a chest out of his room which was on fire with the intention of saving his goods. Chitty joined to assist him and to extinguish the fire. On discovering a body lying close to his room, he suspected 'villainy'. Chitty drew his sword and ran to the governor's lodgings discovering him mortally wounded and being cared for by James Ray. The governor Allan Catchpoole Esq. informed Chitty that he had been 'shot in the breast' and 'stabbed in the belly'. James Cunningham awoken by the commotion and alarmed by the fire was warned of the actions of the Macassars by Peter Hill who came out of the barracks which were burning. Hill explained that he was on guard on a platform not far from the governor's door. He had been asleep as he and his two fellow soldiers from Madras took turns on watch. Hill said he was awoken by the firing of a musket and looking at the governor's door saw two Macassars leaving his quarters, one with a musket, the other with a drawn 'crass', a short sword used by the Macassars. His two companions fled and he dashed into the governor's room to discover he was too late. Pound was informed of the events in the fort, though the use of firearms may have suggested that something untoward was going on in the fort. Pound continued,

> Moses Wilkins, a Writer[25] and I with a black Slave lived in a little house about a quarter of a mile from the Fort: Capt Thomas Dennet wth his Sent[26] and Mr Greenhil above mentioned in a house a little nearer, and Mr Thom[a]s Fuller the Ensigne running by awaked us with *The Fort is on Fire* at which we all leaped out of our beds and without staying to put on our cloths made what hast we could towards the Fort.

Pound and his companions were well protected from the horrors taking place within the palisades. They recognized that the situation was past any kind of assistance. Captain Dennet's servant went over the palisade and heard Lieutenant Rafhel cry out that Ensign Fuller was dead, showing that the massacre was in full flow. Rafhel was heard to shout orders to the defenders. The servant returned informing them of what he had seen and heard. Chitty and Cunningham mentioned the actions of the Macassars who were killing all of the English[27] in the fort. The small party returned to Pound's house. Realizing they were unable to defend themselves from the Macassars, Cunningham decided on going to the Cochin Chinese village about a mile away on the opposite side of the island. The rest of the party seeing a small fishing vessel on the shore abandoned their flight and made their way to the sloop lying at anchor in the bay.

James Ray appeared with two slaves as well as Rafhel's slave who informed them that his master Ridges and others had died before he fled. Aboard the sloop, they found four sailors who had been there all night. A private sentinel who had escaped the fort swam to a pinnace enabling him to reach the sloop. He reported that several of the Englishmen in the barracks were lanced or stabbed in their beds. Several were killed as they left through the door when fire forced them to leave the building. He said that only two others escaped out of the barracks. He had no idea where they went. Going aboard the sloop, they all witnessed the fort and the settlement in flames except for the gatehouse and the powder house which were made of stone. Two houses outside the fort were also fired meaning that if Pound had remained ashore his chances of survival would have been non-existent.

The small party aboard the sloop possessed nine small arms but without ammunition the party agreed they were too close to the shore. They decided to haul further offshore until daylight. Being becalmed, they used the pinnace to tow themselves further out to sea.[28] The party after dawn picked up one of the Swedes, John Peterson, who was employed as a carpenter.[29] He and Will Omens fled through the warehouse, though his companion being wounded died before they had gone a quarter of a mile. They heard the report of four or five muskets which they guessed came from the Chinese village or at the Cambodian settlement up river. So it seems there was no escape for any Europeans on the island. At sunrise they sailed around the island to the north-west harbour seeking any more English or European survivors but none were found. After pondering on their situation, they put several slaves ashore believing they would be safe, and pooling their provisions they decided on a plan. Pound first suggested returning to the south bay but it was agreed that this was too dangerous. All they possessed was a small ration of rice and some fresh water found on 'Flagstaff Island'.[30] If they stayed, they would run out of provisions. The party agreed they would attempt to reach the Dutch settlement at Malacca in Malaya. Pound wrote, 'and finding we had Rice for the rest for 10 days at half a pint each man diem [daily] after we had taken in Water and Ballast we heaved up our Anchor the same 3rd day a little after Sun Set, and resolved to try for Malacca'.

It is revealing to examine Pound's reflections on the possible reasons for the actions of the Macassars. They numbered only sixteen. He compared the Macassars favourably with the English soldiers who had been 'addicted to drunkenness', though the English soldiers spent many days cutting down trees and repairing the palisades. It was because of this hard labour that they were excused night duties. They often fell asleep on guard. For this reason the Macassars usually formed the guard at night. They proved to be reliable and behaved respectfully. Pound was surprised by their actions and was seriously in search of their motivation. He repeated they were trusted because they had always been so reliable and well-versed in 'the exercise of the musket' in the European fashion. He wrote that Mr Chitty was with the governor at Christmas and saw that the Company possessed £5000. This in modern money values was worth well over a million pounds. Whilst arguments are prolonged about money equivalents they should be treated with caution. What can be asserted with certainty is that a considerable amount of money was being held in the safe in the governor's rooms. This would have provided sufficient temptation but Pound's respect for the Macassars suggested he knew some of them well. The rooms used for money transactions had been destroyed and everything was in ashes or held by the Macassars. Even the warehouses were seen in flames so there would be little to salvage. Pound sought other reasons for their treachery in order to explain their actions. He wrote,

> I had almost forgot to mention that 4 or 5 days before that horrid and barbarous fact was committed two Slaves who were put under custody of the Macassar Guard ran away while the Guard slept: Upon which ye Governor soth Mr Lloyd threatened ye Macassars severely telling them they should be served as the Dutch use to serve them in their settlements with other such Menaces if they did not find out and bring back ye Slaves.[31]

The motivation may have been greed but Pound believed the loss of 'face' or honour may have led to the uprising. Contrasting the Macassars with the English soldiers who were so unreliable, their self-perception of trustworthiness but for a single moment's error was met with threats from the governor. This must have rankled with them in Pound's evaluation. What is remarkable about Pound's letter is his attempt to write a dispassionate account and to treat even the perpetrators of a murderous crime with some understanding.[32] Pound does not question the institution of slavery. If he did have any opinions he kept them to himself in a letter addressed to the managers of the East India Company. East African slavery had existed far longer than the three hundred years or so of the Atlantic trade. It can be traced from the time of the Caliphate in the eighth century of the Common Era. Such slavery was practised in Mughal India before the Europeans arrived on the scene. The Honourable Company had few qualms about using slaves in their service. High mortality rates amongst Europeans made the use of native slaves a 'convenience' the Company found impossible to resist. Even so Pound mentioned the fact that it was the escape of two slaves and the governor's reaction to this that may have led to the massacre. In the use of slaves, the English were no different from their Portuguese, Spanish, or Dutch rivals.

The sloop *Rose* had on board basic writing materials[33] which helped Pound form a measured response to the situation they were in. His skill of dispassionate observation and recall was aided by the discovery of these materials. The voyage from Pulo Condore is recorded in Pound's journal,[34] a source full of information on the Indies which could well supplement accounts given by authors like William Dampier[35] who sailed the same waters at almost the same period. This, however, was not a voyage of discovery. It was a voyage of terror and despair during which some members of the party came close to death due to exposure, hunger, or thirst. Throughout his account Pound refers to 'we' and on board all being treated, fed, and provisioned equally including the four black slaves.[36]

On Sunday 4 March 1705 OS,[37] they took in stones for ballast, filled three quarter casks with water, and released three black slaves separately from the four who went on the voyage. Those on board had a box with a little rice, two or three Spanish dollars, and some brass wire. The sloop had no rudder-directed steering, this being applied by the use of an oar.[38] Pound mentions also that a mile from where they boarded the sloop, they sent two children ashore in a boat. They must have come from the Cambodian or Cochin-Chinese communities ashore. This left fifteen people aboard the sloop *Rose*. With the rice and seventy-five pints of water, they allowed each man but half a pint a day enough for ten days. The climate throughout the year is hot and humid so everyone sweated profusely. They gathered rainwater. There was little reported discord on the sloop, testimony to the discipline that was maintained on board. They opened a chest belonging to Mr Wingate who was the original ship's master. There was some clothing, a sea quadrant, a chart of the Straits of Malacca, Banca, and the route to Batavia. There was no compass so they navigated by the sun during the day and by the stars at night. Pound's astronomical knowledge must have been an inestimable boon to the survival chances of those aboard. The party was in a perilous situation as they left Pulo Condore cast adrift in a small vessel in hostile waters with the minimum of navigational aids and very little food and water to support any lengthy sustenance.

On Monday 5 March 1705 the pump broke down and after examination it proved to be beyond repair. They resorted to baling out water with buckets, having found four aboard.[39] The sloop proved to be very leaky, barely seaworthy. The sun was very troublesome not having sufficient clothes to protect themselves. On 6 March 1705 they were buffeted by squalls and it rained heavily. They were far from land heading to the Malay coast. They were unable to keep themselves dry. They bailed out over five hundred buckets of seawater that day. With their near-starvation diet, the heat and humidity probably drained everyone of their physical strength. Wednesday 7 March 1705 saw them frequently becalmed, a dangerous situation as their strength fell away. The next day was dry, hot, and sultry. They were up to ten leagues from land.[40] They saw some fishing boats but were unable to speak with the fishermen. On Friday 9 March 1705 with the monsoon winds shifting, they realized that their rice would soon be spent. Pound mentions that he entered into private discourse (very likely it was with Captain Dennet[41] and Mr Chitty) about which of the black slaves should be put ashore, but at 5pm when they were unable to see the shore, the wind strengthened. Two hours later it was a gale so they had to take shelter behind a rock close in to shore. They went ashore but found no freshwater. Not wishing to lose the aid of a fine wind they set sail with all fifteen on board. On Saturday the 10th of March, a week after the voyage began, with a fresh north-easterly gale they were soon about three leagues offshore. At 9am they dropped anchor. One of the slaves spoke Malay and swam ashore and returned[42] and said he saw human footprints as well as those of hogs. Several men were seen running away evidently in fear. After seeing two fishing boats, they chased one of the occupants ashore and were met with a fearful and aggressive posture which only diminished when 'one of our men spoke friendly to him in Malays'. Although he came aboard, they were unable to combat the resistance ashore. They were unable to acquire rice, fish, or other provisions. At this stage the party was becoming weak and desperate.

Over a week after setting sail from Pulo Condore, they recognized that their situation was getting extremely serious. In the distance astern they observed a junk. They hailed the vessel in Chinese, Malay, and Portuguese without any response. Although the junk was much larger than the sloop its occupants feared them, an indication at how much Europeans were now mistrusted and feared. At 5.30pm, they entered 'the straits of Sincapore' between a swampy island and the Malay shoreline. Following the junk about two miles ahead of them they finally anchored in thirteen fathoms in calm water. They rowed the pinnace to the junk and procured 1¼ bushels or ten gallons of rice. The captain of the junk had no spare fish or other provisions but took two Spanish dollars for the rice which Pound thought extortionate. The crew of the junk were all Chinese sailing from Siam. With enough rice to get them to Malacca, they traded for a few fish from a Malay fishing boat. All were hungry, weak, and faint. They resolved that everyone should have a 'full belly'. With enough rice and the fresh fish they enjoyed 'a cheerful Dinner'. They were able to spend 12 March 1705 largely at repose in the Strait of 'Sincapore'. In his entries for 13 March 1705, Pound remarked that the straits formed a fine passage like a large river a mile or more broad with three or four islands. One day Singapore would supply all of the functions originally intended for Pulo Condore and considerably more. Pound wrote that the villagers asked for opium or tobacco beating their drums all night. On 14 March 1705, they made a good distance with the help of a fine gentle breeze and anchored in the roads of Malacca about a mile from the fort on the following day.

In the evening Pound along with Chitty and Captain Dennet went to 'wait on the Governour' giving him an account of who they were and what had happened informing him that they sought to go to Batavia. However, the ship needed caulking to stop the leaking which had been such a burden during the voyage. Two days later on the Saturday, the sloop was put ashore while carpenters and caulkers worked on the vessel. A week later she was made tolerable to put to sea properly provisioned for the voyage south along the length of Sumatra to the port of Batavia on Java. In Batavia Pound wrote his account of the violence on Pulo Condore and the voyage to Malacca and Batavia. He also listed the names of those known to have perished in the massacre,[43] those for whom it was impossible to give an account,[44] and those who made it to the sloop.[45] A year later, after receiving his reply from the managers of the Company, Pound was recalled to London.

Two months after writing to the managers of the Honourable Company Pound penned a letter to John Flamsteed on 7 July 1705. He informed him of the uprising at the Company settlement adding that he lost everything including all of his observations, records, and his instruments. He explained to Flamsteed that everything including the quadrant had all been in good order. He further added that he built a small house for it and took every opportunity to observe the southern stars with Moses Wilkins who came out to the Indies as Pound's servant before entering the service of the East India Company as a writer. Wilkins took a great delight in astronomy and understood everything very well.[46] When Pound understood that the East India Company was withdrawing the factory at Pulo Condore and settling it at Banjar,[47] he abandoned all hopes of making progress with his observations. He wrote,

> for there being nothing but water at Banjar and commonly thick cloudy weather; I would propose to do nothing there: I was fully resolved not to go to live there, and had it in my thoughts to have spent at least one year in this place tho it had been at my own expenses to have rectified the Catalogue of the fixt Southern Stars:[48] but now having neither Instrument nor Money nor Books, I must be forced to return to England which I design (if please God) by the next shipping from hence in October, November or December if we come back safe from London.[49]

I'm sure that Flamsteed had only the haziest notion of what Pound and his colleagues had gone through. Pound returned to England, arriving in London in July 1706.

Murdin appears to be under the impression that Pound had made himself known 'to men in whose hands rested the power of patronage' citing his election as a Fellow of the Royal Society (FRS) in 1699 and his being made Rector of Wanstead in July 1707 as evidence of this. She interprets Pound's acquisition of the living at Wanstead as providing him with the resources to marry and while some of this may be justifiable she appears unaware that he went to the Indies already a gentleman of some substance in the service of the Honourable Company, along with his servant Moses Wilkins. In July 1707, Sir Richard Child (1680–1750) offered him the newly vacant living at Wanstead as the rector of that lucrative parish. Sir Richard was the third son of the one-time Governor of the Honourable Company, the incredibly rapacious and wealthy Sir Josiah Child (1630–1699).[50] Sir Richard sat as the Tory MP for Maldon from 1708 to 1710. He was a patron of the arts, commissioning the building of the great Palladian mansion of Wanstead House in 1715, and was a patron of the Flemish artist 'Old Nollekens'. Sir Richard was a man of taste and high culture which he could afford on an income of £10,000 a year.[51]

Almost as soon as he took up residence at the parsonage[52] between the small but prosperous village of Wanstead and the Red Bridge over the River Roding, Pound began his own innovative observations with several years of experience behind him in the Indies. Over the next few years he acquired a growing reputation for reliability. His relationship with Flamsteed had cooled, almost in reciprocity with the growing notice of his abilities by Halley and Newton. Seeking to restore good relations with the Astronomer Royal, Pound wrote to him on 14 January 1713 with the results of several observations of eclipses of the satellites of Jupiter. Appreciating that time had passed since these observations had been made, they were intended as an attempt to repair relations. Pound was corresponding with the Astronomer Royal following the unauthorized and scandalous publication of Flamsteed's observations under the unauthorized editorship of Halley. He wrote,

> Revd Sir,
> I cannot conveniently come to Town this week but I here send you those satellite Eclipses which I observed last year, and am sorry I had not leasure and opportunity to make more observations.

August 1712		Mr Gray	Mr Derham
14d 9h 35m 10s	1 ε	14d 9h 40m	Mr Derham
16d 10h 32m 30s	2 ε	16d 10h 36m	
September			
6d 07h 37m 30s	4 ε	6d 07h 43m	6d 07h 39m
16d 09h 47m 20s	3 ε		16d 9h 50m

> I have set down Mr Gray's and Mr Derham's observations of the same eclipses that you may see the Difference. My Observations were made with a 15 foot Glass and the time set down is the true time when I first saw the Emersions, that of the 16th September may not be so exact as the other, and as I remember that of Mr Gray's on Sept. 6 you told me he could not depend on for certain. Mr Gray's in August agree well enough with mine.
>
> I am
> Sr Your Obliged humble Servt
> J Pound.

Flamsteed was well aware that Derham was amongst those fellows of the Royal Society who resisted the dominating presence of Newton in the society's affairs. Pound's domestic affairs were settling into a welcome routine that had been denied him in his various travels. On 14 February 1710, Pound married Sarah Farmer the widow of Edward Farmer. Pound was forty years of age. His years in the Indies may have aged him prematurely. His accounts book reveals a busy household with some long-standing loyal servants. Late in 1710 a son was born who died soon after his birth. On 16 September 1713, a daughter also named Sarah[53] was born. In 1711 came the moment why Pound's story is so central to any account of the work and achievements of James Bradley, his maternal nephew, who came to live with him in Wanstead. Pound took responsibility for Bradley's education with the objective of preparing him for a life in the priesthood of the Church of England.

Given the ravages of the voyage to Malacca, Pound's survival was indicative of his robust constitution. Although we only have scant evidence of Pound's developing skills as an astronomer in the Indies, his work at Pulo Condore from late 1702 until March 1705 would have been extremely valuable had his observations survived. Two years of sustained effort were lost in two hours of unbridled mayhem, but his astronomical expertise and experience was not lost. Pound's many tacit skills and dexterity with his instruments prepared him to perceive astronomy as something more than a gentlemanly pursuit. He had been transformed from a gentleman with an interest in astronomy into one of the finest observers in England with skills approaching those of Flamsteed. The cooling of relations between Pound and Flamsteed may be connected to the Astronomer Royal's belief he was owed a debt. Yet Pound never received the tables he believed to be essential to his work in the Indies. Pound's skills were notable: compare the precision of his observations of the occultations of the satellites of Jupiter sent to Flamsteed with those made by Gray and Derham, both respected observers. James Pound was probably the first observer in England to use a transit instrument, first pioneered by the Danish astronomer Ole Römer.

When Bradley joined Pound in 1711, he was about to enter Oxford University to begin his preparation for a career in the Church of England. Bradley was instructed in the skills required to assist his uncle's observations. It was a fortuitous moment in the history of science.[54] Bradley had no practical experience of astronomy because of the limitations of his domestic situation in Sherborne.[55] Although Bradley went up to Balliol College in 1711, he spent much of his time with his uncle in Wanstead. He graduated with a BA on 15 October 1714 and an MA on 21 June 1717. Bradley was commonly at Pound's side and capitalized on his uncle's knowledge and skills to which he added his own recognized abilities within a small but influential circle.[56] It is difficult to pinpoint what these abilities were but his later work and his remarkable achievements as an astronomer suggests he must have possessed remarkable eyesight.[57] In the era before the use of photography the ability of an astronomer to 'see' was critical.

When I suggest that it was a fortuitous moment when Bradley went to live with Pound, this was when the young man came under the tutelage of one of the finest observers in England. Agnes Clerke's entry in the *Dictionary of National Biography* for James Pound is full of misconceptions and inaccuracies blighting so many accounts of the work of both Pound and Bradley. She writes,

'He was elected a Fellow of the Royal Society on the 30th Nov. 1699, but his admittance was deferred until the 13th July 1713, *when his astronomical career may be said to have begun*'[58] (my italics).

How a good historian of astronomy such as Agnes Clerke could have made such an elementary error is beyond my grasp. This source probably misled Murdin. She was convinced that Pound's astronomical career began on his return to England. It ignores his correspondence with Flamsteed from the Indies, freely available in the archives of the Royal Observatory. We must look for reasons why such a misleading opinion by a trusted source could have been written. I suppose Clerke assumed Pound's observations in England really were the beginning of his astronomical career. Had she examined his correspondence with Flamsteed from 1700 she would have been better informed.

There is circumstantial evidence that Pound was already making serious observations well before he was admitted to the Royal Society.[59] Flamsteed requested Pound to observe on his behalf when he went out to Madras in 1699 the year he was first elected as an FRS. From this it can be assumed he was already an experienced astronomer sufficient for him to come to the notice of both the Royal Society and the Astronomer Royal. To believe his astronomical career began after his return to England in 1706/07 is a mistake. Pound made many observations at Pulo Condore on the positions of the southern stars and the satellites of Jupiter. It may seem to be of little consequence except that a source as important as the *Dictionary of National Biography* succeeded in propagating inaccuracies and maintaining false traditions. They are so often repeated that they become 'established truths'. There will always be room for different interpretations, but historians must report 'matters of fact' as accurately as possible.

Halley was a regular visitor to the Wanstead parsonage as reported on the observations of the total solar eclipse of 3 May 1715. Pound was a source of reliable observations both for Halley's own purposes but also for Newton. All communication with Flamsteed had irretrievably broken down. Pound was essential for the development of the practical and theoretical objectives of both men. Halley remarked that Pound was 'furnished with very curious[60] instruments, and well skilled in the matter of observation'.[61] It was during June 1715 that tragedy affected Pound who hitherto seems to have met all with fortitude and resolution. Pound's wife Sarah died. Pound was forty-six and we can assume Sarah was younger, having given birth two years earlier.[62] Pound was left with a daughter in her second year. By examining Pound's accounts book, it can be discerned that his daughter was placed under the care of a nurse named Mary Lea. She was in the employ of James Pound for three years. I have reason to believe that the Pounds were socializing with the family that owned the Wanstead estate[63] the Wymondesolds, brother and sister, Matthew and Elizabeth.

By 1715 Bradley was coming into his own as an astronomer of worth. He had become his uncle's partner rather than a mere assistant. Pound frustrated by Flamsteed's failure to forward his tables of the motions of Jupiter's satellites formed a growing interest in the use of their motions as a celestial timekeeper. Bradley's first *recorded* observations indicate that at this time he was working entirely independently, his uncle mourning the loss of Sarah. Bradley made many observations between 1711 and 1715 as his uncle's assistant and partner. A month after the death of his wife, Pound began observing again. On 14 July 1715, Pound observed an occultation of a star by Jupiter at a time when he and Bradley were timing Jupiter's occultations of the four Galilean satellites. On 30 October 1715, Pound observed an eclipse of the Moon.[64] The observation books reveal a period of heightened activity, much of it recorded and reproduced in the 'Miscellaneous Astronomical Observations' in the *Miscellaneous Observations and Correspondence of James Bradley*.[65] His observations through a fifteen-foot telescope[66] demanded dexterity and precision, characteristics of Pound's observations. Grief can be buried in work or in other distractions. The Wymondesolds may have become friends in Wanstead from the time Pound became the rector in July 1707.[67] The friendship of Matthew Wymondesold with Pound before his demise,[68] and with Bradley after it reveals a long-standing friendship years in gestation. After his uncle's death Bradley was accepted and treated as family during his years of financial insecurity.[69] Bradley found ready acceptance and friendship with the family lasting most of his life.[70]

Bradley's surviving archive includes some of Pound's original observations which were referred to in the third edition of Newton's *Principia*.[71] The observations were made during September 1716, passing on many of his skills to Bradley who was coming to the notice of Halley as an astronomer of great promise. On 6 March 1716, Pound and Halley were observing an uncommon phenomenon in London, the Aurora Borealis. On 22 March 1716, Bradley reported the same aurora in Oxford where he was detained by his studies. Bradley reported that the aurora tended to the middle between *Castor* and *Pollux* 25° from the zenith. Bradley had no access to instruments. It was on a visit to Pound at Wanstead, knowing that his friend was making various attempts to determine the distance between the Earth and the Sun by observing the planet Mars at opposition, that Halley mentioned the possibility of calculating this distance by observing the transits of Venus in 1761 and 1769. Bradley contracted smallpox. The disease was endemic in London. On 11 March 1717, there is an entry in Pound's accounts book which reads, 'By Cousin Bradley's Nurse in the Small Pox £2 3s 0d'. This came in the year Bradley graduated with his MA degree.

Pound's accounts book is full of entries detailing payments to a poor baronet, a poor clergyman, a poor woman, and other poor people. As a clergyman he would have been expected to care for those less fortunate. Pound lived in a society in which the values of *noblesse oblige* were still evident as the background to a society in which patronage with mutual obligations was given and accepted. Patronage was bestowed because both giver and recipient shared common interests, one of which was the Royal Society itself. I have stated that Pound was a convivial character who wasn't easily offended. During the years after his return from the Indies he attracted the friendship and support of figures of import in early Hanoverian England. One of the reasons why I suggest that Pound was so important in the life and work of Bradley is that he made his nephew's entry into the upper echelons of English society so much easier. He also introduced his nephew to some of the highest circles of English science.

Rigaud wrote that Bradley was fortunate to be related to a prosperous man but his wealth was late in coming. He always possessed the resources required to live the life of a gentleman.[72] He also seemed capable of living within his means evidenced by his leaving for the Indies in 1699 with a servant Moses Wilkins. If Pound really was wealthy, it was in his patrons and friends. He became a trusted friend of Halley, leading to Flamsteed turning his back on him. Adding insult to injury Pound was also admired by Newton who found him a useful source of dependable and accurate astronomical data. Other members of the Royal Society proved important to Pound and Bradley. These included Thomas Parker FRS who was the Regent of Great Britain from 1 August 1714, when Queen Anne died, to 18 September 1714, when George I arrived from Hanover and took the crown.

Parker became Lord Chancellor and in 1721 was raised to the aristocracy as the 1st Earl of Macclesfield. He was impeached in 1725, the year after Pound's death. The Parker family always remained staunch friends to Bradley and his family until his demise in 1762. The first Earl may have been a prominent Whig politician but his son's interests were in astronomy and chemistry. He had little interest in the political scene. The second Earl became President of the Royal Society in 1752 until his death in 1763. Another friend and future patron of Bradley was Samuel Molyneux FRS, the son of William Molyneux FRS, the prominent Irish philosopher and politician who was also a prominent

Whig. Samuel was the Prince of Wales's private secretary and in 1727 was appointed one of the seven Lords of the Admiralty. Another close friend was the Rt Revd Dr Benjamin Hoadly FRS (1676–1761), who became the most controversial churchman of the age and as the Bishop of Hereford ordained James Bradley as a priest in the Church of England. He was a favourite of George I. Hoadly was a rationalist who supported freedom of conscience and the right of all to judge the Bible for themselves. He was a supporter of the political views of John Locke and was referred to by Samuel Adams as a source of American liberties and of the Constitution of the United States of America. Many of the political and religious views of Benjamin Hoadly were possibly shared by Pound and Bradley. In 1716, the then Bishop of Lincoln William Wake was appointed Archbishop of Canterbury. His political views were more ambivalent but his appointment met the approval of the king. William Wake's nephew Martin Foulkes, later elected President of the Royal Society, supported Bradley's advancement. Most of Pound's and Bradley's patrons lay on the Whig side of the political divide. Several Tory figures were also important patrons, including Sir Richard Child of Wanstead who appointed Pound as the Rector of Wanstead and Edmond Halley who had several Tory connections. Bradley, like his uncle, attracted the support of all shades of political and religious opinion including the admiration of non-jurors.[73]

Pound was open to new ideas and flexible in the way he tackled scientific problems with new techniques.[74] About this time he formed a friendship with Samuel Molyneux who was introduced to the Royal Society by Newton in 1712. This was an important association because from 1722 Molyneux and Pound planned to repeat Robert Hooke's 1669 attempt to prove the motion of the Earth through observations of a star at the zenith. During the final years of Flamsteed's life, he was in regular correspondence with Samuel Molyneux. As I hope to show this apparent friendship may have affected Molyneux's scientific judgement during the attempt to measure the parallax of γ Draconis utilizing the method that had been pioneered by Hooke. The most important partner in Bradley's work was George Graham FRS. As well as being the finest clock and scientific instrument maker in London, he was also a scientific investigator of some considerable merit.

It is difficult to determine the exact nature of the relationship between Pound and Molyneux, for the Irishman[75] also maintained a concurrent friendship with Flamsteed. Although his father William had been rejected by Flamsteed, Samuel maintained a respectful relationship with the Astronomer Royal. At the end of a letter Molyneux wrote to Flamsteed, on 19 May 1718, he added a postscript, 'Pray whereabouts in Yorkshire is Mr Sharp Settled'.[76] On 30 August 1718, he wrote to Flamsteed again,

> Sir,
>
> I have been a very long journey into Yorkshire and some other distant parts of England from whence I returned hither but the other day and this hath prevented my acknowledging the favour of yr last very obliging letter wherein you are so good as to communicate to me some things in relation to the Meridionall Instruments I have bespoke. I was in hopes to have found it upon my return so far advanced as that I might have assured many of yr questions concerning it but to my surprize the workman hath done nothing yet but a little of the gross part of the Iron Work. Mr Glyn at the Hercules and Atlas in Fleet street is my Operator, he is a very honest man but I doubt he must have some assistance in dividing the Instrument which it will not be ready for this long time.[77]

This friendship had consequences so far unacknowledged or suspected during the search for parallax. I will discuss this at length in Chapter 4, on the observations made at Molyneux's fine residence by Kew Green, the White House, when in November 1725 he was joined by Bradley and Graham when they began their repeat of Hooke's famous attempt to measure the parallax of γ Draconis in 1669.

When Molyneux was corresponding with Flamsteed, his perceived enemy Halley made a notable discovery regarding the so called 'fixed stars'. In a paper reproduced by R. G. Aitken in 1942,[78] Halley made plain his discovery of the 'proper motion' of the stars. Pound was amongst the first to be apprised of this discovery. Halley wrote (the *italics* are Halley's own),

> Having of late had occasion to examine the quantity of the Precession of the Equinoctial Points, I took the pains to compare the Declinations of the fixt Stars delivered by *Ptolemy*, in the 3rd Chapter of the 7th Book of his *Almag.* As observed by *Timocharis* and *Aristyllus* near 300 years before *Christ*, and by *Hipparchus*, about 170 years after them, that is about 130 years before *Christ*, with which we now find: and by the result of very many Calculations, I concluded that the fixt Stars in 1800 years were advanced somewhat more than 25° in Longitude, or that the Precession is somewhat more than *50″ per ann.* But that with so much uncertainty, by reason of the imperfect Observations of the Ancients, that I have chosen in my Tables to adhere to the even proportion

of five Minutes in six Years, which from other Principles we are assured is very near the Truth. But while I was upon this Enquiry, I was surprised to find the Latitudes of three of the principal Stars in Heaven directly to contradict the supposed greater *Obliquity* of the *Ecliptic*, which seems confirmed by the Latitudes of most of the rest: they being set down in the old Catalogue, as if the Plain [sic] of the Earth's Orb had changed its situation, among the fixt Stars, about 20′ since the time of *Hipparchus*. Particularly all the Stars in *Gemini* are put down those to the *Northward* of the Ecliptick, with so much less Latitude than we find, and those to the *Southward* with so much more *Southerly* Latitude. Yet the three Stars *Palilicium*[79] or the *Bull's* Eye, *Sirius* and *Arcturus* do contradict this Rule directly: for by it *Palilicium* being in the days before *Hipparchus* in about 10gr. of *Taurus* ought to be 15 Min. more *Southerly* than at present, and *Sirius* being then in about 15 of *Gemini* ought to be 20 min. more *Southerly* than now: yet *è contra* Ptolemy places the first 20 Min. and the other 22 more *Northerly* in Latitude than we now find them.[80]

This paper by Halley is taken as the first full recognition that the stars moved independently of one other, that they evinced 'proper motions'. This discovery stemmed from Halley's studies of Ptolemy's star catalogue in the *Almagest*, comparing the positions of important stars in the catalogue with the positions he observed. After making allowance for the precession of the equinoxes, Halley recognized these stars had their own separate motions. The term 'fixed stars' only gradually faded from use.

Halley, it appears, was ever more dependent on Pound and his nephew making observations determining the complex motions of the satellites of Jupiter as a method of determining the longitude at sea. In the enmity between Halley and Flamsteed, Pound attempted to be a friend to both men, though with little success regarding Flamsteed. Halley also made overtures to heal the rift, even visiting Flamsteed with his entire family, only to be forthrightly rebuffed. Pound was a Whig but he could still form friendships with men with Tory sympathies including Halley.

Pound was at the hub of a growing circle of influential people. Mainly Fellows of the Royal Society, they regularly conversed with each other in town. Pound and Bradley were busily engaged in extensive programmes of observation and research bringing them into regular contact with other members of the society. When John Hadley (1682–1744) finally completed his 6-foot reflecting telescope, it was Pound and Bradley who put the instrument through its paces. When, a few years later, Hadley produced a mariner's octant,[81] the forerunner of the mariner's sextant, it was given sea trials on the

Chatham yacht under the direction of Bradley. Molyneux worked with Hadley, Pound, and Bradley developing new alloys to be used as specula in reflecting telescopes. These overlapping friendships led to the commissioning of instruments by George Graham and to advances in instrumental design and the practice of positional astronomy. Some of this will be discussed in Chapter 3, 'An Ingenious Young Man', a study of Bradley's development as an astronomer.

Pound's invaluable accounts book reveals that the acquisition of the living at Burstow cost substantial fees and other outlays. Pound met with Lord Chancellor Parker at the House of Lords. Various fees were paid to a whole line of 'gatekeepers', clerks, bearers, and porters, various officers who received fees every time a new living was bestowed. Some of these offices were lucrative. The outlay for acceptance of the living at Burstow came to £78 14s 0d, equivalent to just under £20,000 in today's monetary values. Little wonder that many public offices, even modest ones, were sought after. They guaranteed solid incomes. More outlays continued with fees of £15 15s 0d for Bradley's small living in Pembrokeshire, with sundry expenses of £21 12s 0d for a formal gown for his nephew's presentation. The acceptance of offices and church livings always involved fees. Almost every office or position in society came with these charges, acting as a barrier to the advancement of any young person without the support of acquired wealth or a patron. Pound was wealthy and settled, with an infant daughter and a nephew he treated like the son he never had. Pound entered marriage once more looking forward to a prosperous life. Matthew Wymondesold, his future brother in law, had been a successful investor and speculator in the City for several years. In his marriage to Elizabeth Wymondesold, a woman with a personal fortune of £10,000, Pound truly became a man of substance.

Due to the circumstances of Bradley's upbringing[82] it had become vital that his education for a career in the Anglican Church be supported by his mother Jane's clerical brother. Pound, however, introduced Bradley to the science of astronomy initially as his assistant. Pound was trusted by Halley and Newton, and Bradley's own skills were soon noticed by Halley who became an influential mentor enabling the young man to master much of the *Principia* with his guidance. With two such mentors Bradley could not have enjoyed better support. It was soon apparent to all concerned that he had great natural abilities. As already discussed quite what Bradley's skills entailed are difficult to comprehend. Was it his eyesight, dexterity, or his intelligence, or was it some other skill or combination of skills? We are left guessing. Halley confidently commissioned the young astronomer to undertake an extended programme of observation of the four Galilean satellites of Jupiter. It is puzzling how an experienced naval captain like Halley ever believed it was a practicable method of determining the longitude at sea given the conditions on board a ship.

Bradley was introduced to several notable instrument makers in London at a time when the London market for scientific instruments was the largest in Europe. Bradley openly confessed his reliance on his instrument makers, most notably George Graham and later John Bird (1709–1776). If Pound had perished on the night of 3 March 1705 what would have become of James Bradley? How much longer would it have been before the phenomenon called the aberration of light was identified? Given Bradley's extraordinary skills as a technician, how long would it have been before the nutation was finally exposed? One thing is certain, without Pound neither the aberration of light nor the nutation of the Earth's axis would have been discovered by James Bradley DD, FRS.

The year 1720 was the year of the infamous South Sea Bubble. At the beginning of the year the directors of the South Sea Company laid before Parliament their plan for paying off the national debt.[83] At this time the debts of the state stood at £30,981,712. On 2 February 1720, the House of Commons resolved that the Company's proposals were advantageous to the country. The City was in a fever of excitement and on that day Company stock stood at 130. The Prince of Wales, with a debt of £40,000, and all sorts of projects for the renewal of London, including those of the Duke of Chandos and the New River scheme, were financed by the returns of the 'bubble'. Experienced financiers were probably ready to sell their holdings on 1 June 1720, the day before the bubble burst, and according to Rigaud, Wymondesold sold all of his South Sea stock just before that event.[84] After Pound married Elizabeth two years later, we find evidence in Pound's accounts book of his stockholdings in the South Sea Company. Some £5000 of South Sea stock may very well have been a marriage settlement on behalf of Elizabeth, a sizable sum worth in excess of a million pounds today. Of important notice was that with Pound's marriage to Elizabeth, any wills made up to that date became null and void. Pound didn't draw up a new will, surprising given the care with which he dealt with most of his financial affairs. It left Bradley in a difficult financial position for several years.

On 9 November 1724, Pound paid a visit 'to town' as it is recorded in his accounts book. He appeared to be in good health after an eventful life of fifty-five years. A week later on 16 November Pound was dead. As he died intestate, any provisions he intended to make for his nephew had disappeared. Bradley was left bereft, his only income being the comparatively modest stipend he received from the Savilian professorship. Even this was subject to various deductions. After a period of mourning, it was Bradley's good fortune that he continued living with the Wymondesolds, being regarded as a member of the family by Elizabeth. With the exception of the 123 foot Huygenian glass that had been lent to Pound by the Royal Society, he was able to continue with the use of all the instruments owned by his uncle. He maintained his observations at Wanstead, a continuation of those he was making of the satellites of Jupiter together with his observations of Saturn and its system. He continued using the transit instrument, one of the finest in England, the reflecting telescope constructed by his friend John Hadley, a fifteen-foot telescope he used generally as well as three other refractors of seven-foot, ten-foot, and twelve-foot focal lengths. Pound's life had been one of varying vicissitudes most of which he coped with. Newly married, a man of substance in his mid-fifties, one final turn in his life brought him to his unexpected demise. He left a wealthy widow and a nephew with an uncertain financial future. He also left Bradley with influential patrons and many friends in the Royal Society and within the Whig ascendancy.

Notes

1. James Pound to the Court of Managers for the Affairs of the Honourable the United Company trading to the East Indies, Batavia, 2 May 1705. I have used Pound's copy of the letter contained in his journal. Bodleian Library, Oxford, Bradley MS24. James Pound's reports of his voyages in the Indies, etc. pp. 1–10.
2. James Pound was the son of John Pound of Bishop's Canning, near Devizes in Wiltshire. He matriculated at St Mary Hall, Oxford in 1687, graduated with a BA at Hart Hall 1694, MA at Gloucester Hall, also 1694, before gaining a medical diploma, with the degree of MB in 1697. He was elected FRS 30 November 1699, though he was not admitted until 30 July 1713. He entered the service of the English East India Company as a chaplain to the merchants at Fort St George at Madras in 1699. Over the next few years he sailed between various company stations in China and the South China Sea, on the island of Pulo Condore (now part of Vietnam and known as Con Son). He was one of just fifteen survivors of a massacre in March 1705 by native troops (Macassars) in the employ of the company. He returned to England in 1706, and was given a lucrative living as the Rector of Wanstead in July 1707.
3. This is the English East India Company, also referred to as the Honourable Company or simply the Company.
4. The modern city of Jakarta, capital of Indonesia.
5. James Pound compiled a list of those who died from natural causes on Pulo Condore. They were John Goodman, Joseph Pishan, Thomas Clark, Robert Shearer, John Shearer, Alex Cutbeard, Charles Tomson, Richard Prat, John Child, Thomas Powel, Thomas Cordial, John Newbold, Nathaniel Hayfield (surgeon), Richard Horn

(surgeon), Henry Swift (surgeon), Francis Boulton, John White, Daniel Doubty, Joseph Helder, Richard Williams, John Smith, Richard Griffith, Richard Hooper, Matthew Peck, Giles Battersby, John Exeter, Andrew Rowlandson, John Derickson, Peter Shirston (Commander of the Safe), George Haughton, William Brown, Mr Woolston (from Bengall in the *Union*), Mrs Boulton, Mrs Linch, Mrs Peck, Mrs Griffith, Mrs Newbold, Mrs Ray, and Florentia Griffith, daughter of Richard and his wife. Given that he fled from the island with little more than what he wore in bed, Pound recalled these deaths from memory, either because he was present or was familiar with those who may have died before he was stationed on the island. Either way, it suggests a man who was personally attached to the people who lived in the settlement. This recall is continued when he describes the deaths of those who perished in the uprising and the way he presented the names of the survivors.

6. Alan Cook, *Edmond Halley: Charting the Heavens and the Seas*, Oxford, 1998, p. 352. Here Cook states that Flamsteed thought poorly of Pound. Given that he had originally thought he was to be trusted before he went to the Indies, it suggests yet another person falling beneath his expectations. There are times when Flamsteed regarded people like Pound with a mercenary attitude.

7. Larry Stewart, Global Pillage: Science, Commerce and Empire, in *The Cambridge History of Science, Volume 4, Eighteenth-Century Science*, edn. Roy Porter, Cambridge, 2003, pp. 825–844.

8. By 'ordinary' I include people who were very often poor or at least from modest backgrounds. Most of those who made fortunes never or rarely left London. Poverty was the driver for so many.

9. This means that the telescope had a focal length of three feet or thirty-six inches. It was probably a shipboard telescope which therefore must have been a Galilean instrument which did not invert the image.

10. James Pound to John Flamsteed, 18 December 1700, Cambridge University Library, RGO 1/37, fol. 91v.

11. James Pound to John Flamsteed, 18 December 1700, Cambridge University Library, RGO 1/37, fol. 91r.

12. I have never been able to determine why Pound never received these essential tables. Given the amount of time Pound lived in the Indies the only conclusion to be drawn was that Flamsteed didn't intend Pound to receive them. Elsewhere I have referred to Flamsteed's mercenary attitude with regards to Pound's observations in the east.

13. It is commonly asserted that Pound never set up a medical practice and this may have been true on his return to England in 1706. I am sure he put his medical knowledge in the service of the East India Company, particularly in the situation he was placed in on the island of Pulo Condore.

14. Lesley Murdin, *Under Newton's Shadow: Astronomical Practices in the Seventeenth Century*, Bristol, 1985, p. 79.

15. Pound's patron was Thomas Parker, who became the Regent of Great Britain during the period between the death of Queen Anne and the accession of George I after his journey from Hanover. Parker later read the 'gracious speech' due to the king's inability to read English and was appointed the Lord Chancellor. He was raised to the nobility as the first Earl of Macclesfield.

16. William of Orange ruled the Dutch Republic as Stadtholder and England, Scotland and Ireland as monarch up to his death on 8 March1702.

17. This book, sent by Flamsteed to Pound, was published in 1700 in Paris. Written by Nicolas Charmot, it is bitter in its denunciation of the Jesuit mission to China. Undoubtedly, Pound sought to acquire a more informed opinion of the theological and political issues that led to his party leaving China. Charmot's entire thesis is fed by error and misinterpretation. See Albert Chan SJ, *Chinese Books and Documents from the Jesuit Archives in Rome*. San Francisco, 2001.

18. James Pound to John Flamsteed, 9 June 1702, Cambridge University Library, RGO 1/37, fol. 100.

19. The three teachings being those of Confucius, Lao Tse, and the Buddha, represented in the Western interpretation as 'religions' which they identified as Confucianism, Taoism, and Buddhism.

20. This was Moses Wilkins, a writer. Wilkins originally travelled to the Indies as Pound's servant before he became a servant of the company. He acted as Pound's assistant when he made astronomical observations.

21. Lesley Murdin, *Under Newton's Shadow: Astronomical Practices in the Seventeenth Century*, Bristol, 1985, p. 79.

22. As reported in Pound's journal.

23. Again reported in Pound's journal.

24. We can assume from this information that company ships were using the Dutch settlement of Malacca from where news passed to Batavia or the main centre of Dutch power, Batavia itself. Either way Pound kept himself informed of the affairs of the company.

25. The East India Company employed many junior clerks, referred to as 'writers'. They recorded accounts, minutes of meetings, made copies of contracts, wrote out copies of Company orders, filed reports, and made copies of ship's logs. Their duties were therefore extensive and necessary to the efficient running of the company's affairs.

26. Sent. An abbreviation of sentinel meaning that he was a private soldier.

27. This term seems to have covered all Europeans, including Dutch and Swedes who worked alongside the English.

28. This reveals just how small the sloop really was for it was possible to use a small rowing boat to tow the vessel further out to sea.

29. His two brothers who were both sentinels also survived. John was originally a sentinel before his skills as a carpenter were employed by the company.

30. A small island close to Pulo Condore.

31. James Pound, Letter to the Court of Managers for the Affairs of the Honourable the United Company trading to the East Indies, Batavia, 2 May 1705. Bodleian Library, Oxford, Bradley MS24. James Pound's reports of his voyages in the Indies, etc. pp. 1–10.

32. In order to inform the company on how not to treat native troops.

33. I infer this for several reasons. The ships log would have been aboard, and as the sloop was prepared for the short voyage to nearby Cambodia it probably had formal writing materials aboard to draw up contracts, treaties, and so forth. In addition, Pound

maintained a log of the voyage upon which he must have based his account of the massacre and the voyage to Malacca.

34. He spent almost a year in the Indies after the conflagration, time to reflect and time to bring forth many vivid memories of his travels, supplemented by discussions with other travellers. His account of the massacre must have been written after interviewing his fellow escapees. The rest is from the sloop's log.

35. William Dampier, *A Voyage to New Holland: The English Voyage of Discovery to the South Seas in 1699*. London, 1729. Edited with an Introduction by James Spencer, Stroud, 2006.

36. This is surely indicative of Pound's character, though as the 'property of the company' they possessed a value to the company, surely of significance in his account to the managers. Some accounts of the number of survivors only count the number of Europeans stating there were eleven survivors, but in fact fifteen people survived, including four black slaves. Pound and the rest of the party appeared to treat the slaves well on board the sloop. All shared in the slim ration of rice and water.

37. Old Style, that is according to the Julian Calendar used in Great Britain, Ireland, and the American colonies until 1752 when the Gregorian or New Style (NS) was adopted.

38. This surely confirms that the sloop *Rose* was a very small vessel.

39. Another reason for believing that at least one of the buckets was used to save rainwater.

40. Ten leagues is 30 nautical miles, that is about thirty-five statute miles or about fifty-five kilometres.

41. Captain Dennet must also have possessed navigational skills but he could not have had the knowledge of the Sun and stars possessed by Pound. As they lacked, a compass Pound's astronomical skills were very important to their own survival. His location on the island well away from the distractions of the fort so that he could observe the skies probably saved his life and now his astronomical knowledge contributed to the further survival of the group.

42. Surely a sign that he was treated humanely or else he might have taken his chances ashore and not returned to the sloop.

43. Those who were seen dead at Pulo Condore were Allan Catchpoole Esq. (Governor), Mr John Ridges (3rd in Council), Mr Joseph Ridges (his son, a writer), Mr Thomas Rafhel (Lieutenant), Mr Thomas Fuller (Ensign), Arthur Aust (Sergeant), Robert Emet (Drummer), John Walton (carpenter's boy), John Pennyman, and John Marefield (both governor's servants). Then came seven private sentinels, Peter Hill, John Bolt, Henry Ormond, William Omery, George Shadford, Thomas Herring, and Richard Bradford, from Madras. Another private sentinel was a Dutchman, Peter Benseley. 'In all 18 besides 4 or 5 blacks'.

44. Those missing included, Mr Solomon Lloyd (2nd in Council), Mr James Cunningham (5th in Council), Mr Henry Pottinger (Secretary), Ambrose Baldwyn, Henry Savage, George Townsend, (all writers), John Hunsdon, (storehouse keeper), Michael St Paul, (surgeon), George Wingate (master), whose effects on board the *Rose* helped those who escaped, John Linch (armourer), Cornelius (the smith, a Dutchman), Alexander Linsey, Henry Slade, and John Watts, (all private sentinels), John Allan, a foremast man belonging to the *Seaford*, plus some slaves.

45. The men who made it to the *Rose* were James Pound (chaplain), Moses Wilkins (a writer), James Ray (a bricklayer), Thomas Emerton, John Hall (both private sentinels), John Peterson (carpenter), Henry and Adrian Peterson (private sentinels), all of these in the service of the East India Company, then Mr Abraham Chitty, Captain Thomas Dennet, with his servant, Mr Henry Greenhil, and a servant of Lt. Rafhel, and two black slaves.

46. This possibly suggests that Moses Wilkins may have been taken as Pound's servant because he was already experienced as his observational assistant in England.

47. Banjar is in southern West Java.

48. This suggests that Pound did indeed escape with more than just the clothes he stood up in. It may also indicate that he was being used by Flamsteed in order to replace the observations made by Halley at St Helena.

49. James Pound to John Flamsteed, 7 July 1705, Cambridge University Library, RGO 1/37, fol. 102v.

50. Several references report that Sir Josiah Child cared about little but money.

51. It could also be argued that Sir Josiah's immense wealth laid the foundation of Sir Richard's enhanced appreciation of the arts as well as the wealth that enabled him to commission the building of Wanstead House from 1715 to 1722.

52. Pound always refers to his dwelling as the parsonage.

53. Sarah Pound never married. She lived in her cousin James Bradley's household in Greenwich and died 19 October 1747, aged 34.

54. Not quite comparable I think to the coming together of Tycho's observations with Kepler's mathematical genius but nevertheless in view of the advances made by Bradley, a pregnant moment indeed.

55. I have no doubt at all that the juvenile James Bradley was familiar with the night sky. Even today the village of Sherborne, bereft of street lighting and situated in the Windrush Valley, offers a vista of the stellar universe and the Milky Way rare in the twenty-first century in England. How much more would the celestial sphere have imprinted itself on the conscious awareness of an intelligent young man in the clearer skies of the eighteenth century.

56. This circle included Edmond Halley and Isaac Newton.

57. Edmond Halley seems to have had no doubts about Bradley's remarkable abilities. These were expressed when Halley introduced Bradley for the Fellowship of the Royal Society in 1718. These abilities will be detailed in the next chapter, 'An Ingenious Young Man'.

58. Agnes Mary Clerke, *James Pound*, Dictionary of National Biography, 1885–1900, Vol. 46, pp. 232–233.

59. The loss of all Pound's records, observations, and calculations during the Pulo Condore massacre may give the casual researcher the impression that Pound simply made a few observations which he sent to the Astronomer Royal on occasion. Even so it is remarkable that Agnes Clerke could have been so mistaken about Pound's astronomical career.

60. The word 'curious' during this period could mean 'precise and accurate'.

61. *Philosophical Transactions*, Vol. XXIX, p. 252.

62. Although I have absolutely no evidence, a possible cause for the premature death of a woman during this period could arise from complications attendant to both pregnancy and/or parturition. Having given birth to a daughter two years before, there is always the possibility that Sarah died as she was expecting another child. Of course any such information would have been withheld or destroyed by Bradley's executors between 1762 and 1776. For information about these latter inferred actions by Bradley's executors refer to Chapter 11 where I discuss the actions of the Peach family.

63. Not to be confused with the estate of Sir Richard Child. Matthew Wymondesold owned an extensive estate bordering on the village.

64. *Philosophical Transactions*, Vol. XXIX, p. 401.

65. James Bradley, 'Miscellaneous Astronomical Observations' in *Miscellaneous Observations and Correspondence of James Bradley*, Oxford, 1832, pp. 339–380.

66. This refers to the focal length of the instrument.

67. I base this on the closeness of the two households at the time of Pound's demise and the manner in which Bradley was accepted as a member of Elizabeth's own household. Without this Bradley's work on the aberration of light might have been jeopardized.

68. As revealed by entries in Pound's Accounts Book.

69. James Pound died intestate. Bradley's only income was a comparatively small stipend from his professorship in Oxford.

70. *Reports of Cases Argued and Determined in the Court of King's Bench*, Vol.4, 1791, reports that Matthew Wymondesold made a will dated 12 June 1748 and died shortly afterwards. His son Francis, who lived with Bradley and his aunt Elizabeth Pound in Oxford in 1732 died in 1759.

71. Isaac Newton, *Philosophiæ Naturalis Principia Mathematica*, Book III, Prop. 42, 3rd edn. London, 1726.

72. During the early eighteenth century an annual income of £300 would suffice to live with all of the appurtenances of a gentleman.

73. Non-jurors were those who refused to take an oath to the new monarchs Mary II and William III after James II had been deposed in 1688. They believed this violated the oath given to James II. They were mostly High Church Anglicans. They were stripped of all offices and livings.

74. Pound was arguably one of the very first astronomers in England to use the transit instrument, first used by the Danish astronomer Ole Römer. Used in conjunction with a good regulator (accurate clock), it enabled the measurement of accurate positions in right ascension.

75. He had been born in Chester but his father William had been a prominent Irish politician and Samuel inherited a substantial estate from his father in Ireland. He attended Trinity College Dublin where his tutor was George Berkeley, who was a prominent critic of the basis of Newton's calculus. Samuel was a member of both Parliaments, in London and in Dublin.

76. Samuel Molyneux to John Flamsteed, 19 May 1718. Cambridge University Library, RGO 1/37, fol. 85r.

77. Samuel Molyneux to John Flamsteed, 30 August 1718. Cambridge University Library, RGO 1/37, fol. 87r.

78. During the 'blitz' in London, from 1940 to 1941, much of Halley's archive was lost to bombing by the Luftwaffe when the Guildhall was badly damaged and many archives held in the Guildhall Library were lost. Any reproductions of the Halley archives made before this catastrophic event are precious.

79. *Palilicium* is another name given to *Aldebaran*, the brightest star in the constellation of Taurus. Bayer designated it as α Tauri.

80. Edmond Halley, *Considerations on the Change of the Latitudes of some of the principal fixt Stars*, Reproduced in R. G. Aitken, Edmund Halley and Stellar Proper Motions, *Astronomical Society of the Pacific Leaflets*, 1942, Vol. 4, No. 164, p. 103.

81. According to principles revealed by Newton in the *Opticks* and two years after Thomas Godfrey of Philadelphia independently devised such an instrument. Godfrey was championed by his fellow Philadelphian Benjamin Franklin FRS.

82. These circumstances will be examined in detail in Chapter 3.

83. Charles Mackay, 'The South-Sea Bubble' in *Extraordinary Popular Delusions and the Madness of Crowds*, London, 1852, new edn., New York, 1980. pp. 49–91.

84. This is according to Rigaud, but I have been unable to determine the exact date when this was, for the stock reached two distinct 'highs', the first on 2 June 1720 when stock reached a high of 890 before falling back to about 650 on 22 June 1720, and the second, reached after various illegal procedures were introduced by the directors of the company, of about 1000 on 1 August 1720, before entering a steady fall to 400 and lower.

John Flamsteed, King's Observator, the first Astronomer Royal 1675–1719.

EDMUNDUS HALLEIUS R.S.S.
Astronomus Regius et
Geometriæ Professor Savilianus.

Edmond Halley. Bradley's mentor and second Astronomer Royal 1720–1742.

Sir Isaac Newton, President of the Royal Society 1703–1727. He, early on, described Bradley as "the finest astronomer in Europe".

The Rt. Honble. THOMAS, Earl of. MACCLESFIELD Lord HIGH CHANCELOR of GREAT BRITAIN.

G. Kneller Eq. Baront pinxit. Pub. by E. Parker & J. Tonson. Geo. Vertue sculpsit. 1722.

Thomas Parker, Regent of Great Britain, Lord Chancellor of England, First Earl of Macclesfield, was patron of both Pound and Bradley.

The Reverend
Mr. Benjamin Hoadly B.D.
Rector of St. Peter Poor. LONDON.

G. Vertue Sculp.

Benjamin Hoadly, progenitor of the Bangorian Controversy. A friend of Pound, he ordained Bradley in 1719.

George Graham, clock and instrument-maker. He worked closely with
Bradley on a number of experiments and epitomized the changing status of
artisans in the early eighteenth century.

The Right Hon:ble George Earl of Macclesfield.

President of the Royal Society &c:

George Parker, Second Earl of Macclesfield and President of the Royal Society 1752–1764. He was a close friend and the most long-standing of Bradley's patrons.

Henry Pelham, Prime Minister of Great Britain 1743–1754. He arranged for an annual pension to be paid to Bradley from 1751 up to his death.

3

An Ingenious Young Man

The Emergence of James Bradley, Astronomer

A well instructed young person who also in his ingenious skills and
enthusiasm reveals an inborn aptitude.[1]

Edmond Halley on James Bradley.

There is no evidence that James Bradley had any prior knowledge of astronomical practice[2] prior to 1711 when he took up residence with his maternal
uncle, the Rev. James Pound, Rector of Wanstead in Essex. Bradley may have
been aware that his uncle was a fine observational astronomer. Later, following the termination of all communications with the Astronomer Royal, Pound
made observations on behalf of Newton that were later included in the third
edition of the *Principia*.[3] Bradley's earliest 'recorded' observations were made
in 1715. When Pound and Bradley's earlier observations from 1711 are examined, it is difficult to decide which of the two made the observations. Stephen
Peter Rigaud, Bradley's first biographer, refers to this confusion when he wrote,

> Some difficulty occurred from Pound's observations having been entered pre-
> vious to 1725, indiscriminately in the same book with Bradley's. They are
> constantly working together, and it is not improbable that either of them
> might have entered what was really the result of their joint labours; but there
> is now no way of ascertaining in what cases this was or was not so, and it
> seemed therefore to be most safe to confine the publication to what Bradley
> had written himself. Fortunately the two handwritings are so different that
> the separation could be made without much danger of mistakes.[4]

Given Rigaud's caution in not accepting the provenance of any observation
unless it was clearly indicated as Bradley's in clear distinction from Pound's,
it is obvious that Bradley's earliest 'recorded' observations evince a degree of
expertise not expected in a beginner. The earliest observation clearly ascribed
to Bradley on 21 October 1715 at 6hrs 55min 20 sec (that is just before 7pm
as the astronomical day begins at noon), was to record the right ascension and
declination of the planet Mars. Bradley wrote, 'Mχ in the short glass = 31.20

The Life and Work of James Bradley. John Fisher, Oxford University Press. © John Fisher (2023).
DOI: 10.1093/oso/9780198884200.003.0004

rev. = 36′ 45″. Xd = difference of declination between M and χ in the same glass was 19.60 rev. = 23′ 23″.

This meant the screw micrometer was turned through 31.2 complete revolutions, corresponding to 36 arcminutes and 45 arcseconds. The entry states that the difference in the declination between the planet Mars and the star χ is 23 arcminutes and 23 arcseconds corresponding to 19.6 complete revolutions of Bradley's micrometer screw. Rigaud commented that the star is χ Sagittarii in Bayer's Catalogue. Mars was more northward and westward than χ Sagittarii. This is an observation made by an experienced, consummate practitioner. It reveals that Bradley's familiarity with the procedures of positional astronomy, were well established. Claims that these were his first recorded observations also reveal that these were not his earliest. Given the instrumentation at his command it shows evidence of a high degree of competence by the youthful astronomer. Later on 29 October 1715 Bradley revealed his burgeoning abilities in the capable way he used several instruments in conjunction with each other.[5]

Bradley had not yet acquired the more remarkable skills of his mature observational practice. Even so, a cursory understanding of his observations during the evening of 29 October 1715 reveals that he was a competent, experienced astronomer. Long before Bradley made his earliest recorded observations, he was not merely working with his uncle as an assistant he was already acquiring the skills and techniques of his highly regarded relative. My chronology of Bradley's work[6] reveals that Bradley made observations only when he was in Wanstead. In his earlier years when he was in Oxford, he had no access to astronomical instruments. By the time of his earliest recorded observations he was reading for an MA with which he graduated on 21 June 1717. Bradley was making observations while his uncle was grieving over the death of his wife Sarah. In June 1715, when Bradley's first identifiable observations were made, his aunt Sarah had recently died. Pound was licensed to practice medicine, graduating with the medical degree of BM on 21 October 1697. Pound was unable to prevent his wife's death which he undoubtedly took grievously. His accounts book[7] reveals a period of withdrawal and mourning. It reveals why Bradley worked alone during this period and why the observations were clearly identifiable as his. During this period of grief, they reveal a fully competent observer able to use several instruments together, these being telescopic sights, micrometers, and a regulator.

Some of Bradley's observations included those of Jupiter's satellites. This extended series of observations of the Jovian system reveal the flowering skills of an astronomer soon to be the equal of his uncle. Rigaud's work has led me to examine these observations in detail, confirming that Bradley had developed the habit of calculating and then observing each occultation of the four Galilean bodies.[8] In turn he developed an observational protocol as he connected each calculation with every observation. These calculations were based on Pound's Tables, adapted from those of Giovanni Domenico Cassini's of 1693.[9] These were improved by Bradley from his many subsequent observations. These provisional tables were then sent to Edmond Halley, who had them printed in the *Philosophical Transactions,*[10] who added them to his growing collection of astronomical tables.

Bradley's papers for 1716 include observations by which Pound determined the positions of the stars recorded in the third edition of the *Principia.*[11] In a letter written from London to Pound in Wanstead dated 6 September 1716, Edmond Halley wrote,

Dear Sir,

Entreating you to pardon the frequent trouble I give you,[12] these are to let you know that by the distances you took the other night I find the place of the nebula in Hercules to be ♍ 25° 6′, with the latitude 58° 1′, whereby it appears that it precedes π of Hercules 33′¼ minutes of time with very little difference of declination. That in the foot of Antinous I find to have long ♑ 8° 55′½ with north latitude 16° 37′½, preceding the bright foot of Antinous (λ Bayero). Bayer has two stars preceding λ, to which he has put no letters, but they are i and k in Mr. Flamsteed's Catalogue;[13] the preceding of which k follows the nebula 6 minutes of time, and is 22′½ minutes more northerly than it. By these Mr. Bradley may at his leisure examine the above situations of the nebulae, which are fitted to Mr. Flamsteed's epocha, or anno ineunte 1690.

On Saturday next, Sept. 8vo, about 5 in the morning, (if you please once more to verify the places of the stars in the beginning of Virgo, to which the comet was applied in November 1680) you will find Mars close to the two contiguous d and e, in Transactions 342, which are but 7 minutes asunder. According to my calculus I expect him nearer to each of them than they are to one another, and about 4 minutes more northerly than they, which have precisely the same declination. I fear it will be too light for me to see so small stars, when Mars gets over the houses, or you should not have had this trouble from, kind Sir,

Your much obliged, faithful servant,

EDM. HALLEY.[14]

It seems that as early as 1716 Halley recognized that his eyesight was deteriorating, sufficiently to become reliant on Pound and Bradley to make observations of small stars, particularly during daylight conditions. Bradley made positional observations of Mars during 1715 and 1719 for papers written by Halley in the following years, as well as observations of various nebulæ.[15] Halley's eventual concentration on his lunar theory led to his reliance on other astronomers to determine stellar positions. It is surprising that when Bradley was in his prime during the 1730s, observing minute variations in the orientation of the Earth's axis by measuring the motions due to nutation, Halley found his 'lynx-eyed' friend's claims impossible to accept. He did not believe it was possible to discern motions as small as half an arcsecond. Bradley aware of his friend's comparative observational shortcomings simply accepted Halley's strictures, apparently without murmur.[16] During the summer vacation in 1716 Bradley observed Jupiter carefully noting its position relative to stars observable with the naked eye. On 12 August 1716 Bradley's entry in the Wanstead Observation Book reads as follows, '14hrs 27′ [2.27am] Jupiter was distant from a star visible without a telescope 37.10 rev.in the long glass,[17] the star was more northerly and easterly than Jupiter. This star is called Propus'.[18]

In 1717, Bradley expected to graduate from Oxford University with a master's degree. He must also have planned to fulfil his growing dedication for the practice of observational astronomy. Under the tutelage of his uncle, he was prepared to take holy orders in the Church of England. This was the original intent of his father William. It was why his education had been placed under the care of his ordained maternal uncle. Any examination of the chronology[19] of Bradley's working life reveals that there was a radical falling away of Bradley's astronomical activity. His only recorded observations for the whole of the year were made on 17 January 1717 when he observed the occultation of Jupiter's third satellite (Ganymede) behind the giant planet. Written on a separate piece of paper Bradley recorded that by the clock he witnessed at 6hrs 4min 10sec the immersion of the satellites behind Jupiter.[20] At 9hrs Bradley wrote 'it was bright as any of the others' referring to the third satellite which in future reports he would describe as being dark. Between 17 January and 21 June 1717 when Bradley graduated with an MA degree there is no evidence of any recorded observations. From Pound's accounts book, it can be discerned that Bradley contracted smallpox. Pound's medical qualifications and his licence to practice medicine allowed him to oversee his nephew back to health whilst protecting his household by isolating him.[21] Bradley seems to have fully recovered. Pound must have tended to his nephew's illness with all expediency, setting aside his own astronomical studies. On 11 March 1717, Pound paid £2 3s 0d 'By Cousin Bradley's Nurse in the Small Pox'[22] thus ensuring round the clock care of his nephew.

More than once Rigaud, Bradley's first biographer, believed he was fortunate in being related to such a prosperous man as Pound. He wrote, 'If money was wanted beyond what his father's limited means could supply, it was the uncle who advanced or gave it to him.'[23]

But this was only a part of the story, for the background seems to have been unknown to Rigaud. The reason why William Bradley was in such financial difficulties was because his income was only intermittently paid, even though he held a position of trust on the prosperous Sherborne estate in Gloucester-shire.[24] His employer was Sir Ralph Dutton Bt. (1642/6–1721) a compulsive gambler. In a handsome history of the village and estate of Sherborne, Sybil Longhurst and others, revealed that Sir Ralph had considerable problems in controlling the management of his estate and monies. His eldest son John Dutton was eventually forced to issue an indenture of assignment dated 23 June 1708,

> between Sir Phillip Meadows of St. Anne's Westminster Knight of the one part and Hans Sloane of St. Giles-in-the-Field M.D. and Richard Husbands of St. Andrews, Holborn Gent of the other part by which Sir Ralph Dutton for £600 to Sloane and Husbands for £270 each. Signed by Phillip Meadows with his receipt for the money.[25]

These were considerable sums of money, and yet most likely only a small proportion of his accumulated gambling debts.[26] Sybil Longhurst and the Tufnalls, her co-authors, state that by 1709, and this is only a short time before Bradley left the village to reside with his uncle in Wanstead, John Dutton took over control of the estate. A further indenture of assignment dated 4 February 1709/10 OS between the Right Worshipful Sir Ralph Dutton of Sherborne Bart and John Dutton Esq. his son and heir states that, 'Sir Ralph assigns to the said John the whole of his real and personal estate in return for £5000 and a life annuity of £400 yearly.[27]'

In a later deed John also 'undertakes to satisfy all debts that Dame Mary [his mother] may contract and provide her and her daughters with maintenance and to indemnify Sir Ralph from all debts due by him to Dr. Barwick and John Prinn'.[28]

The Dutton family evidently feared that Sir Ralph would gamble away much of the value of the estate. His gambling debts must have been considerable for his son to have taken such legal measures to protect his family from the deprivations imposed by his debts, which the family had become aware of. The deleterious effects that the wayward actions of Sir Ralph Dutton had on his family were certainly worrying, but the effects on those dependent upon the estate to provide their livings must have been concerning indeed. Little imagination is needed to comprehend the reasons why the education of James Bradley was given over into the care of James Pound. William Bradley's lack of a guaranteed income was only partially compensated by his ability to gain access to the produce of the estate, of which he was a steward. For a gentleman to have been placed in a position where he was unable to guarantee the education of his third son must have been disturbing.[29] So when Rigaud asserted that Bradley was fortunate to have a prosperous relative, it fails to acknowledge the reasons why he had become dependent on his uncle's fortune. Bradley's father had been failed by his employer. The young James and his father's family must have lived in straightened circumstances with or without a prosperous relative.[30]

Although James Pound was prosperous, he too had experienced life's changing fortunes. In a letter dated 7 July 1705 from Pound in Batavia to the Astronomer Royal at Greenwich, he briefly described the massacre at the East India Company's station at Pulo Condore, described in detail in Chapter 2. Pound had regularly corresponded from the Indies to Flamsteed from 1700 onwards describing experiences and sending astronomical observations. He spent time on Chusan, giving accounts of life around the Imperial Court. Referring to Pound's Journal,[31] we discover that he returned to England in July 1706. Sir Richard Child, the third son of the former governor of the Honourable Company[32] Sir Josiah Child, gave Pound the prosperous living of Wanstead close to his own residence.[33]

It was during 1718, possibly in 1719, that Bradley wrote the notes that were included in the precepts to Halley's Tables.[34] Bradley began with the words,

> In these tables we have determined the mean motions of the satellites, by comparing such of the oldest observations we could procure, as seemed to be the most accurate, with our own taken lately at Wanstead; when Jupiter, after four revolutions, was nearly in the same place in his orbit.[35]

Using older observations from 1671, 1676, and 1677 Bradley discovered that the various computations differed greatly from his and other observations.

By putting back the apsis[36] to $= 14° \ 00'$ at the beginning of 1677 the differences largely vanished. Bradley's work on Jupiter and its satellites and his accurate calculations later formed a foundation to his discovery of the aberration of light during 1728.

Irregularities in the observed timings of a great number of occultations in the Jovian system had led Giovanni Domenico Cassini to briefly suggest that it was a consequence of the progressive motion of light[37] and the relative motions of the Earth and Jupiter. As quickly as he formed the idea, he rescinded it. Cassini was a conservative astronomer as many of the Bolognese school were. Recognizing that he was disavowing Cartesian natural philosophy, the dominant school in France with its insistence on the instantaneous velocity of light, he backtracked. More seriously for a Bolognese astronomer he was contradicting established Roman Catholic doctrine. The Council of Trent (1545–1563) held that the immobility of the earth was to be accepted as Church teaching, following the Church Fathers. Astronomers of a more conservative bent of mind supported this, usually adopting the Tychonic alternative to that proposed by Copernicus, with the Sun revolving around the Earth, while the planets revolved around the Sun. The radical ideas briefly mooted by Cassini were taken up by his assistant at the Paris Observatory Ole Römer, who being a Danish Lutheran had no qualms about admitting the motion of the Earth. When Bradley published his discovery of the aberration of light on 9 and 16 January 1729, it posed a radical challenge to Roman Catholic doctrine, to the work of Cartesian practitioners, and their commitment to the instantaneous velocity of light. The validity of Bradley's work was challenged, but all but the most conservative Catholic authorities bowed to the inevitable. Eventually, in response to Bradley's innovations, the Church of Rome made acceptance of the motion of the Earth 'admissible' to the faithful. Bradley, by then the most eminent and celebrated astronomer in Europe, was later made a member of the Institute of Bologna, an admission that the doctrine of the motion of the Earth had become 'acceptable' to the faithful. A fuller discussion of this will be found in Chapter 5.

Bradley made many observations from March through to May 1718, but there is one that reveals something of the young astronomer's burgeoning abilities. Referring to Cassini's method of observation by applying threads at right angles in the common focus Halley mentions the exactness with which, 'Dr. Pound and his nephew Mr. Bradley did, myself being present in the last opposition of the sun and Mars, this way demonstrate the extreme minuteness of the sun's parallax, and that it was not more than 12″ nor less than 9″'.[38]

The accepted contemporary value is just under nine arcseconds. It appears that Bradley made the observation and Pound entered it into the observation book. The precision of Bradley's observations were now so noticeable that Halley felt compelled to report it.

On 18 September 1717, when Pound was still tending to his nephew's convalescence from smallpox, he began making preparations to use the Huygens 123-foot-long focus glass, paying 4s 6d for tin and brass work to hold the glass in place. During 1718, Pound continued with his preparations. On 25 April 1718, he purchased an eyeglass for 2s 6d. On 13 May 1718, he paid £2 for drink for the men to raise a one-time Maypole given to Pound as a gift by Newton. Three days later on 16 May 1718, he paid the men who finally erected the pole.[39] It is both amusing and concerning to note that the pay of 17 shillings for the labour was less than the 40 shillings that Pound spent on drink for the men.[40] The skills that Bradley displayed when using the 123-foot glass will be referred to later. He proved to be adept with long-focus aerial telescopes, using one with a focus of 212¼ feet.

On 28 October 1718, Pound presented a series of observations he made, assisted by Bradley, on the lunar eclipse on 29 August 1718, to the Royal Society.[41] At the same meeting, Edmond Halley proposed Bradley as a person qualified to be a member of the society. It was referred to the next council meeting due on 6 November 1718. That meeting with Newton in the chair approved Bradley's fellowship. He was elected alongside Nicholas Saunderson (1682–1739) later to be elected Lucasian Professor of Mathematics at Cambridge. Bradley's election was a formality given that his proposer was Halley. Newton was already an admirer of Bradley's abilities. With his growing reputation as an astronomer of repute, Bradley's life was nevertheless moving in other directions, to fulfil his father's expectations. Bradley may well have believed he was in a quandary. His education befitted a man of the cloth, yet his interest and noted expertise in observational astronomy, diverted him to a very different vision of heaven. His uncle had prepared Bradley for a life as an Anglican priest. However, Pound had also introduced his nephew to an interest that had become an all-consuming passion. I agree with Rigaud that a life in the established church given his native intelligence and connections would have led Bradley to high office in the Church of England. At this time there was no obvious career in prospect as an astronomer, the only offices on offer were at the two universities or at the Royal Observatory, none of which were lucrative nor likely to be on offer any time soon.

Bradley's uncle's life as the Rector of Wanstead offered a good living and sufficient time and leisure to pursue his interests as an astronomer. He had sufficient time to allow both Newton and Halley to make use of his services, taking the place of Flamsteed, who was himself the Rector of Burstow in Surrey. Pound's accounts book reveals many gifts and loans to his nephew addressed as 'Cousin Bradley'. The death of Bradley's aunt Sarah and his con-traction of smallpox brought with it, if such a lesson was ever needed, full recognition of how fragile life could be. His father's life as a steward on the Sherborne estate gave him no reason to believe that life owed him a living either, so his financial dependency on his uncle was gratefully accepted. I have been unable to discover anything about the life of Pound's sister Jane, Bradley's mother, other than that she had been born and raised in the village of Bishop's Canning near to Devizes in Wiltshire, the daughter of a gen-tleman. How Jane met William Bradley who lived a day's journey away in Gloucestershire we are given no clues. Although most of the private corre-spondence between family members has not come down to posterity, every insight we can infer suggests his familial life was one of strong affections with regular visits to friends and relatives in Gloucestershire. One observation of a comet was recorded in Cirencester when Bradley was probably visiting friends or relatives.

In the period before Bradley's entry into service with the Church, he busied himself with what may be regarded as his major vocation. From January to June 1719, Bradley's observations reached a high level of activity. The over-whelming majority of his observations were of Jupiter and its satellites, in preparation for tables on behalf of his mentor Edmond Halley, who was still wedded to the idea of using the motions of the satellites of Jupiter to deter-mine the longitude at sea. Later he settled on a more practical method of using 'lunars'. This, however, proved to be elusive due to the complexities of the motion of the Moon. Halley held a commission as a captain in the Royal Navy and must have recognized that it was asking too much of anyone to observe Jupiter and its system with the degree of accuracy required on a ship at sea, even if it was visible at all. Bradley's extensive observations and calculations of the motions of Jupiter's satellites were not, however, in vain. His work would pay immense dividends when he made the fundamentally important discovery of the aberration of light.

Bradley was twenty-five years of age on or about 27 September 1717 OS. At this age he was qualified to take holy orders. He delayed. It may have been because he was still recovering from smallpox, but there is more valid-ity in the belief that he was reluctant at this time to take up his career in the Church. Bradley was working assiduously on observations of the satel-lites of Jupiter for Halley. His observations were so determinedly pursued that any suggestions that his recovery from illness had inhibited him from entering the Church must be treated with scepticism. Aware his time for astronomical observation would soon be cut short, the completion of his planned observations of Jupiter's satellites became important to him. From January to the end of May 1719 Bradley's studies of the Jovian system were only diverted by an occasional measurement of stellar positions in right ascension and declination in order to fix Jupiter's orbit. This enabled him to reduce observations of the satellites in their never-ending dance around the giant planet. Bradley's observations of 9 and 10 March 1719 are highly indicative of the energy of his work. Bradley's entries into the Wanstead Observation Book were reproduced in the *Miscellaneous Works and Correspondence of James Bradley,*

March 9.	7hrs 26' 0"	a bright spot in the northern belt appeared in the middle of Jupiter, it appeared a little black on the north side of it.
	11hrs 28' 45"	mean time, the 3rd satellite of Jupiter [Ganymede] emerged.
	34' 50"	the 1st satellite [Io] touched Jupiter's western limb. Apogee.
	38' 20"	the 1st satellite disappeared.
	14hrs 34' 10"	the 1st satellite began to emerge out of Jupiter's shade.
March 10.	8hrs 44' 8"	the 1st satellite touched Jupiter's eastern limb, perigee.
	49' 50"	it appeared all within.
	9hrs 27' 50"	the shade of the 1st seen entering on Jupiter.
	31' 40"	all within.
	10hrs 58' 38"	the 1st satellite touched Jupiter's western limb.
	11hrs 4' 18"	separated from the limb.
	38' 40"	the shade of the 1st touched Jupiter's western limb.
	44' 0"	the limb free from the shade.
	13hrs 4' 0"	the bright spot on the northern belt appeared again in mid.
	56' 20"	the 2nd satellite [Europa] touched Jupiter's eastern limb.
	14hrs 2' 50"	it appeared all within.

Bradley added a *memorandum* detailing the observations made on 9 and 10 March 1719. He wrote this in the Wanstead Observation Book,

In these observations allow for the versed sine of the parallax of the Earth's orb.

March 9.	11hrs 34' 50"	[See above] the 1st satellite touched Jupiter's western limb, being in apogee.
March 10	8hrs 44' 8"	it touched Jupiter's eastern limb, being in perigee so that
	in 21hrs 9' 18"	it moved 5s^{42} 29° 57', its mean motion in the same time being but 5s 29° 21', the difference being +36'.
March 9	14hrs 34' 10"	the 1st satellite began to emerge out of Jupiter's shade, and
	on 10 11hrs 38' 40"	the first edge of its shade touched the eastern edge of Jupiter,
	whence in 21hrs 4' 30"	it moved 6s 0° 6' 15", its mean motion for that time being 1° 20' 40" less.

to which he adds,

> whence it may be concluded, that if these inequalities proceed from an excentricity in the satellite's orbit, the perigæon part must be nearer the middle between the two last observations than the two first, the motion of the satellite being swiftest between the last observations [made on the 9th of March]. If the place of the apogee be supposed in the beginning of Leo, and the greatest equation one degree, 'twill very well account for these observations.

Rigaud wrote that Bradley's earliest endeavours were on the places of Jupiter's satellites. He adds that he was in the habit of calculating each eclipse he observed, marking the quantity against the observation by which it differed from the times derived from the tables. One object was the correction of the tables made by his uncle, based in their turn on those of Cassini's of 1693 with corrections being introduced as observations revealed them to be necessary. These were communicated to Halley who published them in the *Philosophical Transactions*.[43] Halley inserted them into his own collection of tables published as an appendix to the general tables for the four that were given him by Bradley. Although these tables were due to be published in 1719 Halley kept them for his own personal use. They were not published in a form fully available to the public until 1752 when Dr John Bevis (1695–1771) published them. Bevis was in regular communication with Bradley translating them from Latin into English. It was Bevis, the discoverer of the Crab Nebula in 1731,[44] who revealed that the tables were entirely the work of Bradley. Hodgson of Christ's Hospital published tables of Jupiter's satellites in 1749, 'studiously ignoring all mention of Bradley whose previous labours he must surely have been acquainted.'[45]

Jean Sylvain Bailli (1736–1793)[46] in his *Essai sur la Théorie des Satellites de Jupiter* of 1766 gave Bradley the credit for detecting the greater part of the inequalities which had been recognized in the motions of the satellites of Jupiter.[47] Why these valuable tables were not published for thirty years is difficult to fathom. At the time when they should have been published Bradley was fully engaged with his ordination as a priest in the Church of England and the various responsibilities demanded of him.

On 24 May 1719, Bradley was ordained a deacon by former diplomat John Robinson (1650–1723) the Bishop of London,[48] with testimonials signed by his uncle as Rector of Wanstead, Chishall the Vicar of Walthamstow, and Chisenhall the Vicar of Barking, all of whom were to testify they had known him for six years and could vouch for his character. Two months later on

25 July 1719, Bradley was admitted to priestly orders in the presence of the Rt Revd Dr Benjamin Hoadly FRS, Bishop of Hereford, a friend and long standing Whig associate of James Pound. Two years previously as the Bishop of Bangor, Hoadly initiated the so-called 'Bangorian Controversy'. On 31 March 1717, Hoadly gave a sermon in the presence of George I on the *The Nature of the Kingdom of Christ*. His text was John 18:36, 'My kingdom is not of this world'. Hoadly expressed his interpretation that the church was identical to the kingdom of heaven with the conclusion that it was not of this world. He went further, spelling out his belief upholding the view that Christ had not delegated his authority to any earthly representatives. The full social, theological, and ecclesiastical ramifications of this sermon lasted for over a century. The consequences of this sermon, and the inevitable controversy that followed it, went beyond any theological repercussions or even church affairs. It led to a suspension of Convocation, the government of the Church of England, for many decades, well into the nineteenth century. It does throw some light on the character of Pound's associations as well as the nature of Bradley's later social and political associations which he used with such remarkable acumen in the furtherance of his astronomical and social careers. Bradley remained close to the Hoadly family for the rest of his life. When Bradley made the observations leading to the discovery of the aberration of light at Wanstead, he was occasionally assisted by Dr Benjamin Hoadly, the bishop's eldest son.[49]

It has been suggested to me, more than once, that many of the resources Bradley later acquired when he was the Astronomer Royal were gained because of the importance of the work of the Royal Observatory. There is little evidence to support this view. Neither Flamsteed nor Halley, who were Bradley's predecessors in the office, were ever able to acquire anything like the resources they really needed to administer a national institution. In a manner so often characteristic of English public life, parsimony was the rule of the day even if it was in support of the national interest.

Bradley's brief career as a churchman lasted until late in 1721. There is no evidence that he made any astronomical observations during the period between his ordination by Hoadly at Hereford and his election as the Savilian Professor of Astronomy at Oxford. Rigaud correctly suggests that Bradley's church career promised much, given who his patrons were, combined with his native intelligence. Rigaud argues that he could have reached the highest levels of the Church hierarchy. Two days after his ordination by Bishop Hoadly, Bradley was appointed as his private secretary,[50] and received a living at Bridstow, a small parish near Ross-on-Wye, Herefordshire. The first National Census in 1801 reveals that the population of Bridstow was only 471. From this we can surmise that in 1719 the population didn't much exceeded 400 – a meagre living indeed. It led to Bradley seeking a supplementary income. The patronage of Samuel Molyneux FRS, yet another associate of Pound, led to a source of support that eventually gave Bradley the patronage of George II. Early in 1720, Bradley in addition to his small living at Bridstow gained half of a divided living in Pembrokeshire due to the intercession of Molyneux. As the private secretary to the Prince of Wales, Molyneux acquired for Bradley the patronage of the Prince who had in his gift half of the shared living at Llanddewi felfry in the hundred of Narberth.[51] Bradley had thus acquired the patronage of the Prince of Wales, the future George II. The year 1720 also witnessed radical changes in the lives of Bradley's two mentors, Halley and Pound. They resulted from the death of John Flamsteed on 31 December 1719.[52]

Halley was appointed as the second Astronomer Royal only to discover that the Royal Observatory was almost completely devoid of instruments. All the instruments belonging to her late husband were removed by Margaret Flamsteed. Even if she had been willing to sell the instruments to her husband's successor, there can be no doubt she would have opposed passing them on to a man she believed had caused immense grief to her husband. It is not known what happened to the instruments she removed. She spent her final years completing her late husband's work by seeing through the publication of the catalogue of 2883 stars in the authorized *Historia Coelestis Britannica* in 1725. She published, with the aid of Joseph Crosthwait, the *Atlas Coelestis* in 1729. Thanks to the patronage of the Lord Chancellor, Thomas Parker, soon to become the 1st Earl of Macclesfield, Pound was presented with Flamsteed's former living of Burstow in Surrey. Pound now acquired the income of two substantial livings in Essex and Surrey. On 20 April 1720, Pound's accounts book reveals that he received a gift of fifty guineas[53] from Newton.

The relationship between George I and the Prince of Wales was one of mutual loathing. It led to the creation of rival courts, the King's at St James and the Prince's at Leicester Fields.[54] Pound's connections linked him to both courts. One of his associates was Samuel Molyneux who had been born in Chester on 18 July 1689 and raised in Dublin by his uncle Thomas Molyneux (1661–1733). Thomas was elected an FRS and practised as a physician, becoming President of the Irish College of Physicians in 1713. He provided Samuel with a good education at Trinity College, Dublin, where his tutor was George Berkeley, who dedicated his *Miscellanea Mathematica* to him. Molyneux travelled throughout Ireland as the secretary of the Dublin Philosophical Society, founded by his father, where Samuel published much of his research. This enabled him to build relationships with many of his father's friends and associates. He visited England in 1712, being nominated for membership of the Royal Society by Newton no less. He was elected FRS on 4 November 1712 along with the Swiss mathematician Jean (Johann) Bernoulli (1667–1748).

Although Flamsteed had fallen out with Samuel's father William following the publication of the *Dioptrica nova*,[55] he welcomed Samuel on friendly terms.[56] The correspondence between the two suggests they were on friendly terms for several years. I have transcribed several convivial letters dating from 1717 to 1718 all of which remain friendly and respectful. Flamsteed was in a state of rage over the unauthorized publication of his observations in the *Historiæ coelestis* edited by his former friend Edmond Halley, now perceived by him as a bitter enemy. From his letters and the accounts of others, it appears that Molyneux was a polite, personable man. His firm Whig loyalties and membership of the entourage of the Duke of Marlborough at the Electoral Court of Hanover, led to his appointment in 1715 as the Prince of Wales' personal secretary, thus exposing him to criticism from various quarters. The non-juror and university proctor Thomas Hearne (1678–1735), the second librarian at the Bodleian Library, met Molyneux when he visited Oxford. Hearne thought Molyneux both 'pushy' and 'impatient'. Expressing the opinion that Molyneux's superficial knowledge, a consequence of taking 'Accounts from Conversation with Gentlemen, and not from study', Hearne wrote, 'I had sufficient and full Proof of Mr Mollineux's Confidence and his Ignorance in Antiquities'. This sense of the superficiality of Molyneux's character was later underlined by the naivety displayed in the scandalous Mary Toft affair.[57] Later Hearne would hold equally negative opinions of Bradley, although these were later ameliorated by the passage of time.[58]

Molyneux was already wealthy before he married one of the wealthiest heiresses in England, Elizabeth Capel in 1717, having inherited his father's substantial estate in County Armagh. In contrast to his patron and future astronomical partner, Bradley was financially dependent on his uncle's benevolence, as evident from many entries in Pound's accounts book. The acceptance of a half-living in the gift of the Prince of Wales still involved the outlay of more than £15.[59] An entry for 25 February 1720 reveals another expense of £2 12s 6d paid to 'rectify' Bradley's presentation so Pound's outlay amounted to over £18. If this seems to be a trifling amount, compare it to an entry in Pound's accounts book on 8 April 1720, also for £18, which was a quarter of the annual stipend of Mr Charles Leaver, the curate who took care of the pastoral needs of the parishioners of Burslow on behalf of Pound.[60] It can be seen that the formalities required to gain a modest living for Bradley in Pembrokeshire were too expensive to come out of the pocket of the recipient of that living.

Even modest livings such as this were the preserve of those with wealth or patronage.

On 31 August 1721, the Savilian Chair of Astronomy at Oxford was vacated by the sudden and unexpected death[61] of Dr John Keill (1671–1721), a Scot who supported Newton in the priority dispute with Leibniz over the invention of the calculus. Pound was regarded as the most suitable candidate to take up this prestigious position being described as 'the most suitable person in Europe'.[62] But it was a condition of the office to relinquish all church sinecures. It wasn't in Pound's financial interest to take up the vacant post. Pound with two lucrative livings was settled in his life at Wanstead with a young daughter and was soon to marry again. He was unlikely to uproot himself from his present comfortable position, even for such an eminent post as the Oxford chair in astronomy. A much favoured candidate was John Whiteside FRS (1679/80–1729) chaplain of Christ Church who had been a candidate when Keill was elected to the Savilian chair in 1712. Whiteside had given lectures on experimental philosophy from about 1715.[63] Once Pound had rejected the offer, Lord Chancellor Thomas Parker interceded and supported the candidature of James Bradley. For the rest of his life Bradley was to enjoy the patronage of the Parker family. He also acquired other influential supporters. On 4 September 1721, four or five days after the death of Keill, a letter from a future President of the Royal Society Martin Foulkes, was addressed to Archbishop Wake in support of Bradley's election. He wrote,

My Lord,

I humbly take the liberty of applying to your grace on the occasion of the vacancy at Oxford in the Astronomy chair, occasioned by the decease of the late Dr. John Keill, in behalf of Mr. Bradley, a gentleman who is thoroughly qualified for that part, and will be, I make no doubt, every way recommended to your grace's satisfaction. Your grace, I am informed has the first vote, and the calling of the election; besides that, your countenance will be of the greatest weight with the other electors. Mr. Bradley was of Oxford, and is five years master of arts: he has lived for some years with his uncle, Mr. Pound of Wansted, where he has had great opportunities of joining to his theory the practical part of astronomy, in which he has made himself very eminent, having prepared for the press accurate Tables of the satellites of Jupiter, with some other curious pieces; and I am satisfied his being professor will do honour and service to the science. I shall only take the liberty of adding, that he is perfectly approved, and will be entirely recommended by Sir Isaac Newton, whom your grace knows for the great judge of this sort of learning. I most humbly ask your grace's pardon for this liberty, and begging your blessing, desire the honour of subscribing myself,

My lord,

Your grace's most dutiful nephew and most obliged

M. Foulkes.

Foulkes was familiar with Bradley's work on Jupiter's satellites on behalf of Halley. Foulkes as well as being a friend of Pound knew his nephew to be more than a merely competent astronomer. He added a brief addendum calling his uncle's attention to a growing tendency to confuse Bradley with another of the same name: 'Mr. Bradley is fellow of the Royal Society, but not the same as has published some late books on gardening, &c'.[64]

Rigaud reveals that it was under these auspices that Bradley's election took place at the House of Lords on 31 October 1721, with all the expected expenses being paid by Pound.[65] Bradley relinquished his church sinecures in accordance with the wishes of the founder of the chair Sir Henry Savile[66] in 1619. The immediate cost to Bradley was modest. Both of the sinecures, one in Herefordshire and the other in Pembrokeshire, were unexceptional, though there can be little doubt he would have accepted the chair under any circumstances. Yet the full extent of the consequences of the decision to abandon his career with the Church of England has been commented on by Dr Allan Chapman. He writes,

It is not easy for us today to grasp the magnitude of Bradley's decision to relinquish such powerful ties of clerical patronage in favour of science, for in 1721, astronomy was not a profession, but a scholarly pursuit for men who were either privately well-off or were in receipt of clerical funds.[67]

Rigaud, like Chapman, is at pains to express his opinion that Bradley was almost certain to have risen to a high place within the Church. In contradiction to a common opinion held, particularly in France by many of his admirers there, he did not give up the Church out of any aversion to the clergy.[68] Rigaud asserted that astronomy 'had grown into an all-absorbing passion'[69] and opined that Bradley was so actuated by 'the sacredness of priesthood' that he felt he could not be an astronomer and a churchman simultaneously.[70] In 1751, the vicarage of Greenwich became vacant after the death of the Revd R. Skerret. It was offered to Bradley by Sir Henry Pelham, the Prime Minister and Chancellor of the Exchequer. Pelham along with his wife Lady Catherine, who was the Keeper of Greenwich Park, had formed a friendship with Bradley and his wife Susannah.[71] Bradley refused the preferment. More surprisingly, Rigaud overlooked the consequences of accepting the living at Greenwich, which would have meant abandoning his Savilian Chair at Oxford and probably the Readership as well.[72]

Although elected on 31 October 1721 Bradley did not take possession of the chair until 18 December 1721. He read his inaugural lecture on 26 April 1722. Thomas Hearne claimed that this 'didn't add to his reputation'[73]. Rigaud defended Bradley, stating that Hearne was hardly conversant with astronomical knowledge sufficient to be acceptable as a reliable judge of such matters. Rigaud suggests that he supported the rival candidacy of Rev. John Whiteside FRS of Christ Church. However, 'the elephant in the room' was Bradley's election by a London-based Whig power base that was hardly conducive to gain the favour of the prominent non-juror.[74] There must have been some latent resentment at the appointment, particularly at Tory-leaning Christ Church. Yet a few years later Bradley sought to become the Keeper of the Ashmolean Museum, the location of his lectures on experimental philosophy, when his only supporter was Hearne. For whatever reasons, Bradley seems to have developed a habit of impressing people, be it through the force of his intellect or the charm of his character or the integrity of his standards. He was a man who rarely if ever acquired known enemies.

As James Bradley was coming to terms with his new duties and responsibilities at Oxford, his uncle was preparing for his marriage to Elizabeth Wymondesold. Pound had known his future wife for several years through his friendship with her brother Matthew, a successful financier who owned the Wanstead Estate close by the township of Wanstead.[75] Very shortly before his election Bradley began to make observations of the planets Jupiter and Venus, and mainly of Mars. As would be expected, after his election he began further precise observations and calculations of the motions of the satellites of Jupiter. In a period of just two evenings, Bradley displayed his skill as an observer using an aerial telescope with Huygens' 123-foot glass in conjunction with Hadley's 6-foot focus reflector. His entries in the Wanstead Observation Book included,

1722 June	14 9hrs 45' 00"	the shade of the 3rd sat. of Jupiter all within the limb coming on.
	9hrs 50' 00"	the 4th sat. of Jupiter in conjunction with Jupiter (in apog.)
	10hrs 45' 00"	the 1st sat. began to emerge in the Hugenian glass, and was seen at the same time by Mr. Hadley's reflecting telescope, a very exact observation: the air being serene and quiet.
	11hrs 46' 00"	the shade of the 3rd sat. in contact with Jupiter's limb going off; the air was now very unconstant, and Jupiter's limb did not appear well defined by reason of the undulation of the vapours, so that this last observation is not much to be depended on.

Pound and Bradley compared the performance of Hadley's reflector with Huygens' glass. Bradley wrote that the reflector compared favourably with the refractor, though it lost out in terms of its light grasp.

The reflector won greatly in terms of its utility, even though Bradley was extraordinarily skilful in the use of long-focus aerial telescopes. Bradley's skills with aerial telescopes were so marked that he accurately measured the diameter of Venus across 'its horns' using a telescope with a focal length of 212¼ feet. Rigaud referred his readers to Robert Smith's *Optics*[76] a standard text of the time where mention of Bradley's skill with long-focus aerial telescopes is acknowledged. Bradley's 'Miscellaneous Astronomical Observations'[77] lists the entry, obviously made during the Christmas recess which reads, '1722. Dec. 27. 5hrs.[78] the distance between the horns of Venus was observed 29 rev. = 57.8 in a telescope of 212½ f. — m. diam. 19″.04'.

These observations were made in the evening twilight when Venus was near the horizon. It was a remarkable feat. In spite of supporting a chair in astronomy, the University of Oxford did not at this time possess an observatory. Astronomy was treated as a branch of applied mathematics. Oxford remained tied to the ancient ethos, almost as if the micrometer, let alone the telescope, had never been invented. For several years, Bradley relied mainly on the instruments he had access to at Wanstead,[79] and from 1739 at Shirburne Castle, the seat of the 2nd Earl of Macclesfield and arguably the finest observatory in England. Bradley was instrumental in its construction, which later gave him the knowledge that led to the reconstruction and reform of the Royal Observatory from 1748 to 1750.

In Rigaud's account in the *Memoirs of Bradley* it is easy to be misled by the belief that the Wanstead astronomers acquired Hadley's reflector as late as 1723. Rigaud writes,

Reflecting telescopes had been invented in the middle of the seventeenth century; but the mechanical difficulty of perfecting the mirrors was long an obstacle in the use of them. Hadley applied himself to overcome it, and in 1723 communicated to the Royal Society an account of his success,[80] making them a present of one of his instruments. This Pound compared with the Huygenian telescope of 123 feet, and he tells us that although the focal length of the object-metal was not quite 5½ feet, it bore an equal magnifying power, and represented an object as distinctly as the refractor, although not so clearly and bright. Bradley assisted him in the comparison, and, encouraged by what he saw of it, applied himself afterwards to the construction of mirrors.[81]

Yet, as the observations cited above reveal, Bradley was using one of Hadley's reflectors on 14 June 1722, even though Rigaud left his readers with the impression that Pound and Bradley only acquired the instrument after it had been presented to the Royal Society in 1723. John Hadley, along with his younger brothers George and Henry, built a reflector with a 6-foot focal length speculum from 1719 to 1720. This was tested by Hadley before he passed it on to Halley, the new Astronomer Royal.[82] At that time Halley possessed an observatory devoid of large instruments so it must have been gratefully received.

Bradley in the company of Hadley, and sometimes with the support of Molyneux, investigated various alloys to resolve problems associated with the application of this promising technology. Later Bradley abandoned these pursuits, not because he lacked interest in these problems but because these investigations took him away from his observations, when he was engaged in the work that led to his momentous discovery of the aberration of light. Bradley was skilled in the upkeep of his instruments, cleaning, fitting, and repairing them. Unlike many gentleman astronomers, Bradley was a good mechanic with a working knowledge of current workshop practice, possibly acquired from artisans such as George Graham. I am not in any way suggesting that Bradley was a skilful artisan; I am leaning to the view that he formed an attentive appreciation of those skills. It is one of the keys to understanding the working relationships Bradley had with his instrument makers.

Bradley settled into his new duties at Oxford whilst also maintaining his observations when at Wanstead, particularly on Jupiter's satellites and the orbit of Jupiter. His interests expanded into studies of double stars and the kinematics of the Saturnian system.[83] On 9 October 1723, Halley discovered a notable comet.[84] He immediately contacted Bradley in nearby Wanstead,[85] who began to observe it the following night. Bradley continued observing it, occasionally with Pound's or Graham's assistance until 14 November 1723 when duties forced his return to Oxford.[86] Bradley was not content to offer a qualitative description of the appearances of the comet of 1723. His ideas on natural philosophy, as explicitly expressed a few years later in his lectures in experimental philosophy, made a clear distinction between observational or experimental science and speculative philosophy. This was spelled out in the first lecture in the course presented seventy-nine times between 1729 and 1760.[87] The ideas he presented in the course must already have been formed in his mind before he expressed them at Oxford. Yet Bradley expounded many ideas that were possible only after the publication of Newton's *Principia*. The motions of comets were ordered and regulated by the 'laws of nature'. Bradley, under the tutelage of his mentor Halley, came to master most of the arguments of Newton's masterpiece. By the application of natural laws as they were presented in the *Principia*, the motion of comets was a predictable phenomenon. For many centuries, comets had been accepted as atmospheric phenomena, often seen as portents of impending disaster. This view had been challenged by the parallac-tic observations of Tycho Brahe.[88] Bradley's lecture series in post-Newtonian experimental philosophy[89] was first delivered just six years after observing the notable comet of 1723 when he nevertheless argued that the conclusions to be drawn in natural philosophy were not as certain as those that could be derived in mathematics.

Two weeks after making his final observation of the comet in Wanstead, Bradley delivered an Oxford lecture *De Cometa* detailing his observations at Wanstead, but it was in his paper *Observations of the Comet that appeared in the months of October, November, and December, 1723* published in the *Philosophical Transactions*[90] that Bradley gave a more considered account of the phenomenon. He described the appearance of the comet when visiting friends or relatives in Cirencester, when he observed it on 3 December 1723, four days before observing it for the final time in Wanstead. After detailing his observations in tabular form Bradley wrote,

> In order to determine the orbit of this comet, I supposed it to describe a parabola agreeable to what is delivered in the third book of sir Isaac New-ton's Princip. Math. and then I found the inclination of the planes of the orbit and ecliptic 49° 59'; the place of the ascending node 14° 16'; the place of the perihelion 12° 52' 20"; the distance of the perihelion from the node 28° 36' 20"; the logarithm of the perihelion distance 9.999414; the logarithm of the diurnal motion 9.961007; the time of the comet's being in its perihelion September 16 16hrs 10' equal time. In its orbit thus situated, the motion of the comet was retrograde, or contrary to the order of the signs.[91]

To this Bradley added a table detailing his observations. The table consisted of the following elements:

1. The date and time to the nearest minute of the observation
2. The comet's longitude observed
3. The comet's latitude observed
4. The comet's longitude computed
5. The comet's latitude computed
6. The difference in longitude between the observed and computed position
7. The difference in latitude between the observed and computed position

Differences increased as the comet sank to the horizon when the uncertainties due to the effects of atmospheric refraction were at their greatest. I reproduce *the tables of differences* to allow an appreciation of the degrees of precision achieved by Bradley.

1723	Differ. Long	Differ. Latit.
October 9 8h 5′	+49″	−47″
10 6h 21′	−30″	+55″
12 7h 22′	−21″	+5″
14 8h 57′	−48″	−11″
15 6h 35′	−4″	−4″
21 6h 22′	+11″	−14″
22 6h 24′	−8″	−5″
24 8h 2′	+18″	+9″
Oct. 29 8h 56′	−25″	+17″
30 6h 20′	−8″	+16″
Nov. 5 5h 53′	+7	+26″
8 7h 6′	−18″	−10″
14 6h 20″	−35″	+30″
20 7h 45′	−53″	−32″

The determination of the observed position of an object that appears little more than a fuzzy blob is difficult. The uncertainty between an observed and a computed position was always less than a minute of arc, some considerably closer, indicative of Bradley's observational skills and computing abilities. Early in the eighteenth century, the calculation of cometary orbits was fraught with difficulties. Bradley's calculations for the comet of 1723 fill thirty-two pages of foolscap.[92] His final observation of the comet came on 7 December 1723 when the object was low on the horizon.

In January 1724, James Bradley could not have predicted how pivotal this year would prove to be. His academic career at Oxford was counterpoised by his observational activities at Wanstead. His wealthy uncle had once more settled into the routines of married life. Bradley's relationship with his new aunt Elizabeth, as well as the Wymondesold family in general, was relaxed. Wanstead was a fashionable location, where the well-off settled within easy travelling distance of the City of London and yet remained free of the unhealthy confines of the City itself. It was where Sir Richard Child had recently built his imposing Palladian mansion. Its grounds were intended to emulate those of Versailles.[93] It was built with an inheritance gained from the wealth of the East India Company, in whose service Pound had almost lost his life. The next important estate in the locality was Wanstead Grove, the home of the Wymondesold family. It lay in spacious grounds to the east of the township and was often called the Wanstead Estate. Today this is commemorated by a road called Grove Park, close to where Elizabeth Pound's house in Wanstead was located. Wanstead Grove was built late in the seventeenth-century. According to the Victoria County History, 'It is said to have been built about 1690 by Sir Francis Dashwood, Bt. Son of a Turkey merchant.[94] Matthew Wymondesold, owner in the early eighteenth century, was a successful financier.[95]

On 16 November 1724, James Pound died and his household soon had to relinquish the parsonage where Bradley had lived since 1711. Elizabeth Pound had a modest house in Wanstead, close to her brother's fine estate. Bradley it appears continued living with his aunt and her brother, using her town house as his observatory. As an extremely wealthy woman, she would hardly decide to live in such modest accommodation.[96]

During 1725, Halley took possession of George Graham's newly completed 8-foot mural quadrant. It cost £300 out of the £500 he was granted to equip the observatory. This never allowed Halley to acquire the second required quadrant so essential to allow the observation of stars to the northern horizon as well as to the south. When Halley attempted to gain sufficient finance to acquire such an instrument, he was refused further support.[97] Bradley joined Graham who was a member of the committee of the Royal Society that commissioned the construction of the quadrant that Graham was responsible for, supporting Halley when he took possession of an instrument that was the prototype of all such instruments for much of the eighteenth century. This year was the first time that Bradley was elected a member of the Council of the Royal Society. It was an indication of the respect he enjoyed amongst the fellowship of the society, though it may also have been for the memory of Pound. For several years, the pair had been inseparable co-workers, so respect for Bradley also brought to mind his uncle. Bradley was now about to enter a career that would surely have surprised even Pound's wildest expectations. Before revealing Bradley's work on his first major discovery, that of the aberration of light, I want to examine the exceptional work that led to the communication of his first individually published paper in the *Philosophical Transactions*.

Bradley's paper on the determination of the longitudes of Lisbon and New York, from observations of the motions of Jupiter's innermost Galilean satellite, is a remarkable document. The body referred to as satellite I is now better known as the satellite *Io*, the most volcanic body in the solar system. It demonstrates Bradley's mastery of the data revealed by the motions of the satellites he had observed for almost ten years. This mastery would later be a major factor in the discovery of the aberration of light. Bradley's work depended on the location of his observatory in Wanstead. This he determined from the position of the Royal Observatory, a few miles[98] away on the opposite bank of the Thames. Under the heading of 'Observations made at my aunt Pound's house in Wanstead Town', he determined the position of Elizabeth's town house at

1′ 45″ or 7 seconds in time east of the Royal Observatory, and 5′ 53″ north. Bradley's paper based on calculations made from his observations made in Wanstead, John Machin's (1680–1751) observations in London, and others received from New York by a correspondent of Bradley's, William Burnet the Governor of New York, and finally by John Baptist Carbo an FRS living in Lisbon. Bradley began his paper, 'Some curious[99] astronomical observations having lately been communicated to this Society from Lisbon, among which were several eclipses of the first satellite of Jupiter'.[100]

Bradley examined the observations made in Lisbon pondering if they tallied with those he made in Wanstead. By comparing them, the difference in longitude between these places would be discovered. He discovered only two 'emersions' observed on the same night in Lisbon and Wanstead. There were other observations that might have been of use though they would not be as accurate. This was due to the irregular motion of the satellite due to the gravitational interactions it had with the other Galilean satellites. He had determined the elements of the orbits of Jupiter's satellites from reiterative observations, and discovered the mutual gravitational effects of the satellites and various gravitational resonances which he calculated. The variations in the periods amounted to thirty or forty minutes in time over the space of just seven months. An interesting feature of the paper is that Bradley reduced all of the observations according to the Gregorian Calendar, that is 'new style' over twenty-seven years before it was applied in Great Britain and the American colonies from 1752, avoiding confusion between timings in Lisbon and Wanstead, London, and New York. Bradley as Astronomer Royal from 1742 made the calculations necessary to lay the Bill before Parliament.

Using a Hadley reflector Bradley observed the immersion on 4 August 1725 of the first satellite (Io), 45 seconds after the time calculated in his tables. On 29 August 1725, the immersion was timed at 1 minute and 55 seconds ahead of the time in his tables, which made the immersion at 12h 48m 45s apparent time. The immersion at Wanstead on 28 July 1725 NS was 12h 50m 00s. At Lisbon it was observed at 12h 12m 26s apparent time. The difference therefore was 37m 34s. Comparing observations of emersions made on 14 October 1725 and 16 October 1725 NS, the difference was 36m 20s. Bradley made comparisons on other dates coming to differences between Lisbon and Wanstead of 37m 25s and 37m 20s. The emersion seen at Lisbon on 16 January 1726 led to a difference of 37m 12s. Bradley calculated that the difference of meridians between Lisbon and Wanstead averaged 36m 58s. Bradley referred to other data[101] which included the observations of William Burnet. These eclipses of *Io* made at the fort of New York were compared with Bradley's observations made at Wanstead. By the observation on 25 August 1723 OS described as the most distinct, the satellite emerged at 9h 35m 14s by the clock, about one and a quarter minutes too fast for the expected time at the emersion. The emersion at New York was at 9h 34m apparent time, or 9h 32m 20s mean time. In Wanstead on 27 August 1723 at 8h 57m 50s mean time, Bradley observed the satellite emerge using the Hadley reflector. On 12 September 1723 at 7h 17m 15s mean time, it was observed emerging again in the reflector so that in a period of 15d 22h 19m 35s there were nine emersions of the first satellite. The interval between each of the emersions was 1d 18h 28m 50s. Bradley wrote, 'This subtracted from the time of the emersion observed at Wansted August 27, will give the true emersion at Wansted on August 25, 14h 28m 50s mean time; that is, 4h 56' 30" later than it was observed at New York'.[102]

From another observation made in New York on 10 September 1723 and comparison with a similar observation made at Wanstead on 12 September 1723, Bradley made similar calculations so that it emerged that the difference in mean time between New York and Wanstead was 4h 56m 33s. Bradley concluded his paper, 'The difference therefore of meridians between Wansted and New York, allowing about 15" for the difference of telescopes, is about 4h 56' 45", and between London and New York, 4h 56'¼: so that the true longitude of New York from London is 74° 4' west'.[103]

The use of tables of the motions of the satellites of Jupiter combined with telescopic observations, could determine the longitude. Before Halley determined that he would seek the method of using lunars, the method of determining longitude by observing the motions of Jupiter's satellites was vindicated. Yet as both Bradley and Halley knew, it was one thing to make observations of the motions of the Galilean satellites from a land-based observatory but quite another from the deck of a small wooden ship at the mercy of waves, winds, and the contrary motions of a ship. Even when the Jovian system was in view, given the vagaries of the weather and of the position of Jupiter in its orbit, it was a non-starter. Were the thousands of observations made by Bradley of the orbit of Jupiter and of the motions of the four Galilean satellites a waste of time? It led to a greater understanding of the motions of the satellites of the Jovian system, including the gravitational interactions between the satellites. Bradley was amongst the first to allow for the velocity of light in his tables of the four Galilean satellites. He understood how the inequalities in their motions were interconnected, due to their gravitational interactions.[104] His studies of the Jovian system climaxed the earliest exploratory phase of empirical studies of the motions of Jupiter and its satellites.[105] The payoff for Bradley's knowledge of these motions of the Jovian system came when this knowledge was applied in the development of the hypothesis that became known as the aberration of light.

Notes

1. Expressed by Edmond Halley in support of Bradley's fellowship of the Royal Society of London. The original Latin reads 'eruditus juvenis, qui simul ingenio et industria pollens his studiis promovendis aptissimus natus est'. This translation by John Fisher, Astronomy and Patronage in Hanoverian England: The Work of James Bradley, Third Astronomer Royal of England, PhD thesis, University of London, 2004.

2. The night sky in the location of Sherborne in the Windrush Valley in the Cotswold Hills reveals the stars and the Milky Way clearly. The night sky must have been familiar to Bradley as a child.

3. As I have shown in Chapter 2, James Pound was in regular correspondence with John Flamsteed from 1701 when Pound was in the service of the East India Company in China and the Indies. Cambridge University Library, *Papers of the Royal Greenwich Observatory*, RGO 1/31, fols 90r to 102r.

4. Stephen Peter Rigaud, in 'Miscellaneous Astronomical Observations' in *Miscellaneous Works and Correspondence of James Bradley*, Oxford, 1832, p. 339. Pound's 'e' is so easily confused with his 'a'; indeed it is completely idiosyncratic, and clearly identifies his handwriting.

5. 6hrs Jupiter's diameter = 1.45 rev. = 46″ 6hrs 45min 0sec sp, = 30.13 rev. In the long glass, pJ. being the parallel of Jupiter's nearest limb to the star. 7hrs 5min 30 sec pJ = 43.20 rev. being the diff. right asc. between the star and Jupiter's nearest limb. 11min 15sec sJ = 53.55 rev. being the dist. Of the star from Jupiter's nearest limb. 19min 0sec sp. = 30.00. Hence at 7hrs 11min 15sec by the clock, the star was as follows: clock 1min 40sec too slow for the mean time. 0° 27′ 14″ the distance of the star from Jupiter's centre. 0° 15′ 40″ the difference of declination between the star and Jupiter's centre. 0° 22′ 10″ the difference of right asc. between the same, measured by the micrometer. NB. The star was westward of Jupiter, and nearer the equator than Jupiter.

6. See Appendix 1.

7. James Pound, Bodleian Library, Oxford, Department of Western Manuscripts, Bradley MS.23, *James Pound's Accounts Book*. June 1715.

8. These are the four satellites Io, Europa, Ganymede, and Callisto.

9. Stephen Peter Rigaud, *Memoirs of Bradley*, Oxford, 1832. p. v.

10. *Philosophical Transactions,* Vol. XXX, p. 1023.

11. Isaac Newton, *Philosophia Naturalis Principia Mathematica*, A New Translation by I. Bernard Cohen and Anne Whitman, *The Principia: Mathematical Principles of Natural Philosophy*, 3rd edn. London, 1999, Book III, *The System of the World*, Prop. 42, p. 935.

12. This testifies to the frequency that Halley now referred to Pound, and in consequence Bradley also, given that all referral to Flamsteed was now impossible.

13. This would be in the unauthorized version edited by Halley.

14. Edmond Halley to James Pound, London, 6 September 1716. This letter is reproduced in the *Memoirs of Bradley*, p. iii.

15. Allan Chapman, 'Pure Research and Practical Teaching: The Astronomical Career of James Bradley, 1693–1762', *Notes and Records of the Royal Society of London* Vol. 47, No. 2, 1993, pp. 205–212, see p. 206.

16. One of Bradley's characteristics was the deference he evinced in his relations with friends, patrons, and those in positions of power or influence. He was always modest in his dealings with others.

17. Even during this early period Bradley's skill with long focus telescopes was notable.

18. James Bradley, *Miscellaneous Works and Correspondence of James Bradley*, Oxford, 1832, p. 340. *Propus* was designated as η Geminorum by Bayer.

19. Appendix 1

20. These observations can be found as an addendum in the *Miscellaneous Works* on p. 380. Rigaud writes, 'The following observations, having been written on small separate pieces of loose paper, were overlooked at the time when they might have been inserted according to their respective dates'.

21. Isolation was the common response to the contraction of smallpox for the middling members of society.

22. James Pound, *Accounts Book,* Bodleian Library, Department of Western Manuscripts, Bradley MS 23, 11 March 1717.

23. Stephen Peter Rigaud, *Memoirs of Bradley*, p. ii.

24. The estate which has recently come into the possession of the National Trust was originally the property of Winchcombe Abbey, the monks gathering their flocks at shearing time and washing them in the stream that runs through the village. Sherborne means 'clear brook'. It long came under the ownership of the Dutton family, the Lords Sherborne.

25. Sybil Longhurst, Walter Tufnell, and Alice Tufnall, *Sherborne: A Cotswold Village*, Stroud, 1992, p. 42.

26. Gambling debts were not redeemable in law so such debts were regarded as debts of honour. In modern money values these three debts amount to £285,000!

27. Ibid. p. 42.

28. Ibid. p. 42.

29. There was a long standing oral tradition in Sherborne on the financial difficulties experienced by the Bradley family who apparently lived in what was known in the village as the Monk's House. This building was used by the monks of Winchcombe Abbey in medieval times, when they gathered sheep at Sherborne to wash them in the stream at shearing time.

30. I recognize that all of this is highly conjectural, but my conclusions are based on well-documented circumstances, that must have constrained the life choices of the Bradley family.

31. James Pound, Bodleian Library, Oxford, Department of Western Manuscripts, Bradley MS23, *James Pound's Journal.*

32. A term used to refer to the English East India Company.

33. In 1722, Sir Richard Child took possession of the newly built Palladian mansion house in handsome grounds. The architect of the house Colen Campbell, the author of *The British Vitruvius*, was a cousin of the Duke of Argyll. Sir Richard was a Tory whilst Campbell's family had strong Whig associations.

34. These tables were of the satellites of Jupiter.

35. James Bradley, *Miscellaneous Works and Correspondence of James Bradley,* p. 81, Remarks from Jupiter's Satellites, Mayer's Tables, &c. The remark suggests that

observations were collected and made from a period of just under forty-eight years, four Jovian years. Jupiter's 'year' is equal to 11.86 Earth years.

36. The apsis of an orbit is either of the points which are nearest or farthest from the primary. When discussing both points in the plural they are referred to as the apsides of the orbit.

37. That is, the finite velocity of light.

38. James Bradley, *Miscellaneous Works and Correspondence of James Bradley*, p. 353.

39. I have been unable to discover how such a cumbersome object was transported from Charing Cross to Wanstead, over a distance of some eight miles. It would have been an object of great curiosity.

40. James Pound, Bodleian Library, Oxford, Department of Western Manuscripts, Bradley MS23, James Pound's Account Book 1715 to 1724 with some receipts and bills. See also, Stephen Peter Rigaud, *Memoirs of Bradley*, p. ix.

41. *Philosophical Transactions,* Vol. 30, p. 855.

42. The symbol 's' signifies the word 'sign', referring to the signs of the zodiac. This by convention was 30°, so 5s is the equivalent of $5 \times 30°$ which equals 150°. So the distance travelled was 179° 57'.

43. Vol. 30, p. 1023.

44. Known as M1 in Messier's catalogue.

45. Stephen Peter Rigaud, *Memoirs of Bradley,* p. v.

46. Bailli was guillotined during the Terror of 1793.

47. Ibid. p. xxxi.

48. John Robinson took part in the negotiations leading to the Treaty of Utrecht in 1713.

49. See various entries in Bradley's Wanstead Observation Book included in the *Miscellaneous Works and Correspondence of James Bradley.*

50. This statement gives lie to the suggestion that Bradley was loath to maintaining his correspondence as a way of 'explaining' the lack of surviving correspondence in his name.

51. Stephen Peter Rigaud, *Memoir of Bradley*, Oxford, 1832, p. vi.

52. This is according to the Julian Calendar or 'Old Style'. According to the Gregorian calendar which was in use throughout most of Europe Flamsteed died on 10 January 1720 NS.

53. Fifty guineas was the equivalent of £52 10s. Later in the year Pound received another gift of 50 guineas from Sir Isaac.

54. Now Leicester Square.

55. William Molyneux, *Dioptrica nova*, London, 1692.

56. John Flamsteed and William Molyneux had been in regular correspondence on the science of dioptrics since 1682, well before the publication of the *Dioptrica nova* in 1692. Eric Forbes discusses the letters dated 29 May 1682 and 1 August 1682 in his paper 'Early Astronomical Researches of John Flamsteed' in *Journal for the History of Astronomy*, Vol. 7, 1976, pp. 124–137, pp. 134 and 135.

57. Samuel Molyneux got involved in a ridiculous fraud when a young woman named Mary Toft claimed to have been giving birth to live rabbits. Molyneux was taken in by the fraud before it was finally exposed.

58. When Bradley sought election as Keeper of the Ashmolean Museum in 1731, the location of his lectures in experimental philosophy, his only elector was Hearne.

59. The sum of £15 may seem modest but in terms of modern money values this was the equivalent of £3,750.

60. On 14, 26, and 29 January 1720, Pound paid a series of fees, for the presentation to the Rectory of Burstow, for a dispensation for holding two livings and for the institution and induction, amounting to £32 6s 4d, £32 3s 0d, and £14 4s 8d, a total of £78 14s 0d.

61. Possibly from food poisoning.

62. By Lord Chancellor Thomas Parker, later 1st Earl of Macclesfield

63. Robert Fox, Graeme Gooday, and Tony Simcock, 'Physics in Oxford: Problems and Perspectives' in *Physics in Oxford, 1839–1939: Laboratories, Learning and College Life*, Oxford, 2005.

64. Rigaud writes of this addendum, 'This was Richard Bradley, the first professor of botany at Cambridge in 1724. The caution was not unnecessary. In the Index to the Phil Trans the very first article which is mentioned under the head of James Bradley belongs really to Richard'. *Memoirs of Bradley*, p. vii.

65. Stephen Peter Rigaud, *Memoirs of Bradley*, pp. vii—ix. Rigaud makes reference to entries in Pound's Accounts Book referring to the costs of Bradley's election to the Savilian Professorship of Astronomy at Oxford. On 2 September 1721, Pound enters By coach hire, pocket expenses, & c, about the Oxford professorship £2 12s 0d. On 31 October 1721 we read, By cousin Bradley lent him £4 4s 0d, by pocket expenses in London, & c. £0 11s 3d. Rigaud in footnote a. writes, 'It is probable, therefore, that Pound accompanied his nephew to London on the day of election', and we find from another part of the accounts that the four guineas were supplied '*to give to the door-keepers of the House of Lords*'. This is another example of the many public offices that could be so lucrative to whoever held the post. The fee is equivalent to £1,100 in today's money values.

66. Sir Henry Savile founded two chairs, one in astronomy and another in geometry. This latter chair was held by Edmond Halley conterminous with Bradley's holding that in astronomy. Sir Henry Savile was Warden of Merton College, Oxford from 1585 to 1625.

67. Allan Chapman, 'Pure Research and Practical Teaching: The Astronomical Career of James Bradley, 1693–1762', *Notes and Records of the Royal Society of London*, Vol. 47, No. 2, pp. 205–212, 1993, see p. 206.

68. Anti-clericalism was strong in France both before and after the Revolution. There were marked divisions in early nineteenth century French society which persisted well into the twentieth. Stendahl's novel *Le Rouge et le Noir* reveals how during the Bourbon Restoration the clerical route to social advancement reappeared. In the minds of many scientific practitioners clericalism was regarded as an obstruction to rationalism. This was particularly true in France when Rigaud penned his *Memoirs of Bradley*. The situation in England was not so polarized as it was in France.

69. Stephen Peter Rigaud, *Memoirs of Bradley*, p. viii.

70. I have some doubts about Rigaud's opinion at this point. It didn't seem to bother his uncle who held two highly lucrative livings at Wanstead and at Burstow, nor did it

bother many of his close astronomical colleagues who were often men of the cloth. It certainly didn't bother the 1st Astronomer Royal John Flamsteed.

71. Lady Catherine was famous as a socialite as Lady Catherine Manners.

72. Rigaud believed it was because Bradley was against being both a churchman and an astronomer. I simply believe that Bradley didn't wish to become encumbered with the responsibilities of caring for a parish at the very moment he was beginning the great series of observations that in many respects remains his greatest monument. It may have escaped Pelham's notice that the acceptance of a church living would have disqualified Bradley from the Savilian Chair in Astronomy at Oxford. This in turn would have undermined his legitimacy of holding the readership in experimental philosophy.

73. Stephen Peter Rigaud, *Memoirs*, p. ix.

74. The proctor was a well-known non-juror whose political sympathies were decidedly in support of the claims of the Jacobite cause. This would have placed him in direct opposition to those who supported Bradley's candidature.

75. Not to be confused with Wanstead House, the great Palladian Mansion with its grounds.

76. Smith, *Optics*, p. 892.

77. James Bradley, *Miscellaneous Works and Correspondence of James Bradley*, Miscellaneous Astronomical Observations, pp. 339–380. See p. 354.

78. Seemingly made during winter twilight, an indication of Bradley's skills.

79. There is some suspicion he used one of his telescopes at the Ashmolean Museum.

80. *Philosophical Transactions*, Vol. XXXII, p. 303.

81. Ibid. Vol. XXXII, p. 382.

82. J. J. O'Connor and E. F. Robertson, Entry for *John Hadley*, McTutor, 2004. www-groups.dcs.st-and.ac.uk/history/Biographies/Hadley.html read 25/03/2018.

83. In the 1720s, five of Saturn's satellites had been discovered. These were Tethys, Dione, Rhea, Titan, and Iapetus.

84. Stephen Peter Rigaud, *Memoirs of Bradley*, p. x.

85. Just a handful of miles away but north of the Thames.

86. By 1719 roads between Oxford and London were maintained by turnpike trusts for at least 80% of the total route miles. This suggests that travel between Oxford and London and the Home Counties was efficient. The ease with which Bradley travelled between Wanstead or London and Oxford seems to support this view. See Dan Bogart, 'Turnpike trusts and the transportation revolution in 18th century England', *Explorations in Economic History* (2005), London, p. 19.

87. These ideas will be discussed and analysed in Chapter 6, detailing Bradley's lectures in experimental philosophy at Oxford.

88. Tycho was able to establish that comets were the phenomena of the celestial realm, beyond the orbit of the Moon.

89. What we now term as the science of physics.

90. *Philosophical Transactions*, No. 382, Vol. XXXIII, p. 41.

91. Ibid. p. 41

92. Similar in size to the modern A4 sheet.

93. It is therefore not surprising that Wanstead House became the official residence of Louis XVIII of France during Napoleon Bonaparte's brief reign as the Emperor of France during 1815.

94. Obviously, a merchant trading with the Levant.

95. *Victoria History of the County of Essex*, Vol. VI, p. 319.

96. As well as the £10,000 that she possessed in her name she inherited at least a similar amount from the death of her husband. A sum of £20,000 is the equivalent of about £5 million in current money values. She was a woman of substance and high social standing unlikely to live in the small confines of the house that was in her name.

97. As with Flamsteed, the government never quite seemed able to understand the importance of the work of the Royal Observatory. All further support was denied by the Master of Ordnance, the Duke of Argyle.

98. An English mile is approximately 1609 metres.

99. In this sense the word 'curious' means precise and accurate.

100. *Philosophical Transactions*, Vol. XXXIV, p. 90. This is a report made by John Baptist Carbo.

101. Ibid p.90.

102. James Bradley, 'The Longitude of Lisbon and the Fort of New York from Wansted and London, determined by Eclipses of the First Satellite of Jupiter', in the *Miscellaneous Works and Correspondence of James Bradley*, Oxford, 1832, p. 61.

103. Ibid. p. 61.

104. John North, 'The Satellites of Jupiter, from Galileo to Bradley', in *Old and New Questions in Physics, Cosmology, Philosophy, and Theoretical Biology*, pp. 689–717.

105. Ibid. p. 689.

4

A New Discovered Motion

The Search for Annual Parallax and the Experiment at Kew

'I can calculate the motion of heavenly bodies, but not the madness of
people'.

Isaac Newton.

The first major discovery made by James Bradley was that of the aberration of
light, with a zenith sector suspended at his aunt's house in Wanstead. Before
we can apprehend the work at Wanstead, it is necessary to understand what
took place at Kew. It is important that the sorry tale of attempts to discern the
annual parallax of the stars be first examined. It is difficult to understand the
immense problems that Bradley overcame, or the significance of his discovery,
without some appreciation of the work of earlier astronomers. Thereafter, the
central narrative will remain solidly on the work of Bradley

There are various secondary and tertiary accounts of attempts to measure the annual parallax of a star, but the most informative coming into my
possession is a doctoral thesis written by Mari Williams in 1981, under the
supervision of Marie Boas Hall at Imperial College London. When I began my
research, Mari kindly lent me a copy of her thesis which I gratefully copied. It
has remained in my library ever since.[1] Her invaluable work forms the central
argument of my account up to the work undertaken by Molyneux, Graham,
and Bradley from November 1725. The problem of the lack of observational
evidence for annual parallax had been increasingly urgent since Copernicus
published *De revolutionibus* in 1543. Tycho Brahe abandoned the Copernican system, precisely because he was unable to discern any such evidence for
annual parallax. In January 1587, referring to the motions of the planets, Tycho
proposed that a new system be found. He wrote, 'Either the Earth is whirled
about in an annual motion whereby all the planet's epicycles are eliminated,
or another, hitherto-unconstructed arrangement of the heavenly revolutions
must be sought'.[2]

Before the close of the sixteenth century Tycho substituted his own cosmographical system. He proposed that the planets orbited the Sun, while the

The Life and Work of James Bradley. John Fisher, Oxford University Press. © John Fisher (2023).
DOI: 10.1093/oso/9780198884200.003.0005

Sun in its turn orbited a stationary Earth. Yet a century later, a majority of astronomers were supportive of the Copernican thesis. However, there was not a single convincing observation of the annual parallax of a star. There were claims of course, but one by one each claim met its refutation.

Some astronomers observed a motion, later revealed as annual aberration, mistakenly claiming observation of annual parallax. The phenomenon of the aberration of light was finally identified by Bradley in 1728. It is, therefore, necessary to look at claims for the discovery of annual parallax, before Bradley and his partners came onto the scene. There are commentators who claim that Bradley was fortunate in using instruments made by George Graham. I concede having Graham as his main instrument maker gave him confidence in the dependability of his observations. However, the phenomenon he finally discovered was of such an immense magnitude that any time after the application of micrometers to telescopic sights, it should have been isolated and identified. I argue that the reason Bradley discovered aberration had little to do with his access to remarkable instruments. What is so remarkable about Bradley's achievement is that he quickly perceived that the phenomenon he began to observe in December 1725 was *not* annual parallax. Claimants had for so long persisted in 'seeing what they wanted to see'[3] which was annual parallax. It was the clarity of Bradley's mind, and the discipline of his astronomical practice, rather than the precision and accuracy of his instrument that led to the discovery of the phenomenon of annual aberration. The confidence that Bradley had in the craftsmanship of George Graham and the skilled artisans in his workshops, certainly enabled him to be sure of the veracity of his observations of the phenomenon.

With the publication of *De revolutionibus* in 1543, when Copernicus proposed that the Earth was a planet in orbit around the Sun, some astronomers recognized that the 'fixed stars' ought to exhibit an apparent annual motion that revealed the motion of the Earth around the Sun.[4] The absence of such observed phenomena, so it was argued, was because the stars were so distant that such phenomena were indeterminably small. At the time of the publication of *De revolutionibus*, it wasn't a problem for church doctrine. It met with the blessings of the Church of Rome and was dedicated to Pope Paul II. Copernicus was accepted as 'an adornment of the Church'.[5] It was only after the Council of Trent met from 1545 to 1563[6] that the heliocentric hypothesis by default, became heretical to the Roman Catholic Church. After almost twenty years of deliberation by the leading theologians of the Church, the doctrines not considered, due to the lack of time to clarify the issues, were deferred to the teachings of the Church Fathers. During late classical and medieval times it was universally believed, often with apparent Biblical support, that the Earth was stationary. After the deliberations of the Council of Trent, any belief in the motion of the earth was interpreted as heresy. Copernicus, who had been approached to help reform the calendar, was now implicitly a heretic. The Polish canon[7] had refused to take up the challenge of calendrical reform in the belief that he did not possess sufficient information. Initially, however, it was philosophers and astronomers who rejected his system, not theologians. Tycho argued that the enormous distances involved in the space between Saturn and the sphere of the fixed stars was an unwarranted waste of space. It could have no place in the economy of nature. Mari Williams writes, 'Tycho clearly tried to observe parallax; in a letter to Kepler written in 1598 Tycho said that he had made a series of observations of the Pole Star with the specific aim of measuring its annual parallax.'[8]

By the end of the sixteenth century, it appears that only Tycho had made a serious attempt to measure parallax. On the 'third day' of Galileo's *Dialogues concerning the two chief world systems*, Salviati dismissed Tycho's claim that the diameter of the stars was at least one arcminute. Later observers, including Halley, observed the near instantaneous disappearance of the light even from bright stars when occulted by the Moon, revealing that such estimates of the diameter of the stars were entirely mistaken. Faced with the absence of annual parallax, Galileo's consistent opinion was that it simply hadn't been observed up to that moment.

In 1674, Robert Hooke gave the first of his Cutlerian Lectures[9] entitled, 'An attempt to prove the movement of the Earth from observations'. His claim for the measurement of the annual parallax of an object star *Eltanin* was almost universally rejected. Even so his *methodology* promised to be a means by which annual parallax might possibly be observed. The *method* had few flaws but its execution had failed. Hooke initially presented his plan to the Royal Society in 1666 but was thwarted by the great fire of London lasting from 2 to 6 September 1666. Hooke was encouraged by the Society to renew the project with all possible urgency.[10] The Corporation of the City of London had appointed Hooke as a surveyor in March 1667 to stake out and certify 3,000 foundations over the next five years.[11] Margaret 'Espinasse wrote that the rebuilding of the City of London reached its peak of activity between 1668 and 1674.[12] This must be considered, when examining the reasons why Hooke failed to measure the annual parallax of the object star, which Johann Bayer designated as γ Draconis. The star was chosen because it was the brightest that passed through the meridian close to the zenith at the latitude of London, avoiding many of the problems associated with atmospheric refraction.

Hooke was unable to make any observations until 6 July 1669.[13] In order to diminish the problems associated with atmospheric refraction Hooke had devised a way of observing a star that passed as close to the zenith as possible. The object star passed through the meridian 2 arcminutes from the zenith. Using rooms at Gresham College[14] Hooke suspended a thirty-six-foot telescope that was attached to a chimney stack. The object glass protruded through the roof of the college, with the eyepiece situated on the ground floor, with holes cut through the floors of the intervening storeys. The experiment undertaken in 1669 was not published until after he gave the first of his Cutlerian Lectures in 1674. Hooke was overwhelmingly busy as the Surveyor for the City of London, though I am sure he possessed little confidence in his results. Given Hooke's temperament, he would almost certainly have made a claim immediately if he believed he had confirmed the motion of the Earth. So when, *five years later* he spoke to his results he wrote,

> 'Tis manifest then by the observations of July the sixth and ninth: and that of the one and twentieth of October that there is a sensible parallax of the Earth's Orb to the fixed star in the head of Draco, and consequently a confirmation of the Copernican System against the Ptolomaick and Tichonick.[15]

It suggests that his claim was at best tentative.

Christopher Wren, the former Savilian Professor of Astronomy at Oxford, also had an interest in establishing the annual parallax of a star. During the 1670s, he designed the London Monument.[16] According to John Ward in the *Lives of the Professors of Gresham College,* he wrote, 'The learned architect built it hollow, that it might serve as a tube to discover the parallax of the Earth, by the different distances of the star in the head of the Dragon from the zenith, at different seasons of the year'.[17]

The Monument started in 1671 was completed in 1677, but Wren was unable to use the structure to make zenith observations because it was 'liable to be shaken by the motion of coaches and carts almost constantly passing by'[18] situated as it was at the northern end of the old London Bridge. It was one of the busiest thoroughfares in the entire City of London.

James Gregory was interested in the distance scale of the stars.[19] He made known his opinion on the problem of stellar parallax to Henry Oldenburg, praising Hooke's work. The problem of the distance scale of the stellar universe also concerned Newton. He was perturbed that claims for the measurement of parallax of the order of some thirty arcseconds would place many stars in proximity to the solar system. If so, its gravitational stability would be threatened.

By using a photometric argument that implied that the brightest star *Sirius* was also the nearest, he calculated that it possessed an annual parallax of a single arcsecond or less. Newton suggested that this was too minute to be observed with the technology then available.

Gregory had tried Galileo's proposed method of comparing the position of a star relative to a distant marker.[20] This failed him so he considered Galileo's other suggestion of observing close pairs of stars. This was before the recognition that many close double stars were binary systems bound together by their mutual gravitation. They were useless for estimating the distance of stars because any observed motions would be due to their gravitational interactions. With the publication of Hooke's claim in 1679, Oldenburg sent copies to Christiaan Huygens, Giovanni Domenico Cassini, and Johannes Hevelius, asking for their opinions. Cassini's reply revealed that similar trains of thought had developed in France. Observers in France had noticed annual alterations in the position of some stars, unexplained by the precession of the equinoxes. Cassini mentioned efforts were being made in Paris to follow similar approaches which had been used in London. Cassini added,

> We have a well shaft at the Observatory for this, having an aperture in the vault (built over it) open to the sky which is fully 200 feet from the bottom, to which a convenient ladder leads; but (it has been) hitherto obstructed by the works of the masons constructing the upper vault. And so it has not been possible for us to attempt anything with it so far.[21]

French astronomers had also chosen to observe stars as close to the zenith as possible but unlike Hooke and Wren, they chose to use a well-shaft and have an eye-piece below ground. According to Williams, there were also traces of an attempt to use a well-shaft to house a zenith sector by Flamsteed in the 1670s at the newly instituted Royal Observatory at Greenwich. She mentions a similar approach, using a well-shaft by William Ball (c1631–1690). He wrote to Huygens in March 1674 stating that he was planning to use a well at his home in Devon.[22] Williams writes,

> The difficulties lay in putting the idea into practice, and it is greatly to Hooke's credit that he succeeded at least in building a workable instrument, even if he did not obtain enough observations. It is also to his credit that he published his account of the major practical problems involved and of how it might be possible to overcome them.[23]

Williams adds that in the period after the publication of Hooke's paper there was a general reluctance to detect the elusive phenomenon of parallax. Astronomers tended to lose interest once they began to recognize the difficulties involved. Only the Danish astronomer Ole Römer set out to make a serious attempt in a manner similar to the way that Hooke pioneered, which I will soon discuss.

In 1680 Jean Picard, professor of astronomy at the Collège de France in Paris, published *Voyage d'Uranibourg*. Bartholinus[24] sent him an exact copy of Tycho's observations in 1671 from Copenhagen. Picard travelled with him to the site of Tycho's observatory which the great Danish astronomer left in 1597 to enter the service of the Emperor Rudolf II in Prague. Picard sought the *exact* position of Tycho's observations as accurately as possible so his observations could be correlated with those of other astronomers including his own. In the *Voyage* the Parisian astronomer made a list of observations of *Polaris* over a period of ten years which betrayed an annual positional variation.[25] Picard conjectured this might be evidence of parallax. He considered the effects of refraction, the almost complete elimination of which would be expected when observing at the zenith. Picard eventually gave up, unsure of the causation of the variations in the position of the pole star. A major problem he faced was the lack of dependability of the instruments at his disposal. At least Picard clearly understood the difficulties involved in any attempt to observe annual parallax. In 1693, a similar series of observations of variations in the position of *Polaris* was sent to the Royal Academy of Sciences in Paris by Cassini, this time covering a period of about twenty years. Mari Williams quotes Cassini when he wrote it was unlikely that 'one could attribute that to a real change in the position of the Heavens with respect to the Earth, or to a movement of the stars; and it is far more reasonable to attribute this difference to accidental causes'.

Williams argues that neither Picard nor Cassini was concerned with the detection of parallax, their leading object being the determination of an accurate value for precession. This was only determined after Bradley discovered the nutation of the Earth's axis between 1727 and 1747. They studied changes in the position of the pole star to determine how it differed over time. Williams writes, 'their consideration of parallax as an explanation of their observations was cursory, and their attitude seems understandably to have been one of extreme caution'.[26]

Cassini remained sceptical about the motion of the Earth, as were so many of the Bolognese school. Williams asserts that once they recognized the practical difficulties 'were virtually impossible to solve', such an attitude of 'extreme caution' was justified. Williams interprets the attempt by Hooke as the culmination of a phase in the development of attempts to determine the annual parallax of the stars which came to an end once the difficulties were recognized. This would underline why they lost interest, moving to more 'rewarding projects'. The argument is convincing, but I prefer to expose the nature of the problem. What would its solution mean, particularly to astronomers like Cassini? He was a conservative astronomer, an excellent observer, but one who remained resistant to the Copernican thesis. After working with the wealthy astronomer the Marquis Cornelio Malvasia (1603–1664) at an observatory in Panzano near Bologna, Cassini moved on to observing under the direction of Giovanni Battista Riccioli (1598–1671) and his student, Francesco Maria Grimaldi (1618–1663). Both astronomers were Jesuits and supported geocentric cosmologies. Riccioli especially, was determinedly opposed to the Copernican thesis. This didn't mean Cassini followed Riccioli's every opinion, but he held conservative views throughout his career. It is Römer who has the credit for revealing that the velocity of light was finite. It was Cassini who first proposed the hypothesis, to explain variations in the timings of occultations of Jupiter's satellites, Römer assisting him at the Paris Observatory. As I earlier intimated, Cassini immediately retracted his conjecture, recognizing that acceptance of the progressive[27] motion of light led to a belief that the Earth was in motion, anathema to the Roman Catholic Church. Riccioli and Grimaldi supported the Tychonic alternative for it accounted for Galileo's reported observations, but allowed the earth to remain stationary. To assert that Cassini turned his back on attempts to determine annual parallax because of the difficulties involved may be justified, but such a programme would have been anathema to Cassini. More than observational evidence was at play here.

During the 1690s, two papers appeared in the *Philosophical Transactions,* both concerned with the practicalities of detecting annual parallax. The first, entitled 'Concerning the parallax of the fixed stars, in reference to the earth's annual orbit' was written by John Wallis (1616–1703), Savilian Professor of Geometry at Oxford. He discussed Galileo's proposition of observing pairs of stars, suggesting two *particular* stars[28] β Ursæ Majoris (*Mizar*) and 80 Ursæ Majoris (*Alcor*), which many astronomers have long called 'the horse and rider'.[29] It was never attempted. The other paper included in the *Philosophical Transactions*[30] during the 1690s, written by Francis Roberts (1609–1675),[31] was concerned more generally with stellar distances. He remarked that the measurement of annual parallax would give an accurate value for stellar distances, but that the diameter of the earth's orbit might not offer a base wide enough to provide a discernible angle. Roberts's paper never made an impact. Wallis was the recipient of a letter from the Astronomer Royal John Flamsteed in 1698. He claimed the observation of an annual parallax of forty arcseconds for *Polaris*.[32] The observations were made between 1689 and 1695 with his newly acquired mural arc. Flamsteed measured the altitude of the north celestial pole in response to a request by Newton in preparation for his lunar theory.

Flamsteed's claim to have discerned annual parallax provoked several reactions. The claim was generally accepted as valid by many astronomers in England, but it was heavily criticized by some continental observers on what Williams argues were 'justifiable theoretical grounds'.[33] Jacques Cassini (1677–1756) took exception to Flamsteed's claim on geometrical grounds.[34] Flamsteed retracted his claim in a letter to Christopher Wren, written in November 1702. In spite of the controversy surrounding Flamsteed's claim, none of his contemporaries made their own observations. Flamsteed in his letter to Wren acknowledged that Jacques Cassini was perfectly correct in his criticism and added, 'I shall return to my stock of night observations, to seek out such as are most proper for discovering the Errors of the Instrument; afterward those that are most convenient for showing the parallaxes of the Orb'.[35]

Mari Williams reveals 'there are no subsequent references to show that Flamsteed ever carried out his plan'.[36] The problem of parallax must have been a live issue because of the immediacy of Jacques Cassini's response to Flamsteed's claim. Yet there remained an apparent reluctance for astronomers to undertake the observations required. The attempts made by Ole Römer and Jacques Cassini will be examined, as well as Halley's response to Cassini's attempt. This will reveal the contexts within which the attempt at Kew in 1725 may be more fully appreciated.

Williams argues that in England those who accepted Flamsteed's claim saw little point in seeking to confirm them. It was widely believed that Flamsteed had access to the finest instruments in the country. It would be futile to attempt confirmation with inferior instruments. When Wren received his letter from Flamsteed, he was still thinking about the problems of using a long-focus telescope. In 1692, Huygens presented the Royal Society with a glass with a focal length of 123 feet.[37] By 1703, Wren was giving thought to using 'the hollow in the great staircase on the fourth side' of St. Paul's Cathedral to house Huygens's telescope.[38] Williams ruefully adds that all his plans came to nought. The staircase is a mere 96 feet 10 inches high. Wren abandoned any plans he entertained about resolving the problem of parallax. Wren was fully engaged with the construction of the new St Paul's Cathedral. It appears that the supporters of Flamsteed's claim included John Caswell (1654/5–1712)[39] and John Wallis in Oxford, and William Whiston (1667–1752) in London, though not

David Gregory (1659–1708). Wren also had doubts. The altercation between Flamsteed and Jacques Cassini brought declining enthusiasm for attempts to measure parallax. However, before leaving this sorry tale of failed attempts to observe annual parallax, that made by Römer should be considered. For a historian of astronomy it is an imponderable tragedy that most of Römer's papers, as well as his instruments, were destroyed in the great fire of Copenhagen which burned from 20 to 23 October 1728. It took away over a quarter of the entire city including almost half of the medieval city. One fragment of Römer's archive that survived was a short unpublished piece of work entitled 'The movement of the Earth, or the parallax of the annual orbit from observations of *Sirius* and *Lyra*, 1692–93'.[40] Römer analysed the differences in right ascension in time between *Sirius* and *Vega* (α *Lyrae*). Mari Williams writes, 'Because all that remains of Römer's work on parallax is one small fragment, it is impossible to say whether he made the observations of the right ascensions of *Sirius* and *Vega* in the hope of detecting stellar parallax'.[41]

In the conclusion to his paper, Römer argued that despite detecting variations in the difference in the right ascensions between these two stars, he was convinced that it could not have been caused either by refraction or by the motion of the Earth. He thought it might be a nutation of the Earth's axis the theory of which he hoped to work out later. Bradley and Molyneux were to play with a similar idea, though it was soon abandoned. Römer's method of observing variations in right ascension was to be applied by Eustachio Manfredi (1674–1739) from 1707 until 1719 when he described these motions as 'aberrations'.

In 1714, Jacques Cassini set out to determine the annual parallax of *Sirius,* the brightest star in the heavens. Mari Williams pondered on the reasons why he sought to measure annual parallax at this time. She writes,

> From his own account, in a letter addressed to the Académie Royale des Sciences in 1717 it is apparent that he had developed an interest in the fixed stars as individual objects and was determined to find out whatever he could about them, including therefore their true size and their distance from the Earth.[42]

With considerable insight, Williams adds that in Cassini's search to seek the sizes of the stars he was explicitly rejecting the assumption of uniformity, a common assumption amongst astronomers during the seventeenth century. He was also, by implication, rejecting the notion that the brightness of a star was a function of its distance. In this implicit rejection of photometric methods to determine the distances of stars, Jacques Cassini was forced to accept that the only way of determining the scale of the stellar universe was through measuring the annual parallaxes of the stars. Cassini recognized that all prior attempts to measure parallax had failed. Williams asserts, 'the passage of time had renewed his hope that parallax might in fact be measurable'.[43] Not having experienced the difficulties exposed by his father, he may have thought them exaggerated.

Cassini gave four reasons for choosing *Sirius*. Firstly, being bright it would be easy to observe. Secondly, its brightness *might* indicate proximity and an observable parallax. Thirdly, it lay close to the solstitial colure so its celestial latitude varied similarly to its declination. Fourthly, the declination of *Sirius* due to precession is only about 2½″ so any variation caused by parallax would be recognized. The final reason given by Cassini reveals his expectation that parallax would be a large value relative to the value for precession. Cassini observed *Sirius* during transits until late June 1715. In his 1717 paper, Cassini included a discussion of the theory of annual parallax. Williams introduced a discussion of Cassini's letters of 1702 and 1717. The second letter, the first published, was the first account of the practical difficulties and a theoretical explanation of the phenomenon. He gave data comprising observations collected over a year. Williams states that it has been almost completely ignored by historians.[44] It did appear to have been noticed by Cassini's contemporaries. More importantly, it provoked Halley into considering the problems of measuring annual parallax.

Edmond Halley replied to Cassini's paper in 1720–21 in the *Philosophical Transactions*.[45] Halley pointed out that the effects due to refraction would be so great they would invalidate his observations. The position of *Sirius* would vary by as much as 8″ to 10″. In addition, Cassini took no barometric readings, essential for the evaluation of refraction. Halley took issue with Cassini's estimate of the value of the diameter of *Sirius* calling it an 'optick fallacy'. Following the observations of Halley, Pound, and Bradley they revealed that when the Moon occulted any stars, even the brightest, they disappeared *almost instantaneously*. If the diameter of *Sirius* was anything like that claimed by Cassini (four to five arcseconds) light from the star would disappear over several seconds. According to Halley the 'optick fallacy' was a product of Cassini stopping down his object glass too much. This he did to eliminate scintillation of the star's image. These insights lead to the conclusion that Halley was a perceptive observer, though undermined during the decline of his physical abilities in his later years. He commended Cassini's attempt suggesting he repeat it. There is no evidence that Cassini ever did. Halley, now Astronomer Royal, took possession of a Royal Observatory bereft of instruments[46] and it took him several years before he was in position to undertake any work on the 'parallax'.

From about 1722, possibly from the celebrations of the opening of Sir Richard Child's magnificent new Palladian mansion at Wanstead, the earliest of its kind in England, several figures close to Pound gathered together. Unlike his father Sir Josiah Child, who was described disparagingly by John Evelyn as 'sordidly avaricious',[47] Sir Richard was a socialite, known for his patronage of the Flemish painter 'Old Nollekens'. No doubt during the celebrations, when 'the great and the good' all attended, Pound possibly took the opportunity to discuss scientific issues with friends and acquaintances. I am sure, given the various times that Pound later visited Molyneux at his handsome manor house in Kew,[48] that they were seriously discussing the parallax problem. With Halley's paper fresh in their minds, and their knowledge of the attempt by Robert Hooke in 1669, the problem must have been how to make a practical attempt to resolve the long-standing problem of the parallax. There is no documentary evidence for this. There are entries in Pound's accounts book giving circumstantial evidence that there would be an attempt to observe annual parallax, led by Samuel Molyneux at Kew. It is ironic that Pound's journal of his travels reveals more about his activities in the Indies than any extant evidences of his life in England. I will discuss this in Chapter 11 in some detail when discussing the fate of Bradley's papers after his death in 1762.

The unexpected death of Pound on 16 November 1724 delayed the attempt to measure annual parallax for a few months, but plans were in place before Molyneux commissioned George Graham to construct a 'parallax instrument'. Hooke's attempt was studied closely, particularly by Graham. The instrument maker studied the weaknesses in the design of Hooke's suspended telescope. When Graham constructed the 'zenith sector' for Molyneux, it became the prototype of all such instruments during the eighteenth century. It incorporated various features to ensure very high accuracy and precision, such as a weight to keep the tube in contact with the high precision micrometer. It was rectified to the zenith with a bob, damped in a dish of water. Such were the intricacies that Graham introduced in this remarkable instrument. Unfortunately the sector does not appear to have survived to the present day. Paul Quincey, late of the National Physical Laboratory, has sought the fate of the instrument but so far without success.[49]

Bradley now took an active role in the project, after a period of mourning and assessment of his domestic situation, using Molyneux's instrument and taking up Pound's role. Molyneux led the project, not only did he commission the sector but his house in Kew was capable of accommodating a long-focus telescope. Graham's role in the project was essential. He made the zenith sector but also as a practising astronomer and experimental philosopher,[50] he

fully appreciated the objectives of the partners, in seeking the annual parallax of γ Draconis. Bradley's role was to contribute his outstanding technical
and observational skills in the use of long-focus telescopes. With the finest
instrument maker and the finest observer in England at his disposal, it would
seem that Molyneux had instituted the most serious attempt ever to determine the annual parallax. Yet Molyneux still seemed to display many of the
characteristics of a dilettante. He was a wealthy gentleman with an interest
in astronomy, whose motivations in the attempt only became apparent in the
midst of the experiment. Bradley was a gentleman of limited means, who now
dedicated his life to astronomy, whilst Graham was a leading craftsman of outstanding abilities. Molyneux was someone with the means to put ideas into
action.[51] Hearne's assessment of Molyneux as a man with a shallow understanding, impatient, and 'pushy', would be confirmed by his conduct during
the experiment. The conduct of Molyneux was sometimes so perverse, that the
motivation underlying his interest in the experiment requires close examination. After describing the attempt, I will reflect on what drove Molyneux, and
compare how different his motivations were from the contributions made by
Bradley and Graham. I have few doubts that Molyneux was quite conscious of
his superior social position. He sometimes treated Graham like a tradesman,[52]
and Bradley as though he was still in his debt as his patron, describing him as
'one of the astronomy professors at Oxford'.[53]

In the earlier discussions about the distance scale of the stars in the first
chapter, two quite distinct schools of thought developed during the seventeenth century. I will refer to these as the 'short-distance scale' and the
'long-distance scale'. Those belonging to the latter school were sceptical about
the possibility of determining annual parallax. It must be admitted that the
three partners must have entertained the possibility of the short-distance scale,
or they would not have admitted their support for such an experiment.[54] Newton argued decisively in favour of the idea that even the nearest stars were to
be located at enormous distances, which he estimated to be about a million
times further from the Sun as is the Earth, the equivalent of several light years.
Newton referred to the photometric method devised by James Gregory in his
Geometriæ pars universalis.[55] Newton held two presumptions, first that the
stars all possessed the same intrinsic luminosity as the Sun, and second, that
the luminosity diminished as to the square of the distance. Given these presuppositions, the apparent magnitude of a star is a function of its distance, so
the brightest stars must be the nearest. Newton looked at the Sun through a
pinhole which allowed him to form an image with the estimated brightness of
Sirius.[56] He then calculated the distance and estimated that it would possess

an annual parallax of half an arcsecond, far below the ability of contemporary optics to resolve.[57]

The reason Newton insisted that the stars were located at enormous distances was because it was a necessary prerequisite for the observed stability of the solar system. In Book III of the *Principia*[58] Newton wrote, 'the aphelia and nodes of the [planetary] orbits are at rest'. This leads to two corollaries, 'the fixed stars are also at rest, because they maintain given positions with respect to the aphelia and nodes'. This is an odd assertion, given that eight years prior to the publication of the third edition of the *Principia,* his close confidant and friend Edmond Halley, discovered the proper motion of the stars. Newton adds,

> And so, since the fixed stars have no sensible parallax arising from the annual motion of the earth, their forces will produce no sensible effects in the region of our system, because of the immense distance of these bodies from us. Indeed, the fixed stars, being equally dispersed in all parts of the heavens, by their contrary attractions annul their mutual forces, by Book I, prop. 70.[59]

How Newton was able to reconcile the first corollary that the fixed stars were at rest, with Halley's implied discovery of the proper motion of several bright stars, is not even hinted at in the third edition of the *Principia.*

In opposition to the estimates made by Newton, that the stars were at enormous distances from the solar system, was a tradition of claims made by astronomers for the discovery of annual parallax, with implied annual shifts of thirty or forty arcseconds. These alterations in the positions of stars were regarded differently by astronomers committed to geocentric or heliocentric models. Geocentric astronomers, a minority in early eighteenth-century Europe, tended to interpret such motions as 'anomalies' or 'aberrations' or 'monsters'. Observational or systemic errors were suspected. Heliocentrically inclined astronomers often interpreted the motions as evidence of annual stellar parallax. These claims were invariably challenged by those with a better appreciation of parallax theory, whether committed to geocentric or heliocentric cosmologies.

Molyneux, Bradley, and Graham began their attempt to measure the annual parallax of the second magnitude star γ Draconis, which at the latitude of London passes through the meridian within two arcminutes of the zenith. They did not expect to observe any motion. It was the wrong time of the year to expect any motion due to annual parallax. They expected to use this period to adjust the instrument and iron out the inaccuracies and idiosyncrasies they expected from such a new instrument.

In his discovery paper,[60] Bradley writes,

> Before I proceed to give you the history of the observations themselves, it may
> be proper to let you know, that they were first begun in hopes of verifying
> and confirming those that Dr. Hooke formerly communicated to the pub-
> lic; which seemed to be attended with circumstances that promised greater
> exactness in them, than could be expected in any other that had been made
> and published on the same account.[61]

Great attention is required here. The account may not be as transparent as
it may at first appear. As Chapter 4 has already established, Bradley's practice emphasized rigorous precision, accuracy, and reiteration. When Bradley
stated that the object of the investigation was to verify and confirm Hooke's
observations, a problem immediately asserts itself on anyone who recognizes
the perversity of this statement. Hooke claimed that he had discovered annual
parallax on the basis of just four observations made over a period of three
and a half months. Hardly an investigation that would have impressed an
astronomer noted for his reiterative observational methods. There is further
perversity in the account, suggesting Bradley was not as candid as he might be.
Hooke's claim proposed that γ Draconis was only about 13,755 astronomical
units[62] away from the Sun. If this is combined with the common presumption that the magnitude of a star was a function of its distance, then it would
infer that if a second magnitude star was less than 14,000 astronomical units
away, the distance of the brightest stars could be as little as 4,000 astronomical units or less. Bradley was aware of Newton's belief that the stars were at
immense distances, implying annual parallaxes that were beyond the ability of
contemporary technology to measure. A confirmation of Hooke's *result* with
the implication that γ Draconis was less than 14,000 astronomical units away
from the Sun posed real problems for Newton's natural philosophy. It would
mean that hundreds of stars could be between 400 and 1,000 times the distance
of Saturn, the outermost planet then known.

Bradley appears to have reflected about this problem. I do not think he
thought Hooke's *result* was valid. The leading figure in this attempt to measure
the parallax of the star was Samuel Molyneux. More attention should be paid to
the motives behind the experiment. Halley's recent work on the proper motion
of certain bright stars, derived when he compared their positions with those
recorded in the star catalogue in Ptolemy's *Almagest,* contrasted starkly with
Newton's insistence that the stars were stationary relative to each other. Newton had written as we have already learned that, 'the fixed stars, being equally

dispersed in all parts of the heavens, by their contrary attractions annul their mutual forces.'[63]

There is sufficient ambiguity in Bradley's reasoning to lead us to query his own motives in asserting confidence in Hooke's result. It was written in December 1728 when he was seeking to justify his participation in an experiment based on a belief that Hooke's long-abandoned claim was valid. Yet as revealed, any attempt to measure annual parallax in the full expectation that it would lead to a result in the vicinity of thirty arcseconds would pose real difficulties for the continuing veracity of Newton's universal law of gravitation.

Molyneux was the prime motivator, and it was he who commissioned Graham to construct the twenty-four-foot zenith sector. Bradley was a junior partner, invited after the death of Pound because of his outstanding technical and observational abilities. Molyneux wrote an account of the instrument which was seemingly intended for publication.[64] It was published by Rigaud from the first part of *The Kew Observation Book*.[65] It was the last of three versions. The first was roughly written out in Molyneux's hand whilst the second was written in an unknown hand, possibly Molyneux's trusted servant Giles. The instrument was suspended at Molyneux's residence, the White House, on 26 November 1725.[66] As I have already mentioned, no observable motion of the object star γ Draconis was expected at this time of the year, so the next few days were devoted to adjusting the instrument.

Elsewhere I have briefly touched on Manfredi's work on 'aberrations', studying the relative motions of stars in right ascension, a more demanding task than making observations of motions in declination. It is reliant also on an accurate regulator.[67] In contrast to this, the Kew experiment, based as it was on the method pioneered by Robert Hooke, observed motions in declination.[68] Observation in declination was more efficient and precise because observations were not so time critical. It allowed accurate calibration of the object star as it passed through the observational field. In Bradley's mature practice, he used an array of hairlines above and below the central horizontal line as well as either side of the central vertical line.[69] It enabled the partners to observe the passage of the star with great confidence as they evaluated the distance the star had shifted north or south with each succeeding passage. The turns of the micrometer screw determined the shift in declination as it passed along the central hairline.[70]

The star *Eltanin*[71] designated by Bayer as γ Draconis, passes through the southern meridian at noon in December. On 3 December 1725, its position in declination was recorded by Molyneux.[72] In the *Memoirs* Rigaud wrote,

'The motion which might be produced in it by parallax towards the pole would at this time have been very gradual.'[73] Observations of the object star were repeated on 5, 11, and 12 December 1725 to verify its position in declination.[74] On 17 December 1725, Bradley repeated the procedure. Observing its passage through the field of view he perceived that the star had passed 'sensibly' south of its former position. His response was it was an anomaly as no sensible motion was expected this time of the year. Bradley characteristically examined the equipment minutely.[75] Molyneux recorded the incident in the *Kew Observation Book*.[76] He wrote,

> This day, Mr Bradley, one of the astronomy professors at Oxford, who had been acquainted with the whole contrivance of this instrument, and was present when it was first set up in its place, did with great care examine the situation thereof. He found the index at 10.3, where it had been left on the 12th of December, and adjusting the mark to the plumbline, about 11h.15' mean time, he found the index stood at 9.7, so that from the 12th of December to the 17th, it had been altered but a little more than ½ a second.[77]

This passage clearly reveals that Molyneux regarded Bradley as a social inferior and junior partner in the enterprise. He obviously respected Bradley but the remark that he was 'one of the astronomy professors at Oxford' rather than the Savilian Professor of Astronomy was not only patronizing, but ranked him as a subservient partner. Discussions of the Kew experiment must never lose sight of Molyneux's perception that he was the leading voice in the experiment. Any discussion of the motives, to repeat and verify Hooke's *results*, must recognize they were inordinately Molyneux's. This was evident when Bradley constructed his account of the work undertaken at Kew and Wanstead on the 'new discovered motion'. The passage quoted above recorded the precision of Bradley's work and the accuracy of Graham's instrument. It was a continuation of an outstanding partnership between Bradley and Graham, working to the limits of the technology then available. Bradley perceived a motion of 0.6″ which Molyneux referred to as ½″. This at a time when other observers including Halley did not believe any motion less than four or five arcseconds could be validly recorded.[78] Rigaud politely suggests Halley lacked Bradley's observational skills. There is a revealing passage which Rigaud included in the *Miscellaneous Works,* in the *Zenith Observations at Wanstead*.[79] Bradley recorded Halley's observations with the new zenith sector made for Bradley by Graham, suspended in Elizabeth Pound's house in Wanstead. The entry for 2 September 1727 reads,

Dr. Halley observed *Capella;*[80] The star fluttered much, and he said that as it went out[81] it appeared to be more southerly in reality than the thread; he therefore fancied from this observation that the direction of the thread was not right; but as all other stars going out appear on the other side of the wire, this appearance must be owing to his not being able to bisect the star at the cross because of its fluttering.[82]

With reference to this passage Rigaud comments, 'Though Halley's was certainly a very powerful mind, it was not always free from error; for he not only had too much contempt for what appeared to him to be trifles, but he was often too rapid in drawing his conclusions'.[83]

Some commentators have stated that without Graham's remarkable instruments Bradley could not have made his discoveries. Bradley never denied this. In his later paper on the nutation, he sings loudly his praises for his friend. What the passage also revealed was that no matter how accurate the instrument, it required a competent practitioner to make the best use of it. In 1818, Friedrich Wilhelm Bessel (1784–1846) referred to Bradley's work as 'incomparable', appreciative of the difficulties he overcame, and how he recorded every conceivable cause of error for every one of over 60,000 precise and accurate observations at Greenwich from 1750 to 1762. Halley's observational practice was maintained with the view that calibrations of less than five arc seconds were dubious. Halley never accepted Bradley's observations of the nutation of the Earth's axis, for he considered observations of motions of under a second of arc to be deluded. In December 1725, Bradley adjusted the instrument at Kew satisfied it could be rectified to within half a second of arc.[84]

The declination of γ Draconis was varying at a rate of about an arcsecond every three days[85] which was entirely unexpected and unattributable. On 20 December 1725, according to his account[86] and 21 December 1725 according to the *Kew Observation Book,*[87] Bradley discovered the star passed even more southerly. The discrepancy between these dates is easy to explain. The star had passed conjunction on the intervening days between 17 and 20 December 1725. Civil dates are reckoned from midnight but astronomical dates are reckoned from midday. γ Draconis having passed conjunction, passed across the meridian after noon on 20 December 1725 according to the civil date 21 December 1725 according to the astronomical calendar. Rigaud and others refer to this observation as being *decisive.*[88] However, the observation could only be treated as an anomaly, like thousands of others in the sorry tale of attempts to measure annual parallax. Rigaud seems to have interpreted this observation as the discovery of aberration. He believed he had to defend Bradley's claim for priority over earlier observations made by Jean Picard.

The claim was made on Picard's behalf by Jean-Baptiste-Joseph Delambre in his *Histoire de l'Astronomie au 18e Siècle*[89]. Rigaud continues,

> Picard was the first to bring forward an undeniable objection to Hooke's supposed discovery of the parallax of the fixed stars. He also detected some extraordinary annual motion of the pole star, during a portion of the year when it could not be accounted for either by parallax or by refraction; but instead of his observations having given the direction to Bradley's, or having formed the foundation of his discoveries (which were necessary to explain them) that they were recalled into notice, and that their value was established.[90]

Picard's observations gained significance mainly in the light of Bradley's work. In the history of science, there are many instances of anomalies that only achieved significance in the context of a new theoretical framework. Picard's observations, with those made by other astronomers, including those made by Bradley on 17 and 20 December 1725 were *anomalies*.[91] They were not validated by reference to any known phenomenon, either parallax or atmospheric refraction. The observation of 20 December wasn't decisive, it was perplexing. Yet the assertion that the observation of 20 December 1725 was 'decisive' left a permanent legacy. The memorial to commemorate the discovery of the aberration of light was erected at Kew. The Kew observations only acquired importance in the context of the work performed by Bradley at Wanstead from 19 August 1727 to 21 December 1728.

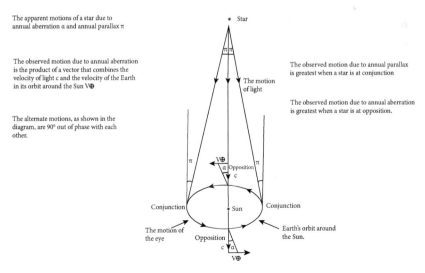

Annual Aberration and Annual Parallax

After a year of observations at Kew, and another year or more at Wanstead with a different zenith sector, designed on similar principles but with a greater range of observable angles from the zenith, Bradley was able to propose his hypothesis of the new discovered motion. He constructed a new theoretical framework to distinguish the motion from that of annual parallax. The theory was a product of Bradley's astronomical practice, based very largely on principles drawn from Newton's *Principia* and the *Opticks*. Bradley's methods were revealed in his lectures in experimental philosophy. These lectures were presented from 1729 to 1760 emphasizing rigor, reiteration, precision, and accuracy, with an understanding of the importance of determining the parameters of the *errors* of his observations. The discovery of the aberration of light was a singular achievement, validated in the context of the shared values of a scientific community. These values and assumptions led to the rapid acceptance of his paper on 'the new discovered motion', soon to be called the aberration of light, throughout much of Europe, including even Papal Italy where less than a century earlier Galileo had been condemned for promulgating the assertion that the Earth was in motion. The aberration of light validated Galileo's attestation, although the trial had been more about the sources of authority. The hypothesis of the new discovered motion met with resistance from many devout Roman Catholics. It was also rejected by various upholders of Cartesian theory, particularly in France. It was a critical element of Cartesian physics that the velocity of light, interpreted as a pulse or a wave passing through a universal plenum, was instantaneous. Acceptance of Bradley's hypothesis was contingent on the acceptance that the velocity of light was finite.

As Delambre wrote, Picard indicated the shortcomings of Hooke's claims for the discovery of annual parallax.[92] I have conveyed the view that it is difficult to maintain that Bradley believed Hooke's *result* was acceptable. His own stated commitment to reiterative rigor, as well as precision and accuracy, as revealed in his mature observational work, would have ruled it out. Bradley remarkably states,

> Before I proceed to give you the history of the observations themselves, it may be proper to let you know, that they were first begun in hopes of verifying and confirming those that Dr. Hooke formerly communicated to the public; which seemed to be attended with circumstances that promised greater exactness in them than could be expected in any other that had been made and published on the same account.[93]

Bradley's account[94] dates from December 1728. His erstwhile partner Molyneux died on 13 April 1728 of what was described as a 'brain seizure'.[95] In my master's dissertation,[96] I held an opinion that Molyneux was prompted by anti-Newtonian sentiments. I reflected on the possible influence of his tutor George Berkeley at Trinity College Dublin, a perceptive critic of the basis of Newton's calculus. After further post-doctoral research, I now think that Molyneux's opposition to Newton was emotive rather than philosophical. Studying Molyneux's correspondence with Flamsteed, before the first Astronomer Royal's death, reveals a genuine friendship. Molyneux's antipathy towards Newton was probably derived from his distaste of the way he believed Newton had abused his friend. The 'anti-Newtonian' aspects of the Kew experiment are conspicuous. Bradley discovered that he had to justify his work at Kew, well known to members of the Royal Society, yet present it in terms deflecting suspicion that he was associated with work that might be perceived as being contrary to Newton's natural philosophy. He also undoubtedly sought to protect Molyneux's reputation, which had been damaged by his association with the Mary Toft affair. He had claimed that he had witnessed Toft giving birth to live rabbits. His credulity about such an obvious fraud made him a 'laughing stock' in the press.

What is unexpected about Molyneux's 'anti-Newtonian' stance is that when he was elected an FRS in 1712, he was introduced to the Royal Society by no less than Sir Isaac Newton. This was certainly because of the reputation of Samuel's highly respected father William, who corresponded regularly with the philosopher John Locke, an acolyte of Newton's. William Molyneux had been forthrightly rejected by Flamsteed, with whom he had corresponded since 1680, even though he had included acknowledged work by the Astronomer Royal in the *Dioptrica Nova* of 1692.

Samuel's mother Lucy Domville, the youngest daughter of Sir William Domville the attorney-general of Ireland, became blind and lived in pain for her short life following an illness contracted after her marriage to William. It led to the discussion of a famous philosophical conundrum still referred to as 'Molyneux's Problem'. In its simplest form it can be stated, 'Suppose a man is born blind but can still feel the differences between shapes such as spheres and cubes, could he if given the ability to see (but without recourse to touch) be able to distinguish those objects by sight alone with reference to the tactile schema?'

The problem was brought up in correspondence with John Locke who referred to it in his book *An Essay Concerning Human Understanding*. Locke concluded that there was no necessary connection between objects known by touch and those known by sight. After the death of William in 1698 and Locke in 1704, the problem was advanced by George Berkeley in his book *A New Theory of Vision*. He agreed with Locke that the connection could only be established by experience and not a priori. Samuel Molyneux's philosophical, social, and political connections made his entry into the Royal Society a formality. When Samuel was proposed by Newton, it coincided with the moment when an unauthorized publication of Flamsteed's observations as the *Historia Coelestis* was printed under the editorship of Halley but directed by Newton. Flamsteed angrily resented Halley as much as Newton. Samuel Molyneux must have been aware that Bradley was a protégé of Halley's. He would have carefully avoided any show of resentment towards Newton or Halley in the presence of his partners. It is my conjecture that the Kew experiment, with an expectation that a 'repeated' observation of annual parallax, confirming the 'short-distance scale' of the cosmos, was intended to produce a problem for the 'Newtonian school'. It may explain Molyneux's peculiar and precipitate actions early in 1726.

After 20 December 1725, Bradley discerned a continued southerly motion of γ Draconis. The partners were perplexed about the possible causes of this anomalous motion of the object star.[97] It moved south in declination up to 5 March 1726 by when it had moved 20″ further south.[98] By 1 April 1726, the motion had reversed, the star moving north,[99] so by 19 June 1726 it had returned to the same distance from the zenith as it was on 20 December 1725.[100] It kept moving to the north until about 11 October 1726 by when it began moving to the south again.[101] In September the star was 39″ further north than it had been in March. On 26 December 1726, it had returned to the same position as the previous year.[102] The predictable but counterintuitive motion of γ Draconis suggested this might be the result of a hitherto unknown natural cause. Bradley knew it was irreconcilable with annual parallax. The phenomenon of the annual parallax of a star always appears to take the object body in the opposite direction from that of the Earth against the background of more distant stars. However, the observed motion appeared to take the object star γ Draconis as though moving *with* the Earth against the background of more distant stars, an entirely counterintuitive and utterly perplexing phenomenon.

From December 1725 to December 1726, conjectures were proposed by the partners to explain the cause of the phenomenon. Bradley's account in the version published on 9 and 16 January 1729 by the Royal Society[103] revealed that one of the conjectures was that the motion was the product of a nutation[104] of the Earth's axis. Bradley wrote,

> A nutation of the earth's axis was one of the things which first offered itself upon this occasion, but it was soon shown to be insufficient; for though it might have accounted for the change in declination in γ Draconis, yet it would not at the same time with the phenomena in other stars, particularly a small one almost opposite in right ascension to γ Draconis, at about the same distance from the north pole of the equator: for though this star seemed to move the same way as a nutation of the earth's axis would have made it, yet, it changed its declination but about half as much as γ Draconis, in the same time, (as appeared upon comparing the observations of both made upon the same days, at different seasons of the year) this plainly proved that the apparent motion of the stars was not occasioned by a real nutation, since, if that had been the case, the alteration would have been equal.[105]

The problem presented to Bradley was that in order to determine whether the motion was due to an annual nutation, it was necessary to observe a control star separated by twelve hours in right ascension.[106] The star chosen was constrained by the design of the zenith sector[107] for it was only capable of being directed twelve minutes of arc from the zenith.[108] This limited the choice of a suitable star to act as a control. It had to be bright enough to be observable in daylight. From 3 January 1726, Bradley, Molyneux, and Graham sought a suitable control star.[109] They found one that Molyneux called *Telescopica in Auriga* and Bradley designated *Anti-Draco*. Bessel later confirmed it as the star 35 in Flamsteed's Catalogue in the constellation of Camelopardalis.[110] Between the death of Bradley and the early nineteenth century there was debate over the identity of this star, for Flamsteed only gave its declination. The exact position of the star was indicated by Bessel in 1818.[111]

The control star was observed on 7 January 1725/26 OS and 3, 12, and 20 February 1725/26 OS. Its motion was consistently northward while γ Draconis was moving south. While the motion of *Anti-Draco* between 7 January 1726 and 20 February 1726 was five arcseconds, the opposing motion of γ Draconis between 3 January 1726 and 13 February 1726 was 9.1 arcseconds.[112] The annual nutation hypothesis was therefore extremely short-lived. Both stars were observed throughout 1726.[113] Using the evidence revealed by the *Kew Observation Book*, it is patently obvious that the nutation hypothesis was soon abandoned. It could not have been sustained much into March 1726 because the motion of the control star *Anti-Draco* or 35 Camelopardalis was incompatible with the hypothesis.

This makes the resultant precipitate actions of Samuel Molyneux difficult to understand unless he was highly motivated to disconcert Newton. Bradley believed a nutation of forty arcseconds was irreconcilable with Newtonian theory.[114] He treated the hypothesis with reticence and caution. In conformity with his usual practice he preferred to observe the motion of both stars for several months before coming to any final conclusions. In contrast to this, Molyneux made an immediate, ill-judged, and premature approach to an elderly and infirm Isaac Newton via John Conduitt (1688–1737). Newton at this time appeared to live at Conduitt's residence at Cranbury Park near Winchester.[115] Molyneux informed Newton that he, Graham and Bradley had discovered 'a certain nutation in the earth which they could not account for and which Molyneux told me *he thought destroyed entirely the Newtonian System* (my italics) and therefore he was under the greatest difficulty how to break it to Sir Isaac'.[116]

Throughout the Kew experiment, there is a lingering suspicion that Molyneux was working to a different agenda from that of Bradley and Graham. Molyneux's precipitate approach to Newton reveals an inordinate emotional need to cause distress to the elderly Newton. From the entries in the *Kew Observation Book,* the hypothesis of an annual nutation contradicting the *Principia*[117] could not have been supported for longer than four weeks and probably less. Molyneux's approach was unwarranted and needless. It confirms Thomas Hearne's earlier assessment of Molyneux as shallow and impatient. He made the approach without the knowledge of Graham or Bradley who would have been greatly disconcerted by Molyneux's actions. Against my assessment of Molyneux it could be argued otherwise, for Conduitt also stated 'he was under the greatest difficulty how to break it to Sir Isaac'. Molyneux was a gentleman and a prominent politician skilled in the arts of persuasion and diversion. It is the haste of his approach that betrays his motives. There are other instances during the Kew experiment when these motives are suggested by his actions.

Molyneux was the prime mover of the Kew experiment. The attempt to repeat Hooke's result of 1669 'in hopes of verifying and confirming those that Dr. Hooke formerly communicated to the public'[118] was under the direction of Molyneux. If an annual nutation of some forty arcseconds was damaging to the validity of Newtonian theory, then any confirmation of Hooke's *result* would also have been regarded as a threat to the validity of Newtonian natural philosophy.[119] It is possible to assume that the Kew experiment was driven by Molyneux specifically to cause discomfort to Newton's acolytes, though I do not believe he was quite so calculating.[120] A positive result posed problems for Newtonian physics. Bradley thought so because he attempted to reconcile such a result with the validity of Newtonian theory. He made various calculations in the back of his copy of the *Kew Observation Book* and sought to reconcile Hooke's result with Newton's concerns over the stability of the solar system.

The approach by Molyneux to Newton would not have received any encouragement from Bradley even if observational evidence had appeared reconcilable with such an annual nutation. Bradley was a cautious, patient, and thorough observer. His methodology was such that he would have suppressed any acknowledgement of an annual nutation until he had confirmed the motion, ruling out any other explanation of the phenomenon.[121] The zenith sector had been designed only to observe stars at the zenith or at least within five or six arcminutes of it. This led to difficulties when the partners sought a control star to test the hypothesis of an annual nutation. Bradley's apparent acceptance of the Hooke result[122] is not what it seems. Bradley is seeking to justify himself when he recognized that the whole project was tinged with an anti-Newtonian bias.[123] More than anything else he sought to argue that his partnership in the investigation was undertaken in a spirit of pure inquiry free from bias or prejudice. Yet after the abandonment of the nutation hypothesis another was proposed. This conjecture was even more detrimental to Newtonian theory than either a confirmation of Hooke's result or an annual nutation of forty arcseconds.

What was proposed was nothing less than a hypothesis that a French savant committed to Cartesian natural philosophy might propose. The hypothesis envisaged that the motion of γ Draconis was due to refraction. It proposed the Earth was moving through a dense medium. To account for a motion of forty arcseconds this matter would affect the Earth and all the planets in their orbits around the Sun. Familiarity with Book II of the *Principia* and Bradley certainly was[124] would not have allowed any such suggestion.[125] According to Rigaud this hypothesis was dated 18 June 1726.[126] This was evidenced by Molyneux's endorsement of the paper, although I have been unable in the time I had available to locate this source.[127] It is difficult to understand how this medium would account precisely for the differences in the relative motions of γ Draconis and *Anti-Draco*. It is difficult to conceive how someone as knowledgeable of Newtonian experimental philosophy as Bradley could for a single moment have entertained the idea, unless he felt some moral obligation towards his former patron Molyneux. Bradley was deferential to a fault. It seems that this strange hypothesis was abandoned in a moment. According to Rigaud's account in the *Memoirs of Bradley* Samuel Molyneux ended his paper on refraction with the words, 'I have wrote down this hypothesis this day, not as a thing which I look upon as certain and demonstrative, but barely as a probable conjecture, to be examined by further observations'.[128]

This was an example of the sort of speculative natural philosophy that was so repellent to Bradley. His fierce rejection of such procedures was evident in his lectures in experimental philosophy at Oxford, as I describe them in Chapter 6. Rigaud asserts that another paper was written, offering reasons for the abandonment of both hypotheses – annual nutation and refraction – possibly as late as September or even November 1726.[129] He mentions that the partners, recognizing that the two stars were moving very differently[130] raised other possibilities. In Rigaud's words,

> It had been found that the quantity of motion in declination was somehow connected with the latitude of the star. The precise nature of this connection does not yet appear to have been distinctly made out, but it was now clearly seen, that the few stars that passed within the narrow range of Molyneux's instrument were insufficient for the complete examination of the question.[131]

Once the motions of γ Draconis and *Anti-Draco* (HR 2123) failed to support evidence for annual parallax, and the inadequate hypotheses proposed did not explain the observations, I have little doubt that Bradley found the entire experience frustrating. He realized that the motion he was observing had nothing to do with annual parallax and that the entire experimental set-up was hopelessly inadequate for any other purpose. He surely discussed this with Graham who must also have recognized the shortcomings of the design of his earlier commission. A zenith sector is a modified transit instrument. The function of the zenith sector is to measure the drift in declination of a star as it passed through the meridian close to the zenith. The function of a transit instrument is likewise, to observe the passage of a star through the meridian. Bradley needed an instrument that combined the advantages of observation at or near the zenith with the adaptability of a transit instrument, capable of observing many bright stars as they passed through the meridian. Bradley's observational practice emphasizing reiteration and precise analysis led to his commission[132] of a sector that could be directed several degrees from the zenith in order to observe as many bright circumpolar stars as was possible in order to isolate a universal natural law.

Notes

1. It was Jim Bennett who put me in touch with Mari Williams – one of the various actions Jim took on my behalf when I was setting out on my studies of Bradley's work. I have never really thanked him enough for all the help he gave me at the start of my studies.
2. Brahe to Landgrave Wilhelm 18 January 1587, Quoted from Robert Westman, *The Copernican Question: Prognostication, Skepticism, and Celestial Order*, London, 2011. p. 287.
3. This process of 'seeing what you want to see' I identify as *perceptual bias*.
4. I have long taken issue with this proposition. Parallax is discerned by the motion of nearer stars against the background of more distant objects. But the common belief at the time of the publication of *De revolutionibus* was that the fixed stars were attached to the surface of a single crystalline sphere making the discernment of parallax impossible. It was when astronomers and philosophers began to think of stars being differentially distant from the earth that the search for annual parallax made sense. It was probably the Englishman Thomas Digges (c1546–1595) who was the first to represent the concept of the stars being distributed throughout space beyond the solar system and thus made the problem of annual parallax a real issue.
5. Mentioned in many references.
6. The Council of Trent was called by Pope Paul II so that the Church could reach agreement about church doctrine in response to the challenge posed by the Reformation. All matters not resolved were referred to the teachings of the church fathers up to the fifth century. They universally believed the Earth did not move. It is therefore somewhat ironic that the pope who called the Council which led to the doctrine that the motion of the Earth was heretical was also the pope to which *De revolutionibus* was dedicated.
7. Copernicus was a canon, that is a lawyer learned in Church law. He was not an ordained cleric as is so often mistakenly asserted.
8. G. R. Fockens, 'Exponantur observandi et computandi rationes quibus astronomi stellarum fixarum parallaxin annuam definire conati sunt', in *Annales of the Academia Lugduno Batavia*, Leyden, 1834. Quoted in M. E. W. Williams, *Attempts to Measure Annual Stellar Parallax: Hooke to Bessel*, PhD Thesis, London University, 1981, p. 7.
9. Robert Hooke, *Lectiones Cutlerianæ* or *A Collection of Lectures: Physical, Mechanical, Geographical & Astronomical, etc.*, London, 1679.
10. J.-B.-J. Delambre, *Histoire de l'Astronomie au XVIII Siècle*, Paris, 1819, p. 419.
11. M. A. R. Cooper, *'A More Beautiful City': Robert Hooke and the Rebuilding of London after the Great Fire*. Stroud, 2003.
12. Margaret 'Espinasse, *Robert Hooke*, London, 1956, pp. 85–86.
13. Bodleian Library, Bradley MS20*, fol.68. Here Bradley recorded the four observations made by Robert Hooke in pursuit of the measurement of the annual parallax of the circumpolar star γ Draconis. These were made on 6 and 9 July, 6 August, and 21 October 1669.
14. Located in present day Broad Street in the forecourt of the building formerly known as the NatWest Building. Gresham College was the largest building in the City of London

to escape the ravages of the Great Fire of London in 1666. It was where the Royal Society held its meetings. The building was used for the governance of the City of London immediately after the Great Fire.

15. Robert Hooke, *Lectiones Cutlerianæ*, Lecture 1, 'An attempt to prove the movement of the Earth from observations', London, 1679.

16. The Monument is more properly 'The Monument to the Great Fire of London'. It is 62 metres tall and is situated 62 metres from the location of the bakery in Pudding Lane where the fire began.

17. John Ward, *Lives of the Professors of Gresham College*, London, 1740. p. 104, Quoted in M. E. W. Williams, *Attempts to Measure Annual Stellar Parallax: Hooke to Bessel*, PhD Thesis, London University, 1981, p. 23.

18. Ibid. p. 23.

19. Ibid. p. 23.

20. J. Gregory to H. Oldenburg, 8 June 1675; printed in T. Birch, *The History of the Royal Society*, iii, London, 1757, pp. 225–226, and subsequently in H. Oldenburg, *The Correspondence of Henry Oldenburg, ix*, pp. 336–339. The first half of the letter is published in S. P. Rigaud, *Correspondence of Scientific Men of the Seventeenth Century*, ii, Oxford, 1841, pp. 262–263, and subsequently in H. W. Turnbull (ed), *James Gregory Tercentenary Memorial Volume*, London, 1939, pp. 306–307. The mathematical contents of the letter were sent on by Oldenburg to Huygens; see C. Huygens, *Oeuvres*, vii, pp. 475–476, and H. Oldenburg, xi, pp. 375–376. Quoted in M. E. W. Williams, *Attempts to Measure Annual Stellar Parallax: Hooke to Bessel*, PhD Thesis, London University, 1981, p. 23.

21. Gregory attempted to derive stellar distances using photometric methods in addition to direct methods. See J. Gregory, *Geometriæ pars universalis*, Padua, 1668, p. 147.

22. William Ball to Christiaan Huygens, 10 March 1674, printed in C. Huygens, *Oeuvres*, vii, pp. 428–429, Quoted in M. E. W. Williams, *Attempts to Measure Annual Stellar Parallax: Hooke to Bessel*, PhD Thesis, London University, 1981, p. 27.

23. Ibid. p. 27.

24. Probably Rasmus Bartholin (1625–1698).

25. Ibid. p. 28.

26. Ibid. p. 29.

27. Finite.

28. Ibid. p. 29.

29. At one time it was used as an eye-test for those seeking service in the Persian army. Short-sightedness was endemic throughout the Middle East.

30. Ibid. p. 30.

31. Ibid. p. 30.

32. Ibid. p. 30. At this time, the annual parallax was measured from the mean diameter of the Earth's orbit around the Sun. Modern practice is based on the mean semi-diameter, which is referred to as an astronomical unit (AU). Subsequently, a parallax of $40''$ would in modern parlance become $20''$.

33. M. E. W. Williams, 'Flamsteed's Alleged Measurement of Annual Parallax for the Pole Star', *Journal for the History of Astronomy*, x, (1979).

34. Jacques Cassini was the son of Giovanni Domenico Cassini.

35. M. E. W. Williams, 'Flamsteed's Alleged Measurement of Annual Parallax for the Pole Star', *Journal for the History of Astronomy*, p. 32, (1979).

36. M. E. W. Williams, *Attempts to Measure Annual Stellar Parallax: Hooke to Bessel*, PhD Thesis, London University, 1981, p. 32.

37. Ibid. p. 33. Reported in James Pound, 'A rectification of the five satellites of Saturn', *Philosophical Transactions.* Vol. XXX, 1718, pp. 768–774, see p. 768.

38. Ibid. p. 33.

39. He was Savilian Professor of Astronomy at Oxford.

40. Ibid. p. 34 .

41. Ibid. p. 34.

42. Ibid. p. 37.

43. Ibid. p. 38.

44. Ibid. p. 40.

45. Edmond Halley, 'Some remarks on a late essay of Mr. Cassini, wherein he proposes to find, by observation, the Parallax and Magnitude of Sirius'. *Philosophical Transactions*, Vol. XXXI, (1720–1), pp. 1–5.

46. Following the death of her husband, Margaret Flamsteed removed the instruments he used, virtually all of which were his own personal property. The fate of these instruments is unknown.

47. John Evelyn, *The Diary of John Evelyn*, ed. Guy de la Bédoyère, Woodbridge, 1995, p. 258.

48. As revealed repeated times in his accounts book. See Bradley MS23, Bodleian Library, Oxford.

49. Paul Quincey has taken an interest in my research on Molyneux and Bradley for some time.

50. George Graham discovered the diurnal variation of the terrestrial magnetic field in 1722/23.

51. As the Prince of Wales's private secretary and a Member of Parliament at London and Dublin, the time he had available for his personal interest would have been severely eaten into at various times.

52. Evidenced by the times he summoned Graham from London to Kew to make quite minor running repairs.

53. A description that is somewhat patronizing.

54. This may have been true of Pound. Bradley was incorporated after the death of Pound. Bradley may have been reticent to admit to the 'short-distance scale' with his understanding of the arguments in the *Principia*.

55. James Gregory, *Geometriæ pars universalis*, Padua, 1668.

56. In fact *Sirius* which is designated α Canis Majoris has an absolute magnitude of −1.5, which makes it twenty-three times more luminous than the Sun.

57. James Bradley, *Miscellaneous Works and Correspondence,* p. 208. Edmond Halley's observational practice was predicated on the belief that angles of less than 4 or 5 seconds of arc could not be observed. Bradley developed observational techniques which allowed him to calibrate motions of less than 1 arc second. Rigaud expressed views that were highly critical of Halley's observational abilities, and Halley's expressed

amusement at his protégé's claims of the observation of such fine angles. See S. P. Rigaud, *Memoirs of Bradley*, p. xxix.

58. Isaac Newton, *Philosophiæ Naturalis Principia Mathematica*, Book III, Proposition 14, Theorem 14, 3rd edn. London, 1726, trans. I. Bernard Cohen and Anne Whitman, London, 1999. p. 819.

59. Ibid. p. 819.

60. Written as a letter addressed to Edmond Halley.

61. Bodleian Library, *Bradley MS20**, fol. 18r—18v.

62. An astronomical unit (AU) is a fundamental measure of distance. It is the mean distance between the Sun and the Earth.

63. Isaac Newton, *Philosophiæ Naturalis Principia Mathematica*, Book III, prop. 14, Theorem 14, Coro. 2, 3rd edn. London, 1726. Trans. I. Bernard Cohen and Anne Whitman, London, 1999.

64. Samuel Molyneux, 'A Description of an Instrument set up at Kew, in Surrey, for Investigating the Annual Parallax of the Fixed Stars, with an Account of the Observations Made Therewith', included in the *Miscellaneous Works and Correspondence of James Bradley*, pp. 93–115.

65. Bodleian Library, *Bradley MS14*, marked K14, *The Kew Observation Book*.

66. Samuel Molyneux, ibid. p. 99.

67. Eustachio Manfredi, *De annuis inerrantium stellarum aberrationibus*, Bologna 1729. Caput 1.

68. James Bradley, Bodleian Library, *Bradley MS14*, fol. 2v

69. James Bradley, 'Directions for Using the Common Micrometer' in *The Miscellaneous Works and Correspondence*, pp. 70–74.

70. Ibid. pp. 70 –71.

71. Richard Hinckley Allen, *Star Names Their Lore and Meaning*, (formerly *Star Names and Their Meaning*), London 1899, reprinted Dover Publications, New York, 1963. pp. 207–209. The name of the star *Eltanen*, is derived from Uleg Beg's *Al Ras al Tinnin* (the Dragon's Head). Riccioli called it *Ras Eltanim*, whilst Bayer referred to *Rastaben*. In the Alphonsine Tables it is called *Rasaben*. It passes within two minutes of the zenith at the latitude of London and is occasionally referred to as the Zenith Star. In 2000 BCE, it was the nearest bright star to the pole and was of great significance to the Egyptians and was variously referred to as *Apet, Bast, Mut, Sekhet*, and *Taurt*. It was associated with Isis and the worship of that deity. It was designated by Bayer as γ Draconis. It is a double star, a bright component of magnitude 2.4 and a companion of magnitude 13.2. The primary star possesses a distinct orange hue.

72. S. P. Rigaud, *Memoirs of Bradley*, p. xvi.

73. Ibid. p. xvi.

74. James Bradley, 'A Letter to Dr. Edmond Halley, Astronomer Royal and etc. Giving an Account of a New Discovered Motion of the Fixed Stars' in the *Miscellaneous Works and Correspondence*, p. 2.

75. Samuel Molyneux, 'A Description of an Instrument set up at Kew, in Surrey, for Investigating the Annual Parallax of the Fixed Stars, with an Account of the Observations Made Therewith'. In James Bradley, *Miscellaneous Works and Correspondence*, pp. 109–110.

76. Bodleian Library, *Bradley MS14,* K14, *Kew Observation Book,* fol. 10r.

77. Samuel Molyneux, ibid. p. 109.

78. S. P. Rigaud, *Memoirs of Bradley,* p. xxix.

79. James Bradley, 'Observations of the Fixed Stars made at Wansted in Essex by J. Bradley' in the *Miscellaneous Works,* pp. 201–286.

80. Designated by Bayer as α Aurigæ. It is a brilliant yellow-white star of magnitude 0.3. It is significant as the brightest near zenith star at the latitude of London. Although much more luminous than the Sun, it belongs to the same spectral class on the main sequence. It passes within 6½° of the zenith.

81. The terms 'went out' and 'going out' refer to the star moving out of the field of vision.

82. James Bradley, 'Observations of the Fixed Stars made at Wansted in Essex by J. Bradley' in the *Miscellaneous Works* p. 208.

83. S. P. Rigaud, *Memoirs of Bradley,* p. xxix.

84. Shortly after Nevil Maskelyne was appointed as the fifth Astronomer Royal in 1764, he tested the zenith sector constructed for Bradley by Graham in 1727 and was confident that it was accurate to half a second of arc. It was used to collimate the instruments of the Royal Observatory before being sent to Cape Town from 1837 to 1850. It is now mounted on the west wall of the Transit House at the Old Royal Observatory at Greenwich.

85. James Bradley, 'A Letter to Dr. Edmond Halley, Astronomer Royal and etc. Giving an Account of a New Discovered Motion of the Fixed Stars' in the *Miscellaneous Works and Correspondence,* p. 3.

86. Bodleian Library, *Bradley MS20*,* *Drafts of Papers by James Bradley on Aberration.* fol. 19.

87. Bodleian Library, *Bradley MS14,* K14, *Kew Observation Book,* fol. 10v.

88. Rigaud believed this so firmly that he had Bradley's record of it reproduced and included in the *Miscellaneous Works* as a facsimile.

89. J.-B.-J. Delambre, *Histoire de l'Astronomie au 18e Siècle,* p. 427.

90. S. P. Rigaud *Memoirs of Bradley,* p. xvii.

91. From the Greek, anomalia, meaning an irregularity.

92. Ibid. p. 427.

93. Bodleian Library, *Bradley MS20*,* 3rd unpublished version, fol. 18r.

94. Bradley wrote at least four versions of his account. One version was published by the Royal Society, 'A Letter to Dr. Edmond Halley, Astronomer Royal and etc. Giving an Account of a New Discovered Motion of the Fixed Stars', No.406, Vol. XXXV, pp. 637–661. There are three other versions at the Bodleian Library, *Bradley MS20*,* all of which were transcribed and included as appendices to my MSc dissertation, 'James Bradley and the New Discovered Motion: The Origins, Development and Reification of James Bradley's Hypothesis of the New Discovered Motion of the Fixed Stars', University of London, 1994. The third version was included as appendix 3, and is in my considered opinion the intended final account. This is now included as Appendix 2 in this work.

95. Molyneux died in his thirty-ninth year.

96. John Fisher, 'James Bradley and the New Discovered Motion: The Origins, Development and Reification of James Bradley's Hypothesis of the New Discovered Motion of the Fixed Stars', University of London, 1994.

97. This perplexity indicates that the so-called discovery of the aberration a few days earlier was merely a recognition of an anomaly which could not be explained.

98. Bodleian Library, *Bradley MS20**, 'Drafts of papers by James Bradley on Aberration'. fol. 19r.

99. Ibid. fol. 19v.

100. Ibid. fol. 19v.

101. Ibid. fol. 20r.

102. With allowance for precession.

103. James Bradley, 'A Letter to Dr. Edmond Halley, Astronomer Royal and etc. Giving an Account of a New Discovered Motion of the Fixed Stars' in the *Miscellaneous Works and Correspondence*, pp. 1–16.

104. Nutation literally means 'nodding'. It refers to a regular wobble in the Earth's axial rotation.

105. Ibid. p. 3

106. If the control star was in the opposite area of the sky, any motion in declination due to nutation would be similar for both the object and the control stars. In fact the motion of the control star was only half that of the object star. The hypothesis of annual nutation therefore failed.

107. Samuel Molyneux, 'A Description of an Instrument set up at Kew, in Surrey, for Investigating the Annual Parallax of the Fixed Stars, with an Account of the Observations Made Therewith', included in the *Miscellaneous Works*, p. 98. There are some formerly loose papers in George Graham's hand (now in the *Kew Observation Book, K14*) that were titled by Samuel Molyneux as 'Mr Graham's dimensions of part of the parallax instrument', which includes the data, Radius = 24 feet 3½ inch, nearly, Screw 42 threads in one inch. Molyneux adds that the head was divided into 17 equal parts and each part was equal to one second on the radius of 24 feet 3.15inches.

108. Ibid. p. 98.

109. James Bradley, 'A Letter to Dr. Edmond Halley, Astronomer Royal and etc. Giving an Account of a New Discovered Motion of the Fixed Stars' in the *Miscellaneous Works and Correspondence*, p. 11. Also in 'Observations on the Fixed Stars made at Wansted in Essex, by J. Bradley', also in the *Miscellaneous Works*, p. 212.

110. Due to boundary changes the star is now located in the constellation of Auriga. It is now designated as the star HR2123.

111. Friedrich Wilhelm Bessel, *Fundamenta Astronomiæ*, p. 176.

112. Bodleian Library, *Bradley MS14, Kew Observation Book, K14.* fol. 17 passim.

113. Ibid. fol. 17 passim.

114. Isaac Newton, *Philosophiæ Naturalis Principia Mathematica*, Book III, Prop 21, Theorem 17. 3rd edn. London, 1726. Trans. I. Bernard Cohen and Anne Whitman, London, 1999. Newton wrote, 'The equinoctial points regress, and the world's axis, by a nutation in every annual revolution, inclines twice toward the ecliptic and twice returns to its former position. This is clear by Book I, Prop. 66, Corol. 20. *This motion of nutation, however, must be very small – either scarcely or not at all perceptible*'. (My italics).

115. Conduitt was married to Newton's niece and adopted daughter Catherine Barton. They took care of Newton in his declining months.

116. John Conduitt, 'Memorabilia', King's College Library, Cambridge, Keynes MS 130.5. This approach must have been some time in late in January or during February for by the end of February the hypothesis of the annual nutation was simply unsustainable. The memorandum is dated 13 May 1730, over four years after the incident recalled, and over two years after the death of Samuel Molyneux. Conduitt's motive in recalling Molyneux's approach was to reveal Newton in a good light. Conduitt recalled that Newton's response was generous and magnanimous, for he is reported to have said, 'there is no arguing against facts and experiments'.

117. Isaac Newton, *Philosophiæ Naturalis Principia Mathematica*, Book III, Prop 21, Theorem 17. 3rd edn. London, 1726. Trans. I. Bernard Cohen and Anne Whitman, London, 1999.

118. Bodleian Library, Bradley MS20*, *Drafts of papers by James Bradley on Aberration,* fol. 18r.

119. Bodleian Library, Bradley MS14, *Kew Observation Book,* K14. Bradley evidently considered that a positive result would pose some difficulties for Newtonian theory, and in his copy of the observation book he makes various calculations attempting to reconcile the possibility of a 'short-distance scale' with Newtonian physics.

120. Bodleian Library, Bradley MS 23, *James Pound's Account Book 1715 to 1724.* Pound lists expenses for various visits to Kew in order to confer with Molyneux during much of 1724 in the months before his demise. The fact that Pound travelled to Kew and not vice versa can be regarded as circumstantial evidence of Molyneux's primacy in the attempt to repeat Robert Hooke's attempt and to achieve Hooke's result.

121. Later Bradley was to suppress all public discussion of the lunar-associated nutation for several years, keeping knowledge of it to a few trusted co-workers and friends like Graham, Bevis, and Machin. Though he observed the phenomenon from 1727, he did not publicize it until 1737 and did not publish his paper until 1748. Bradley would surely have disapproved of his senior partner's approach to Newton of all people.

122. James Bradley, 'A Letter to Dr. Edmond Halley, Astronomer Royal and etc. Giving an Account of a New Discovered Motion of the Fixed Stars' in the *Miscellaneous Works and Correspondence of James Bradley,* p. 1.

123. Molyneux had died over six months before Bradley wrote his account of the discovery of the newly identified motion. He was able to give an account free of the suspicion of anti-Newtonian intrusions.

124. During 1726, whilst Bradley was working with Molyneux on the motion of γ Draconis, Isaac Newton personally presented James Bradley with one of twelve special copies of the 3rd edition of the *Principia* bound in Morocco leather.

125. Isaac Newton, *Philosophiæ Naturalis Principia Mathematica* Book II, Section 1, Prop. 4, Problem 2, reveals that the amount of resistance encountered in a fluid medium to create such an effect would definitely lead to the earth and the other planets spiralling into the sun.

126. S. P. Rigaud, *Memoirs of Bradley,* p. xxi.
127. Including archives of the Molyneux family at Southampton Records Office.
128. S. P. Rigaud, *Memoirs of Bradley,* p. xxii.
129. Ibid. p. xxii.
130. As was very obvious by the end of February 1726.
131. Ibid. p. xxii.
132. Quite how Bradley was able to afford the instrument is still a topic of contention. Possibly it was financed by his aunt, or just as likely Graham deferred payment. He was supporting John Harrison financially with his work on his chronometers at this time.

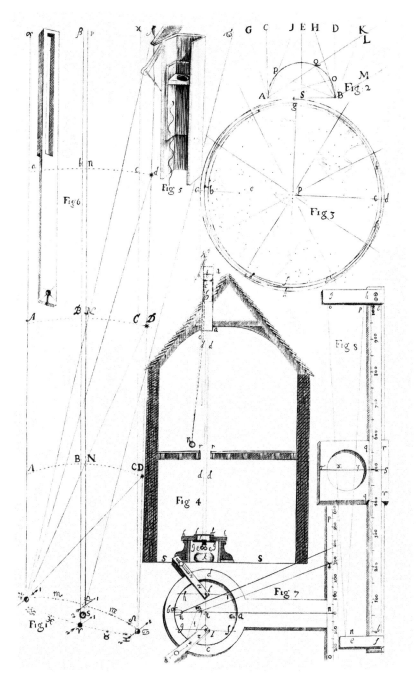

Robert Hooke's celebrated experiment of 1669 to observe the parallax of a star using a suspended telescope with a focal length of thirty-six feet attached to a chimney.

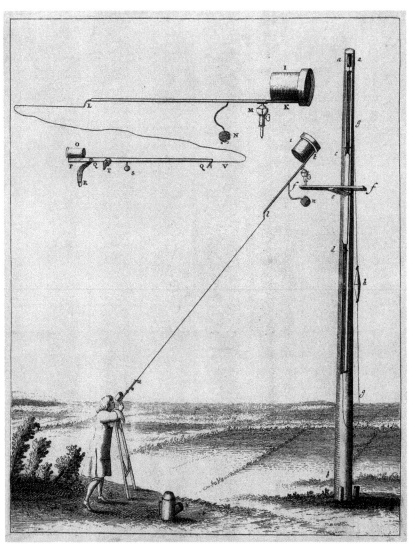

Huygens' aerial telescope. An example of the kind of aerial telescope that Bradley used. The invention of the tubeless telescope is attributed to the Huygens brothers. This one, sixty-five metres (210 feet) long, was described by Christiaan Huygens in 1684.

5

And Yet It Moves

The Discovery of the Aberration of Light

'therefore after all, [as there is] no sensible parallax in the fixed stars,
the Anti-Copernicans have still room on that account to object against
the motion of the earth;... But ... I do not apprehend that [this] will
be denied me by the generality of astronomers and philosophers of the
present age'.[1]

James Bradley.

James Bradley's paper on a 'new discovered motion of the fixed stars' transformed him from someone admired by a small but select circle of English
virtuosi in London, into a highly regarded astronomer known throughout
Europe.[2] His isolation of a phenomenon later called the aberration of light
was indubitably the most important single astrometrical discovery since the
application of the micrometer to the telescopic sight. It transformed the entire
science of positional astronomy making obsolete almost all prior meticulous
positional observations, including most of those made by Flamsteed, whose
catalogue and celestial atlas did, however, serve astronomers well for many
decades. It vindicated Galileo who has been attributed with the phrase that
forms the title of this chapter, which in the original Italian was 'e pur si muove'.[3]
In spite of various advances in the precision and accuracy of astronomical
observations, the whole science had been held back by the attribution of this
long-observed, sensible motion caused by aberration, to observational error or
as erroneous claims for the discovery of annual parallax. It seems so implau-
sible that an annual motion in declination of up to forty arcseconds of every
object in the sky had never been identified during the previous decades, when
so many notable astronomers had observed its effects. It was Bradley's perception that the motion could not be reduced to the phenomenon of annual
parallax that led to his research programme at Wanstead, in order to isolate
and identify the causes of the newly discovered general motion of the stars.
It was almost universally accepted as the first hard empirical evidence of the
motion of the Earth.

The Life and Work of James Bradley. John Fisher, Oxford University Press. © John Fisher (2023).
DOI: 10.1093/oso/9780198884200.003.0006

After the unexpected and sudden demise of James Pound on 16 November 1724, Bradley absorbed himself with his academic work at Oxford. His first recorded observations after the death of his uncle were on 24 July 1725, when observing Jupiter's first satellite (Io) as it was eclipsed by Jupiter. He used his friend John Hadley's six-ft focal length reflector. His next observations were made on 18 August 1725 when he was visiting Samuel Molyneux at Kew. Both sets of observations were of Jupiter's innermost satellite. His work was in support of Edmond Halley, his mentor and friend. Bradley constructed accurate tables of the satellites of Jupiter in 1719. These showed the places of the satellites in degrees and minutes of arc to which he added ecliptic tables of the first satellite as calculated by Pound. His uncle had determined the mean motions of the satellites with characteristic accuracy. Pound compared the observations made by Bradley and himself with those made by predecessors, most particularly Giovanni Domenico Cassini and his nephew Giacomo Maraldi (1665–1729) that were the most reliable up to that date. Both of the naturalized French astronomers did not include an 'equation of time' when they reduced their observations. This would imply an acceptance of the motion of the Earth or of the progressive motion of light. Both were anathema. To admit to either contravened their conservative instincts. Yet it was the accuracy of Cassini's observations of Jupiter's satellites that gained the attention of Jean Picard who recommended him to Jean-Baptist Colbert, the Minister of Finances under Louis XIV. Cassini was recommended as a suitable candidate to become acting director of the Paris Observatory that had been completed in 1672.[4] In 1694, Halley insisted that an equation of time should be applied to the motions of the satellites. Bradley was the first to apply it to all four satellites. For this he settled on 14 minutes, adding 3 minutes 30 seconds, accounting for the eccentricity of Jupiter's orbit. Robert Grant expressed surprise[5] that Bradley never applied the maximum value assigned to the aberration of light amounting to 16 minutes 26 seconds for the greater equation. This was more in conformity with the observations. Grant wrote, 'It is surprising that he should have overlooked the importance of his great discovery in furnishing an independent means of calculating this equation.'[6]

For a historian of astronomy as knowledgeable as Grant, he does not seem to have appreciated that Bradley tendered his tables to Halley in 1719, ten years before the appearance of his epoch changing paper on the aberration of light. The tables were the personal possession of Halley. They reappeared in 1744, two years after Halley's death but were not published until 1752. Later in 1744 Bradley received a letter from Nicolas-Louis de Lacaille (1713–1762) regretting that the tables had remained unpublished. Why Halley didn't publish the tables is not known. He regularly published papers in the *Philosophical Trans-actions*. A highly perceptive observer, Halley was occasionally rather careless, undervaluing the precision of his protégé.[7] The motions of Jupiter's satellites absorbed much of Bradley's attention before he shifted his emphasis to the motions of γ Draconis.

Bradley's work on the motions of seventy circumpolar stars in Wanstead fol-lowed the many difficulties he experienced, observing the satellites of Jupiter. Bradley's work on Jupiter's satellites was later applied when working on 'the new discovered motion of the fixed stars'. The inclusion of an equation of time for all of his observations implied both the motion of the Earth and the finite motion of light. The precision of Bradley's studies showed that the second satellite (Europa) revealed anomalies unaccounted for by circu-lar or elliptical orbits. Bradley recognized that the three inner satellites[8] were interacting gravitationally with each other. These interactions created com-plex irregularities in their mutual motions around Jupiter. Bradley perceived that these bodies were in gravitational resonance with each other, motions now commonly referred to as *Laplace resonances*. Robert Grant referred to these irregularities: 'He [Bradley] discovered that the three interior satellites passed through the irregularities of their motions in 437 days; the errors returning at the close of this period in the same order and magnitude as before.'[9]

Bradley assessed that the period of the inequalities corresponded to that which brought the satellites to their mutual positions relative to each other. From this he calculated that the inequalities were due entirely to the mutual attractions of the satellites. It required the age of interplanetary probes to reveal that the tidal effects of these repetitive perturbations had remarkable conse-quences for all three bodies.[10] Bradley revealed that the orbit of the fourth satellite (Callisto) was eccentric. He fixed the maximum of the equation of the centre. This is the angular difference between the actual position of a body in an elliptical orbit and the position it would be located in if its motion was uniform.

Soon after his work observing the Jovian system on behalf of his mentor was concluded, Bradley was invited to assist Molyneux with the suspension of the zenith sector constructed by Graham at Kew. Bradley had chosen to live as much as circumstances would allow as an astronomer, the real vocation of his life. His Oxford teaching duties sometimes intruded. His later courses in experimental philosophy reveal his commitment to an open-ended pursuit of natural knowledge. His astronomical practice rested not only on a thorough understanding of the materials and instruments at hand, but many of the current developments in calculation. His work on Jupiter and its satellites revealed his increasing mastery of the contents of Newton's *Principia*.

Few men[11] in eighteenth-century Great Britain could conceive of the pursuit of 'science' or natural philosophy as a paid career. If Bradley had relied solely on the stipends he received as the Savilian Professor of Astronomy, later adding his wholly inadequate income as the Astronomer Royal, then his meal table would have been scarce indeed. Portraits of Bradley suggest that his meals were substantial. Rigaud asserts that Bradley was 'abstemious' in his habits. I tend to agree with Allan Chapman[12] that I see little evidence at all for this. His relaxed attitude to religious doctrine and his social ties with the Parkers, the Hoadlys, and the Pelhams all suggest that Bradley was very much a man of the world. His acquisition of the extra-curricular post of lecturer in experimental philosophy at Oxford became a lucrative source of income, sufficient to exceed his requirements.

Bradley was widely celebrated as the person who had established beyond reasonable doubt that the Earth really was a planet in motion around the Sun. In educated circles this belief in the motion of the Earth had been accepted with increasing regularity for over a century. Establishing it, however, had proved elusive. Hooke's method of observing how far from the zenith a star moved through successive observations, the alignment of the telescope being at all times trained on the zenith, promised to be the best means of observing annual parallax, whilst avoiding most of the problems associated with atmospheric refraction.

As seen in Chapter 4, Bradley, Molyneux, and Graham were confused by an observation that was inexplicably opposed to all rational expectations. It is little wonder that the partners set upon a course of testing the instrument and trying to fathom the origins of this apparent motion. Yet to assert as did Rigaud, that this was the moment when Bradley discovered the aberration of light is simply absurd. Rigaud also argued that the time of the earliest observations at Kew was at the most favourable time. He suggested that if the observations had begun three months earlier the star, by the effects of aberration, would have moved southward, in the direction expected by annual parallax. Rigaud suggested that the rate of variation would have been too great for annual parallax, though precisely how the three partners in the investigation would have been able to discern this was not explained.[13] He concedes that a continuation of the observations beyond conjunction would have contrasted the effects of the two phenomena, or a false conclusion might have been drawn. Familiarity with Bradley's approach to observation, and certainly Rigaud must have been fully apprised of this, could not suggest this, except as an aside not meant to be taken seriously.

There persists a tradition that suggests that Bradley was 'fortunate' to initiate his observations of the object star late in December. Yet regardless of the time of the year the partners, led by Bradley's consistent systematic approach to observation, would certainly have observed the phenomenon over the course of an entire year at the very least. Any suggestion that Bradley might have been confused fails to comprehend how he undertook investigations, always completed with utmost thoroughness. He always delayed drawing any conclusions until completing his observations over the full course of the investigation. This I might contend was a reflection of his modest demeanour, surely revealed in his deferential attitude to his 'social betters'. If some catastrophe had overcome the instrument, Bradley would have set everything aside until the experiment could be replicated, this time taken to its conclusion. As it was, the partners had more than enough to ponder on, trying to follow the newly discovered motion over the course of the year. Only then could a reasoned explanation be sought for a phenomenon that was entirely inexplicable and counterintuitive.

It is difficult to work with Bradley's surviving papers, because so many of the documents that would have allowed researchers to clarify Bradley's intentions are absent. When Rigaud began writing the *Memoirs of Bradley*, he found the task inestimably more difficult than it should have been. This was due to the way that much of the surviving archive had been thoughtlessly dispersed when under the custody of Bradley's successor at Oxford, Thomas Hornsby. Without Rigaud's intervention, for which we should be very grateful, more of Bradley's already depleted archive would have been hopelessly lost, increasing the difficulties of the task of later authors.

I am firmly of the opinion that Bradley was forming in his mind the expectation that the object star was about to move in a complete ellipse. What he couldn't comprehend was the completely counterintuitive phase and direction of the phenomenon. He determined to resolve the problem by expanding the enquiry in order to examine the motions of as many different stars as was possible. This reiterative methodology is characteristic of Bradley's entire approach to scientific investigation. In this he was a pathfinder, who along with other contemporary workers, such as Lacaille in France, established many of the procedures of modern natural science. It reflected his knowledge of Newtonian natural philosophy, through his approach to the resolution of the problems of the motions of Jupiter's satellites.

The observed motion of γ Draconis could not be reduced to annual parallax, but as the months went by the observed motion seemed nevertheless to reflect that of the motion of the Earth, but 90° out of phase and in an opposing direction to that expected from parallax theory. He remained confused about the reasons for these extraordinary and unexpected motions of γ Draconis. He was sure the observed motion was a natural phenomenon due to some as yet unascertained natural law. Some of these expectations will be examined when discussing Bradley's lectures in experimental philosophy at Oxford in Chapter 6.

During his experiences at Kew Bradley had complained that 'he had in several journeys to Kew been successively hindered by clouds from making any observations'.[14] Bradley claimed that having his own instrument at Wanstead he would always have the means to record observations at hand. He no longer had access to the parsonage, but he observed at 'aunt Pound's house in Wansted town' having gained access here from July 1725.[15] Bradley and his aunt Elizabeth were in some form of close relationship for several years, an association that was maintained even after Bradley moved his chief abode from Wanstead to Oxford in 1732. By the standards of Molyneux's fine mansion or even the Wanstead parsonage, Elizabeth's town house was modest. We are informed that the ground room was only 7½ feet high and above it was a small loft or garret.[16] Her house was close to the extensive grounds of her brother's handsome residence. The modest dimensions of Mrs Pound's town house imposed limits on the length of Bradley's zenith sector. There were calculations by Bradley of the length of a sine of a second on an arc whose radius was thirteen feet. There was also mention of 13½ feet but he settled on a shorter length of 12½ feet. Bradley drew up a short catalogue of stars passing within one degree of the zenith, another for those which passed within three degrees. Then he drew up one to take in *Capella*, a first magnitude star that passed within 6¼° of the zenith. The instrument was contrived by Graham and completed by him. Bradley specifically mentioned that the critical apparatus upon which the instrument was suspended was made by George Hearne[17] an optical and mathematical instrument maker who had a shop in partnership with Joseph Jackson in Chancery Lane, London.

Although the length of Bradley's tube was only about 12½ feet[18] he found that he could depend on the adjustment of it to ¼ of an arcsecond. Referring to events on 19 August 1727 when Molyneux and Graham joined him at his aunt's town house in Wanstead to begin observations with the new zenith sector, Bradley wrote,

When we had fixed up the instrument and put on the wire and the heaviest ball, Mr Molyneux and myself tried with what certainty we could set the instrument, by bringing the same spot several times to the plumbline; and from the trials which we then made with the apparatus for enlightening the wire, &c. we concluded that when the plummet was dead or still, the instrument might be rectified to ¼ of a second.[19]

The weight of the plummet was less than half of that of the instrument at Kew. They believed this would avoid the difficulties faced at Kew when the plumbline regularly broke. The shorter tube was much firmer, an important prerequisite in order to observe *Capella* 6¼° from the zenith as it passed through the meridian. All flexure in the tube had to be avoided, particularly in the plane of the meridian. When out of the perpendicular, it had to be supported by a screw that pressed against the lower part of the tube. In view of the significance of this instrument, some gratitude must be expressed that it has survived to the present day. The instrument is on display in the Transit House at the Old Royal Observatory at Greenwich, on the west wall of the building. Unfortunately its location means that it is usually missed by members of the public as it is to the right of the entrance into the hall, when attention is immediately attracted to the two imposing 8-foot mural quadrants. There is every justification to regard the zenith sector made for Bradley, as one of great significance in the history of astronomy. It was the instrument with which Bradley discovered the aberration of light, the earliest firm empirical evidence of the motion of the Earth, as well as the nutation of the Earth's axis, which was integral to the most far-reaching transformations in the foundations of positional astronomy during the eighteenth century. The instrument needs to be highlighted so that its significance can be appreciated by visitors. It was the instrument that finally established once and for all, that the Earth really was a planet in orbit around the Sun.

Bradley's sector gave him the choice of more than 200 bright stars[20] from which he selected seventy. There were eleven stars bright enough to be seen the year round even in full daylight. In a manner that was distinctively Bradley's, he observed his chosen stars systematically and with diligence.[21] When he began observing γ Draconis on 19 August 1727 it was 18″ further north than it had been when it was observed on 26 December 1726. The star continued moving until it was another 1.3″ further north during the period from 7 to 14 September 1727, when (as fully expected) γ Draconis began moving to the south. By 10 December 1727, it had returned to the position it was in at Kew a year earlier. By 7 March 1728, it had moved 19.3″ south of this location before returning to the position where it was observed exactly a year earlier. To complete the year γ Draconis returned to the position it was initially observed at Wanstead on 19 August 1727. This confirmed that the motion of γ Draconis at Kew from 20 December 1725 was not the product of some instrumental peculiarity. It dispelled all doubts that the phenomenon observed at Kew was due to instrumental or systemic error.

This was as expected. Bradley was observing several other bright stars throughout the year. These included a bright star with a similar right ascension to γ Draconis, a group that were in opposition to γ Draconis and two groups in quadrature, one group six hours ahead in ascension and another six hours behind. The 'object group' included γ and β Draconis, a 'control group' made up of α Aurigae (*Capella*) and 35 Camelopardalis (HR 2123) and two groups in quadrature. The stars β Cassiopeiæ, α Cassiopeiæ, τ Persei, and α Persei were six hours in arrears, and γ Ursæ Majoris, ε Ursæ Majoris, and η Ursæ Majoris six hours ahead. After observing these stars throughout the period of a complete year from 19 August 1727 a distinctive pattern became evident. To make sense of these results I will briefly rehearse the evidence that was presented at Kew on the motion of the object star γ Draconis. Recall the bewilderment of the partners as the object star moved southward on 20 December 1725, the reversal around 20 March 1726, the motion northwards until 27 September 1726, when it reversed its motion until 26 December 1726. The other stars followed similar patterns varying according to their right ascension. I present the observed data in tabular form:

Star	RA.	Dec.	Stationary periods
β Cassiopeiæ	00h 09m 20s	59° 10′ 09″	Late Nov. & late May
α Cassiopeiæ	00h 40m 40s	56° 33′ 23″	Early Dec. & early June
τ Persei	02h 54m 28s	52° 46′ 35″	Late Dec. & late June
α Persei	03h 24m 32s	49° 52′ 23″	Early Jan & early July
α Aurigæ (*Capella*)	05h 16m 55s	46° 00′ 55″	Early Feb. & early Aug.
35 Camelopardalis	06h 04m 35s	51° 34′	Mid. Mar. & mid. Sept.
γ Ursæ Majoris	11h 53m 58s	53° 40′ 34″	Late June & late Dec.
ε Ursæ Majoris	12h 54m 08s	55° 56′ 30″	Early July & early Jan.
η Ursæ Majoris	13h 47m 37s	49° 17′ 50″	Mid. July & mid. Jan.
β Draconis	17h 30m 27s	52° 17′ 52″	Early Sept. & early Mar.
γ Draconis	17h 56m 38s	51° 29′ 22″	Mid. Sept. & mid. Mar.

It is soon apparent that the stationary points coincide with periods when the Earth is in quadrature to each star in turn. This pattern is referred to in Bradley's account of the 'new discovered motion of the fixed stars'.[22] He wrote,

after I had continued my observations a few months, I discovered what I then apprehended to be a general law, observed by all the stars, viz. That each of them became stationary, or was farthest north or south, when they passed over my zenith at six of the clock, either in the morning or evening. I perceived likewise, that whatever situation the stars were in with respect to the cardinal points of the ecliptic, the apparent motion of every one tended the same way, when they passed my instrument about the same hour of the day or night; for they all moved southward, while they passed in the day, and northward in the night; so that each was the farthest north when it came about six of the clock in the evening, and farthest south when it came at six in the morning.[23]

Bradley determined that the motion he first observed in γ Draconis conformed to a general law obeyed by all of the stars he observed.[24] Early in his investigation Bradley remained uncertain about the parameters of the motion he observed in these stars. He first presumed that the magnitude of each star's motion was connected with the declination of the star. This suspicion proved to be unsustainable. With this in mind, Bradley's initial presumption that the magnitude of each star's motion was connected with its position (in this case) north of the equator, was quickly rejected. I present the data in order of increasing declination to make plain that there is no connection between the declination and the maximum of aberration.[25]

Star	Declination (north)	Max. of aberration
α Aurigæ (*Capella*)	46° 00′ 55″	8.07″
η Ursæ Majoris	49° 17′ 50″	17.81″
α Persei	49° 52′ 23″	11.43″
γ Draconis	51° 29′ 22″	19.30″
35 Camelopardalis	51° 34′	9.44″
β Draconis	52° 17′ 59″	19.80″
τ Persei	52° 46′ 35″	12.87″
γ Ursæ Majoris	53° 40′ 34″	16.67″
ε Ursæ Majoris	55° 56′ 30″	18.03″
α Cassiopeiæ	56° 33′ 23″	16.55″
β Cassiopeiæ	59° 10′ 09″	17.51″

The positions of the stars are in accordance to epoch 1950. It does not change the argument. The positions of even the fleetest of stars, due to their proper motions, are mostly indistinguishable to the naked eye over a period of two centuries. The maxima of annual aberration are to a degree of precision not possible with the technology at Bradley's disposal. Again, this does not affect the argument. Moreover, during the early eighteenth century motions were recorded according to the *Orbis Magnum,* that is to the mean diameter of the Earth's orbit. Modern practice relates such data to the astronomical unit, that is the mean semi-diameter of the Earth's orbit. Bradley would have measured the motion of γ Draconis in declination over the course of the year, being about 39″; modern practice based on the AU calculates the annual aberration at 19.30″. Whether the standards of the eighteenth century are used, or those of contemporary practice it is clear that there is no connection between the maximum of a star's annual aberration and its declination.

It was while Bradley was in the midst of this extensive inquiry into the annual motions of about seventy stars that his partner Molyneux intended to observe the object star together with *Telescopium in Auriga,* the star Bradley usually referred to as *Anti-Draco,*[26] to act as a control for the observations taking place at Wanstead. However, just seventeen days before he assisted Bradley and Graham with the installation of the Wanstead sector Molyneux was appointed one of Lords of the Admiralty.[27] This deprived Molyneux of almost all of his free time. His recorded observations fell away radically during the rest of the year. He was only able to observe on six nights on 3 and 4, and 23 and 27 September 1727, then 1 October 1727 and finally on 29 December 1727. The sector then fell into disuse. Apparently Molyneux was thwarted by technical problems and he lost touch with the work that Bradley was now undertaking at Wanstead. He did not visit Wanstead again. Not only did his workload intrude, but he may also have lost interest in the project now that he could no longer achieve his earlier objective of invalidating Newtonian natural philosophy. Bradley was visited and supported at various times by Edmond Halley, Lord Cavendish, Dr Benjamin Hoadly, the son of the bishop, and other notable figures. There can be little doubt that the experiment interested many members of the Royal Society.

The death of the Honourable Samuel Molyneux took place on 13 April 1728, a few days after he suffered a fit and collapsed in the House of Commons in the middle of a debate. For a few days Bradley did not make any observations. It seems he may have entered into a period of mourning for his friend and former patron.[28] Molyneux's funeral was not attended by his widow, Lady Elizabeth Diana Molyneux, nee Capel, eldest daughter of Algernon Capel the 2nd Earl of Essex. Her father, a member of the Kit Kat Club, was a man who courted scandal. His eldest daughter Elizabeth maintained the family's reputation. The day following her late husband's death she eloped with her lover Nathaniel St Andre who had been Molyneux's carer, to Southampton. Molyneux was badly served by St Andre. He entangled him in the Toft affair. It was claimed that a young illiterate woman from Godalming was giving birth to live rabbits which Molyneux under St Andre's prompting naively thought he had observed. It opened him to public ridicule even before it was exposed as a fraud. St André was convinced that it was genuine, believing they were supernatural births.

The result for Molyneux not only opened him to derision in the press, but as a prominent Whig politician, to the acerbic pen of Tory authors such as Jonathan Swift, the Irish satirist and wit. Because Molyneux died in suspicious circumstances St André was placed on trial for his murder though he was found not guilty.

On 23 April 1728, in the month of Molyneux's death, after more than six months of continuous observation, the patterns of the motions of the stars were becoming apparent. Bradley extrapolated their motions relative to the observed motions of γ Draconis at Kew from 20 December 1725. He began to discern the parameters of 'the laws of the counterintuitive motion' with greater clarity. However, he remained perplexed by the causes of the phenomenon. The earlier conjectures concerning annual nutation or atmospheric refraction had long been abandoned.

Before continuing with my account of Bradley's discovery of the aberration of light I think it necessary to examine the unwarranted but influential passage included in Thomson's *History of the Royal Society*,[29] which unfortunately was included in Rigaud's *Memoirs of Bradley*,[30] giving it an importance it has never deserved. It reduces Bradley's discovery as the result of 'an accidental observation'. In order to counter the unjustified importance of this celebrated passage I have gone to great lengths to emphasize the importance of the work that Bradley undertook over many years, observing the motions of the satellites of Jupiter.

Thomson refers to a pleasure trip by boat on the Thames. The passage, included in the *Memoirs*, continues,

> The boat in which they were was provided with a mast, which had a vane at the top of it. It blew a moderate wind, and the party sailed up and down the river for a considerable time. Dr. Bradley[31] remarked that every time the boat put about, the vain at the top of the boat's mast shifted a little, as if there had been a slight change in the direction of the wind. He observed this three or four times without speaking; at last he mentioned it to the sailors, and expressed his surprise that the wind should shift so regularly every time they put about. The sailors told him that the wind had not shifted, but that the apparent change was owing to the change in the direction of the boat, and assured him that the same thing happened in all cases. This accidental observation led him to conclude, that the phenomenon which had puzzled him so much was owing to the combined motion of light and the earth.[32]

This all too often repeated passage has troubled me for many years. I first discussed it in my master's dissertation on Bradley's discovery of the aberration of light, completed in 1994. Scholarship demands a balance between receptiveness and scepticism. This account of a boating party is one that elicits scepticism. The events described supposedly took place during the summer of 1728, yet Thomson's passage was published in 1812. Was there an oral tradition that neither I nor other authors have ever traced? What is its value if it was written eighty-four years after the event? With so many unresolved questions concerning the credibility of this passage I turned to the only primary source. This is Bradley's own published account. I scrutinized not merely the published paper but also the three unpublished versions that were preserved in Bradley's aberration papers. I transcribed and edited all three of them. Though time consuming, involving several visits to the Bodleian Library, this proved to be invaluable.

In the final unpublished version of Bradley's discovery of 'the new found motion of the fixed stars'[33] which is included as Appendix 2 to this work, I have strong documentary reasons to believe it was written later than the version that was published.[34] The published version has a postscript detailing observations made by Flamsteed. The third unpublished version dated 5 January 1728/29 OS incorporates Flamsteed's work in the body of the account, indicating that it was written later than the hurried version published by the Royal Society. I regard the last unpublished version as Bradley's intended final account. The printed version was published under time pressure from Halley, who was fearful that his protégé might lose priority for the discovery of such a major discovered motion of the stars. With the words 'At last after several fruitless conjectures I hit upon one which if it be allowed me',[35] he began with the conjecture that the communication of light was not instantaneous.

In the third unpublished version Bradley wrote,

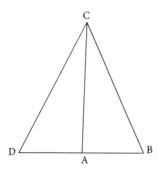

Bradley's geometrical argument that demonstrates the aberration of light

fol. 22r

At last

fol. 22v

I conjectured that all the phenomena hitherto mentioned proceeded from the progressive motion of light and the Earth's annual motion in its orbit about the Sun. For I perceived that if light was propagated in time, the apparent place of a fixed object would not be the same when the eye is at rest, as when it is moving in any other direction than that of the line passing through the eye and object, and that when the eye is moving in different directions, the apparent place of the object would be different.

I considered this matter in the following manner. I imagined CA to be a ray of light proceeding from an object and falling perpendicularly upon the line BD; then if the eye is at rest in A the object must appear by that ray in the direction AC, whether light is propagated in time or an instant. But if the eye is moving from B towards A, and light is propagated in time; the velocity of light will be to the velocity [of] the eye in a given ratio (suppose as AC to AB) and whilst the eye moves from B to A light will move from C to A so that particle of light by which the object will be discerned when the eye by its motion comes to A, is at C when the [eye] is at B. Joining the points C, B, we may suppose the line BC to be a tube (inclined to the line BD in the angle CBD) of such a diameter as to admit of but one particle of light at C (by which the object will be seen when the eye in its motion arrives at A) will pass through the tube BC if it is inclined to BD in the angle CBA, and accompanies the eye in its motion from B to A. But it cannot come to the eye placed behind such a tube, if

fol. 23r

It has any other inclination to the line BD. If instead of supposing CB so small a tube we imagine it to be the axis of a larger, then the particle of light at C cannot pass through that axis unless it is inclined to BD in the angle CBD. In like manner, if the eye moved the contrary way from D towards A with the same velocity, then the tube must be inclined in the angle BDC. So that, though the true or real place of an object perpendicular to the line in which the eye is moving, yet the visible place thereof will not be so, since that, no doubt, must be in the direction of the tube; but the difference between the true and apparent place will be (caeteris paribus) greater or less according to the different proportion between the velocity of light and that of the eye; if light is propagated in an instant, then indeed there can be no difference between the real and visible place of an object although the eye is in motion; for in that case, AC being infinite with respect to AB, the angle ACB vanishes. But if light is propagated in time (which I presume will readily be allowed by most of the philosophers of this age) then it is evident that there will be always a difference between the true and visible place of an object, unless the eye is moving directly towards or from the object; and in all cases the sine of the difference between and the real and visible place of the object, will be to the sine of the visible inclination of the object to the line in which the eye is moving, as the velocity of the eye to the velocity of light.[36]

Bradley continued with his argument in simple geometry, contending that the angle will vary according to different proportions between CA and AB, the velocity of light and that of the eye (as it is carried by the earth in its orbit around the Sun). He added that if the communication of light from the object to the eye be instantaneous, which he confessed to be an extraordinary sup-position,[37] there would be no difference between the *real* and the *visible* place of an object. Bradley recognized that the motion of the Earth was based on the supposition that the velocity of light was finite and that the finite velocity of light was based on the supposition that the Earth was in motion![38] Neither proposition had been established independently of the other with certainty, although Bradley included an 'equation of time' in his tables of the motions of the four Galilean satellites of Jupiter. In view of this, he concluded his paper with the following challenging remarks addressed to Edmond Halley,

> There appearing therefore after all no sensible parallax in the fixed stars,[39] the Anti-Copernicans have still room on that account to object against the motion of the Earth; and they may (if they please) have a much greater objection against the hypothesis by which I have endeavoured to solve the forementioned phenomena by denying the progressive motion of light as well as that of the Earth. But as I do not apprehend that either of these postulates will be denied me by the generality of the astronomers and philosophers of the present age, so I cannot doubt of obtaining their assent to the consequences which I have deduced from them, if they are such as have the approbation of so great a judge of them as yourself.[40]

Bradley's own account is so lucid and straightforward that there is no need for any unsupported accounts like Thomson's. All that is regaled by Thomson is a vague account that can leave a reader perplexed. One may as well say the sailors have a claim! At least the simple geometrical account given by Bradley revealed how the object star moved in the opposing direction expected from parallax. My real objection to the Thomson passage is that he ascribes Bradley's discovery to an 'accidental observation'. It sits well with the commonly found statement in some secondary and tertiary sources, that Bradley's discovery of annual aberration was a 'lucky' consequence, following his 'failure' to observe annual parallax. Bradley's discovery of the aberration of light wasn't the consequence of some fortuitous observation. It was the result of years of disciplined, accurate, and precise observation, the culmination of which was a clearly designed programme of observation of over seventy circumpolar stars at Wanstead, leading to the convincing conclusion he fully expected.

This reiterative approach enabled Bradley to determine the laws and parameters of the observed phenomenon. It was a consequence of a wider programme of research utilizing skills and insights derived from years of observation and calculation, including the motions of the satellites of Jupiter. Bradley determined an equation of time which was the product of the Earth's and Jupiter's motions around the Sun and the progressive motion of light. Römer initially determined the finite velocity of light in 1676 though it met with little favour in Paris where Cartesian natural philosophy upheld the fundamental attestation of the instantaneous velocity of light. Römer calculated that light travelled the distance between the Sun and the Earth in eleven minutes. From his discovery of the aberration of light Bradley was able to calculate a more accurate and precise parameter when determining that light took 8 minutes and 13 seconds to travel a distance equal to the mean distance from the Sun to the Earth, a distance now defined as an 'astronomical unit'.

The recipient of Bradley's paper was the man who set him the task of observing the satellites of Jupiter. Although I cannot always agree with the effusive language used by Bradley's earliest biographer, his appraisal of the achievement of first isolating the phenomenon later called the aberration of light, before deducing its physical causes, must at least lead to some agreement with Rigaud when he stated that,

We naturally feel anxious to observe all the various workings of a mind which seized the clue as soon as the scarcely perceptible extremity of it presented itself, which with unremitting diligence pursued the path to truth, and gradually subdued every difficulty, till the whole was brought out in that clear and brilliant light which leaves us almost in wonder at its having remained so long in obscurity.

It is a common characteristic of so many important scientific breakthroughs that they evince a simplicity that often surprises the inquiring mind. In retrospect, they can appear obvious. Of course this view of hindsight is misleading. Bradley's perceptive mind avoided 'seeing what people wanted to see': annual parallax. It is why the phenomenon led to various unfounded claims of annual parallax or was ascribed to observational error. Astronomers had observed aberration for decades, but until Bradley recognized what he and his partners were observing, a 'new discovered motion of the fixed stars', different from annual parallax, the phenomenon lay hidden in open sight.

Bradley's first response was to attempt to understand this motion, for whatever else it was it could not possibly be annual parallax. Bradley, as we have seen, was not a man who jumped to rapid conclusions. His understanding included a sound working knowledge of the geometry of motions due to parallax. Not surprisingly the partners, Molyneux, Graham, and Bradley, first suspected instrumental or systemic error. This degree of modesty by Graham and scepticism by Bradley speaks much for the 'apparent' disinterested approach pursued by Graham and Bradley. However, were Bradley and Graham as impartial and unbiased as they appeared? As their responses to the hypotheses of the annual nutation or atmospheric refraction demonstrated, they were firmly committed to the arguments contained in the *Principia*. It is difficult to determine where Molyneux's commitments lay, beyond a sense that he sought some form of discomfort or embarrassment for Newton. That Molyneux initiated an experiment that might have caused distress to the President of the Royal Society reveals something about his motives. Did he really comprehend this? Was he fully apprised of the issues concerning the debate about the distances to the stars, and the possible consequences with regards to the validity of Newton's *Principia*?

There is a passage in Bradley's paper which I have referred to. I quote from the third unpublished version of his paper,[41] written later than the version published in the *Philosophical Transactions*[42] which has since remained the only version available to the public. Bradley writes as will be found in Appendix 2,

> Before I proceed to give you the history of the observations themselves, it may be proper to let you know, that they were at first begun in hopes of verifying and confirming those that Dr. Hooke formerly communicated to the public; which seemed to be attended with circumstances that promised greater exactness in them, than could be expected in any other that had been made and published on the same account.[43]

Earlier I asked why Bradley of all people, whose methodology was based on reiterative and precise observation, entertained hopes of verifying and confirming Hooke's long-abandoned claims for the discovery of annual parallax? But now I propose examining why Molyneux, the initiator of the Kew experiment, sought to verify and confirm Hooke's claim. Given the sorry history of the relations between Hooke and Flamsteed, it may seem surprising that Molyneux might support Hooke. It leads to a supposition that he was largely ignorant of the role Hooke played in tormenting his late friend Flamsteed. However, the words are Bradley's and it is his account that is under question. So what is Bradley seeking to accomplish by expressing words that he couldn't possibly have believed and which, if they had been fulfilled, would have undermined some of his most fundamental philosophical commitments? The reasons lie elsewhere. When two members of the Council of the Royal Society join another respected fellow of the society in an attempt to observe the annual parallax of a star using Robert Hooke's method, possibly confirming the former Curator of Experiments claim, it will have attracted the attention of many in the society. Advancing age and increasing infirmity now prevent me from examining the relevant archives of the Royal Society for documentary evidence to support my argument. I must contend, however, that circumstantial evidence can suggest that the Kew experiment attracted widespread attention within the Royal Society. When Bradley wrote his account he was in a sense hoisted on his late friend's petard. He had to remind his fellow members of the validity and promise of using Hooke's method. With this, Bradley could have nothing but approval. Many within the society and the astronomical community would have been aware that it was Molyneux who was the lead investigator and the person who commissioned Graham to construct a 'parallax instrument' in order to confirm *Hooke's result*. Bradley and Graham believed *Hooke's method* was the most likely to lead to a positive result, that is, of observing annual parallax at the zenith.[44]

Circumstantially, it would appear that Molyneux was more concerned with the intention of confirming Hooke's result in order to embarrass Newton by confirming the 'short distance scale' of the stellar universe, which would threaten the validity of Newton's 'system' as he termed it. I have real doubts whether he had any solid grasp of the arguments in the *Principia* but I'm sure he understood that if Hooke's result was confirmed, it would pose a threat to Newton's theories. Flamsteed, his late friend, held a low opinion of Newton's 'theories' which he held to be of less consequence than his own observations. Molyneux possibly shared something of Flamsteed's modest opinion of Newton's great work. In support of this it is well to remember that one of Newton's most perceptive critics George Berkeley, was Molyneux's tutor at Trinity College, Dublin. Molyneux was quickly disabused of his hopes of observing annual parallax when Bradley observed the motions in declination of γ Draconis from 17 December 1725. But once Molyneux believed that the observed motions could be reduced to an annual nutation that solidly contradicted Newton's calculations, his object of causing distress to the President of the Royal Society was quickly rekindled. That is why Molyneux acted so precipitately when the nutation remained a viable hypothesis during the first two months of 1726. When Bradley wrote his account, the objects of the experiment were common knowledge to many in the Royal Society, as were the hypotheses suggested and abandoned with alacrity. I have no wish to denigrate Molyneux's character. I am sure, both from his experience, and by reputation, he thought Newton was capable of vicious behaviour which I surmise he greatly disapproved of.

I can discuss something that is central to my account of Bradley's astronomical career. A comparison of the observational abilities of Flamsteed and Bradley suggests that they were at the cutting edge of their science during their lifetimes, but Bradley was much more successful in gaining patronage and support. In character the men were almost polar opposites. Flamsteed was defensive and argumentative. Bradley was more open. Flamsteed had few supporters. Bradley gained many important patrons, friends and supporters: he was as Sir William Jones was to say later, a man without enemies. Even a non-juror like Hearne altered his opinions of the man, from dismissal after his inaugural lecture as Savilian Professor of Astronomy at Oxford in 1722, to becoming his sole elector when Bradley sought election as the Keeper of the Ashmolean Museum in 1731. Bradley was deferential and unlikely to take offence, whereas his predecessor at Greenwich was, unfortunately, a man who easily took offence and subsequently retained few friends. Therein lies the tragedy of Flamsteed's life. He spent the final decade of his life defending his reputation. If he had garnered friends in the way that Bradley did, his reputation would never have been questioned. Much of Flamsteed's worth was vindicated after his demise, by friends and supporters including Abraham Sharp. The strength of character and determination of Margaret Flamsteed was surely admirable. She saw the publication of the authorized edition of her husband's observations, and the star atlas based upon them. The latter remained important for almost a century.

The social and psychological contrasts between Molyneux and Bradley were quite pointed. Following the unexpected death of his uncle, who died intestate, Bradley had been left in a fragile financial position. Bradley was reliant on his uncle's widow Elizabeth, who accepted him as a member of her household. Bradley came from a social and economic background where nothing was certain. After the sudden death of his uncle, his life, in spite of his chair at Oxford, became even more uncertain. This rather unbalanced relationship between Molyneux and Bradley must be acknowledged before examining the way the Kew investigation was conducted. This is why I have queried Molyneux's motives, the conjectures that were made, and why Bradley appeared to accept them so passively, even though he must have realized they would not stand up to any extensive analysis or investigation. Herein lies the surest guarantee that regardless of the various conjectures that were made, Bradley would have insisted that the investigation would always be pursued to its conclusion.

Molyneux, unfortunately, did not survive to the resolution of the problems that the partners faced just before Christmas 1725. Three years later before Christmas 1728, Bradley was composing the paper on what was then referred to as 'the new discovered motion of the fixed stars'. Without Molyneux's insistence that Graham construct and then suspend a 24-foot 'parallax instrument' at the White House,[45] his mansion on Kew Green, who knows how long it would have remained before the aberration of light was isolated and accounted for? Would it have been discovered by Bradley? Little wonder then that Halley chivvied Bradley into writing his account as quickly as possible, lest another equally perceptive, but as yet unknown astronomer, claim the prize. It explains why the published paper was concluded hurriedly with a short postscript referring to observations made by Flamsteed. This information was possibly passed on by Halley whose knowledge of claims for parallax was, as his paper on Jacques Cassini's attempt reveals, well informed. In the third unpublished version (Appendix 2), this reference to the work of the first Astronomer Royal is included in the body of the paper, proof if any was needed, that the final unpublished version was as near to being Bradley's definitive account as we are ever likely to read.

I will conclude my examination of Bradley's discovery of the aberration of light, with a discussion of its reception, particularly in the Papal States in Italy where it was impossible to publish works that supported the Copernican thesis. The reception of Bradley's paper amongst the learned in this most Catholic of domains must have been disturbing. Yet Bradley, forever open to other interpretations of his observations,[46] laid it open for the 'Anti-Copernicans' to challenge his own. In Italy there was an astronomer with the required skill and experience to challenge Bradley's interpretations. Eustachio Manfredi had been observing the motions of stars in right ascension that he called 'aberrations' since 1707. A treatise entitled *De annuis inerrantium stellarum aberrationibus* (*The annual aberrations of the fixed stars*), completed in 1721 by Manfredi, remained in the possession of the censors in Rome, who were fearful, according to O'Connor and Robertson,[47] that 'his results might be used by those wishing to show the earth was not stationary'. If this is what the censors thought, I can only surmise that they were rather confused.[48] In the introduction to his treatise, Manfredi asserted that his investigations success-fully challenged Copernican theory. There can be absolutely no doubt about Manfredi's intent. He wrote,

> There are some however who, noticing that these strayings of stars recur at fixed times of the year, think there is no doubt the matter has to do with an annual movement which with Copernicus they attribute to the earth, and so they proclaim that a true universal system is disclosed and demonstrated by this very piece of evidence, in which however it is easy to be deceived, since the Copernican system is tottering together with the arguments by which they think it is supported, as I am confident that I shall show in this very work.[49]

Before he discussed his observations of aberrations[50] in right ascension in the final chapter of his treatise, Manfredi examined the observation of aberrations in declination, the subject of his penultimate and eighth chapter. To mea-sure motions in declination, as Bradley did, requires the use of an instrument capable of measuring motions north and south with great precision. What is required is a single instrument of great reliability and accuracy: a transit instru-ment or a zenith sector. To measure motions in right ascension, as Manfredi did, requires the use of two instruments: a quadrant or an instrument capa-ble of duplicating some of the functions of a quadrant and an accurate clock or regulator, a timepiece of reliability.[51] Bradley possessed the advantage of a precision instrument constructed under the supervision of a man who was regarded as the finest instrument maker in Europe. Manfredi had access to a sizable fixed quadrant of 2.5 metres[52] diameter and a clock of indeterminable reliability. To measure celestial positions in right ascension a reliable time-piece is a minimal requirement. Manfredi's observational abilities are clearly displayed in his treatise. The accuracy of the instruments he had at his com-mand may have been less reliable. French astronomers of this period regularly complained about the quality of the work of the artisans who provided them with their instruments. If this was the situation in France, I see no reason to believe the situation was any different in Italy. In both countries, the social status of artisans was little better than that of common labourers. In London and probably in Great Britain as a whole, the social status of skilled artisans was much higher, several being Fellows of the Royal Society, as was George Graham.

It was through the agency of Flamsteed's letter of 20 December 1698 to John Wallis[53] that Manfredi obtained the data of Hooke's four observations of 1669.[54] He conceded that Hooke's results did not obviously contradict the theory of annual parallax.[55] Manfredi analysed the observations made by astronomers of 'aberrations' of the pole star including those made by Picard, Flamsteed, and Cassini with whom as a fellow Bolognese he had been in correspondence for many years. Manfredi expressed a forceful criticism of Flamsteed's claim for the discovery of annual parallax, based on observations of *Polaris* up to and including 1696.[56] Manfredi concluded,

> in the observation of the 14th or 16th, the actual direction of the aberration is at variance with annual parallaxes, since from that day to December 2nd, the star moved according to the observations 48″ to the north... So it seems that a reason must be sought for the movements of this star other than the annual rotation of the earth around the sun.[57]

Although he mentions that Picard had noted the fact that these 'aberrations' of *Polaris* did not appear to conform to the theory of annual parallax in the *Voyage d'Uranibourg*[58] and that Jacques Cassini had successfully challenged Flamsteed's claim in a paper addressed to the Royal Academy of Sciences in Paris in 1699.[59] Manfredi did not successfully determine the possible causes 'for the movements of this star'.

In the final chapter of his treatise, Manfredi discussed the results of his investigations, turning his attention to the observations of 'aberrations' of stars in right ascension. He wrote,

> Rather more laborious is the consideration of those aberrations which happen to the fixed stars in ascension, since of course they hardly maintain the same straight ascension or arc of distances from the colures[60] at all, but wander somewhat to the east or west. I am sure that this was noticed by no astronomer before the celebrated Maraldus,[61] on comparing the times at which *Sirius* and *Arcturus* cross the meridian, found a *varying* difference between those times on various days of the year, and warned me about this in his letter in the year 1707. He also mentioned that he had observed wanderings of this kind not only in the case of these stars, but in others besides,

which first induced me too to attempt the same investigation at the Marsilian Observatory with Stancarius[62] and in this public observatory of the Institute of Sciences.[63]

Working with several assistants including Francesco Algarotti (1712–1764),[64] Manfredi observed the aberrations of many so-called 'fixed' stars at various latitudes and succeeded in identifying many such motions that ironically for Manfredi were used as corroborative evidences in support of Bradley's hypothesis of the new discovered motion of the fixed stars. In explaining his method, Manfredi revealed that he did not appreciate the *universality* of the phenomenon he was observing. This lack of recognition may have stemmed from the potential inaccuracy of the method he used. The instrumentation available to Manfredi was inadequate to meet the immense precision that was required before it was possible to recognize that the motions were common to all of the stars in right ascension. However, the real problem with Manfredi's approach lay in his presumptions.

In Manfredi's interpretation of the evidence of such 'aberrations', he was correctly declaring that they invalidated or falsified claims for the discovery of annual parallax by revealing that these observations did not conform to parallax theory. He was convinced that the stars were 'fixed' for the use of the term 'aberrations' indicated that the behaviour of many stars was not 'normal'. Bradley's 'new discovered motion' was interpreted as a *universal* phenomenon. Most of Manfredi's research had been completed by about 1719 before Halley published his discovery of 'proper motions', the separate motions of individual stellar systems. When referring to the motions he was observing, Manfredi wrote,

> we are forced to acknowledge that it is impossible to judge which of the two stars the error[65] is to be attributed, to what those aberrations we are seeking, or to the other which we are comparing it, or perhaps to both. In this uncertainty nothing better occurs to me than if we choose two or more of the fixed stars, whose intervals of arrival at the meridian are found by very many precise observations made at different times of the year [to be] perpetually the same.[66] ... I do not doubt that this happens to many among the vast number of stars... and refer to these or to some one of them, the times of that star, whose aberration we have undertaken to investigate and define by measurements.[67]

In contrast with Manfredi's belief that many stars really must be 'fixed' or at least mutually stationary, and that aberrations were not universal, Bradley presumed that the phenomenon of 'the new discovered motion' *was* universal. Bradley wrote,

> At last I conjectured that all the phenomena mentioned proceeded from the progressive motion of light and the earth's annual motion in its orbit. For I perceived that, if light was propagated in time, the apparent place of a fixed object would not be the same when the eye is at rest, as when it is moving in any other direction than that of the line passing through the eye and object; and that when the eye is moving in different directions, the apparent place of the object would be different.[68]

Bradley's far-reaching perception was based on a distinctive way of thinking about 'the laws of nature'. To Bradley, the universe was constructed on cogent and lucid grounds.[69] His initial presumption when he first discovered the counter-intuitive motion of γ Draconis on 20 December 1725 through to 25 December 1728 was to seek a universal explanation or a 'law of nature'. His observations at Kew up to December 1726 confirmed that the phenomenon though inexplicable, was nevertheless predictable. He initiated a new investigation at Wanstead which involved the observation of the motions of about seventy circumpolar stars which passed through the meridian of his zenith sector within $6\frac{1}{4}°$ of the zenith.[70] From these extended observations of the motions of many stars Bradley was able to deduce a law that would account for the observable motion of any object in the sky.[71] So far as it is possible to determine, Manfredi remained until his death on 15 February 1739, a firm supporter of a geocentric conception of the cosmos, whether out of personal conviction which I believe to be the case, or out of loyalty to the teachings of the Church of Rome. Even so, there remains much confusion about the nature of Manfredi's work on aberrations.

It is unfortunate that some authors have confused Manfredi's work on aberrations with Bradley's work on the new discovered motion. A recently published paper asserted that Manfredi provided, 'the first demonstration, though unsought, of the revolution of the Earth around the Sun, and thus the reality of a heliocentric system'.[72]

This clearly contradicts Manfredi's *own words* in the treatise that was finally published in Bologna in May 1729. I think I have demonstrated that the Italian astronomer never appreciated the universality of the phenomenon he was observing. Moreover, he believed that his treatise clearly revealed that, 'The Copernican system is tottering together with the arguments by which they think it is supported, as I am confident that I shall show in this very work'.[73]

O'Connor and Robertson refer to Manfredi and write, 'he discovered that the stars appeared to move in circular orbits which he realised could not be due to parallax. He had come to these conclusions in 1719 but he did not publish these results until ten years later in *De annuis inerrantium stellarum aberrationibus*'.

Later in the passage they add, 'Although Manfredi had, earlier than Bradley, discovered the same phenomenon and given it the name "aberration", he could not accept Bradley's explanation. Manfredi remained throughout his life a believer in a stationary earth in the centre of the universe'.

The point at issue here is, *were they were observing the same phenomenon?* Manfredi was observing 'aberrations' in right ascension. Bradley was observing motions in declination. Manfredi was observing these abnormal 'aberrations' in order to disprove all prior claims to the discovery of annual parallax, in order to invalidate the Copernican system. Bradley was observing these counter-intuitive motions precisely *because* they contradicted the theory of annual parallax. He sought to resolve the problem posed by the phenomenon because he was certain that the Earth was in orbit around the Sun. In seeking to resolve this problem, he discovered the aberration of light. It was widely accepted as the long-sought after direct observational evidence of the motion of the Earth around the Sun. Initially of course Bradley called the phenomenon 'the new discovered motion' and left it open to the 'anti-Copernicans' to challenge his conclusions. No one was ever able to explain the phenomenon in geocentric terms.

In conclusion, I will refer to Bradley's discovery, explaining why the term 'aberration of light' was used to describe the 'new discovered motion of the fixed stars'. I will refer to two *transitional aberration* documents that are included as Appendix 3 of this book. In these documents, Bradley referred to 'parallax on account of distance' which is what we now term 'annual parallax'. He referred to 'the parallax of the stars on account of the motion of light' which is what he provisionally called 'the new discovered motion'. Bradley clearly recognized that this terminology was confusing, particularly as 'parallax on account of the motion of light' was nothing of the sort. This document, judging by the ink and paper used,[74] was written at the end of 1729 or during 1730, which was after Manfredi's treatise was finally published. Bradley, being aware of Manfredi's terminology, realized that it enabled him to distinguish between *annual parallax* and what he soon termed *annual aberration*. Folio 56 of Bradley MS20* is entitled 'Rules for computing the Aberration or Parallax of the Stars arising from the Motion of Light'.

The 'aberrations' observed in right ascension by Manfredi were made with the clear purpose of invalidating all claims for the discovery of annual parallax. He demonstrated that these claims were false and he believed thereby he had demonstrated that the Copernican system was 'tottering'.[75] Yet nearly all his observations were compatible with those made by Bradley in declination. Indeed, Bradley was later to reveal that not only were Manfredi's observations a confirmation in right ascension of his own in declination, he absolutely agreed that they contradicted the theory of annual parallax. Bradley must have possessed a high regard for his older Italian contemporary. Manfredi was elected a Fellow of the Royal Society in 1729. This was through an introduction to the Society by the Italian based Roman Catholic non-juror Sir Thomas Dereham (1678–1739) a Fellow since 1720 who was an informal representative of the 'Old Pretender', known to his supporters as James III (1688–1766). A close confidant and friend to Pope Clement XII, Dereham was actively involved in reporting scientific developments in Italy to the Royal Society, as well as interpreting *Philosophical Transactions* into Italian from 1729. From 1722, he was in correspondence with James Jurin, secretary of the Royal Society, as well as occasionally with Newton. Manfredi was a supporter of Newton's theories of light and did much to propagate many Newtonian ideas in his native land. Bradley and Manfredi never met nor as far as I am aware did they ever correspond.

Notes

1. James Bradley, A Letter to Dr Edmond Halley, Astronomer Royal and etc. Giving an Account of a New-Discovered Motion of the Fixed Stars, in *Philosophical Transactions*, Vol. XXXV, 1729, pp. 637–661.
2. There was some hesitation within the domains of Papal Italy. Acceptance of the theory came with an acceptance of the motion of the Earth, the very thing for which Galileo Galilei had been found guilty of, just 96 years earlier. Jacques Cassini and others supporting Cartesian physics resisted Bradley's theory because they were unable to accept the finite or the progressive motion of light. Cartesians like Cassini postulated that the velocity of light was infinite or instantaneous.
3. Galileo Galilei, 'and yet it moves' or in the original Italian, 'e pur si muove', words attributed to Galileo after he had been found guilty of the vehement suspicion of heresy by the Holy Office of the Inquisition in Rome in 1633.
4. The Paris Observatory did not have a director as such at first, but as an Academician Cassini spent much of his time directing the activities of the observatory. Later his grandson Cassini de Thury was appointed as the first official director.

5. Robert Grant, *History of Physical Astronomy from the Earliest Ages to the Middle of the Nineteenth Century*, London, 1852, p. 82.

6. Robert Grant adds a footnote, 'Strictly speaking, the equation is equal to only half this quantity' but in the tables of the satellites the coefficients are doubled in order to render the results always additive.

7. Edmond Halley never accepted James Bradley's claim that he could observe motions as small as ½ an arcsecond. Halley, as revealed in Bradley's observations of aberration at Wanstead, when he occasionally acted as his protégé's assistant, revealed serious shortcomings in his observational technique, which were recorded in the Wanstead Observation Book. See Halley's entries during September 1727.

8. I refer of course to the three inner Galilean moons, not to the various smaller bodies discovered since the nineteenth century, and particularly during the period at the turn of the twentieth and the twenty-first centuries, when space probes explored the system of Jupiter.

9. Ibid. p. 82.

10. The first satellite (Io) is the most volcanically active body in the solar system, the second satellite (Europa) has what appears to be a global ocean capped by pack ice, and the third satellite (Ganymede) also now appears to show evidence of a sub-surface oceanic layer as well as a magnetic field. Most of these phenomena are the consequence of tidal flexing induced by mutual gravitational forces, being endlessly repeated.

11. No woman at this time would ever be considered for any such public position.

12. In remarks shared whilst attending a day conference at the Royal Society.

13. The first measurements of annual parallax date from 1837, all of which were less than a second of arc.

14. Stephen Peter Rigaud, *Memoirs of Bradley*, p. xxiii All three transcriptions are included in John Fisher, *James Bradley and the New Discovered Motion: The Origins, Development and Reification of James Bradley's Hypothesis of the New Discovered Motion of the Fixed Stars*, appendix 1, The First Unpublished Version, pp. 53–64, appendix 2, The Second Unpublished Version, pp. 65–99, appendix 3, The Third Unpublished Version, pp. 100–134. London University, 1994.

15. James Bradley, *Miscellaneous Astronomical Observations* in *Miscellaneous Works and Correspondence of James Bradley*, Oxford, 1832, p. 356.

16. James Bradley, Memoranda Respecting the Instrument at Wansted, in *Miscellaneous Works and Correspondence of James Bradley*, Oxford, 1832, p. 195.

17. A. D. Andrews, Cyclopedia of Telescope Makers, Part 2 (G—J), p. 42, in *Irish Astronomical Journal*, Vol. 21, No. 1, 1993, p. 1

18. James Bradley, *Miscellaneous Astronomical Observations* in *Miscellaneous Works and Correspondence of James Bradley*, Oxford, 1832, p. 195.

19. James Bradley, Observations on the Fixed Stars made at Wansted in Essex by J. Bradley in James Bradley, *Miscellaneous Astronomical Observations* in *Miscellaneous Works and Correspondence of James Bradley*, Oxford, 1832, p. 202.

20. By this we mean stars down to magnitude eight, as evidenced by the catalogue of Bradley's stars at Greenwich, included in Bessel's *Fundamenta Astronomiæ* published in 1818.

21. Refer to the evidence included in the Chronology in Appendix 1

22. James Bradley, A Letter to Dr Edmond Halley, Astronomer Royal and etc. Giving an Account of a New Discovered Motion of the Fixed Stars, in *Miscellaneous Works and Correspondence of James Bradley*, Oxford, 1832, pp. 1–16.
23. Ibid. p. 5.
24. Ibid. p. 5.
25. Of course at this stage, the term 'aberration' lay in the future. I use the term for the clarity it offers.
26. The star is now generally referred to as HR2123 and is located in the constellation of Auriga.
27. Many authorities state that Molyneux was appointed as one of the Lords of the Admiralty, but this term was not fully in use at the time, coming into vogue later during the eighteenth century. The term 'Commissioner' was in common usage at this time. Henceforward Molyneux was addressed as the 'Honourable'.
28. After Bradley's death all of his observations, correspondence and other personal and public items passed to the executors of his Will. Samuel Peach was the subject of a legal dispute with the Board of Longitude over the ownership of Bradley's effects, particularly the observations made by Bradley at Greenwich. None of Bradley's *personal* correspondence has survived. Subsequently it is impossible to trace anything about Bradley's response to the death of his erstwhile partner and former patron. Circumstantial evidence can be inferred from the lack of observations during 10–16 April 1728. Refer to the Chronology in Appendix 1.
29. Thomas Thomson, *History of the Royal Society*, London, 1812, p. 346.
30. Stephen Peter Rigaud, *Memoirs of Bradley*, Oxford, 1832, pp. xxx–xxxi.
31. Bradley didn't receive his doctorate until 1742 when he was appointed as the third Astronomer Royal.
32. Ibid. p. 346.
33. John Fisher, *James Bradley and the New Discovered Motion: The Origins, Development and Reification of James Bradley's Hypothesis of the New Discovered Motion of the Fixed Stars.* Appendix 3, The Third Unpublished Version, pp. 100–134. See Appendix 2 of this book.
34. James Bradley, A letter to Dr Edmond Halley, Astronomer Royal and etc. Giving an Account of a New Discovered Motion of the Fixed Stars, in *Philosophical Transactions*, Vol. XXXV, 1729, pp. 637–661.
35. Neither the progressive (finite) motion of light nor the orbital motion of the earth had been conclusively established at the time of Bradley's paper, even though both were generally accepted. However, his work on the satellites of Jupiter on behalf of Halley, had established beyond reasonable doubt, that the velocity of light was indeed finite, having to include an equation of time to account for it. Halley is the recipient of the paper, who is fully aware of the issues Bradley is discussing.
36. James Bradley, Bodleian Library, Department of Western Manuscripts, Bradley MS20*, fol. 22r–fol. 23r
37. This proposition was supported by many Cartesians, including Jacques Cassini and his supporters, in the Royal Academy of Sciences in Paris. Bradley's account is a polemic directed against current practice in France and many other parts of Europe.

38. Here may be discerned one of the foundations of special relativity. It is an optical phenomenon explained by, or is the consequence of, a kinetic cause.

39. Bradley asserted very confidently that if the annual parallax of the fixed stars was as great as a single second of arc he would have detected it. His assertion concurred with the long-distance estimates made by Newton, rather than the short-distance scales favoured by Hooke, Flamsteed, Picard, and others. The annual parallax of the nearest star, *Proxima* Centauri, part of the α Centauri system, is about 0.75″ vindicating Bradley's confident remark.

40. James Bradley, Bodleian Library, Department of Western Manuscripts, Bradley MS20*, fol. 29v—29r.

41. This is included as Appendix 2 of this study.

42. James Bradley, A Letter to Dr Edmond Halley, Astronomer Royal and etc. Giving an Account of a New Discovered Motion of the Fixed Stars, in *Philosophical Transactions*, Vol. XXXV, 1729, pp. 637–661.

43. James Bradley, Bodleian Library, Department of Western Manuscripts, Bradley MS20*, fol. 18r.

44. Bradley favoured the 'long distance scale' and was aware that Graham's zenith sector might have the capacity to observe motions of less than a second of arc, confirming Newton's argument about the scale of the stellar universe.

45. A profile of the White House can be found on the first page of Rigaud's *Memoirs of Bradley*. It is Rigaud who seriously entertains the view that the aberration of light was discovered in the White House at Kew. This is the reason why Bradley's first biographer includes the illustration in his account of Bradley's life.

46. Bradley was nescient in two distinctive ways. First, he did not know – and this is why he was an empirical investigator. Second, he believed there was a disjunct between the results derived from observation and experiment and the various theories, hypotheses, conjectures and forms of knowledge derived from the results. He believed it was always possible to interpret results in different ways.

47. J. J. O'Connor and E. F. Robertson, School of Mathematics, University of St. Andrews, 2012 July, Biographical sketch of the life of Eustachio Manfredi, 20 September1674 to 16 February 1739. https://mathshistory.st-andrews.ac.uk/Biographies/Manfredi. [Accessed 9 May 2023.].

48. Like many before it appears that the censors believed that any evidence of any motion of the stars was evidence of annual parallax.

49. Eustachio Manfredi, The Latin text reads, 'Sunt tamen nonnulli qui cum illas Stellarum evagationes statis anni temporibus recurrere animadvertant, minime dubitandum existiment quin ea res ad annuum, quem Telluri cum Copernico tribuunt, motum pertineat, atque adeo vel hoc, ipso indicio detectum, demonstratumque verum Mundi systema depraedicent, in quo tamen decipi proclive est, labante Copernicano Systemate una cum argumentis, quibus illi sussultem arbitrantur, in quod in hoc ipso opere demostraturum me esse confide'. For this translation I have to thank my wife's late uncle David Simonson, who was the classics master at the King's School at Gloucester.

50. All of the random observed motions of the fixed stars, whether due to observational or instrumental or systemic errors, or claimed evidences of annual parallax, indeed any observed motion at all, was described by Manfredi as an 'aberration'. His use of

the term 'aberration' must not be confused with the phenomenon that was later called annual aberration or the aberration of light. Manfredi utilized the term 'aberration' in the sense of a moral failing. A few years later Bradley's use of the term 'nutation' was contested by his friend Lacaille. He preferred to use the more equivocal term 'deviation' until he was sure the phenomenon was due to the gravitational effects Bradley ascribed to it. The term 'deviation' also, like 'aberration' possessed moral overtones.

51. By this I mean a timepiece that was *predictable*. Whether the clock gained or lost time is less important than having a clock that lost or gained time predictably. If it loses for instance 3 seconds a day it doesn't matter provided it *always* loses 3 seconds a day *so that all observations may be reduced with confidence*. Whether the clocks in the possession of Manfredi were as predictable as those available to Bradley is doubtful.

52. According to O'Connor and Robertson. Whatever its diameter, it would not have been measured in metres in the period under discussion. However, I will rely upon their information as an approximation.

53. John Flamsteed to John Wallis, 20 December 1698. It was translated into Latin by Wallis and published in *Opera mathematica,* Vol. III, pp. 701–8, Oxford, 1699.

54. Ibid. p. 702.

55. This opinion appears to have been shared by Bradley prior to the Kew observations initiated by Samuel Molyneux in 1725.

56. Eustachio Manfredi, *De annuis inerrantium stellarum aberrationibus,* Caput VIII, Sectio 150, p. 63.

57. Ibid. p. 63.

58. Ibid. p. 64.

59. At the conclusion of Caput VIII, Sectio 150, Manfredi concludes that Picard recognized that this observed motion did not conform to the theory of parallaxes, and that Jacques Cassini 'brilliantly demonstrated' this in 1699.

60. Colures are great circles that pass through both celestial poles. They are found earliest in Ptolemy's *Almagest*. An equinoctial colure passes through both poles and the two equinoxes, the vernal and the autumnal, whereas the solstitial colure passes through both poles and the two solstices. These are the principal meridians of the celestial sphere.

61. Jacques Maraldi, cousin of Jacques Cassini, who worked at the Paris Observatory.

62. Vittorio Francesco Stancari.

63. Ibid. Caput IX, Sectio 153, p. 68. The Latin text reads, 'Paullo operosior est earum aberrationum contemplatio, quae secundum ascensiones inerrantibus Stellis contingent, cum scilicet eandem ascensionem rectam, sive arcum distantiae a coluris, minime servant, sed nonnihil ad ortum, vel ad occa sum evagantur, id quod nemini Astronomorum certo animadversum mihi constat, ante celeberrimum Maraldum, qui collatis Sirii, atque Arcturi temporibus, quibus Meridianum pertranseunt, alium allio anni diebus inter ea tempora differentium invenit, deque eo literis suis me admonuit anno 1707; neque in his tantumodo Stellis, sed in aliis praeterea, ejusmodi excursiones ab se deprehensas significavit, quae res me quoque impulit, ut & in Marsiliana Sepecula cum Stancario, & in hoc publico Scientarum Instituti Observatorio cum Sociis, eandem indaginem aggrederer'. Again I thank the late David Simonson for this translation. See note 49.

64. Francesco Algarotti was the author of one of the best known popularizations of the philosophy of Newton in *Newtonianismo per la dame*. He was a frequent visitor of Voltaire and Emelie du Chatelet at Cirey. He was an Anglophile who was associated with the court of Frederick the Great of Prussia.

65. From this it is easy to perceive that 'aberrations' are 'errors', they are shortcomings, but what of? Is it our ability to perceive the world as it is? Therefore, are 'aberrations' a product of the shortcomings of our senses or our instruments or both?

66. Ibid. Caput IX, Section 155, p. 69.

67. Ibid. Sections 154–155, p. 69. The Latin text reads, '…sateri cogimur ambiguum esse judicium, utri Stellarum, illine, cujus aberrationes quaerimus, anne alteri, cum qua hanc ipsam comparamus, aut forte utrique, error fit tribuendus. Section 155. In hac dubietate nihil mihi fatius occurrit, quam si duas, pluresue ex innerrantibus Stellis seligamus, quarum intervalla adventus ad meridianum plurimis observationibus, iisque exquisitis, ac per diverto Stellarum numero multis earum contingere non dubito, atque ad hasce ipsas, vel ad unam aliquam earum, Stellae illius tempora resenius, cujus aberrationem investigare, ac mensuris praesinire suscepimus'. I refer to the late David Simonson's translations on my behalf. These were undertaken when I was preparing my MSc dissertation in 1994.

68. *Philosophical Transactions,* No. 406, Vol. XXXV, p. 646. It is also to be found in S. P. Rigaud, *Miscellaneous Works and Correspondence of James Bradley,* p. 6.

69. During the eighteenth century natural law was often perceived in terms of providential or rational grounds.

70. Ibid. p. 643.

71. Ibid. p. 644.

72. Andrea Gualandi and Fabrizio Bonoli, Eustachio Manfredi e la prima conferma osservativa della teoria dell'aberazzione anna della luce In *Atti del XXII Convegno SISFA – Università degli Studi di Genova, Genova-Chiavari, 6-8 giugno 2002,* edited by M. Leone, A. Paoletti, and N. Robotti, IISF, Napoli, 2003, pp. 476–481.

73. Eustachio Manfredi, *De annuis inerrantium stellarum aberrationibus,* Caput I. p. 1.

74. Bodleian Library, Department of Western Manuscripts, James Bradley's Manuscripts *MS20*,* fols. 56r and 60v.

75. Manfredi's term in his treatise.

6

The Laws of Nature

James Bradley's Lectures in Experimental Philosophy at Oxford 1729–1760.

> Nature hath no goal, though she hath law.
>
> John Donne.

From 1722, when he delivered his inaugural lecture as Savilian Professor of Astronomy at Oxford, James Bradley spent much of his time in Wanstead, five miles north-east of the City of London. There he lived at the parsonage that came with his uncle's living as the Rector of Wanstead. Later, he apparently worked at his aunt Elizabeth Pound's more modest town house.[1] The Savilian chair entailed modest demands, the delivery of two 45-minute Latin lectures a week during term time. From 1722 to 1725, he was primarily engaged on his extensive observations of the satellites of Jupiter towards the compilation of a table of ephemerides on behalf of his mentor Edmond Halley. From 1725 to 1728, he was engaged largely at Kew, later at Wanstead, on the work that led to the discovery of the aberration of light, the most important innovation in the science of positional astronomy since the micrometer had first been applied to the telescopic sight. From 1728, Bradley was engaged in a more exploratory investigation within the residuals of his observations of aberration, that led to his discovery of the nutation of the Earth's axis.

In 1729, Bradley acquired the post of lecturer in experimental philosophy at Oxford[2] in succession to John Whiteside (1679–1729). Bradley delivered three courses of twenty lectures in experimental philosophy each year. These lectures provided his financial security.[3] After the publication of his paper on 'the new discovered motion' in January 1729 and the acquisition of his new lectureship at Oxford, the increased demands on his time and intellectual energy forced Bradley to move from Wanstead, his home for twenty-one years. He moved more permanently to the dwelling in New College Lane that came with the Savilian chair. It is worthy of note that Elizabeth Pound moved to Oxford with Bradley, continuing to live with him until her terminal illness forced a return to Wanstead and the care of her brother Matthew.

The Life and Work of James Bradley. John Fisher, Oxford University Press. © John Fisher (2023).
DOI: 10.1093/oso/9780198884200.003.0007

It is difficult to determine the exact nature of the relationship that existed between Bradley and his aunt Elizabeth. Under Hanoverian family law, women were forbidden to marry their nephews. I believe it may partially explain why Bradley's personal effects and correspondence was never released into the public domain.

Bradley evinced a mastery of the content of Newton's *Principia* and the *Opticks* as well as many of the ruling texts of contemporary natural science. It suggests that when unable to engage in observation due to inclement weather, he busily engaged in the exchange of scientific ideas and knowledge with fellow astronomers and natural philosophers, though all too little of this survives. Bradley was well versed in contemporary natural philosophy and mathematics and did much to marry these two disciplines that had been separated from the time of the Peripatetics; Johannes Kepler had long previously led the assault on the eradication of this epistemological divide in the *Astronomia nova* of 1609.

Bradley's investigation of the suspected lunar-induced nutation necessitated regular journeys to Wanstead where the 12½-foot zenith sector designed by George Graham remained in *situ* at Elizabeth Pound's town house, although from 1734 the house was occupied by a tenant, Mrs Elizabeth Jenkins. I am sure Bradley stayed with Matthew Wymondesold[4] as a member of the family. Bradley also on occasion travelled to Chatham and London, the former when conducting sea trials of John Hadley's octant, the forerunner of the mariner's sextant, and the latter attending meetings of the Royal Society, occasionally being elected a member of the Council. His work was progressively focused on his University activities. Of these, the most demanding were his lectures in experimental philosophy.

Bradley did not formally write out his lectures in natural and experimental philosophy, although he sketched out the content of what he was to deliver.[5] As well as Bradley's notebooks and loose notes, a student's notebook has come to light, revealing the content and something of the presentation of the lectures. On the first page of his own notebook Bradley wrote out the express purpose of his lectures,

> The design of this Course being to explain several of [the] principle Phenomena of Nature and to give some Account of their Causes as far as they depend upon the Situation. [The] motion of Bodies and the general laws of Nature according to which the motion of Bodies are performed. It may be proper to begin by considering Body or matter in general with respect to its most obvious Properties and then proceed to inquire into the Laws of Nature by Examining the Natural Phenomena from whence they must be deduced.[6]

There is a pencil note in this notebook written in 1893 which estimated that it was written between 1740 and 1750.[7] Comparisons of Bradley's penmanship, paper, and ink with other sources written during the early 1730s and those definitely written in the mid 1740s strongly indicate that Bradley's notebook was written during the early 1730s, and may even date from the time he acquired the post of lecturer of experimental philosophy late in 1729. The text is a preliminary working out of a course in rough form and not a coherent presentation as would be expected if it had been written after the course had been presented for ten years and about thirty presentations.

A brief examination of Bradley's notes suggests that he conceived of natural and experimental philosophy as an activity reducible to mathematical analysis. In this he follows Newton's lead in the mathematization of natural philosophy. As his research with Molyneux on the new discovered motion clearly indicated, he was averse to unsupported speculation of the kind that was sometimes proposed by his senior partner. This distinction between unsupported speculative hypotheses and knowledge claims based on observation and experiment is quite fundamental in Bradley's *modus operandi*. In his notebook Bradley refers immediately to the 'laws of nature', to be discovered *only* 'by experiments and observation & examining the Phaenomena & finding from them by what laws their motions are ordered and regulated which is properly the Business and scope of Natural and Experimental Philosophy'.[8]

This is a fair description of the procedures that led to Bradley's discovery of the aberration of light after extended observation and analysis. In this way only, are the laws of nature revealed. This is a very English procedure in that it bears a resemblance to the way that the Common Law is said to be 'revealed' in English jurisprudence, being established on the disclosure of precedents. The foundations of this method of deriving a knowledge of the laws of nature in England are in part located in the works of Francis Bacon, himself a lawyer. This 'Baconian tradition' was writ large in the early ethos of the Royal Society and made explicit in many contributions to the *Philosophical Transactions*. It wasn't without its detractors. In writing of the discourse between Robert Boyle and Thomas Hobbes in the *Leviathan and the Air Pump*, Shapin and Schaffer ask 'why does one do experiments in order to arrive at scientific truth?'[9] This groundbreaking book revealed that the acceptance of an empirical approach to 'truth' cannot be divorced from a 'social philosophy'. Bradley's heuristic was constructed on procedures to be found in English jurisprudence, itself a form of socialization. From this we can discern that even in the pursuit of natural knowledge there has to be agreement on what is valid for knowledge claims to be acceptable. From the introductory paragraph Bradley proceeds,

> But then our principle endeavour must be to learn the true & real manner in which the operations of Natur are actually performd & not content ours[elves] with framing Hypoth to explain how such Phaenom may be performd tis on this account that Reasoning much from Hypotheses in Natural Phil is apt to lead people into mistakes and there is no likelier a method to avoid error than having recourse to experiments and trials.[10]

In the English science of his time, evidence was commonly subjected to trials. For Bradley it was an activity in which observations were interrogated so that the 'laws of nature' could be 'revealed'. From these, predictions could be made and tested against further observations. These are the procedures that Bradley applied when he 'revealed' both the aberration of light and the nutation of the terrestrial axis later in his career. Bradley's lectures are illustrated by practical demonstrations, revealing some underlying law or principle. The work of James Bradley strongly suggests that scientific theories are social constructs, and that successful theories are those that most well accord with the demands of the dominant requirements of the social milieu in which they originate. This Bradley intuitively grasped, for he clearly understood that there was a disjunct between the results of observations and experiments and the theories derived from them. The cosmologist P. J. E. Peebles recently wrote that, 'It is sometimes said that the laws of physics were 'there', waiting to be discovered. I would rather put it that we operate on the assumption that nature operates by rules we can discover, by successive approximations.'[11]

He also adds that cosmology [science] is a social construction. Both views are in my opinion perfectly valid. All scientific knowledge is socially validated. The view that laws of nature are 'revealed' by 'trials', similarly to the ways that English Law based on precedents is also revealed by trials, leading to the establishment of case law, is a social construct. It validates itself by a process of social perception and acceptance.

It is difficult to decide whether Bradley had an assistant or not. I suspect he may have gained the help of students to help him use various pieces of apparatus in order to illustrate his lectures; that may well have been one of the reasons why they were so popular and well attended for so long. Larry Stewart recognizes that it was the proselytization of men like John Desaguliers and John Harris that placed Newtonian natural and experimental philosophy before an inquisitive public to whom the *Principia* would forever remain an arcane mystery. Stewart writes, "In order to appreciate the social impact of natural philosophy it is necessary to admit that a wider society exists – in this case, beyond the Royal Society."[12]

It would be easy to underestimate the influence of Bradley's lectures in experimental philosophy but his audiences were made up largely of members of Oxford University, some of whom indubitably achieved high offices both in church or state. His notes reveal a sound introduction to contemporary natural philosophy and experimental physics.

The authenticity of Bradley's sketched out lecture notes are verified by reference to lecture notes written by at least one of his students. Of particular interest is the notebook written by Roger Heber in 1754.[13] There can be little doubt that these notebooks were a reasonably faithful record of Bradley's lectures.[14] In addition there are also various notes made by Heber concerning the concurrent lecture. The lecture notes are mainly written on the recto side of each folio and the personal notes are largely to be found on the verso side. It appears that Bradley expounded freely, including experimental demonstrations to illustrate the phenomenon he was discussing. Whether these were performed by Bradley unassisted or with the help of a student or an assistant is not indicated. With reference to Heber's notebook the opening of the first lecture of the course presented in March or May 1754[15] Heber records the following,

> The purpose of this course of Lectures is First to prove by Experiment what properties belong to matter or Body – Secondly to show how the Phenomena or Appearances of Nature, that is, the Constant settled Rules according to which Effects are produced in the Material World.
>
> That these are such a Law is sufficiently apparent from a survey of Nature, either in the Animate or inanimate World: Similar causes still producing similar Effects, according to what one may call stated Laws, to which by the Creator's appointment they conform to themselves.[16]

What is clear from this account is that as well as being rationally apprehensible, the entire material universe is dependent on God's volition.[17] That the material world is 'rational' is really the expression of the common contemporary notion of providence, or the protective care of nature, coming into being.

As I have indicated, Bradley's earliest notebook for lectures was probably penned sometime during the early 1730s. This was when many of the issues dividing those who adopted a 'post-Newtonian' physics and those who upheld the principles of 'Cartesian' natural philosophy had not been resolved.

To clarify the meanings of the word 'Cartesian' I refer to Mary Terrall's study of the life and work of Pierre Louis Moreau de Maupertuis (1698–1759).[18] She writes,

> When someone was called a Cartesian in early eighteenth century France, it might imply any or all of the following: belief in a material plenum with vortices of subtle matter carrying the celestial bodies; commitment to a 'system', which usually involved hypothesizing about causal mechanisms beyond the reach of observation; matter defined by extension, with no added fundamental properties; 'rationalism' rather than 'empiricism'; conservation of quantity of motion (mv) in the universe; measuring the force of bodies by their quantity of motion rather than their 'living force' (mv^n where n = 2); refusal to accept forces acting across empty space; equating such forces with 'occult qualities' of scholastic philosophy; observable effects of gravity caused by pressure of matter in celestial vortices. Many of these terms were rhetorically laden, and many of the concepts were conflated together.

To these I think I would add another contrast between those who followed the precepts detailed above and those who valued the concepts found in Newton's *Principia*. The fundamentally important work of Rene Descartes' (1596–1650) *La Géométrie* unified algebra and geometry and was included as an appendix to his *Discours de la method*,[19] but his natural philosophy was largely devoid of mathematical analysis, and so differed radically from the mathematization of nature in Newton's *Principia*. Descartes intended his approach to be a replacement for Aristotelian natural philosophy, but he retained the fundamental Peripatetic distinction between physics and mathematics. Many Cartesian natural philosophers thereby interpreted Newton's *Principia* as a brilliant exercise in geometry rather than an account of 'the real world'. This was a consequence of the maintenance of the ancient distinctions between physics and mathematics.

From 1726 to 1729, the French philosopher, playwright, and author Voltaire (1694–1778)[20] lived in London having fled from Paris. When he returned to France he penned his *Lettres philosophique*. In this work he contrasted the many differences, including those between the natural sciences in England and France. Voltaire wrote,

> A Frenchman arriving in London finds things very different, in natural science as in everything else. He has left the world full, he finds it empty. In Paris they see the universe as composed of vortices of subtle matter. In London they

see nothing of the kind. For us it is the pressure of the moon that causes the tides of the sea; for the English it is the sea that gravitates towards the moon[21]

Within the Cartesian milieu the Newtonian approach with its appeal to 'forces' represented a reversion to the occult influences and affinities of Renaissance philosophy. Followers of Newton believed it was the theory of vortices that was open to question as they had not been observed.

Book II of the *Principia* is largely a fundamental critique of the theory of vortices which mathematically demonstrated its inadmissibility. Voltaire wrote, 'It is the theory of vortices that might be called an occult quality, since nobody has proven their existence. Attraction on the contrary is a real thing, since its effects can be demonstrated and its proportions calculated'.[22]

Descartes, one of the co-founders of the mechanical philosophy, explicitly denied that there were any powers or forces in nature. Matter was entirely passive and inert and set in motion by God at the creation of the universe in a perfect clockwork mechanism.[23] There was no need of divine intervention after the creation.[24] This contrasts strongly with the theological voluntarism so characteristic of the practice of natural philosophy in England. When Bradley first prepared his lectures from 1729 and 1730, these were live issues. In France, Voltaire's *Lettres philosophique* was burned in public by the public executioner in 1734. By 1754, when Roger Heber wrote his notebook recording Bradley's lectures, any lingering support for the now esoteric Cartesian natural philosophy was fast fading.

Bradley initially dealt with the long-established Peripatetic philosophy which he contended has been overtaken by contemporary natural philosophies, even though it remained on university curricula. This was surely a source of discontent amongst many in the University teaching classics and divinity, being a cause of opposition to the pernicious influence of the perceived godlessness of Newtonian natural philosophy. Bradley is reported by Heber asserting that,

The Ancient Philosophers dealt little in Experiments in order to find out the properties and Laws of Matter. And to explain the appearances that happen in it. And yet they were as forward to undertake the Explanation of Phenomena, as the Moderns are. This they did by making abundance of suppositions or Hypotheses, which, without proving them, and laid down as principles, which they required to be admitted for certain truths, and these being admitted they could readily assign the Reason of any Phenomenon that happened. Thus if you asked them why water rises in a Pump upon your working it, they would first make you swallow an Hypothesis. Nature they would say abhors a Vacuum, and this being granted the cause of the waters rising in the Pump appeared plainly to be Nature's abhorrence of a Vacuum, to prevent which she sent up the Water into the Pump all along.[25]

This aggressive assault on Peripatetic philosophy[26] was a preparation for a more vigorous denial of many of the presumptions of contemporary non-Newtonian natural philosophies.[27] In early eighteenth-century England, the term 'experimental philosophy' was really a synonym for 'Newtonian natural philosophy'. A course in experimental philosophy was commonly understood as a presentation of the central concepts of Newtonian physics as disclosed in the *Principia* and the *Opticks*. In the preliminary to his dismissal of the claims of Cartesian natural philosophy Bradley asserted, 'To know anything of the Nature of Body we must not guess and suppose, but make trials and experiments which diligently pursued have led Men to discover many Articles of Natural Knowledge, which the most ingenious Hypothesis could never have brought to Light'.[28]

Bradley argued that Cartesian natural philosophy was based on ingenious but misleading hypotheses which were not supported by experiment or even common experience. In a section of Bradley's first lecture on the properties of matter, he contrasted the nature of matter as revealed by experimental philosophy with the speculations of what he understood to be Cartesian natural philosophy. In Heber's notebook under the heading of 'On the Properties of Matter' Bradley asserted,

> The Universal Properties of Matter are
>
> 1st Extension which is threefold, magnitude of length, breadth and thickness.
>
> 2nd solidity or that Property by means of which any Body resists any other Body from coming into its place while it keeps its Place. This Property is prov'd to belong to Air and Water as much to Adamant.
>
> The Cartesians maintaining that Nature is full of Matter make Space to be matter and consequently solid contrary to common sense and Experience.[29]

Bradley contrasted the Newtonian concept of matter with the Cartesian view. The seventeenth-century French philosopher Descartes had conceived of a dualism of mind and matter. These two substances are distinct from one another: matter is characterized as the possession of extension (or a spatial dimension) whereas mind is associated with non-extension.[30] Just as mind was associated with thought then so matter was associated with extension.[31] However, as Daniel Garber reveals, Descartes assertions were perhaps less settled than they have usually been represented, for in the *Meditations* and the *Principles of Philosophy* he sometimes reverted to scholastic terminology with the expressions of 'ratio', 'accident', 'quality', 'attribute', 'property', and 'mode' to represent his views about matter and extension.[32] Bradley presented the accepted Newtonian viewpoint because his presentation of the laws of motion was also a polemic. It was directed against the style of speculative natural philosophy that Bradley identified with the Peripatetic philosophy that Cartesian natural philosophy purported to replace, but merely transmogrified. Bradley was a painstakingly diligent empiricist whose contingent natural philosophy also supported a voluntaristic natural theology. He directed his ire against any notion of a rationalistic system, ancient or modern.[33] To avoid any confusion, whilst Bradley believed in a 'rational' universe his voluntarism opposed any form of systemization. It is the essence of his commitment to empirical investigation.

To Descartes, the purpose of empirical observation was to *confirm* the conclusions of reasoned argument. To Bradley, the purpose of empirical observation was to *establish* new knowledge. Bradley cited experimental evidence and associated intuitive insights to contrast the Newtonian concepts of matter, with those proposed by Descartes *prior* to his presentation of the laws of motion. To Bradley, the concept of extension certainly included the notions of length, breadth, and height (or thickness), but it also included solidity and resistance. In his presentation of the Newtonian conception of matter, Bradley emphasized the radical distinction between matter and void. Matter is not identifiable merely with extension.[34]

According to the Heber *Memorandum,* Bradley discussed the properties of divisibility and mobility.[35] When alluding to the property of mobility, Bradley discussed the distinctions between velocity and momentum. This he illustrated by reference to a thought experiment. He referred to the analogy of a horse and a dog moving with an equal velocity, but emphasized that the consequences of a collision with a horse are likely to be more far reaching than colliding with a dog. Bradley refers to the momentum or the quantity of motion as the velocity combined with the 'quantity of matter'.[36] The first lecture concluded with a brief introduction to the fifth property of matter the 'vis inertiæ' of a body, referring to Newton's first law of motion, its 'propensity to continue in its state of Rest till acted upon by some outward force'.[37]

After elucidating the nature of matter with an emphasis on its passivity and the implications of this, Bradley continued the themes in his second lecture when introducing his students to the 'laws of motion'. As reported by Roger Heber he opened his lecture with the words, 'We are next to set forth and explain the Laws or Rules which Matter is found to observe[38] whether it is at rest or in Motion, or suffers a Change from the one to the other, or from one line of Motion to another Line of Motion'.[39]

The first and second lectures of Bradley's course on experimental philosophy were an introduction to the laws of motion. The Axioms or Laws of Motion are to be found in the preface to Newton's *Principia*. Bradley then proceeded to discuss the first law of motion as expounded in the *Principia*.[40] They state,

> Law 1, Every body perseveres in its state of being at rest or of moving uniformly straight forward, except insofar as it is compelled to change its state by forces impressed.
>
>   Law 2. A change in motion is proportional to the motive force impressed and takes place along the straight line in which that force is impressed.
>
>   Law 3. To any action there is always an opposite and equal reaction; in other words, the actions of two bodies upon each other are always equal and always opposite in direction.[41]

After stating that a body at rest will remain at rest whilst a body in motion will continue in motion, he stresses that this is based on the supposition of there being no 'external impediments'. Bradley implies this cannot be proved directly by experiment because no bodies in the observable universe are ever free of such external impediments.[42] Bradley ends his discussion of the first law of motion before introducing the central ideas of the second law. Heber reports Bradley saying, 'The Change that happens in the Motion of a Moving Body, is proportionally to the Quantity of new force impressed, and this force tends to Urge the Body towards that right line in which it acts upon the Body'.[43]

This paraphrase of the second law of motion was used by Bradley as an introduction to the parallelogram of forces that Newton discussed as the first corollary of the third law of motion:[44] 'A body acted on by [two] forces acting jointly describes the diagonal of a parallelogram in the same time in which it would describe the sides if the forces were acting separately'.[45]

In Heber's notebook vector motion is illustrated by reference to the motions of billiard balls before and after contact on a flat billiard table.[46] It appears that a practical demonstration of this was given by Bradley or someone assisting him, though this is not stated. With reference to Roger Heber's notebook and to James Bradley's own outline, here it is salient to note that Heber wrote his in 1754 and Bradley around 1730; there is little difference in the content. Bradley concluded the second lecture with a discussion of the nature of the circular motion of bodies.[47]

This led to a discussion of the second proposition and the second theorem of the first book of the *Principia* which reads,

> Every body that moves in some curved line described in a plane and, by a radius drawn to a point, either unmoving or moving uniformly forward with a rectilinear motion, describes areas around that point proportional to the times, is urged by a centripetal force tending to that same point.

Circular motion is also expressed as the product of two motions – an application of the parallelogram of forces: one rectilinear and tangential to the centre of motion and one directed to the centre of the motion. Bradley attempted to demonstrate this experimentally. Heber reported his words,

> for two impulses acting together and at the same time, as it hath been proved, make the Body move in the middle path in a Curved Line. This is shewn experimentally by a Ball being whirled round a round Table where the vis inertia keeps it some time from motion. But motion being at last communicated to the Ball, the Ball keeps its Motion for some time after the Table is stopped. The Ball being tied with a string to the Center of the Table.[48]

It appears that his lectures were regularly illustrated by experiments and practical demonstrations, performed in conjunction with the lecture itself. Few of Bradley's lectures in experimental philosophy were formally written out. He used the practical demonstrations as his means of making the central points of each lecture. Some of the experiments were direct demonstrations, and others were analogies or thought experiments. It is difficult to be sure of the exact nature of some of the practical demonstrations. One thing is fairly certain, and that is the experiments were performed in the midst of the lectures, rather than after the formal lecture given by Bradley and performed by an assistant in another room, as was common practice with some of Bradley's predecessors in the post. When discussing the parallelogram of forces, it is possible to perceive that Bradley used collisions between billiard balls on what must have been a small, easily transportable table that was sufficiently large enough to show the effects before an audience of some fifty or sixty persons.

After dealing with the nature of matter and the laws of motion, the third lecture began with the action of fluids. Bradley argued from the content of the second book of Newton's *Principia* before discussing the concepts of attraction, repulsion, and cohesion progressing onto magnetism and electricity. Referring to these later phenomena which he distinguished from gravity, he refused to be drawn into making empty conjectures or proposing occult 'explanations'. Bradley wrote in his notes,

> The foregoing experiments [are] sufficient to prove that there are such Laws as Attraction and etc., and if the Effects are in the same circumstances always produced in the same manner, but the causes not known, we do not reason about occult Qualities. The difference between our present method and that which ascribes particular occult Qualities, is that we by experiment find some general cause that is certain though by us inexplicable, yet because tis evident we may properly be said to know the cause of an effect when we have traced its dependence on some known law of nature or the motion of Bodies.[49]

This methodology of reducing general natural phenomena through processes of observation and mathematical analysis to 'laws of nature' enabled observers to make mathematically reducible projections from them. This, as opposed to seeking hypothetical explanations for inexplicable processes, was entirely characteristic of Bradley's work. It is reflected in his experimental philosophy and his lectures at Oxford.

Bradley's next subject in the course of experimental philosophy was the science of 'mechanicks'. In Roger Heber's notebook he recorded Bradley asserting that, 'In general the Application of Geometrical Reasoning concerning the motion of bodies and [this] makes up that mixed part of mathematics which is called Mechanicks. The name seems to confine it to the Effects of Powers and Motion produced by machines'.[50]

Bradley continued with a discourse on simple and compound machines. The discussion involved what he called 'perfect machines' in which the effects of matter are ignored. The behaviour of machines is first discussed as a branch of pure geometry.[51] The philosophical implications of this discussion demonstrate just how far experimental philosophy under the auspices of a Newtonian milieu was related to an increasing mathematization of nature. The contending forces were applied in theoretical terms of reference. Although much of the analysis is presented in Bradley's notebooks, Heber's *Memorandum* makes explicit the lines of Bradley's presentation and their implications.

The final discussion of the principles of mechanics in Bradley's lectures was an examination of how the smallest body or force can give motion to the most massive bodies.[52] The investigation then moved on to an extended study of the lever. This was preceded by a consideration of the concept of weight, both in the sense of lifting and carrying and in a theoretical, mathematical sense.[53] At the end of his presentation of the theory of the lever Bradley put forward an aside to discuss the practical problems of the precise requirements necessary to make scales exact. In itself, it was not of the greatest importance, but it threw light on Bradley's appreciation of the principles underlying the suspension of the zenith sector that he was at that moment using in his continuing observations at Wanstead.[54] The practical skills required in the accurate suspension of scales, or indeed a suspended telescope, involved the overcoming of five critical problems. These were,

Firstly, that the points of suspension and the centre point should be placed in the same line.

Secondly, that the weights are equidistant.

Thirdly, that the weight of the brachia or the arms should be equal.

Fourthly, that the centre of gravity should be a little below the centre of motion.

Fifthly, that there should be a fine edge to avoid friction.[55]

The principle of the lever leads into an examination of inclined planes prior to dealing with the mechanical principles of the wedge and the screw, conceived of as an inclined plane around a cylinder. All five simple machines were presented to the students being firstly the lever, secondly the wheel and access,[56] thirdly the wedge, fourthly the pulley, and fifthly the screw. They disclosed how they transformed various forms of energy or power into mechanical power. As Larry Stewart writes, 'in the natural philosophy of Newtonians like Desaguliers, there was a residual Baconianism hidden and submerged by the contemporary reputation of Newton himself'. [57]

This Baconianism informed much of the ethos of the early years of the Royal Society. Yet there were no *necessary* connections between the increasing mathematization of nature and the beginnings of the socio-economic transformations which are usually described as the 'industrial revolution'. The quantification of human labour is at least one consequence of this form of natural philosophy. In his book, *Britain's Industrial Revolution: The Making of a Manufacturing People, 1700–1870*, Barrie Trinder wrote, 'The broad thrust of change which so excited the attention of contemporaries was that increasing numbers of workers had come to be tied to factory machines and to their employers' systems of timekeeping'.[58]

There followed an examination of the strongest beams and joists obtainable with the least amount of wood. Heber's memorandum states, 'A beam cut out of a given tree will be strongest with the least Timber when the Sides are to one another as $\sqrt{2}/e : \sqrt{\tau}/e$ on the longest side $= d\sqrt{2/3}$ when d = the diameter. That is, the sides are to one another as 5 to 7 or rather 12 to 17.[59]

The symbol τ is a variable signifying different parameters. Building on the introduction to simple machines, there followed a section on how the force of compound machines could be estimated. The purpose of this section of the course was to contrast machines as mathematical constructs with machines implemented in practice.

The sixth lecture of the course, *On Friction*, underlined this shift from the theoretical to the practical. Roger Heber wrote,

> Machines in fact never answer what they ought to do in Theory. This drag back arises from the Friction of the parts of the Machine. Friction proceeds from the Roughness of the Surfaces which rub upon Each other in the Machine, and is generally equal to one third of the weight, whence the power must be increased to a force beyond what is sufficient in theory. The force requir'd to draw a Body along a Horizontal plane is wholly owing to Friction, without which the least force imaginable would draw it along as we see in a hanging body.[60]

Bradley demonstrated simple practical experiments on friction with blocks of wood and various rollers including experiments on the different effects of equal power applied and experiments of two or more power sources. The motion of different carriages on different surfaces was illustrated by the use of wheelbarrows. Bradley had recourse to timbers, ropes, twisted ropes, crossbows, laces, and fibres as well as glass, all to illustrate the different effects of friction on different surfaces and different agents.

The seventh lecture, *On Accelerated and Retarded Motion*, was initially related to the earlier discussion of the second law of motion. For ease of comprehension I will now repeat it. It states, 'A change in motion is proportional to the motive force impressed and takes place along the straight line in which that force is impressed.'[61] Heber reports Bradley saying that, 'The Motion of a Body is said to be uniformly accelerated when a fresh impulse is impress'd Upon it in every Instant of time.'[62]

From this Bradley concluded that a body's velocity increased proportionately to the time in which it is moving as for instance under the force of gravity. According to Roger Heber, various mathematical analyses were offered of the rate of fall of a body through time as it accelerated constantly in which the analysis is dealt with incrementally. The topic was developed as an introduction to the discussion of the retardation of motion. Heber then wrote, 'A heavy Body when thrown perpendicularly upwards (sic) ascends to such a height that when it has fallen back again to the Point from whence it was thrown, it will have acquir'd a Velocity equal to that with which it was thrown'.[63]

The course continued with an extrapolation of the topic, with an extended discussion of the oscillation of pendulums, an expansion of the examination of constantly accelerated and retarded motion.[64] There was no apparent reference to anything other than a continuously acting force so the reference in Heber's notebook[65] may be nothing more than a misapprehension when he 'appears' to assert that Bradley expressed an impetus model[66] of accelerated motion.[67] Bradley developed this section of the course when he was working on the isochronal pendulum experiment with George Graham in London and Colin Campbell in Jamaica. The lecture examined a few of the considerations met with during that work. The method of determining the length of a 'seconds pendulum'[68] was demonstrated as was the manner by which the centre of oscillation was determined. Bradley presented the problems associated with the coefficient of expansion of different metals and how these affected the oscillations of a pendulum as the metal expanded or contracted with changing temperatures. He revealed that the alteration due to expansion and contraction between summer and winter in England for a pendulum of thirty inches is about a fortieth part of an inch. Bradley discussed how clocks slowed as they approached the equator.[69] When Heber recorded this lecture Bradley had published the paper on the 'nutation' a product of the Moon acting differentially on the Earth's equatorial 'bulge'. The implications of the pendulum experiments affirmed that the Earth was indeed an oblate spheroid in conformity with Newton's calculations.

The next lecture examined *The general motion of the heavenly bodies*[70] beginning with a brief examination of Kepler's second law of planetary motion 'that all planets sweep out equal areas in equal times'. Newton stated this in similar terms in the *Principia* Book I, Proposition 1, Theorem 1, which states, 'The areas which bodies made to move in orbits describe by radii drawn to an unmoving centre of forces lie in unmoving planes and are proportional to the times'.[71]

From this Bradley asserts,

Hence Sir Isaac Newton demonstrates that if a Body be acted upon by a force tending to a point or a centre, and the force decreases as the square of the Distance of the Body increases, in that case the Body must describe one of the conic sections and the centre of the force will be the focus of the section.[72]

Now there is but one Conic Section which returns into itself which is the Ellipsis (under which the Circle is a particular form).[73]

Bradley states that the primary planets do in fact describe ellipses about the Sun, one of whose two foci[74] are near the Sun's centre,[75] consequently acted upon by a force tending to the Sun's centre, which decreases as the squares of the distances of the planets from the Sun. He did not conjecture as to the nature of this force and is reported by Heber to have stated, 'There is therefore a known power in nature sufficient to retain all the several Heavenly bodies in their respective orbites we ought therefore conclude that they are retained by this power till we meet with Phenomena that contradict these deductions'.[76]

This avoidance of the propagation of occult causes to explain natural phenomena is a *leitmotiv* of Newtonian natural philosophy. It is entirely characteristic of a practitioner of Newtonian experimental philosophy to assert that the search for causes should be set aside, in favour of determining the behaviour of bodies and forces in space described analytically and in mathematical terms.

There are fifteen folios of text in the ninth lecture[77] located in Bradley's notebook.[78] Although there are various differences between the earlier versions of his lectures and those he penned in the 1740s, the topics remained largely untouched. A comparison of Bradley's lecture notebooks written in the 1730s with those he wrote during the 1740s and 1750s and Roger Heber's notebook of 1754 reveals a consistent continuity. The lecture ranges from a discussion of the shape of the Earth, contrasting the opinions of the Cartesians with the empirical evidence supporting the ideas of Newton. In his discussion of Cartesian physics, Bradley examined the laws of percussion and its effects, again contrasting it with Newton's generalized approach to the subject in the second book of the *Principia*. The lecture concluded with a presentation of the motion of projectiles.

Notes for the tenth lecture are surprisingly brief being written on just six folios.[79] It is an introduction to Newton's *Opticks*.[80] When Bradley started giving his lectures on experimental philosophy at Oxford the fourth edition of 1730 was coming into common use. Bradley presented demonstrations of Newton's experiments with prisms and glasses. This lecture was an introduction to the next two, the eleventh on the principles of 'catoptricks' or reflection[81] in which he displayed developed knowledge of the theory of reflecting telescopes. He had continuous practical experience of one of the world's first viable and efficacious reflectors constructed by John Hadley and his brothers, an instrument he had used from 1722.[82] Bradley demonstrated how many of the problems associated with chromatic aberration, a phenomenon which had for so long been the bane of astronomers using refracting telescopes, were avoided in reflecting telescopes. In the twelfth lecture, on the principles of 'dioptricks' or refraction,[83] his experience of long-focus telescopes, having once used one with a focus of 212¼ feet, is also explained as a means of combating chromatic aberration. Bradley's experience of various optical systems was possibly without equal in England except for Dr Robert Smith at Cambridge. There was a discussion of the theory and construction of microscopes as well as telescopes, using both refraction and reflection. The problems associated with chromatic and spherical aberration were discussed before moving on to the theory of colours.

The next three lectures were a further exposition of the second book of the *Principia*. The thirteenth lecture[84] was on 'hydrostatics or the doctrine of fluids',[85] itself an ancient science dating back to the sixth century before the Common Era. Newton's *Principia* distinguished fluids from solids. Fluids, which included liquids and gases, are unable to resist distorting forces unlike solids. Newton developed a precise theoretic of the continuity and fluidity of liquids. This enabled Newton to construct mathematical proofs that relied on spherical surfaces coming into contact with planes or other spheres. Bradley moved onto the practical consequences of Newtonian hydrostatic theory revealing the empirical evidences supporting the theory. It relied on the work of Robert Boyle (1627–1691) by showing that within any body of liquid that is in equilibrium, there is a pressure exerted on its bounding surfaces that is transmitted as the same force per unit area to any surface within the body of liquid. Introducing the effects of gravity, Bradley disclosed that Newton revealed how the pressure in a liquid is proportional to the depth below its surface. In lecture fourteen, 'of solids immers'd in fluids'[86] Bradley gave a full exposition on the purpose and use of the hydrostatic balance.[87] It makes reference to Galileo's *La Bilancetta* which allowed him to weigh objects in air and water. The final lecture of the three, lecture fifteen, was of 'the motion of fluids',[88] an extended exposition of the content of the second book of the *Principia*, together with an implied criticism of the Cartesian version of the mechanical philosophy. In Section IX of Book II of the *Principia* with direct reference to the Cartesian concept of the motion of vortices Newton discussed the 'circular motion of fluids'. This section is translated by Andrew Motte in his contemporary 1729 English translation of the *Principia* as 'The resistance arising from the want of lubricity in the parts of a fluid, is, other things being equal, proportional to the velocity with which the parts of the fluid are separated from one another'.

The expression 'want of lubricity' has been translated otherwise as 'lack of slipperiness'. The friction that takes place between adjoining layers of a circulating fluid will soon lead to a loss of velocity. Any bodies circulating in the fluid will begin to spiral to the centre of rotation.[89] As discussed when Bradley first wrote his lecture notes in the early 1730s, Cartesian natural philosophy was still supported as the primary orthodoxy within the Royal Academy of Sciences in Paris. There was a growing regard for Newtonian models that were becoming more influential not only in France, but also in the Low Countries where a distinctive Newtonian school developed.

The next four lectures were concerned with an examination of the pneumatic sciences. The sixteenth lecture was an introduction to pneumatics, the sciences of the air and gases.[90] The seventeenth lecture[91] was of a practical vein with various experimental demonstrations. Using pumps and siphons, Bradley demonstrated the effect of the weight and spring of the air before discussing and demonstrating the barometer, an instrument that was becoming critical in his astronomical work. It was important in the evaluation of atmospheric refraction, an important factor in high precision positional astronomy. Bradley referred to the fact that as the Earth spun on its axis, sunrise preceded the actual position of the Sun relative to the observer's horizon. Likewise sunset takes place several minutes after the Sun has in fact disappeared below the observed horizon. These phenomena, according to Bradley, make the accurate positional observations of stars near the horizon impractical. This lecture served as an introduction to the eighteenth lecture, *Of the causes of the rising and sinking of the mercury in the barometer,*[92] that properly concluded the discussion begun in the seventeenth. The lecture was both an exposition of the air pump and a demonstration of Boyle's Law. At a constant temperature the volume of a gas is inversely proportional to the pressure. It is a 'law of nature' that was 'revealed' from experiments with the air pump. It described how the pressure of a gas increased as the volume within which it is contained decreased. Mathematically, the law can be represented by the simple equation $PV = k$ where P is the pressure of the gas, V is the volume of the gas, and k is a constant. This equation revealed that the product of the pressure and the volume is a constant which holds true provided that the temperature of the contained gas is constant.

The penultimate lecture was an exposition of several experiments using the air pump.[93] The air pump, a device invented by Robert Boyle and Robert Hooke at some time between 1656 and 1660,[94] was used by Boyle to demonstrate the various qualities and characteristics of the air contained within a pressurized flask. He revealed the importance of air for combustion, for respiration, and for the transmission of sound. Bradley performed various experiments with an air pump, an expensive item of scientific apparatus, possibly with the assistance of one or more of his students or an assistant.

Bradley's final lecture, the twentieth, was a summary of the leading points of the course, offering an overview of contemporary natural and experimental philosophy, particularly the strong Newtonian strain that directed so much of modern physics in England. Bradley offered his lectures in experimental philosophy at Oxford from 1729 to 1760 when declining health led him to resign the then readership in experimental philosophy. I cannot believe that Bradley did not keep a careful record of the students who enrolled on his courses from 1729 to 1745 particularly as his financial situation in 1729 was so fragile. I am sure that such a record must have been a part of his archive that went missing, particularly during the years when Thomas Hornsby his successor as Savilian Professor of Astronomy carelessly and negligently treated Bradley's archive after it was presented to Oxford University.[95] The registers containing the record of his students by college from 1746 to 1760 are located at the back of Bradley's lectures on the air pump.[96]

When Bradley's opening lectures in experimental philosophy are examined, his adherence to many of the precepts presented in Newton's *Principia* and *Opticks* seem evident. It would be incorrect to think he followed Newton slavishly. Bradley was first and foremost an empiricist. Like Newton, he disdains the use of 'occult' causes to explain natural phenomena. It was a regular accusation by Cartesian natural philosophers that the Newtonians ascribed to 'action at a distance' because they believed in Descartes' assertion of the passivity of matter and the associated belief that motion could only be transmitted by material contact. The Newtonians subscribed to the observation of natural 'forces' and being unable to give explanation of what these forces consisted of, refused to provide 'occult' explanations. The Newtonians, instead, adopted a more modest programme, that of observation and experiment leading to mathematical analyses, enabling investigators to calculate the behaviour of bodies in a void.

This is evidenced by Bradley's attack on Peripatetic philosophy. His rejection of the procedures of the Peripatetic schools leads also to a non-acceptance of some contemporary schools of thought. These apply occult material or mechanical properties to explain phenomena like gravity, such as Descartes' theory of vortices of unobservable fine matter. Cartesians asserted that gravity was the product of the pressure applied by these hidden vortices of fine matter. In Bradley's work there is a comparative absence of concerted attempts to court controversy. This is evident in my account of the construction of his aberration paper, inviting criticism of his own solution to the new discovered motion.

The evidence suggested by his philosophical practice reveals that this was a conscious decision. It is defined most assertively in his astronomical practice. When in December 1728, Bradley wrote his account of the discovery of the 'new discovered motion', he must have reflected on his late partner's lukewarm acceptance of Newton's natural philosophy. As a man of his perception, even without any knowledge of Molyneux's approach to John Conduitt, he surely recognized that his partner was sceptical of at least some of Newton's work. Molyneux's characterization of Newton's work as a 'system' reveals the profundity of his misconceptions. Bradley acquired his own knowledge of natural processes by having two outstanding 'Newtonian' practitioners as his mentors. James Pound was one of the finest astronomical observers in England, and Edmond Halley motivated Newton to write the *Principia* and paid for its publication. Both men were empirical practitioners. No better teachers were available in England. It is no surprise to discover that Bradley's lectures in experimental philosophy bore much of the imprint of Newton's mind. However, it is essential to appreciate that he was not a supporter of a Newtonian 'system'.[97] Bradley's perception of his approach to experimental philosophy was less to support Newton's 'philosophy' rather to allow observation, experiment, and the inferences drawn from them to determine his conclusions.

Bradley's espousal of the methods of experimental philosophy did not meet with universal support either in Oxford or elsewhere in certain Tory and High Church circles. There was a common apprehension in such quarters that it led to atheism and godlessness. There were opposing currents that regarded Bradley's lectures as a threat to the moral wellbeing of the young. One such movement was that which followed the publication of John Hutchinson's *Moses' Principia* in 1724. It propagated a form of natural philosophy that was not only opposed to Newton but also the whole development of heliocentric cosmology from the publication of Copernicus's *De revolutionibus* in 1543. What is so interesting about this opposition to the otherwise overwhelming ascendancy of Newton's influence in England is the way it focused attention on growing misgivings in society. This was particularly so amongst some High Church Anglicans, some of whom were resident in Oxford itself.[98]

An enlightening recent master's dissertation[99] combats a commonly expressed view that Oxford was in some sense 'anti-Newtonian' during this period. It explores the important influence of Bradley's lectures in experimental philosophy and some of the reactions to them, particularly by those who perceived of them as a pernicious influence. Turning to alternative forms of natural philosophy. Stedman Jones's abstract concludes,

> By demonstrating the forceful presence of new styles of teaching, thus far ignored by historians, I situate Hutchinsonianism as not so much a pure defence of high church principles, as some have claimed, but rather as a response to methods of university teaching which endangered young minds, and the pedagogical and educational reforms which they presaged. The Hutchinsonians worried not only about the authority of scripture, but also about the implications of the new science for the future careers of tradition-ally trained men at the University of Oxford and beyond.[100]

The author reminds us that it has been widely accepted that Oxford was a scientific backwater during the eighteenth century. He is critical of the influ-ence of authorities such as Christopher Wordsworth, who in 1877 compared Oxford's scientific backwardness relative to Cambridge. Stedman Jones char-acterizes the opposition to Bradley's lectures in experimental philosophy as an indication or a measure of their importance in the academic life of Oxford University.

The natural philosophies of men such as Hutchinson are taken to be repre-sentative of Oxford academic opinion, supposed by authors like Wordsworth to be 'anti-Newtonian' but as Stedman Jones writes, 'In 1753 Horace Walpole, the English Whig politician saw Hutchinsonianism in Oxford as an esoteric group'.[101]

Stedman Jones forcefully argues that it is possible to reconstruct the intel-lectual aims and influences of the group, in part by the fortuitous publication of William Jones's *Memoirs*[102] of the life of George Horne, along with the writ-ings of some other Hutchinsonians. Stedman Jones asks what drew the Oxford Hutchinsonians to each other and identifies an acceptance of the Mosaic creation of the world. Julius Bate wrote,

> every Coffee-House ...sounds all Day long with the Nonsense, &c. of the Mosaic Account; and every Bookseller's Shop, almost in London, swarms with Books, new and old, to ridicule it, and who have demonstrated their and many other Things, to be contrary not only to the stated Laws of Nature, but of Probability and common Sense also, ...learned Men have shewn this supposed Repugnancy to be imaginary and groundless .... we are contented to the Jest of Unbelievers, but must we be made so of Christians too [?].[103]

According to men such as Bate and Spearman, Hutchinson's 'system' if such it could be called, constituted 'essential articles of our faith' established 'as on a rock'.[104] Citing a modern text, Stedman Jones states that most early follow-ers of Hutchinson believed in a self-sufficient nature activated by light, spirit and fire.[105] Stedman Jones points to works such as George Horne's *Somnium Scipionis* which was a lampoon of 'Whig Newtonianism' claiming its debt to Ciceronian paganism. Various anti-vacuist and ethereal positions were iden-tifiably Hutchinsonian. Horne ridiculed the 'Newtonian vacuum' and asked whether, 'thine hair stand on end, at the excessive effrontery, and absurdity of the man?'[106]

Horne attacked the Newtonian concept of 'action at a distance'. In a vitu-perative aside he added 'Experiment and logic failed in the demonstration of a vacuum' evidently lacking sufficient knowledge of Boyle's work with the air pump or Newton's *Principia* even to instigate an appreciation of the central arguments of Book II. The second book of the *Principia* was a full frontal assault on all systems that proposed either a plenum or a substantial ether. Unless the Earth and the other planets moved in a medium that was close to a vacuum, they would soon spiral into the Sun. Stedman Jones continues his account of Horne's assault on Newtonian natural philosophy, 'He con-sidered whether attraction was a cause or an effect, detailing contradictions in the works of Newtonian apologists, contrasting Newton's claim that the cause of gravity was unknown, with Roger Cotes's assertion that gravity was the simplest cause, not requiring explanation'.[107]

In so doing, Stedman Jones revisited the Clarke-Leibniz disputes over God's function and the occultism of gravity.

What is so interesting about this opposition to the otherwise overwhelming ascendancy of Newtonian natural and experimental philosophy in England is the way that the writings of the Hutchinsonians focused attention on the growing misgivings that many held about the entire thrust and tenor of contemporary natural philosophy. It even brought about something of a rap-prochement between many High Church Anglicans and certain elements in the more evangelical and non-conformist confessions of the faith. John Wes-ley for one, the founder of the Methodist movement within the Church of England, confessed that the 'new ideas' tended the believer 'toward infidelity'. Amongst the polite such misgivings must have been experienced, but the approbation accorded to the rise of the new philosophy ensured that such movements remained muted amongst the leaders of public opinion.

Bradley was the product of a particular milieu and it would be misleading to make overassertive claims on his behalf. For instance Freeman J. Dyson in an article in the journal *Nature* called James Bradley 'the founder of modern science',[108] particularly in connection with his commitment to high precision. But to assert as Dyson does that a complex human activity like science can have a 'founder' fails to acknowledge that this recognition involves a sharing of mutual values. Bradley was a man who shared the values of his own age. Bradley was a rarity in the eighteenth century: he was a man who in part at least earned his living from his science both from his research and his teaching. He was the inheritor of a tradition that was to lead to the basis of so much of modern natural science. It was an approach that was developed by Newton and his acolytes, an approach that Bradley was to develop in his practice, both as an astronomer and as a lecturer in experimental philosophy. Bradley's work was given high value in his day because it reflected the values of an optimistic, expansive, inquisitive, and acquisitional society that invested in 'reason'.[109]

I will conclude my brief discussion of Bradley's lectures in experimental philosophy in only the broadest terms, for the purpose of this examination was to demonstrate their connection with his astronomical practice. It is noticeable that, unlike some other 'followers' of Newton's work, he is not interested in its imputed metaphysical or theological consequences.[110] Although he shares much of the common vocabulary of his age, he avoids any meaningful use of the concept of 'God' in his analyses or discussions. I am not suggesting he was an atheist or even an agnostic, for I am sure that Bradley was a believer. He held a 'longitudinarian' confession of the faith that sought to accommodate itself to modern circumstances.[111]

Although his lectures caused comment in certain High Church quarters in Oxford, Bradley studiously avoided religious controversy, even though he served briefly as the personal secretary of the most controversial churchman of his age, the Rt Revd Dr Benjamin Hoadly. As his work on the law and hypothesis of the new discovered motion revealed, Bradley possessed a perceptive appreciation of the distinctions between the determination of natural law and the various schema, hypotheses, and theories that were applied to account for natural law. Bradley's opposition to unnecessary or insupportable speculation resulted in an intellectual stance that made him suspicious not only of 'systems', but also any connections between established natural philosophical knowledge and religious or theological beliefs; because alternative theoretical frameworks were capable of 'explaining' empirical data, Bradley tended to base any hypothetical conclusions on what was apparently evident in his own practice or what his audiences were ready to accept. As a result, his papers tend to possess a modesty that sometimes disguises his astute understanding of the epistemological and ontological issues that might otherwise be evident.

In comparison to the earlier lectures of William Whiston, for example, his lectures were almost entirely devoid of controversy, unless of course you were a latter-day Cartesian or a passionate Hutchinsonian.

Bradley's lectures made their mark on Oxford life. There is little doubt that he had acquired what could be described as an established status in the life of the University. When he was appointed as the Astronomer Royal on 3 February 1742 in succession to his mentor Edmond Halley, who also held the Savilian chair in Geometry, the University awarded Bradley with the degree of Doctor of Divinity by diploma. This was apparently a rare procedure and obviously reflected the pride that the University had in Bradley. He had been associated with the academic life of Oxford from when he matriculated in 1711, entering Balliol College as a Commoner. This action is firm evidence against the opinions of men such as Christopher Wordsworth.[112] As is plain to see the assault by the Hutchinsonians and some High Churchmen and Tories on the 'fraudulence of experiment'[113] was directed primarily against the 'Bradleian programme'.[114] But such was the authority of Bradley and his lectures in experimental philosophy that he was unassailable. Stedman Jones asks a fundamental question as to why it was that Oxford was the centre of Hutchinsonian resistance against experimental philosophy? He writes,

> The crucial factor was Oxford's role in [the] training of future churchmen, and increasingly lay professionals. For the commercial Whig London to play host to charlatanry was one thing; for the University of Oxford to tolerate it, was another. With the experimental style of knowledge transfer gaining traction, a reaction developed by necessity. The Hutchinsonian method was a way of expressing instinctive distrust of these trends, a dislike which was widespread, but had often been articulated in an anti-intellectual way. If the founder of the 'Holy Club' in Oxford and English Anglican cleric John Wesley opposed Socinianism,[115] his Methodist system was directed to too lowly an audience as to attract [the] assent of high-minded scholars. The Hutchinsonians could never have claimed for themselves the words attributed to Wesley, that the world was his parish. Hutchinsonians thought of their method as a cerebral way of disavowing practices that many disdained, but whose critique few could articulate. Horne could easily attack the long dead Newton, and faraway Martin,[116] but for a junior university member, an overt attack on Professor Bradley would have been daring indeed. Given the tidal wave of the Bradleian lectures, Hutchinsonian attacks on the fraudulence of experiment can be plausibly seen as covert attacks on the Bradleian programme.[117]

Stedman Jones rightly gives expression to the nature of the fundamental challenge posed by the new philosophy to the traditional sources of knowledge. At the time of Galileo, the status of mathematics was far below that of philosophy and theology and this is evidenced by the low salaries commanded by professors of mathematics, who had to accept the low academic status of the discipline. With the publication of Newton's *Principia*, the challenge posed by the new philosophy became explicit, for its subject really was the '*mathematical* principles of natural philosophy', a concept that would have been deemed nonsensical in a previous generation. From the inception of the first universities during the twelfth-century, theology was truly 'the queen of the sciences', and mathematics was held to be of little consequence. Bradley's lectures in experimental philosophy were perceived by many of those who opposed, or were aggrieved by their influence, as a threat to the traditional hierarchies of knowledge as well as the faith of many believers. The new natural philosophy was seen as a threat. The overwhelming importance of mathematics in the new dispensation was regarded by many with alarm, for many were influenced by Aristotle's insistence that there was no meaning in mathematics, a view vehemently and consciously opposed by Bradley who wrote,

> The purpose of this course of Lectures is First to prove by Experiment what properties belong to matter of Body – Secondly to show how the Phenomena or Appearances of Nature, that is, the Constant settled Rules according to which the Effects are produced in the Material World.
>
> That these are such a Law is sufficiently apparent from a survey of Nature, either in the Animate, or inanimate World: Similar causes still producing similar Effects, according to what one may call stated Laws, to which by the Creator's appointment they conform themselves. – For the making the Discoveries above mention'd Experiments are to produce.
>
> ...and this we can't argue so closely and conclusively as in ye Mathematicks, for that the Nature of the Subject will not admit of.[118]

With these words as recorded by Roger Heber, Bradley opened his course of lectures. The divine is emptied of all content and referred to as 'the Creator'. Natural philosophy is dictated by experimental and observational evidences, the more precise and accurate the better, it being the key to the understanding of the laws of nature. Mathematics even more than observation and experiment is the key to the real understanding of nature. The 'laws of nature' are framed mathematically. Through the apprehension of the mathematical laws of nature, it is possible to predict the motions of all the forces and bodies in nature. To such as the Hutchinsonians and Wesley, this was seen as an assault on their belief in divine purpose and helps explain what their concerns were. Even to our own age there remains a sense of distress amongst many of 'the religious', particularly those for whom the mysterious and the transcendental lie at the core of their beliefs. Yet Bradley would have denied that he was an atheist, quite the contrary. For many who accepted the new science, it was the very existence of natural law and the predictability of the 'laws of nature' that revealed the existence of God. Natural law was seen in some circles as providential and this was seen as the action of God in nature matching the actions of God in history.

Stedman Jones points to more subtle effects of the influence of Bradleian pedagogy. He asserts that the rise of experimental science, teaching, and learning devalued the areas of scholarship and expertise that many who followed Hutchinson 'laboured to develop'.[119] Events turned full circle. Stedman Jones writes,

> Horne and Wetherell spent most of their later life in the mainstream. Horne befriended Lord North,[120] and was invited by North to become the preacher to the House of Commons in 1780. Wetherell's lasting contribution as Vice-Chancellor of Oxford was to invite Lord North to be Chancellor in 1772, which finally created a nexus between Whig Hanoverian London and Tory Jacobite Oxford.[121]

I say full circle, because it was to Lord North that the Peach family finally surrendered Bradley's papers in 1776, before he passed them on to the university with instructions to publish Bradley's Greenwich observations. And it was soon after 4 July 1776 that Lord North's administration was faced with the American Declaration of Independence. The new knowledge led to new values. According to John Locke everyone's mind at birth was a clean slate, a *tabula rasa*. All knowledge was based on experience. Nothing could be more opposed to the notion of 'innate ideas' that were so familiar to Cartesian adherents. With post-Newtonian scientific knowledge, all was based on observation and experiment. Little wonder that those of a more conservative bent opposed the new sources of knowledge. Like some of the good citizens of Venice in 1610, who on reading the *Sidereus Nuncius* and hearing the *Vespers* of Monteverdi, many may indeed have thought the world was coming to an end.

To return somewhat more prosaically to the lectures of James Bradley, these presentations and demonstrations of experimental philosophy not only displayed the central tenets of Newtonian natural philosophy, but also attracted a substantial proportion of the academic body of the university. By examining Bradley's records of the students who enrolled on his courses from 1746 to 1760, when he resigned as the Reader in Experimental Philosophy, the perception is that as annual matriculations in the university fell, the number of students enrolling on Bradley's courses increased. During the whole of the period from 1746 to 1760, there were 2,788 matriculants at the University of Oxford. During the same period there were 1,833 enrolments for the courses in experimental philosophy, of which 1,214 attended one course and 619 attended two. Of those attending Bradley's lectures, over 40% of the student body attended once and over 20% twice. Conservative High Church members feared the influence of Bradley's lectures not only for their supposed 'atheistic' content, but also for their evident popularity. The influence of the lectures reached into many sections of English society. One of his students was George Austen, the future father of Jane Austen the novelist, whose radical credentials are clearly expressed in her works to a mind prepared to perceive them.[122]

Notes

1. I have increasing doubts whether Elizabeth Pound or James Bradley actually took up residence in the house in Wanstead, preferring to reside at Matthew Wymondesold's fine residence close by. I believe that the small house in Wanstead was an address of convenience allowing Bradley to make his observations at various times of the night or day. It must be acknowledged that Elizabeth was a woman of substance, worth about £20,000, and unlikely to reside in a small town house.
2. In addition to his post as Savilian Professor of Astronomy.
3. He was earning up to £400 a year from these lectures in addition to about £150 a year, after various deductions from the income he derived from the Savilian chair in astronomy.
4. I am sure Elizabeth and James usually travelled together between Oxford and Wanstead.
5. James Bradley, Bodleian Library, Oxford, Bradley MS 1, Bradley's notebook for lectures at Oxford and Bradley MSS. 3 to 13, Lectures delivered by James Bradley at Oxford as Savilian Professor of Astronomy and as lecturer in experimental philosophy.
6. James Bradley, Bodleian Library, Oxford, Bradley MS 1, fol. 1r.
7. During the 1740s and 1750s, Bradley was inordinately busy dealing with the problems and later the programmes of the Royal Observatory, so he would have found it difficult if not impossible to possess the energy needed to write out a course of experimental philosophy he'd been teaching since 1729/30.
8. James Bradley, Bodleian Library, Oxford, Bradley MS. 1, fol.1r.
9. Steven Shapin and Simon Schaffer, *Leviathan and the Air Pump*, London, 1985. p.3.

10. James Bradley, Bodleian Library, Oxford, Bradley MS. 1, fol. 2r.

11. P. J. E. Peebles, *Cosmology's Century, An Inside History of our Modern Understanding of the Universe*, Oxford, 2020. p. 348.

12. Larry Stewart, The Rise of Public Science: Rhetoric, Technology and Natural Philosophy in Newtonian Britain, 1660–1750, Cambridge, 1992, pp. xxxii, passim.

13. Roger Heber, Bodleian Library, Oxford, MS. Eng. Misc. e. 15. *Memorandums of Doctor Bradley's Course of Experimental Philosophy* [by R. Heber]. There is a pencilled note on the manuscript to show that it had been purchased at auction. It states 'Phillips sale. June 1893. Lot 268'.

14. It was my intention to transcribe and edit Bradley's notebook of his lectures in experimental philosophy and Roger Heber's 1754 notebook, with a commentary that compared each text. However, I am currently in my 81st year and both age and infirmity may yet deny me the opportunity to fulfil this task.

15. James Bradley gave two courses of lectures in 1754. It is not certain whether it was the course given from March or from May 1754 that was the course attended by Roger Heber.

16. Roger Heber, Bodleian Library, Oxford, MS. Eng. Misc. e. 15. Fol. 1r.

17. Throughout his work Bradley rarely or never refers to the works of the Almighty. In Bradley's lectures it has to be recognized that there is a hidden script at play for he was conscious of authorities within the university who were implacably opposed to the 'godlessness' of Bradley's lectures and the whole influence of Newtonian natural philosophy.

18. Mary Terrall, *The Man Who Flattened the Earth: Maupertuis and the Sciences in the Enlightenment*, London, 2002.

19. René Descartes, *Discours de la method*, 1637

20. Voltaire was the nom de plume of François-Marie Arouet.

21. In Great Britain, Ireland and the American colonies it was published as 'Letters from England'.

22. Voltaire, *Letters from England*, London, 1734.

23. This was opposed by Bradley who supported a voluntaristic theology. Descartes' philosophical position made the efficacy of prayer entirely delusional. I realize of course that Malebranche's occasionalism is a deft way of sidestepping this impasse.

24. Rene Descartes, *Le Monde,* contained in Adams's and Tannery's complete edition of Descartes works, Vol. I, p. 90, note a. Quoted in Daniel Farber, *Descartes' Metaphysical Physics*, London, 1992, p. 198.

25. Roger Heber, Bodleian Library, Oxford, MS. Eng. Misc. e. 15. fol. 3r.

26. These assaults on Aristotelian natural philosophy, still maintained in the curricula of the University, must have been resented by many who were still committed to the traditional values promulgated by the teaching of Classics and Divinity, giving succour to reactionary natural philosophies such as that published by Hutchinson in his book, *Moses Principia.*

27. Again, I must stress that when this course was originally framed Cartesian natural philosophy was still very much a live issue on much of the continent, particularly in France.

28. Ibid. fol. 3v.

29. Ibid. fol. 4r.

30. In the *Meditations,* the *Principles of Philosophy,* and *Le Monde,* passim.

31. James Bradley, Bodleian Library, Oxford, Bradley MS. 1. Bradley's notebook for lectures at Oxford. fol. 5r.

32. Daniel Farber, *Descartes' Metaphysical Physics*, London, 1992, pp. 64–65.

33. At this point some of my readers may justifiably feel confused. I have asserted that Bradley believed that the universe was constructed on rational grounds and yet he insisted that the only way to acquire sure natural knowledge was on the basis of empirical observations. A belief in the possibility of a divinely ordered world is not coincidental with the assertion that the only way to knowledge about the natural world is via the construction of principles based on a single fundamental clear and distinctive idea such as the 'cogito ergo sum'. The entirety of Descartes' metaphysical system, including his natural philosophy, is constructed on his fundamental metaphysical division between mind and matter.

34. Ibid. fol. 5r—5v.

35. Ibid. fol. 5v.

36. Ibid. fol. 5v—6r.

37. Ibid. fol. 6r.

38. Interestingly Bradley's description of the laws of nature finds reference to the laws of jurisprudence according to English legal practice based on the recognition of precedents.

39. Ibid. fol. 6r.

40. Isaac Newton, *Philosophiæ Naturalis Principia Mathematica,* Axioms or Laws of Motion, 3rd edn. London, 1726, trans. I. Bernard Cohen and Anne Whitman as *The Principia: Mathematical Principles of Natural Philosophy*, London, 1999, pp. 416–430.

41. Translation by I. Bernard Cohen and Anne Whitman.

42. Ibid. fol. 9r.

43. Ibid. fol. 9r.

44. Isaac Newton, *Philosophiæ Naturalis Principia Mathematica*, 3rd edn. London, 1726, trans. I. Bernard Cohen and Anne Whitman as *The Principia: Mathematical Principles of Natural Philosophy,* London, 1999, Axioms or Laws of Motion, Law 3, Coro. 1, p. 417.

45. Ibid. p. 417.

46. Ibid. fol. 10r.

47. Ibid. fol. 11v.

48. Ibid. fol. 11v.

49. James Bradley, Bodleian Library, Oxford, Bradley MS. 1, *Bradley's notebook for lectures at Oxford.* fol. 21r. This neatly encapsulates the exact procedure Bradley followed after the initial observations of the 'new discovered motion'.

50. Roger Heber, Bodleian Library, Oxford, MS. Eng. Misc. e. 15. *Memorandums of Doctor Bradley's Course of Experimental Philosophy,* fol. 27r.

51. Ibid. fols. 27r–30v.

52. Flamsteed was sceptical about interplanetary perturbations, believing that the influence of the Sun was so overwhelming that such interactions would be swamped and eradicated.

53. Ibid. fol. 29r.
54. Bradley was observing a minute phenomenon that required absolute precision and accuracy in the suspension of his zenith sector. It led to the discovery of nutation.
55. Ibid. fol. 29r.
56. Bradley himself used the term 'wheel and access'.
57. Larry Stewart, *The Rise of Public Science: Rhetoric, Technology and Natural Philosophy in Newtonian Britain, 1660–1750*, Cambridge, 1992, p. 213.
58. Barrie Trinder, *Britain's Industrial Revolution: The Making of a Manufacturing People, 1700–1870*, Lancaster, 2013. This process impinged directly on the present author's life for he worked for several years as a machinist on a night shift manufacturing auto parts. He was the human extension of a machine being paid by the quantity of parts manufactured within each night's work. His work was entirely dictated by the speed of the machines.
59. Roger Heber, Bodleian Library, Oxford, MS. Eng. Misc. e. 15. *Memorandums of Doctor Bradley's Course of Experimental Philosophy,* fol. 27r.
60. Ibid. fols 32r and 33r.
61. Isaac Newton, *Philosophiæ Naturalis Principia Mathematica*, 3rd edn. London, 1726, trans. I. Bernard Cohen and Anne Whitman as *The Principia: Mathematical Principles of Natural Philosophy*, London, 1999, Axioms or Laws of Motion, Law 2. p. 416.
62. Roger Heber, Bodleian Library, Oxford, MS. Eng. Misc. e. 15. *Memorandums of Doctor Bradley's Course of Experimental Philosophy,* fol. 35r.
63. Ibid. fol. 40r.
64. James Bradley, Bodleian Library Oxford, Bradley MS. 1, *Bradley's notebook for lectures at Oxford*. fol. 35r.
65. Roger Heber, Bodleian Library, Oxford, MS. Eng. Misc. e. 15. *Memorandums of Doctor Bradley's Course of Experimental Philosophy,* fol. 40r.
66. This is a medieval concept, which ironically was developed in part by scholars at Merton College, Oxford during the fourteenth century.
67. It is a persistent model still intuitively used in common speech in the twenty-first-century.
68. This is a pendulum that beats seconds.
69. Ibid. p. 36r.
70. Ibid. p. 36r.
71. Isaac Newton, Philosophiæ Naturalis Principia Mathematica, 3rd edn. London, 1726, trans. I. Bernard Cohen and Anne Whitman as *The Principia: Mathematical Principles of Natural Philosophy*, London, 1999. p. 444.
72. Isaac Newton, *Philosophiæ Naturalis Principia Mathematica,* 3rd edn. London, 1726, trans. I. Bernard Cohen and Anne Whitman as *The Principia: Mathematical Principles of Natural Philosophy*, London, 1999. Book I, Proposition 11, Problem 6, states, 'Let a body move in an ellipse; it is required to find the law of the centripetal force tending toward a focus of an ellipse'. p. 462.
73. James Bradley, Bodleian Library, Oxford, Bradley MS. 1. *Bradley's notebook for lectures at Oxford*. fol. 37r.
74. This mathematical term the 'foci' was first implemented by Johannes Kepler when he discovered that the planets all 'swept out equal areas in equal times' and that they all move in ellipses which have two 'foci' or centres.

75. In fact the centre of the orbits of the primary planets is the centre of gravity of the entire solar system. The Sun possesses almost 99.9% of the entire mass of the system so the centre of gravitation is very close to the centre of the Sun.

76. James Bradley, Bodleian Library, Oxford, Bradley MS. 1. *Bradley's notebook for lectures at Oxford.* fol. 40r.

77. James Bradley, Bodleian Library, Oxford, Bradley MS. 17. *Papers relating to James Bradley's lectures on natural philosophy at Oxford from 1740-1760.* fols. 60v to 75r.

78. Ibid. fols. 60v–75v.

79. Ibid. fols. 75r–81r.

80. Isaac Newton, *Opticks*, London, 1704.

81. James Bradley, Bodleian Library, Oxford, Bradley MS. 17. *Papers relating to James Bradley's lectures on natural philosophy at Oxford from 1740-1760.* fols. 81r—87v

82. In his Will, James Bradley left the Hadley reflector to his sister Rebecca. This is surely a sign that his sister possessed more than a passing interest in astronomy. How many more women observed the heavens without reference to a republic of letters generally closed to them?

83. Ibid. fols. 87v—108v.

84. Ibid. fols. 108r—119v.

85. Ibid. fols. 108r—119v.

86. Ibid. fols. 119v—130v.

87. Ibid. fols. 125r—128v.

88. Ibid. fols. 130r—135r.

89. Which is why in the field of the study of exoplanets, many 'Hot Jupiters' have been discovered. These are gas giant planets that formed in the outer domains of the available gas and dust, before the effects of friction gradually forced such bodies to spiral in towards the primary. Jupiter's inward motion was arrested by the formation of the planet Saturn, another gas giant.

90. Ibid. fols. 135r—143r.

91. Ibid. fols. 143r—154v.

92. Ibid. fols. 154v—164r.

93. Ibid. fols. 164r—176v.

94. There is an illustration of Boyle's air pump dated 1661. Various sources including Columbia College New York, Core Curriculum, Drawing of Robert Boyle's Air Pump, 1661.

95. This is attested to in various places by Rigaud who barely disguises his anger.

96. James Bradley, Bodleian Library Oxford, Bradley MS. 3, *Lectures on 'the more known properties of the Air published...' and notes on members attending his lectures.* The notes on Bradley's students from 1746 to 1760 are located at the back of the notebook.

97. Bradley would have denied that any such 'system' ever existed, it being a 'construction' made by those who failed to understand the Newtonian approach to natural philosophy. I am not asserting that this was necessarily so, but rather that this was Bradley's own apprehension of how Newton approached natural knowledge.

98. This opposition to heliocentricism and experimental philosophy because of its so-called moral consequences brings to mind the contemporary opposition to evolution by natural selection also because it is believed to be both Godless and atheistic even though there is no necessary connection between evolution and atheism.

99. Joseph Stedman Jones, *Moses Principia or Principia Mathematica: Hutchinsonianism and Natural Philosophy in Mid-eighteenth Century Oxford*, MSc Dissertation, Oxford University, 2017. I thank Professor Rob Iliffe for access to this insight into the influence of Hutchinsonianism at Oxford during the period when Bradley was teaching Newtonian experimental philosophy.

100. Ibid. p. 2.

101. Joseph Stedman Jones, *Moses Principia or Principia Mathematica: Hutchinsonianism and Natural Philosophy in mid-eighteenth century Oxford*, MSc Dissertation, Oxford University, 2017, p. 19.

102. William Jones, *Memoirs of the Life, Studies and Writing of the Right Reverend George Horne*, London, 1795.

103. Julius Bate, *A Defence of Mr. Hutchinson's Plan: Being an Answer to the Modest Apology, &c. In a Letter to the Country-Clergyman*, London, 1748. p. 9. Quoted from Joseph Stedman Jones, *Moses Principia* or *Principia Mathematica*: Hutchinsonianism and Natural Philosophy in mid-eighteenth century Oxford, MSc Dissertation.

104. John Hutchinson, *An Abstract from the Works of John Hutchinson, Esq, Being a Summary of his Discoveries in Philosophy and Divinity*, J. Bate & R. Spearman. London, 1755, p. 1.

105. D. Gurses, The Hutchinsonian Defence of an Old Testament Trinitarian Christianity: The Controversy over *Elahim*, 1735–1773, *History of European Ideas*, Vol. 29 (2003), p. 395.

106. George Horne, *The Theology and Philosophy in Cicero's* Somnium Scipionis, *Explained*, Oxford, 1762. p. 48.

107. Joseph Stedman Jones, *Moses Principia or Principia Mathematica: Hutchinsonianism and Natural Philosophy in Mid-eighteenth Century Oxford*, MSc Dissertation, Oxford University, 2017, p. 21.

108. *Nature,* Vol. 400, No. 6739, 1999 July 1, p. 27.

109. What is characterized as being 'reasonable' is refracted through the underlying values of a culture. The results of reason are perceived through a medium of cultural values and historical associations.

110. An example of the cultural impositions on the products of reason.

111. Many early Puritans at the turn of the sixteenth and seventeenth-centuries sought to align their faith with the new knowledge. A doctrine often referred to as 'accommodationism' asserted that when God revealed himself to the various people in the Bible, he did so by accommodating himself to their knowledge and understanding.

112. Christopher Wordsworth, *Scholæ Academicæ: Some Account of the Studies at the English Universities in the Eighteenth Century*, Cambridge, 1877, p. 71.

113. Joseph Stedman Jones, *Moses Principia or Principia Mathematica: Hutchinsonianism and Natural Philosophy in Mid-eighteenth Century Oxford*, MSc Dissertation, Oxford University, 2017, p. 39.

114. Ibid. p. 39.

115. It met the ire of many Christian believers for its nontrinitarian Christology.

116. Benjamin Martin (1704–1782) was a scientific instrument maker with premises in Fleet Street in London. He was also a lexicographer. He was celebrated for his scientific public lectures.

117. Joseph Stedman Jones, *Moses Principia or Principia Mathematica: Hutchinsonianism and Natural Philosophy in Mid-eighteenth Century Oxford*, MSc Dissertation, Oxford University, 2017, p. 39.

118. Roger Heber, Bodleian Library, Oxford, MS. Eng. Misc. e. 15. *Memorandums of Doctor Bradley's Course of Experimental Philosophy,* fol. 1r.

119. Joseph Stedman Jones, *Moses Principia or Principia Mathematica: Hutchinsonianism and Natural Philosophy in Mid-eighteenth Century Oxford*, MSc Dissertation, Oxford University, 2017, p. 45.

120. It was to Lord North that Bradley's executors finally surrendered Bradley's works and correspondence as Chancellor of Oxford University.

121. Ibid. p. 45–46.

122. In both *Sense and Sensibility* and *Pride and Prejudice* Jane Austen's ire is directed against the laws of entailment, whilst *Mansfield Park* is set openly against the background of the contemporary conflicts over slavery. Even the name of the house refers to the Somersett case placed before Lord Mansfield in 1772. *Emma* more subtly, refers to the social and economic consequences of the enclosure acts. *Persuasion* gives witness to the fundamental changes of values in society. Even the earlier novel *Northanger Abbey*, which pokes fun at the popularity of the Gothic novel, places reason above that of the imagination. During a period of political repression, Jane Austen reveals her radical sensibilities, both subtly and assertively. Though her novels express her wit, and often disguise her intentions, her characters so often challenge the values that she was so often set against, by making fun of them.

7

On the Figure of the Earth

James Bradley's work on isochronal pendulums and the geodesic conflict in Paris

> The great obstacle to discovering the shape of the Earth, was not ignorance, but the illusion of knowledge.
>
> Daniel J. Boorstin.

Even at an early stage of James Bradley's astronomical career, George Graham was recognized by many as an outstanding exponent of the skills needed to design and construct the high precision instruments needed in the science of observational astronomy. It should be regarded as a fortuitous moment when Graham and Bradley met and quickly recognized each other's abilities. One man created astronomical instruments of unrivalled design, facility, and rigor. The other was an observer who took these instruments to the very limits of what they were capable of. Such was the respect each had for the integrity of the other that they began to work with each other sharing each other's interests. Bradley was able to help Graham appreciate problems experienced in the practice of observational astronomy, but he in turn was taught to appreciate the problems experienced by the artisans at the workbench who constructed the instruments he used.[1] This led to them working as partners on the problems posed in the development of isochronal pendulums.

Bradley continued making observations with the zenith sector at Wanstead after the publication of his paper on the new discovered motion. He began to suspect the existence of another phenomenon hidden within the residuals of his observations of the aberration of light. He suspected that he was observing a possible nutation of the Earth's axis of rotation, a wobble induced by the Sun's 'tugging action' on the Earth's equatorial 'bulge'. Athough little more than a working conjecture, he eventually recognized that it was due to the Moon's influence as the nodes of the lunar orbit processed over a period of over eighteen years. He realized he would be committed to observing the phenomenon for over the length of the period of a complete regression of the nodes of the Moon's orbit.

The Life and Work of James Bradley. John Fisher, Oxford University Press. © John Fisher (2023).
DOI: 10.1093/oso/9780198884200.003.0008

The increased demands of his teaching duties had forced Bradley to take up a more permanent residence in Oxford. He must have pondered on arrangements for the continuation of his observations with the zenith sector at Wanstead. Elizabeth Pound also uprooted herself from Wanstead to Oxford taking up residence with Bradley in the accommodation provided by the Savilian legacy in New College Lane. It was at least two years before the town house in Wanstead acquired a tenant, Mrs Elizabeth Jenkins, whom they believed could be trusted not to disrupt the zenith sector. Bradley seems to have come to an arrangement with the tenant not to make observations during unsociable hours. At least twice during the following years, Bradley returned to discover that, in his words, the instrument had been manually disrupted.

It was not until 1736 when the nodes[2] of the Moon's orbit had regressed half way around the Earth that Bradley was confident that he was observing a nutation. Newton understood that the Earth was an oblate spheroid, a spherical body greater in diameter through the equator than through the poles. He calculated that the effect of gravity would decrease at the equator relative to the force felt at the poles. This he surmised would be true of all the planets, including the Earth. In the *Principia,* Book III, Proposition 9, Theorem 9 Newton wrote,

> In going inward from the surfaces of the planets, gravity decreases very nearly in the ratio of the distances from the centre. If the matter of the planets were of uniform density, this proportion would hold true exactly by Book I, Section 12, Prop. 73, Theorem 33[3]. Therefore the error is as great as can arise from the nonuniformity of the density.[4]

When Newton completed his discussion of the precession of the equinoxes, he offered a suggestion that its motion may not be uniform, due to the effects of the Earth being an oblate spheroid and not a perfect sphere. He argued that if the 'height of the earth at the equator' exceeded 'its height at the poles' by 17.15 miles, matter would be rarer at the circumference than at the centre and the precession of the equinoxes would be increased because of the excess in 'height' and diminished because of the lesser density.[5] When the motion due to the precession of the equinoxes revealed an apparent inequality, Bradley took seriously the possibility of determining the degree of oblateness of the figure [6] of the Earth, by joining Graham in an experiment with a pendulum clock comparing results in London with those recorded at the tropics. This presumption was strongly supported by the work Bradley undertook during the 1720s with Graham on isochronal pendulums, long an interest of the great instrument maker.[7]

It was with reference to another proposition in the *Principia* that Bradley directed his immediate attention. He examined the possible consequences of the Earth being confirmed as an oblate spheroid.[8] In the recent translation of the *Principia* by Cohen and Whitman,[9] Newton asserted in the third edition that the lengths of pendulums oscillating with equal periods are determined by the gravities. Using data gathered by Jean Richer (1630–1696) during the 1670s in Cayenne and at Paris, where the length of a seconds pendulum (oscillating in one second) is 3 Parisian pieds (feet) and 8½ lignes (lines),[10] Newton computed a table. At the equator the length of a pendulum oscillating in seconds will be shorter to an amount equal to 1.087 lines.[11]

Latitude (degrees)	Length of pendulum (pieds & lignes)		Length of a degree meridian (toises)
0	3	7.468	56,637
5	3	7.482	56,642
10	3	7.526	56,659
15	3	7.596	56,687
20	3	7.692	56,724
25	3	7.812	56,769
30	3	7.948	56,823
35	3	8.099	56,882
40	3	8.261	56,945
41	3	8.294	56,958
42	3	8.327	56,971
43	3	8.361	56,984
44	3	8.394	56,997
45	3	8.428	57,010
46	3	8.461	57,022
47	3	8.494	57,035
48	3	8.528	57,048
49	3	8.561	57,061
50	3	8.594	57,074
55	3	8.756	57,137
60	3	8.907	57,196
65	3	9.044	57,250
70	3	9.162	57,295
75	3	9.258	57,332
80	3	9.329	57,360
85	3	9.372	57,377
90	3	9.387	57,382[12]

Newton's calculations were based on Jean Richer's questionable observations made over half a century earlier. It was entirely characteristic of Bradley to undertake an experiment in partnership with Graham, in order to produce data that was as rigorous, precise, and accurate as contemporary instrumentation and practice could determine. More than once Bradley had learned how unreliable earlier data could be, whether provided by Hooke or other more 'reliable' observers. Graham had long been interested in the study of isochronal pendulums so it seemed opportune that the two friends undertake a joint experiment.

Bradley's paper on the behaviour of isochronal pendulums stated that it was 'a well established truth' that pendulums of the same length did not perform their oscillations in equal times in different latitudes, as predicted by Newton. His ostensible purpose was to accurately establish the difference in the vibrations of a pendulum in London from those in Jamaica to determine how accurate Newton's calculations were. A demonstration might offer confirmation of Newton's calculations. The suggestion that the precession of the equinoxes might be affected by the oblateness of the Earth would support Bradley's current work on what he was sure was a nutation[13] produced by the retrogression of the nodes of the Moon's orbit around the Earth. When the motion due to the precession of the equinoxes revealed an apparent inequality Bradley must seriously have considered the possibility of determining the degree of oblateness of the figure [14] of the Earth by joining Graham in an experiment with a 'seconds pendulum'. They showed that pendulums swinging seconds were in general shorter as the equator was approached. The real difference between the lengths in different latitudes,

> does not seem to have been determined with sufficient exactness by the observations that have hitherto been communicated to the public. This may be gathered from the twentieth proposition of the third book of the *Principia* where the observations are compared both with each other as well as with Newton's theory.[15]

What impresses about Bradley's paper is the thoroughness of the preparations together with the careful planning that went into setting the experiment up. Bradley's commitment was doubly focused. This horological experiment was intended to clarify whether Newton's calculations bore any corroboration to the observations of the suspected nutation he was making with the sector suspended at the town house in Wanstead. The experiment would also determine whether it supported the validity of his work on an inferred nutation. Bradley evidently believed he needed this corroboration. He probably discussed his work on the nutation with his mentor Edmond Halley and had been met with scepticism regarding the precision achieved.[16] Bradley claimed he could observe motions as small as half an arcsecond whereas Halley believed that four or five arcseconds was the limit possible with contemporary instruments.[17] Although Bradley admired much of Newton's work, he firmly believed that there was no substitute for accurate and precise data, no matter how accurate or valid the calculations made. Newton would probably have agreed with this Bradleian characteristic for John Conduitt reported Newton saying that, 'there is no arguing against facts and experiments'.[18]

Although the construction of this immensely accurate clock was entirely due to Graham's extraordinary mechanical skills and those of the artisans working in his workshops, Bradley appears to have been the originator of this particular experiment. Bradley and Graham were equal partners in the execution of it. Both sought solutions to different but associated problems. It is commonly forgotten that as well as being a clock and instrument maker of the highest order Graham was also a first-rate scientist[19] in his own right. Several years after his death in 1751 John Canton wrote,

> The late celebrated Mr. George Graham made a great number of observations on the diurnal variation of the magnetic needle, in the years 1722 and 1723 but declared himself ignorant of the cause of that variation, in No. 383 of the Philosophical Transactions, where many of those observations are to be found.[20]

Like Bradley, he relied on continuous programmes of reiterative observation and experiment as the way to identify and determine the parameters of natural law. Graham was more concerned with the determination of mathematical laws rather than with vain attempts to elucidate 'occult causes' to explain them. Graham's skills and achievements had included the invention of the dead-beat escapement, a major advance in the accuracy of time-pieces, though it must be accepted that Bradley's faith in Graham's consummate craftsmanship was based on several years of continuous experience of his immense abilities.

Bradley was confident that the experiment would provide evidence that would support his work on the as yet barely identified nutation and counter the scepticism of his mentor Edmond Halley. In his account Bradley wrote,

> there is a clock whose pendulum vibrates seconds, made by our ingenious member Mr. George Graham, justly esteemed for his great skill in mechanics; who judging that an opportunity was now offered of trying with the utmost exactness what is the true difference between the lengths of isochronal pendulums at London and Jamaica, readily embraced it; and in framing the parts of the clock, carefully contrived that its pendulum might at leisure be reduced to the same length, whenever there should be occasion to remove the clock from one place, and set it up in another.[21]

The clock was adapted so that all of the appurtenances associated with a clock of the time such as striking parts and so forth were omitted. Everything was directed to the accurate timing of astronomical observations.[22] The clock oscillated seconds of sidereal and not solar time to increase its facility as an astronomical time-piece. Various tests were made by Graham whilst the clock was at his premises in the Strand in London. Through observations made of transits of α Aquilæ (*Altair*),[23] Graham was able to determine that the clock gained twelve seconds in ten apparent revolutions of the star.[24] The immense thoroughness of the preparations was demonstrated in the way the pendulum was tested in two different ways. It was tested for the effects of temperature on the length of the pendulum and on the effects that various weights might have on the capacity to keep 'true time'. To determine how the pendulum might be affected by the greater heat of the Jamaican climate, a thermometer was also fixed close to it. Whilst Graham observed and timed the passages of the test star *Altair* through his meridian, he also read off the temperature from the height of the spirits in his thermometer. All of these tests and observations were performed between 20 and 30 August 1731.[25] After these tests Graham altered the clock-weight from one of 12lb 10½oz[26] to one of 6lb 3oz[27] (the weight of the pendulum being 17lb[28]) and noted that the pendulum now oscillated only 1°15′ each way compared with its former 1°45′.[29] There was no difference in the accuracy of the clock. Bradley concluded,

> This experiment shews, that a small difference in the arcs described by the pendulum, or a small alteration in the weight that keeps it in motion, will cause no great difference in the duration of the vibrations; and therefore a little alteration in the tenacity of the oil upon the pivots, or in the foulness of the clock, will not cause it to accelerate or retard its motion sensibly; from whence we may conclude, that whatever difference there shall appear to be, between the going of the clock at London and in Jamaica, it must wholly proceed from the lengthening of the pendulum by heat, and the diminution of the force of gravity upon it.[30]

For a mind as perceptive as Bradley's, this conclusion was unusual. On reflection, such an experimental set up must have been subject to many other sources of error than those mentioned including barometric pressure.

After obtaining the results of the observations in Jamaica[31] and comparing the thermometer readings, Bradley and Graham calculated the effects that the lengthening of the pendulum might have on the slowing of its oscillations. From the records of the performance of the clock in London, it was concluded that the clock was one second a day slower when the thermometer was two divisions higher. From the Jamaican records, the thermometer varied from between fifteen and twenty divisions higher than those recorded in London. It led to the conclusion that on average the clock went some 8½ seconds slower each day due to the greater air temperatures in Jamaica than when it was situated in London.[32] The clock lost 2 minutes and 6½ seconds a day in Jamaica compared to its performance in London.[33] By taking away the effect due to the expansion of the pendulum, Bradley deduced that the clock lost 1 minute 58 seconds a day due to the diminution of gravity. Bradley continued,[34]

> the increase of gravity as we recede from the equator, is nearly the square of the sine of the latitude; and that the difference in the length of pendulums is proportional to the augmentation or diminution of gravity. Upon these suppositions, I collect from the forementioned observations, that if the length of a simple pendulum (that swings seconds at London) be 39.126 English inches,[35] the length of one at the equator would be 39.00,[36] and at the poles 39.206.[37]

When the clock was sent to Colin Campbell in Jamaica, care was taken to avoid the effects of bias. Bradley reported that Graham sent full directions describing how the clock was to be set up and how the pendulum might be reduced exactly to the state as when it had been situated in England. Immense care was taken in not revealing how the clock had performed in London. Bradley and Graham wanted to avoid all possible bias or prejudice in favour of any hypothesis or any former observations that Campbell may have been aware of.[38] After such close attention to every detail, Bradley came to a remarkable conclusion that must have exceeded all expectations. He was aware of the capacity of even the most diligent of scientific investigators to be misled by favourable results. Bradley writes, 'the difference of the squares of the sines of 51½° and 18°, the latitudes of London and Black-River being to the square of the radius as 118 to 228¼, the clock will go 1′58″ in a day slower at Black-River than at London, as was found by observation.'[39]

Precise and diligent workers like Bradley and Graham, having taken so much trouble to avoid bias, must have suspected that the result, in which the observations exactly equalled the prediction of theory, was better than could be hoped for. Whilst Graham gave explicit instructions to Campbell on how to reduce the pendulum to the state that it was in at London, there must have been many other factors involving various forms of tacit skill, knowledge, and practice that came into play. There was a significant variation between the results of this experiment and the predictions contained in the *Principia*.[40] Yet the postulation that the Earth was an oblate spheroid was fully supported by the results of the experiment.

There was at this time an ongoing conflict taking place in Paris over whether the Earth was a prolate or an oblate spheroid. A modest refracting or reflecting telescope will reveal that the gas giant planets, Jupiter and Saturn, are both oblate spheroids in contradiction to the predictions of Cartesian theory. According to upholders of the theory of vortices, with the belief that gravity was the result of the pressure of matter in the vortices, the figures of Jupiter and Saturn were never explained. Bradley measured the diameters of both Jupiter and Saturn across the poles and equators to confirm the apparent oblateness of both planets. Bradley expressed concern about the problems associated with observational uncertainty. His mature work after he took up residence at Greenwich as the third Astronomer Royal in 1742 is characterized by his practice of always recording possible sources of error. A result such as that derived from Campbell's observations should have led to Bradley entertaining misgivings about a result that was so in agreement with prediction. Bradley did, however, have great faith in the precision and accuracy of Graham's handiwork[41] and this may have alleviated a few of the concerns he may well have entertained.

Bradley's reservations were not recorded, but he must have hoped that the example of this experiment with Campbell[42] and Graham, to whom he gives all due respect in his account, might be repeated by others.[43] Bradley believed that experimental results required repeated confirmation or possible refutation. His own reiterative practices were such that he was possibly relieved that the results of the pendulum experiments supported his continuing observational work on what he was increasingly convinced was a nutation of the Earth's axis. He was, as always, concerned about the possible effects of observational bias. Bradley knew that the phenomenon he was observing was of extraordinarily minute parameters and was subject to Halley's scepticism about the observations of a suspected nutation of such minute proportions of the Earth's axis. In spite of any possible reservations he may have held Bradley concludes his paper with confidence. He wrote,

> For these reasons, I esteem Mr. Campbell's experiment to be the most accurate of all that have hitherto been made, and properest to determine the difference of the gravity of bodies in different latitudes; and therefore I will subjoin a table which I computed from it, containing the difference of the length of a simple pendulum swinging seconds at the equator, and at every

fifth degree of latitude, together with the number of seconds that a clock would gain in a day in those several latitudes, supposing it went true when under the equator; by means of which, any one may readily compare other the like observations with his, and thereby discover whether the alteration of gravity in all places be uniform, and agreeable to the rule laid down by sir Isaac Newton, or not.[44]

Bradley was very certain that the Earth was an oblate spheroid for his experiments with isochronal pendulums not only supported his observations of the 'nutation', but also confirmed the observations made with his uncle James Pound of the figures of Jupiter and Saturn. I include Bradley's Computed Table.

Bradley's Computed Table[45]

Lat. Deg.	Diff. Inch.	Gain. Secs.	Lat. Deg.	Diff. Inch.	Gain. Secs.
5	0.0016	1.7	50	0.1212	134.0
10	0.0062	6.9	55	0.1386	153.2
15	0.0138	15.3	60	0.1549	171.2
20	0.0246	26.7	65	0.1696	187.5
25	0.0369	40.8	70	0.1824	201.6
30	0.0516	57.1	75	0.1927	213.0
35	0.0679	75.1	80	0.2003	221.4
40	0.0853	94.3	85	0.2050	226.5
45	0.1033	114.1	90	0.2065	228.3

Lat. = The latitude of the place; Diff. = The difference of the length of the pendulum in parts of an English inch; Gain. = Seconds gained by a clock in one day.

The publication of Bradley's paper on isochronal pendulums in 1732[46] coincided with another paper published in Paris by Pierre Louis Moreau de Maupertuis entitled, *Discourse on the Shape of the Heavenly Bodies*. It led to a major debate in the Royal Academy of Sciences in Paris. Mary Terrall writes,

geodesy became a major topic of discussion in the academy. This later metamorphosed into a new version of the debate about gravity, but initially contention centred on measurement and calculation techniques, rather than on theory. Interpretations of cosmology were overshadowed by arguments about how to perform and evaluate observations and calculations, about who should be entrusted to do this work, and about how they should present it to the public.[47] To demonstrate its ability to enhance the glory of the crown out to the ends of the earth — one to the equator and one to the Arctic circle. These expeditions generated intense public interest in what might seem an arcane scientific question: how much did the shape of the earth deviate from the spherical?[48]

The publication of these papers by Bradley in London and by Maupertuis in Paris coincided with the continuation of a lively debate over 'le figure de la terre'. Although Bradley's work with Graham and Campbell appeared to give experimental support for Newton's theory, most French natural philosophers and astronomers ignored the result.

Any study on the debate about 'the figure of the earth' may possibly contrast the practices of natural philosophy in England and France. It is a diversification apparent from the earliest years of the Royal Society in London, and the Royal Academy of Sciences in Paris. In England, natural philosophy was an activity undertaken by gentlemen motivated by curiosity. In France, natural philosophy, equally motivated by curiosity, was so often undertaken by savants and academicians as servants of the state. The Royal Academy was an extension of a bureaucracy that served to glorify Le Grande Nation. So when Bradley, Graham and Campbell published their experiment and its results in the Philosophical Transactions they did so as Fellows of the Royal Society. It supported an ethos where many of the issues of what qualified as experimental evidence had been socialized differently from that within France. London was fortunate in possessing a vibrant market for scientific instruments with many artisans protected by a robust patent system, providing instruments for a growing market made up of gentlemen amateurs. The situation in France was rather different. There were contentious schools of thought, while artisans generally possessed a low social status, Mary Terrall states,

> In the battle over the shape of the earth, conceptual issues associated with the labels 'Cartesian' and 'Newtonian' were embedded in concerns about other features of scientific practice. As the dispute unfolded, cosmological or theoretical were no more essential to the dispute than claims about authority and expertise, the role of analytical mathematics in astronomy and cosmology, the design and use of astronomical instruments, and the appropriate presentation of quantitative results. Personal animosities also entered into the mix, as time went on.[49]

Throughout the final decades of the seventeenth and the first decades of the eighteenth century several French astronomers, including Jacques Cassini and his supporters, performed extensive geodesic surveys which appeared to support the Cartesian theory that the Earth was a prolate spheroid.[50]

In 1728, Maupertuis visited London and like his compatriot Voltaire[51] discovered that he had moved into an entirely different world.[52] Prior to its publication, his paper on rotating fluid bodies in Paris was read to the Royal Society after a short delay in July 1731. Maupertuis, sensitive to the suspicion that he may have offended his audience, wrote to Hans Sloane expressing his concern.[53] The paper was in the possession of Bradley's friend John Machin (1680–1751) at Gresham College who probably discussed it with Bradley and Graham. There is no doubt that Maupertuis was offended by the delay in the publication of his paper by the Royal Society. In a letter to Pierre des Maiseaux[54] in April 1732 Maupertuis complained that it had not been included in the *Philosophical Transactions*.[55] It had been written by a leading academician supportive of Newton's stated methods, so contrary to much that was published in Paris. Machin wrote to Des Maiseaux who passed on the response to Maupertuis, He sought to reassure him. He wrote that, 'it was so far from being thought slight or trivial or unintelligible that on the contrary every member who has any knowledge in these matters was extremely well satisfied with it as a performance which discovered a great skill and address in the author'.[56]

It appears that Machin pointed out some degree of confusion in Maupertuis's argument, where the law of gravity is assumed as a given, determining the shape of a rotating body. When Maupertuis understood this criticism, he agreed his paper provided 'mathematical solutions rather than true and exact physical explanations'.[57] Mary Terrall writes, 'Maupertuis had hoped to impress his English colleagues, and was relieved that his paper finally appeared in the *Philosophical Transactions*, although the response from mathematicians was modest at best. Machin was certainly polite, but did not see anything path-breaking in Maupertuis's work'.[58]

Bradley, Graham, and Campbell had decided to undertake their experiment on isochronal pendulums in order to subject Newton's predictions to observational verification, partly in order to determine the validity of Newtonian theory. In Bradley's case, he sought to determine whether it gave support for the hypothesis he was assessing, a lunar-induced nutation. The contrast of the philosophical milieu in which Maupertuis worked with that in England was marked. By the time his paper had been published in the *Philosophical Transactions* in the spring of 1732, Maupertuis had decided to reveal to his compatriots where his loyalties lay. In addition to the contrasting philosophical arguments at play in Paris, there were also currents of patriotic loyalty at work.

Mary Terrall mentions an issue I discussed with Eric Aiton. Johann Bernoulli had been awarded the Academy's prize for an essay written in 1730. It criticized Newton's mathematical critique of vortices. Bernoulli's commitment to Cartesian natural philosophy was so marked that he believed it was possible to reconcile the motions of the vortices with Kepler's laws of planetary motion. This was a complete nonsense. How is it possible to reconcile a motion of something that had not been seen or was even possible to see with Kepler's laws based on Tycho Brahe's accurate observations? Bernoulli informed Maupertuis that he was planning to construct a 'system of the world' combining the 'best' of both Descartes' and Newton's works. It would be based on a mechanical percussive physics and not on 'forces' acting across a void. He would challenge Newton's objections to vortices on mathematical grounds. It was challenging the entirety of Book II of Newton's *Principia*. Bernoulli was viscerally opposed to Newton's gravitation across empty space. Maupertuis must have been fearful of giving offence to Bernoulli, his mentor. Mary Terrall neatly sums up Maupertuis's position. She writes,

> He accepted Newton's critique of Descartes, based on the incompatibility of Kepler's planetary laws with fluid dynamics, explaining it in relatively simple terms, and slid almost imperceptibly into a defence of the Newtonian position on void, space and action at a distance. In doing so he rejected Bernoulli's compromise which had impressed him originally.[59]

Maupertuis had no need to worry when his mentor read his book in Basel. Indeed his response was famously magnanimous. He responded to Maupertuis's book as 'a hypothetical presentation of gravity', accepting the author's profession of ignorance about the physical cause of gravity.[60] He took no offence and replied, 'If you appear English to me, Monsieur, you appear to be a reasonable Englishman, who does not push the principle of attraction beyond geometry into physics, as the rigid followers of Newton do, more than Mr. Newton himself did.'[61]

But if the relationship between Maupertuis and his mentor was conducted with a high degree of mutual respect, the atmosphere within the Academy was nothing if not positively venomous at certain times in stark comparison. The disputes were so contentious that any appeal to mere precision and accuracy in observation was 'a matter to be interpreted, attacked, defended, and represented'.[62] The dispute over the figure of the Earth in Paris during the 1730s and later was a vindictive affair. It involved philosophical and intellectual politics, the fashions of the court and salons, and personal antipathy in an atmosphere that ventured on the poisonous.

Terrall claims that the only empirical support came from pendulum measurements taken near the equator, probably referring to Richer's observations in Cayenne in the 1670s, apparently ignoring the more precise measurements made by Campbell, Bradley, and Graham in 1732. The origins of the dispute stemmed from geodesic observations made by Jacques Cassini during the project to produce an accurate map of France. Evidence produced from the measurement of arcs of longitude suggested that the Earth was a prolate spheroid in conformity with the predictions of Cartesian theory. These results were published in 1718. Within the Academy, this result lay unchallenged until the 1730s.

Jacques Cassini took any criticism of his measurements personally. But since Bradley had discovered the aberration of light in 1728, any positional observations made of the stars prior to this, no matter how precise or accurate, were open to challenge, at least according to those who accepted the progressive motion of light. This was denied by Jacques Cassini on philosophical and personal grounds based on his own perceived integrity as an observer. Within the 'English camp', the discovery of aberration made earlier observations invalid until demonstrated otherwise. Jacques Cassini refused to accept this. He held to the Cartesian insistence that the velocity of light was instantaneous. Cassini refused to accept Bradley's 'discovery' or rise to the English astronomer's simple challenge to explain annual aberration in terms that did not rely on the motion of the Earth or the acceptance of the progressive motion of light. The growing acrimony led to calmer voices seeking a resolution of the dispute. Voices within the Academy conceded that measurements made entirely within France were inadequate to decide the issue. Two expeditions were commissioned, one to be sent to equatorial regions in Peru[63] and one to the polar regions in Lapland.

If the determination of the shape of the Earth was free from controversy in England, it was because of the prestige of the recently deceased Sir Isaac Newton. In France, there were rival camps ranged against each other like contending armies. For some the conflict was conducted in gentlemanly terms. Maupertuis and Alexis Clairaut spent the academic vacation in September and October 1735 with the Cassinis at their country estate at Thury.[64] Both sides in this altercation were convinced that they would be vindicated by the results of the two expeditions, often greeting fellow academicians with cordiality. Besides making astronomical observations, each expedition was to undertake pendulum experiments. Maupertuis was determined to acquire the finest instruments available. His familiarity with the London market for scientific instruments led him to approach James Bradley. Referring to the preparations for the pendulum experiments Mary Terrall writes,

> Eighteenth-century investigators, both English and French, used new clocks, paying careful attention to variables affecting their performance, in order to refine measurements of effective gravity.[65] George Graham, the London instrument maker had raised the standard of precision with his measurements of the effect of temperature on the length of the pendulum.[66]

I will discuss these expeditions only sufficient to throw light on Bradley's labour during this period on what he was confident was a nutation of the Earth's axis. This phenomenon was perceived by Bradley to be connected with the establishment of the figure of the Earth as an oblate spheroid. Bradley's indirect involvement with the Lapland expedition came from Maupertuis's request to the English astronomer to commission Graham to construct a zenith sector similar to the instrument he used to discover the aberration of light. Bradley's theory was still not accepted by the Cassinis and their followers because it challenged the Cartesian 'system' on its fulsome insistence on the instantaneous velocity of light. Any theory dependent on an acceptance of the finite velocity of light would challenge the entire rationale of Cartesian theory, the notion of the universe being a plenum, the transmission of light being explained as 'an instantaneous pulse' transmitted through the universal medium.

The opposing Newtonian model of light as a stream of corpuscles moving through a void also threatened the Cartesian nexus. Rather like the Aristotelian system it purported to replace, the Cartesian system was an integrated whole based on mutual necessity which could not be challenged or replaced in part. The Newtonian contingent approach to natural knowledge, which abandoned hypothetical causalities in favour of the determination of mathematical laws, revealed a different heuristic, challenging and undermining the long-standing epistemological tradition that separated mathematics from physics for over two thousand years.

Descartes' *Géométrie* was the third appendix to his *Discours de la methode* supporting his rationalist approach to the study of nature, yet he barely applied it to his natural philosophy. The Cartesian rejection of Aristotelian physics was not complete, for it had maintained the fundamental separation between mathematics and physics. Descartes was a first-class mathematician, the author of a great work revealing how geometrical problems can be reduced to algebraical formulations, a fundamental development in mathematical facilitation. The Newtonian approach to natural philosophy involved the abandonment of qualitative conjectures with the adoption of mathematical analysis in order to predict the future behaviour of natural phenomena. To the Newtonians, 'scientific' problems were open to analysis based on empirical data and the mathematical inferences drawn from them. The Cartesian approach was based on a predetermined rationalistic approach to nature. To Descartes, the problems of nature were simple, and he believed that the outstanding problems would soon all be resolved.

Jacques Cassini had deduced from measured variations of the length of degrees of latitude along a meridian through Paris that the Earth was shaped in accordance with Cartesian theory. It conformed to the Cartesian prediction that the Earth was shaped like a prolate spheroid. Cartesians like Jacques Cassini sought confirmation of the expectations of Cartesian theory. It was an example of what Bradley always sought to avoid: 'seeing what you wanted to see'. Those who took their lead from the more open-ended approaches to natural knowledge developed by the followers of Newton were unsure about the possibilities of establishing theory without reiterative observation or experiment. As in Bradley's work, constant vigilance was always the key to the establishment of knowledge claims.

As Cassini's results were widely regarded as the most accurate ever made, they were accepted as firm empirical evidence in support of the Cartesian system.[67] The leading figure of French science during this period, the long-lived Bernard le Bouvier de Fontenelle (1657–1757) published a hostile review of Maupertuis' *Discours*. He asserted that measurements made by reliable geodesists such as 'the worthy and experienced' Jacques Cassini, were much to be preferred to any theoretical demonstrations of geometry. In this he was very much in agreement with the Newtonian school. Yet Fontenelle directed his ire not merely to Maupertuis, but to the 'Newtonians' in general, most specifically at the *Principia* itself.[68] Fontenelle admired the *Principia* but regarded it as a geometrical exercise that had little or nothing to do with the 'real world'.

The Academy had planned an expedition to Peru in a location in what is presently Ecuador. In a *mémoire* written in 1735, Maupertuis argued that a second expedition was required to travel to Lapland.[69] Some authors have argued that these expeditions took on the character of an *experimentum crucis*. Amongst the authors who have favoured the interpretation that the Peruvian and Lapland expeditions resolved the conflict between the Newtonians and the Cartesians are Rene Taton,[70] Valentin Boss,[71] and David Beeson.[72] Yet the research and writings of Eric Aiton,[73] John Greenberg,[74] Mary Terrall,[75] and Rob Iliffe[76] suggest that the resolution of the conflict was not achieved by these experiments and observations.

Although there were methodological approaches to knowledge claims that radically divided the two schools, there was sufficient overlap to allow discourse between them. The Lapland and Peruvian expeditions were largely inconclusive because the results were open to many challenges that created further debate. Greenberg goes further, and asserts that the conflict between the Cartesian and Newtonian world views would never have been resolved by the geodesic expeditions; both schools in the main supported a flattening of the poles, although the Cassinis adamantly refused to accept this view.[77] Much of the rivalry was coloured by the reception of the *Principia* in France, delayed by its impenetrability, and exacerbated by the conflict between the supporters of Leibniz and those of Newton. The *Opticks* received a more sympathetic treatment, but in France the work of Newton faced a continuing resistance well beyond any clearly discernible reasons.

Bradley appeared to express a degree of reticence in getting entangled in the debate when he stated that he would prefer not to get involved with it.[78] He must have regarded the dispute with a degree of detachment and certainly with a measure of bemusement. Bradley appreciated the intractability of the world and was critical of the observational methods utilized by Maupertuis and his colleagues in Lapland. Against his better judgement, it was a subject that would indeed involve him. When the account of the experiment on isochronal pendulums came to the notice of French natural philosophers and astronomers, some of the French Newtonians perceived Bradley as a supporter in their ongoing disputes. They asked the English astronomer to confirm that their application of the aberration of light was correct, although this was really a courtesy. Clairaut and the other expedition members were highly competent mathematicians. On 27 September 1737, NS Maupertuis wrote to Bradley,

Sir,

The rank you hold among the learned, and the great discoveries with which you have enriched astronomy, oblige me to give you an account of the success of an undertaking wherein the sciences are much interested, even if I were not inclined to do it through the desire which I have of the honour of your acquaintance, on account of the part you bear in our work itself; we being indebted for a principal part of the accuracy thereof to an instrument made upon the model of yours, and towards the construction of which, I understand, you were pleased to give your assistance.[79]

This was a translation made by Bradley in Oxford on behalf of his friend George Graham in London, who had little or no French. Bradley wrote,

Dear Sir,

Mr. Maupertuis having done me the honour to communicate to the Royal Society, through my hands, the observations which the French gentlemen made in the north, as I apprehend his account can be to none more agreeable than to you, who had so much trouble in the affair, and to whom we are principally indebted for the accuracy of the experiment.[80]

I have persistently argued that an instrument, no matter how accurate, is only as good as the practitioner using it. One of the most remarkable of all partnerships, possibly unique in its consequences in eighteenth-century astronomy, was that between Graham and Bradley. I have likened it to the relationship between a great violin maker and a great virtuoso performer. The problem with the observations made by the Lapland expedition is that they did not capitalize on the flexibility of the instrument provided by George Graham. In a passage from his letter to his friend, Bradley wrote,

The method they took to verify the quantity of the arc, by making observations on different stars, and carrying the instrument twice backwards and forwards, does certainly remove all scruples relating to that point. But as they were at all that pains, might they not, with a little additional trouble, have been yet better assured of the true quantity of their arc, by turning the whole apparatus on which the instrument hung from west to east, &c, by which means they would have had on the arc of the sector the double distance of the star from the zenith at each station, and likewise have removed all objections relating to the alteration of the line of collimation, upon carrying the instrument from one station to another.[81]

By the simple expedient of reversing the sector at each station, any possible errors in the collimation of the instrument, as well as the measured position of each star, would be negated. To Bradley it seemed so extraordinary that something so simple and obvious appears to have escaped the notice of such an eminent party. If the expedition had followed such a procedure all objections concerning the collimation, as well as the measurement of each star's position at each station would have been nullified. Maupertuis considered such a procedure but was convinced that the construction of the instrument made it impossible to reverse it and rectify it as Bradley envisaged. This puzzled

Bradley for as he understood it the instrument made by Graham was modelled on the one he was using at Wanstead, which he knew to be entirely reversible. Bradley conferred with Graham suggesting the French party might use it on 'their famous meridian'[82] for they would be able to demonstrate the accuracy of the instrument in a place and in circumstances open to examination by all concerned.

The prime reason for the letter from Maupertuis to Bradley was to give an account of the expedition to the Royal Society. As a courtesy it sought to inform Bradley of the results and to confirm that the aberration of light had been correctly applied. Bradley conferred with Graham discussing whether he should reveal his observations of the suspected 'nutation' and apply a further corrective to the observations made by the Lapland expedition. Although a few friends and colleagues were familiar with Bradley's continuing observations at Wanstead, he had not yet made public his work on what he considered was evidence of such a nutation. Bradley's future correspondent in France, the positional astronomer Lacaille, insisted in later correspondence that the phenomenon be called a 'deviation' until it be fully established as a nutation. Lacaille had enormous respect for his English correspondent and had no doubt that some such phenomenon existed, whatever the cause. In this he was quite unlike Halley who privately considered his friend's observations delusional, convinced as he was that such high levels of precision were beyond the bounds of possibility. Halley died in 1742, six years before Bradley's work on nutation was published in a paper addressed to George Parker the 2nd Earl of Macclesfield. The story of Bradley's work on nutation will be told Chapter 8.

Bradley made known his work on the nutation and applied it to the observations made by the Lapland expedition. Even though I will discuss this letter in greater detail in Chapter 8. I make no excuse for quoting from Bradley's letter to Maupertuis, which is both lucid and informative. He wrote,

> In my account above referred to, I have taken notice that, so far as I could then judge from the observations of a single year, the annual precession of the equinox was at that time greater than 50″, the stars near the equinoctial colure (between the years 1727 and 1728) changing their declination 1″½ or 2″ more than a precession of 50″ required. On the other hand, I observed that stars near the solstitial colure altered their declination less than they ought if the precession was 50″. Which seeming contradiction in the phenomena, with regard to the stars in those different situations, I now conceive to arise from the unequal action of the moon upon the equatorial parts of the earth, which, varying on account of the different inclination of her orbit to the

equator, will cause a small nutation in the earth's axis, as also an acceleration and retardation in the precession of the equinoxes.[83]

Bradley gave details of his observations to Maupertuis revealing the 'whole quantity of this nutation' amounted to $9''$ each way from the mean. He explained that it produced an inequality in the precession so it can vary from $42''$ to $57''$ in any one year. It is the first announcement of Bradley's second great discovery. The two men continued to correspond, each understanding that their researches were mutually supportive at this time, although Bradley was sure that the explanation of what his future *copain* Lacaille referred to as a deviation, was in fact a nutation.

The resolution of the problem of 'the figure of the earth' was never going to be achieved by the highly publicised expeditions to the equator and the polar regions. There were too many disagreements as to what was acceptable as evidence. Many of those who questioned the results of the Lapland expedition awaited the arrival from Peru, the one that in the eyes of those who supported the Cassinis possessed the greatest credibility. However, unlike the speedily organized and effectively led expedition to Lapland the expedition to the Viceroyalty of Peru under the 'protection' of the King of Spain bordered on the farcical. The weak leadership of Louis Godin and his early replacement by the *de facto* leader Pierre Bouguer set the expedition back from the start. The third academician was the young and inexperienced Charles-Marie de La Condamine, a friend of Voltaire's. Yet it was the almost complete unpreparedness of the leading figures of the expedition that hampered it throughout. Larrie D. Ferreiro underlined the incompetence of the party: 'This dichotomy between expectations and reality was already becoming a hallmark of their mission. Plans that seemed to be ideal on first inspection were plagued with overwhelming problems when actually executed. These problems went far beyond the normal expected setbacks of any scientific expedition.'[84]

In addition to this organizational incompetence, the expedition's members were often guilty of treating the native population with contempt, and they in their turn regarded the expedition with suspicion. To a population long inured to exploitation by their Spanish masters, French protestations that they were there as an expedition to seek the figure of the Earth was met with the belief that they were in fact seeking gold or silver. Sometimes the expedition was under siege with the lives of its members in jeopardy. An expedition envisaged to last three years, including the travel to Peru and back to France, delayed some members by about ten years.

No matter what the rival claims within the Academy were, the opposing factions usually maintained their commitments, some now resolutely suggesting that this was hardly a scientific dispute at all, if by that we understood it to be an argument to be resolved by empirical or even rational evidence. No matter how much the two schools of thought were arrayed against each other there was insufficient agreement about how the conflict might be resolved. But it was the Cassinis, and a small number of their supporters, such as Johann Bernoulli, who found it difficult to accept the Newtonian approach, or adamantly refused to accept the arguments contained in the *Principia*. Jacques Cassini was by many accounts, a generous and kindly man, as well as being a good astronomical observer with an acute understanding of the issues. Yet he could never be deflected from his commitment to what he believed he had 'proved' in his geodesic surveys of France, that the Earth was a prolate spheroid. To deny this was to suggest that he was incompetent. Rather than admit it he retired from the debate.

In the rivalry between France and the two naval powers of England and the Dutch Republic, the French state financed and sponsored natural science so that the leading natural philosophers in France effectively became agents or employees of the state. In England, the Royal Society was an association of gentlemen motivated by curiosity in the pursuit of independent enquiries into the natural world. The two rival models are to some degree still played out in modern life. Some institutions support a more open-ended approach to scientific research undertaken in the belief that scientific advances come from entirely unexpected quarters. This is opposed to the notion that science can best be supported by national and commercial budgets with investment in particular areas of research to achieve required objectives. The latter approach makes sense in a Cartesian ethos where the objectives of empirical science were to 'fill in the gaps' or clarify outstanding issues. This was opposed to the more open-ended approach to empirical science so characteristic of Bacon, Newton, and their acolytes.

Jacques Cassini was committed to the maintenance of what he believed to be a thoroughly sound approach to research and was unable to perceive that there were many shortcomings in its acquisition of natural knowledge. At stake were issues far greater than the resolution of an apparent conflict over the figure of the Earth. At stake were opposing models of scientific knowledge claims. During the eighteenth century, there developed a degree of methodological fusion between the contending parties, as can be seen today when enormous international investment is required in order to construct a vast instrument such as the large hadron collider in order to resolve fundamental questions about issues produced by open-ended scientific endeavour.

By the end of the 1730s, when the Peruvian expedition was still at work, the situation in Paris had altered markedly, such that Cassini de Thury recalibrated his father's original survey[85] and had come to the conclusion that his results rather favoured the notion of a flattening of the poles which had been so adamantly denied by his father. In 1740, Pierre Bouguer, the *de facto* leader of the Peruvian expedition, was still involved with the difficult and time-consuming survey in the Andes. In 1742, Maupertuis was elected as the director of the Royal Academy of Sciences, a radical transformation in the position of the Newtonian school in Paris.

In the meantime Alexis Clairaut began working independently of the dispute between those who followed the Newtonian and Cartesian approaches to natural philosophy, working on his own resolution of the conflict. In 1743, he published *Théorie de la figure de la terre, tirée des principles de l'hydrostatique* in which he expressed his ideas in what has since become known as Clairaut's theorem. It connects the effective gravity at various points of a rotating ellipsoid with the compression and centrifugal force at the equator. It revealed that a mass of a homogenous fluid which had been set in rotation around a line through the centre of its mass would through the mutual attraction of its particles adopt the form of an ellipsoid. Yet it was very much a sign of the times that even here the politics of the Academy intruded. Greenberg writes,

> In the introduction to his treatise of 1743, Clairaut stated that Bouguer's Paris Academy *mémoire* of 1734 is what had induced him to investigate the conditions for equilibrium. Clairaut declared that Bouguer had made him wonder whether all of the conditions necessary for equilibrium had been determined.[86] Perhaps what Clairaut stated here is true, but it does seem more likely, based on the chronology of the events told so far in this story, that he first found the letters to him from MacLaurin, whom he neglected to mention at all in the 'introduction' to be more troublesome than Bouguer's *mémoire*. MacLaurin's letters, not Bouguer's *mémoire*, are what originally caused Clairaut to look into the question of conditions for equilibrium.[87]

In the years between the return of the Lapland expedition in 1737 and the publication of Clairaut's treatise in 1743, it had become fashionable in the salons of Paris to abandon the Cartesian accounts of the world, such as Fontenelle's lucid *Entretiens sur la pluralité des mondes*, to be replaced by works like Francesco Algarotti's *Il Newtonianismo per la dame*, a popular presentation of Newtonian theory for the public. Bouguer had been equivocal before he travelled to Peru refusing to side with either party. He returned to France armed with data that would find favour with the supporters of the Newtonian party. He returned with results of the geodesic observations and his gravity experiments in the vicinity of Chimborazo, widely accepted at this time as the highest mountain on Earth.[88] These experiments were the first confirmation of Newton's theory of universal gravitation.

This has taken the narrative far from Bradley and his observations at Wanstead, but they must be placed in the scientific and social contexts within which he undertook his work up to 1748 when he published his papers on pendulums and the nutation of the Earth's axis. There was an appreciation that he was observing a phenomenon of extraordinary minuteness, given the observational limits of the instruments of his time. Many in the astronomical community regarded his work with awe. To spend years observing a motion of about 1 arcsecond a year seemed to be pushing the very limits of what was technically possible. Bradley was aware that in France Maupertuis' observations were often treated with scepticism due to his commitment to Newton's universal law of gravitation. Bradley responded,

I could not without concern hear of the objections that were raised against the certainty of your observations; but the world, I imagine, must soon be convinced of their invalidity; and you, I doubt not, will have the satisfaction of finding the truth of this saying:[89] 'Magna est veritas et prævalebit'.[90]

Given the accuracy and precision of the experiments on isochronal pendulums conducted with Graham and Campbell, he was convinced of the validity of the observations made by the Lapland expedition. Bradley must have regarded the results reported by Maupertuis as a confirmation of his own. But for Bradley having revealed some information about his observations of what he was now sure was a nutation of the Earth's axis, his work of several years of sustained effort, he was sure was not in vain.

Notes

1. This mutual appreciation helped Graham to construct instruments of great facility and Bradley to acquire instruments not only of immense precision and accuracy, but also within the capabilities of current workshop practice. Compare this with Hooke's often quite brilliant but entirely impractical designs.

2. The points where the plane of the Moon's orbit around the Earth intersect the plane of the Earth's equator. They regress in a period of over eighteen years.
3. Isaac Newton, *Philosophiæ Naturalis Principia Mathematica,* Book I, Section 12, Prop. 73, Theorem 33. 3rd. edn. 1726. Translated by I. Bernard Cohen and Anne Whitman, *The Principia: Mathematical Principles of Natural Philosophy,* London, 1999, p. 593, reads 'If toward each of the separate points of any given sphere there tend equal centripetal forces decreasing in the squared ratio of the distances from those points, I say that a corpuscle placed inside a sphere is attracted by a force proportional to the distance of the corpuscle from the centre of the sphere'.
4. Ibid. p. 815.
5. Ibid. pp. 887–888.
6. The term 'figure' was used in the sense of its 'shape'.
7. James Bradley, 'An Account of Some Observations Made in London by Mr. George Graham, F. R. S.; and at Black-River in Jamaica, by Colin Campbell, Esq. F. R. S. Concerning the Going of a Clock, in Order to Determine the Difference between the Lengths of Isochronal Pendulums in Those Places', in *Miscellaneous Works and Correspondence of James Bradley,* Oxford, 1832, pp. 62–69.
8. Isaac Newton, *Philosophiæ Naturalis Principia Mathematica,* 3rd. edn. 1726. Translated by I. Bernard Cohen and Anne Whitman, *The Principia: Mathematical Principles of Natural Philosophy,* London, 1999, Book III, Prop. 20, Prob. 4. 'To find and compare with one another the weights of bodies in different regions of our earth'. p. 826.
9. Ibid. p. 828.
10. The standard unit of the measurement of length in France, prior to the French Revolution after which the metric system was introduced, was the *toise* of six *pieds* (feet) of 12.86 English inches or 12 French *lignes.*
11. Isaac Newton, *Philosophiæ Naturalis Principia Mathematica,* 3rd. edn. 1726. Translated by I. Bernard Cohen and Anne Whitman, *The Principia: Mathematical Principles of Natural Philosophy,* London, 1999, p. 828.
12. Ibid. p. 828.
13. A 'nutation' which literally means a 'nodding' is a term to describe a wobble in the direction of the Earth's axis. In this case a motion barely measured in fractions of an arc-second a year.
14. The term 'figure' was used in the sense of its 'shape'.
15. James Bradley, 'An Account of some Observations made in London by Mr. George Graham, F. R. S.; and at Black-River in Jamaica, by Colin Campbell, Esq. F. R. S. concerning the Going of a Clock, in Order to Determine the Difference between the Lengths of Isochronal Pendulums in Those Places', in *Miscellaneous Works and Correspondence of James Bradley,* Oxford, 1832, p. 62.
16. I am sure of this conjecture. Bradley published his paper on nutation in 1748 whilst Halley died in 1742. As Bradley supported Halley's work, it seems incredible that he would not have shared a phenomenon he identified in the residuals of his observations of the aberration with the very person to whom he had addressed his paper on the aberration in 1729. Halley never accepted his protégé's observations of nutation, believing that the observation of motions of such minute proportions was impossible.

17. Stephen Peter Rigaud, *Memoirs of Bradley*, Oxford, 1832, p. xxix.

18. John Conduitt, *Memorabilia*, King's College Library, Cambridge, Keynes MS 130.5, fol 3r. I think the authenticity of this remark should be qualified for I am sure Conduitt sought to reflect Newton in the best possible light, even though it was an opinion that Newton certainly supported.

19. I clearly recognize that this term, applied to the eighteenth century, is an anachronism, the term not becoming current until the 1830s. More accurately Graham would have been familiar with the terms 'natural or experimental philosopher'.

20. John Canton, An Attempt to Account for the Regular Diurnal Variation of the Horizontal Magnetic Needle and also for its Irregular Variation at the Time of Aurora Borealis, *Philosophical Transactions*, Vol. XXXVIII, 1759, p. 398.

21. James Bradley, 'An Account of Some Observations made in London by Mr. George Graham, F. R. S.; and at Black-River in Jamaica, by Colin Campbell, Esq. F. R. S. Concerning the Going of a Clock, in Order to Determine the Difference between the Lengths of Isochronal Pendulums in Those Places', in *Miscellaneous Works and Correspondence of James Bradley*, Oxford, 1832, pp. 62–63.

22. Ibid. p. 63.

23. *Altair* is the twelfth brightest star in the sky with an apparent visual magnitude of 0.77. It is 16.7 light years from the solar system (5.13 parsecs) so it is a nearby star. RA 19h 50m 47m / Dec. +08° 52' 06".

24. James Bradley, 'An Account of Some Observations Made in London by Mr. George Graham, F. R. S.; and at Black-River in Jamaica, by Colin Campbell, Esq. F. R. S. Concerning the Going of a Clock, in Order to Determine the Difference between the Lengths of Isochronal Pendulums in Those Places', in *Miscellaneous Works and Correspondence of James Bradley*, Oxford, 1832, p. 63.

25. Ibid. p. 63.

26. 5.74 kg.

27. 2.81 kg.

28. 7.71 kg.

29. Ibid. p. 64.

30. Ibid. p. 64.

31. Colin Campbell observed the passage of α Canis Majoris (*Sirius*) and β Canis Majoris through his meridian at Black-water in Jamaica.

32. James Bradley, 'An Account of Some Observations Made in London by Mr. George Graham, F. R. S.; and at Black-River in Jamaica, by Colin Campbell, Esq. F. R. S. Concerning the Going of a Clock, in Order to Determine the Difference between the Lengths of Isochronal Pendulums in Those Places', in *Miscellaneous Works and Correspondence of James Bradley*, Oxford, 1832, p. 67.

33. Ibid. p. 67.

34. Ibid. p. 68.

35. 99.38 cm.

36. 99.06 cm.

37. 99.58 cm.

38. James Bradley, 'An Account of Some Observations made in London by Mr. George Graham, F. R. S.; and at Black-River in Jamaica, by Colin Campbell, Esq. F. R. S.

Concerning the Going of a Clock, in Order to Determine the Difference between the Lengths of Isochronal Pendulums in Those Places', in *Miscellaneous Works and Correspondence of James Bradley*, Oxford, 1832, p. 64.

39. Ibid. p. 68.

40. Isaac Newton, Philosophiæ Naturalis Principia Mathematica, 3rd. edn. 1726. Translated by I. Bernard Cohen and Anne Whitman, *The Principia: Mathematical Principles of Natural Philosophy*. Book III, Prop 20, Prob. 4, 'To find and compare with one another the weights of bodies in different regions of our earth'. London, 1999, p. 826.

41. James Bradley, Bodleian Library Oxford, Bradley MS 20*, *Drafts of Papers by James Bradley on Aberration*, fol. 18v.

42. Colin Campbell was first cousin to the incumbent Duke of Argyle who was the Ordnance General who refused Halley's request to supply sufficient resources to acquire a second 8-foot mural quadrant.

43. James Bradley, 'An Account of some Observations made in London by Mr. George Graham, F. R. S.; and at Black-River in Jamaica, by Colin Campbell, Esq. F. R. S. concerning the Going of a Clock, in Order to Determine the Difference between the Lengths of Isochronal Pendulums in Those Places', in *Miscellaneous Works and Correspondence of James Bradley*, Oxford, 1832, pp. 68–69.

44. Ibid. p. 69.

45. Ibid. p. 69.

46. James Bradley, 'An Account of some Observations made in London by Mr. George Graham, F. R. S.; and at Black-River in Jamaica, by Colin Campbell, Esq. F. R. S. concerning the Going of a Clock, in Order to Determine the Difference between the Lengths of Isochronal Pendulums in Those Places', in *Miscellaneous Works and Correspondence of James Bradley*, Oxford, 1832, pp. 62–69.

47. This raises the question about the identity of the 'public'. This must surely mean the educated classes.

48. Mary Terrall, *The Man Who Flattened The Earth: Maupertuis and the Sciences in the Enlightenment*, Chicago, 2002, p. 88.

49. Ibid. p. 130.

50. That is the circumference was greater over the poles than around the equator.

51. François-Marie Arouet (1694–1778).

52. The 4th Earl of Chesterfield, a well-known wit and raconteur, later to introduce a Parliamentary Bill to reform the calendar in 1750, found Maupertuis to be very amusing and much to his liking.

53. Pierre Louis Moreau de Maupertuis to Hans Sloane 17 September 1731, British Library, Sloane MS. 4052, fol. 14r.

54. Pierre Louis Moreau de Maupertuis in Paris to Pierre des Maiseaux in London, 29 April 1732. Des Maiseaux was a Huguenot living in London who acted as a go-between.

55. British Library, Additional MS4285, fol.212.

56. John Machin to Pierre des Maiseaux 29 March 1732, British Library, Additional MS 4285, fol. 214.

57. Pierre Louis Moreau de Maupertuis in Paris to Pierre des Maiseaux in London, 29 April 1732. British Library, Additional MS4285, fol.212.

58. *Philosophical Transactions,* No.422, 1732, pp. 240–256. 'De figures quas fluida rotate induere possunt problemata duo'.

59. Mary Terrall, *The Man Who Flattened The Earth: Maupertuis and the Sciences in the Enlightenment*, Chicago, 2002, p. 68.

60. This was Newton's own philosophical position.

61. How ironic therefore that gravitation in the General Theory of Relativity is precisely that, a consequence of the geometry of the space–time continuum.

62. Mary Terrall, 'Representing the Earth's Shape: The Polemics Surrounding Maupertuis's Expedition to Lapland'. *Isis* Vol. 83, No. 2, pp. 218–237

63. In what is modern day Ecuador.

64. Ibid. p. 104.

65. The term 'effective gravity' was used by John L. Greenberg, *The Problem of the Earth's Shape from Newton to Clairaut: The Rise of Mathematical Science in Eighteenth-Century Paris and the Fall of 'Normal' Science*, Cambridge, 1995, p. 1.

66. Mary Terrall, *The Man Who Flattened The Earth: Maupertuis and the Sciences in the Enlightenment*, Chicago, 2002, p. 105.

67. Eric Aiton, *The Cartesian Theories of the Planetary Motions*, PhD Thesis, London University, 1958. Eric very kindly lent me a copy of his doctoral thesis, with the warning that he wished to disavow it. However, Eric was always a very modest, private man, whom I found to be a delightful person to know. He was a fine musician playing regularly in a string quartet.

68. Bernard le Bouvier de Fontenelle, Review of Maupertuis' 'Discours sur les differentes de aster', included in *Histoire de l'Académie Royale des Sciences, 1732*. Paris, 1735.

69. John L. Greenberg, *The Problem of the Earth's Shape from Newton to Clairaut: The Rise of Mathematical Science in Eighteenth-Century Paris and the Fall of 'Normal' Science*, Cambridge, 1995, p. 81.

70. René Taton, 'Sur la diffusion des theories Newtoniennes en France: Clairaut et le problem de la figure de la terre', *Vistas in Astronomy*, Vol. 22, pp. 485–450, 1978.

71. Valentin Boss, *Newton and Russia: The Early Influence, 1698–1796*, London, 1972.

72. David Beeson, *Maupertuis: An Intellectual Biography*, Oxford, 1992.

73. Eric Aiton, *The Vortex Theory of Planetary Motions*, London, 1972.

74. John Greenberg, *The Problem of the Earth's Shape from Newton to Clairaut: The Rise of Mathematical Science in Eighteenth-Century Paris and the Fall of 'Normal' Science*, Cambridge, 1995.

75. Mary Terrall, *The Man Who Flattened The Earth: Maupertuis and the Sciences in the Enlightenment*, Chicago, 2002.

76. Rob Iliffe, '"Aplatisseur du Monde et de Cassini": Maupertuis, Precision Measurement, and the Shape of the Earth in the 1730s' in *History of Science*, Vol 31, 1993, pp. 335–375.

77. John L. Greenberg, *The Problem of the Earth's Shape from Newton to Clairaut: The Rise of Mathematical Science in Eighteenth-Century Paris and the Fall of 'Normal' Science*, Cambridge, 1995, pp. 83–88.

78. James Bradley, 'An Account of some Observations made in London by Mr. George Graham, F. R. S.; and at Black-River in Jamaica, by Colin Campbell, Esq. F. R. S. concerning the Going of a Clock, in Order to Determine the Difference between

the Lengths of Isochronal Pendulums in Those Places', in *Miscellaneous Works and Correspondence of James Bradley*, Oxford, 1832, p. 67.

79. Pierre Louis Moreau de Maupertuis in Paris to James Bradley in Oxford, 27 September 1737 NS *Miscellaneous Works and Correspondence of James Bradley*, Oxford, 1832, pp. 404–406.

80. James Bradley to George Graham, December 1737. *Miscellaneous Works and Correspondence of James Bradley*, Oxford, 1832, pp. 406–408.

81. Ibid. p. 408.

82. James Bradley to George Graham, December 1737. *Miscellaneous Works and Correspondence of James Bradley*, Oxford, 1832, p. 407.

83. James Bradley to Pierre Louis Moreau de Maupertuis, 27 October 1737 OS, *Miscellaneous Works and Correspondence of James Bradley*, Oxford, 1832. pp. 408–409.

84. Larrie D. Ferreiro, *Measure of the Earth: The Enlightenment Expedition that Reshaped our World*, New York, 2011, p. xiv.

85. In reality, Cassini de Thury's data was gained from Lacaille's hard-won observations. Lacaille long resented his expedition leader's usurpation of his work.

86. Alexis Clairaut, *Théorie de la figure de la terre, tirée des principes de l'hydrostatique*, Paris, 1743, pp. xxxii–xxxiii.

87. Ibid. pp. xxxii–xxxiii.

88. Although Mount Everest (*Chomolungma*) at 8,848 metres above sea level is now acknowledged as the highest mountain on Earth, if mountain heights were measured from the centre of the Earth, then the summit of Chimborazo at 6,263 metres above sea level would in fact be the furthest point from the centre of the Earth due to the equatorial bulge. It is also the point on earth that approaches nearest the Moon. The highest point on the Moon on the side facing the Earth in the Lunar Apennines, is *Mons Bradley*.

89. I translate this epigram as 'Truth is mighty and will prevail' or more colloquially 'The truth will greatly prevail'.

90. James Bradley to Pierre Louis Moreau de Maupertuis, 28 October 1738 OS. In *Miscellaneous Works and Correspondence of James Bradley*, Oxford, 1832, p. 412.

8

The Triumph of Themistocles

The Reception and Consequences of Bradley's Discovery of the Nutation of the Earth's Axis

'The unequal action of the moon upon the equatorial parts of the earth, which, varying on account of the different inclination of her orbit to the equator, will cause a small nutation in the earth's axis, as also an acceleration and retardation in the precession of the equinoxes'.

James Bradley.

In Chapter 7, I referred to James Bradley's discovery of the nutation of the Earth's axis. There is much yet to say about the discovery of this phenomenon. The discovery of the aberration of light made Bradley's name known throughout the entire astronomical community in Europe and America, but it was his discovery of the nutation of the Earth's axis that was received with some degree of amazement, when it was first revealed to Maupertuis in a letter dated 27 October 1737 OS. [1] The motion was by eighteenth-century standards infinitesimal, so much so that Bradley's close friend and mentor Edmond Halley refused to accept the validity of his friend's claims that he could observe motions as small as a half-second of arc. Halley's limit was about four or five seconds.

On 19 August 1727, George Graham's zenith sector, constructed to Bradley's commission, was suspended at his Aunt Elizabeth Pound's house in Wanstead. It led to the discovery of the aberration of light, but it is his isolation of the nutation of the Earth's axis that now concerns us. The sector was later used to collimate the astronomical instruments at the Royal Observatory. It now occupies part of the west wall of the transit house of the Old Royal Observatory. Given the fundamental discoveries made with the sector, it should be recognized as a critically important instrument in the history of observational astronomy, laying the new foundations of eighteenth-century astronomy. Without the discoveries made by it the exacting science of positional astronomy could not

The Life and Work of James Bradley. John Fisher, Oxford University Press. © John Fisher (2023).
DOI: 10.1093/oso/9780198884200.003.0009

have advanced beyond the faltering steps made by Bradley's predecessors. It revealed the Earth's place in the Universe.

If the discovery of annual aberration demanded a clear understanding of the geometry of annual parallax and the application of that knowledge, the discovery of the nutation of the Earth's axis required levels of consistent, reiterative precision that defied the skills and practice of almost all of his contemporaries, Lacaille his French counterpart being an exception to this. Once Bradley understood the cause of the phenomenon, he recognized that like Halley before him, he was committed to observing a phenomenon through a complete retrogression of the nodes of the Moon's orbit, a task that would eventually demand twenty years of precise observation.

The zenith sector that made all of this possible whose construction was overseen by Graham, was specifically designed to observe circumpolar stars passing within 6¼° of the zenith.[2] It took unprecedented observational skills and geometrical understanding to isolate the lunar induced nutation, observed from 19 August 1727 to 3 September 1747. Bradley composed the paper written, as a letter addressed to Lord Macclesfield dated 31 December 1747 (OS), to be read before the Royal Society. It was a paper some of the leading astronomers in Europe had awaited with anticipation. Bradley replied to Maupertuis on 27 October 1737 OS informing him of his observations of a newly identified phenomenon believed to be a lunar-induced nutation of the Earth's axis. Bradley comprehended that there was a consistent pattern, found within the residuals of his observations of aberration. Bradley wrote to Lord Macclesfield,

> I have already taken notice in what manner this phænomenon discovered itself to me at the end of my first year's observations, viz. By a greater apparent change of declination in the stars near the equinoctial colure, than could arise from a precession of 50″ in a year; the mean quantity now usually allowed by astronomers.[3]

Bradley admits that by 31 August 1728 he was sure that the residuals of his observations of aberration betrayed evidence of another natural phenomenon. Such were his technical skills, and such the precision and reliability of the instrument made by Graham, that these residual motions were soon isolated.

Before this he formed suspicions that his observations were not quite conforming to the expectations of the phenomenological law of the 'new discovered motion'. This in itself was remarkable. In an age when most observers, including his mentor Edmond Halley, hesitated to perceive motions smaller than four or five arcseconds, Bradley had observed a residual as small as 1 or 1½ arcseconds. His trust in the instrument constructed under the direction of his friend and research partner George Graham, meant that he was certain these minute differences were not likely to be owing to instrumental error, but to an unidentified natural phenomenon. As was his procedure before when he discovered annual aberration, he sought to determine the laws of the phenomenon. Yet at no time between 1727 and 1742 did Bradley's mentor and friend Edmond Halley ever accept the validity of his claims that he was observing motions as small as one arcsecond. This did not appear to discomfort Bradley unduly; he was personally acquainted with his mentor's comparative lack of observational dexterity. By way of example, on 2 September 1727 Halley used the zenith sector during a visit to Wanstead. Bradley's entry in the Wanstead Observation Book, reproduced in the *Miscellaneous Works and Correspondence of James Bradley* as 'Zenith Observations at Wansted'[4] reads,

Sept. 2. Manè. Dr. Halley observed Capella; index 24 15,2; before 19 32¼ after 19 32¾. Hence star 4 16,7 south of 44 15. The star fluttered much, and he said that as it went out it appeared to be more southerly in reality than the thread; he therefore fancied from this observation that the direction of the thread was not right; but as all other stars at going out[5] appear on the other side of the wire, this appearance must be owing to his not being able to bisect the star at the cross, because of its fluttering.[6]

Perhaps because of this and other observations made by his friend, Bradley decided to introduce a system of recording the dependability of observations. Those which he regarded as 'good' he marked with a ‖ so that the more horizontal and vertical lines the better the observation. Dubious observations he marked with::. He later modified this by entering a number in square brackets, like so: [7]. The greater was the value of this digit, the better the observation. Bradley's observational technique records his drive to greater precision and clarity.

Bradley became aware that he was observing another phenomenon conflated with the 'new discovered motion'. He also sought any evidence of annual parallax. He wrote,

it must be granted that the parallax of the fixed stars is much smaller than hath been hitherto supposed by those who have pretended to produce it from their observations. I believe I may venture to say, that in either of the two stars last mentioned it does not amount to 2″. I am of the opinion, that if it were 1″ I should have perceived it, in the great number of observations that I made, especially of γ Draconis; which agreeing with the hypothesis (without allowing any thing for parallax) nearly as well when the sun was in conjunction with, as in opposition to, this star, it seems very probable that the parallax of it is not as great as one single second; and consequently that it is above 400,000 times farther from us than the sun.[7]

When Bradley joined Samuel Molyneux at Kew on 17 December 1725, a widespread expectation was that the annual parallax of the stars was of the order of thirty or forty arcseconds. This expectation had been created by the various observed motions of the stars, usually of the pole star, or as in the case of Robert Hooke's observations of γ Draconis, the observation of a star passing close to the zenith at the latitude of London. Most of these motions were reduced by Bradley to observations of annual aberration, later identified by him, which at the time of the Kew experiment most of the reports were asserted to be annual parallax and in Robert Hooke's case claimed as such. It seemed reasonable at the time that annual parallax would prove to be of the order of thirty arcseconds. This is why Samuel Molyneux was so confident that the observation of annual parallax was within easy reach.

By August 1728, one year into his observations at Wanstead, Bradley was confident that he possessed evidence of a 'motion' distinct from that of 'the new discovered motion'. He was unable to reduce it to the theory of annual parallax. Bradley, cautious by nature and nurture,[8] may have expected this motion to be identifiable with some form of systemic or instrumental error. He then accepted that he was observing two different natural phenomena, one of which would be called the aberration of light. The other, which he was unsure of, would be identified later as the nutation of the earth's axis. The differences between the predictions of the law of the 'new discovered motion' and Bradley's precise observations never exceeded more than 1½ or 2 seconds of arc. Almost every other contemporary investigator would have ignored such small differences, sure that this was due to observational error. Bradley's trust in the handiwork of Graham and the artisans who worked in his workshops, gave him great confidence in the accuracy his observations.

Bradley continued observing the unidentified, considerably smaller annual motion than that of aberration. On reflection, it is remarkable that Halley believed that Bradley might be forestalled in his claim for the discovery of the aberration of light. The micrometer had been used with the telescopic sight from the late 1630s and almost continuously since the late 1650s. For decades no one made a claim, for the many observations of the motions of the stars were attributed to annual parallax or to instrumental error. Eustachio Manfredi's observations of 'aberrations' in right ascension was interpreted as evidence of the invalidity of claims for annual parallax. Manfredi was an astronomer wedded to an acceptance of geocentric models of the cosmos. He interpreted these aberrations as clear evidence for the falsity or invalidity of the Copernican 'hypothesis'. These mutually corroborative observations of motions in right ascension by Manfredi and motions in declination by Bradley is a classic example of corroborative empirical evidence leading to diametrically opposed conclusions. Manfredi perceived his observations as evidence of *the invalidity of heliocentricity* whilst Bradley presented his observations as empirical evidence of the motion of the Earth around the Sun, *a confirmation of heliocentricity.* For Bradley this was no surprise, believing that there was a disconnect between observations and experiments and the hypotheses that purported to explain them. Manfredi and Bradley held differing presuppositions about the interpretation of knowledge claims. Empirical evidence alone is insufficient to decide between rival interpretations of evidence, as the conflict concerning the figure of the Earth makes abundantly clear.

Bradley was characteristically reticent about what his data was suggesting. Yet it is obvious from his actions that he felt sure he was observing two quite distinct phenomena.[9] One referred to as 'the new discovered motion' and the other identified as a 'deviation' from the expectations derived from the theory of 'the new discovered motion' which he determined did not conform to the theory of annual parallax. Bradley gave voice to his difficulties when he stated,

> The apparent motions of the heavenly bodies are so complicated, and affected by such a variety of causes, that in many cases it is extremely difficult to assign to each its due share of influence; or distinctly to point out what part of the motion is the effect of one cause, and what of another: and whilst the joint effects of all are only attended to, great irregularities and seeming inconsistencies frequently occur; whereas when we are able to allot to each particular cause its proper effect, harmony and uniformity usually ensue.[10]

Bradley was not anything but persistent. One of the reasons why he was such a first-rate scientific worker was his refusal to accept apparent evidence until he was sure that it had been firmly established. He was not a person who was inclined to see what he wanted to see. He was suspicious if his results were 'too correct', supposing his observations had been affected by psychological bias. It explains why he went to great lengths to avoid bias in the experiments on isochronal pendulums with Campbell and Graham. George Graham was of an equal cast of mind in his concerns for the avoidance of bias, especially when observing phenomena of a small magnitude.

By the standards of astronomical measurement in the 1730s, motions as small as a ½ or 1 or even 1½ arcseconds were regarded as infinitesimally small.[11] It was not backward on Halley's behalf to remain sceptical about Bradley's claims. It remains surprising, however, that up to his death in 1742, he never appeared to come round to believing that his protégé had been observing a 'real' natural phenomenon for fifteen years. During the 1770s the fifth Astronomer Royal, Nevil Maskelyne, confirmed that the zenith sector constructed by George Graham was indeed accurate to within ½ an arcsecond and used it to collimate the other instruments of the Royal Observatory. It indicates a remarkable strength of mind by Bradley to insist that he was indeed observing such minute angles, even in the face of consistent scepticism from his friend and mentor. Bradley was forthright in his praise for his chief instrument maker, stating that,

> A mind, intent upon the pursuit of any kind of knowledge, will always be agreeably entertained with what can supply the most proper means of attaining it: Such to the practical astronomer are exact and well-contrived instruments; and I reflect with pleasure on the opportunities I have enjoyed of cultivating an acquaintance and friendship with the person that, of all others, has most contributed to their improvement. For I am sensible that, if my own endeavours have, in any respect, been effectual to the advancement of astronomy, it has principally been owing to the advice and assistance given me by our worthy member Mr. George Graham; whose great skill and judgment in mechanics, joined with a complete and practical knowledge of the uses of astronomical instruments, enable him to contrive and execute them in the most perfect manner.[12]

From around August 1728, Bradley began to make observations specifically of this suspected and unknown phenomenon as opposed to inferring it from the residuals of his observations of the newly discovered motion. He believed the zenith sector was true and dependable. It was Bradley's practice to rectify his instruments before he used them, and indeed whilst he was observing with them, as it is possible to infer from his observation books. There was little chance of the observed motions being due entirely to instrumental error, though cautious as he remained, he must have kept it in mind as a possibility. He began to revisit the hypothesis that he and Molyneux toyed with at Kew. Bradley resurrected the possibility that it was an annual nutation which he had perfunctorily abandoned when seeking a hypothesis to explain the counter-intuitive motions of γ Draconis and 35 Camelopardalis (now designated as HR 2123 in the constellation of Auriga) at Kew. He now adopted this hypothesis again. He must at this point have suspected that this motion, if it could explain his observations, would be resolved in a year or two. Little did he then realize at this time that this motion would engage him for over twenty years. When at Kew, various hypotheses had been projected to explain the aberration, they included a possible nutation mentioned by Newton in the *Principia*.[13] As Bradley understood it, Newton suggested that it would probably be too small to be observed, but by this he probably meant that it was considerably less than about five arcseconds, beyond the limits that men like Halley were capable of discerning with the instruments and techniques at their disposal. But Bradley worked to a much more exacting standard. He revived a hope that the newly identified motion could be this mooted annual nutation. Bradley studied the phenomenon through 1728 and 1729 and into the 1730s, still uncertain about what he was observing.

Bradley eventually let go of the notion that the phenomenon he was observing was an annual 'nutation', as suggested in Book III of the *Principia*[14]. Bradley was sure that the cause of the 'deviation' was another natural phenomenon as yet unidentified. By May 1732 when Bradley moved his chief dwelling from Wanstead to Oxford he was beginning to apprehend the probable cause of the phenomenon he was observing. Within a year or so he was sure.

> It appeared from my observations, that, during this interval of time, some of the stars near the solstitial colure had changed their declinations 9″ or 10″ less than a precession of 50″ would have produced; and, at the same time, that others near the equinoctial colure had altered theirs about the same quantity than a like precession would have occasioned: the north pole of the equator seeming to have approached the stars, which come to the meridian with the sun, about the vernal equinox and the winter solstice; and to have receded from those which come to the meridian with the sun autumnal equinox and the summer solstice.[15]

All of this accorded with the expectations of a lunar-influenced nutation, a consequence of the fact that the plane of the lunar orbit around the Earth is inclined by about 5° to the plane of the Earth's equator. This induced a differential interaction between the Moon and the Earth's equator evidenced by a nutation. The retrograde motion of the nodes of the lunar orbit takes over eighteen years to complete an entire revolution, meaning that Bradley was committed to observing the newly suspected phenomenon for this entire period. The nodes of an orbit are the points where an orbiting body crosses the plane of the equator of the primary body, one rising and the other falling. More technically, it is where an orbit intersects any plane of reference to which it is inclined. Bradley's account continues,

> When I considered these circumstances, and the situation of the ascending node of the moon's orbit, at the time when I first began my observations; I suspected that the moon's action upon the equatorial parts of the earth might produce these effects: for if the precession of the equinox be, according to sir Isaac Newton's principles, caused by the actions of the sun and moon upon those parts, the plane of the moon's orbit being at one time above ten degrees more inclined to the plane of the equator than at another, it was reasonable to conclude, that the part of the whole annual precession, which arises from her action, would in different years be varied in its quantity; whereas the plane of the ecliptic, wherein the sun appears, keeping always nearly the same inclination to the equator, that part of the precession which is owing to the sun's action may be the same every year: and from hence it will follow, that although the mean annual precession, proceeding from the joint actions of the sun and moon, were 50″, yet the apparent annual precession might sometimes exceed, and sometimes fall short of that mean quantity, according to the various situations of the nodes of the moon's orbit.[16]

If the application of Newtonian principles transcended those of the Cartesian school, no better example can be cited in order to display the real advantages of the former over the latter. Newtonian principles were capable of being applied in so many different ways. Here Bradley uses them to connect a barely perceptible motion in declination to the motions of the nodes of the lunar orbit in order to explain why the observed annual precession of the equinoxes appeared to vary as they did. This was yet barely a working hypothesis, a long way from being demonstrated. Bradley recognized like Halley before him, that he was committed to observing a phenomenon through an entire retrogression of the nodes of the Moon's orbit, a period approaching some nineteen years.[17] He made his final observation of nutation at Wanstead with the zenith sector on 3 September 1747, less than four years before the great instrument maker died.[18] Graham must have reflected on the achievements made possible by his remarkable craftsmanship and the great appreciation of it expressed by his friend James Bradley.

In May 1732, Bradley moved from Wanstead to Oxford after taking up the post of Lecturer in Experimental Philosophy from late in 1729. Bradley regarded Wanstead as his home, having lived there from 1711.[19] After moving to Oxford Bradley made regular visits to Wanstead to observe the declinational motions due to the nutation at Elizabeth Pound's town house.[20] Bradley probably travelled between Oxford and Wanstead accompanied by Elizabeth Pound when visiting her brother. The small house may at first have been used by a servant who kept it in good order, looking after the sector, at least until a tenant took up residence from 1734. It did not make sense that a woman of substance would choose to live in such modest circumstances only a brief walking distance from her wealthy brother's handsome mansion.

Bradley moved to Oxford accompanied by Elizabeth Pound. Whatever the living arrangements were in Wanstead, it is certain Elizabeth lived with Bradley in Oxford until her terminal illness forced her to return to the care of her brother Matthew Wymondesold. Elizabeth died in 1740 leaving half of her extensive fortune[21] to her stepdaughter Sarah Pound and half to James Bradley, Sarah's cousin. Bradley did not marry until 1744, four years after Elizabeth's death. It suggests that a close relationship existed between Elizabeth and James, and given that in Georgian family law they were forbidden to marry, it may offer reasons as to why the 'respectable' Peach family either withheld or destroyed almost all of Bradley's private effects, only passing into the public domain items relating to his public affairs and duties. When he was fifty-two years of age, he married Susannah Peach by whom he had a single daughter.

By 1732, certainly by 1734 at the latest, Bradley clearly recognized he was observing a lunar-influenced nutation of the terrestrial axis. In the retrogression of the nodes of the Moon's orbit around the Earth, it was inducing a differential interaction between the Earth's equatorial regions and the changing latitude of the Moon. He chose only to share this knowledge with a few close friends such as George Graham and John Machin. Whether Bradley chose to share it with Edmond Halley is debatable, for his friend genuinely believed that his observations were beyond the capability of his instruments. Once Bradley was approached by Maupertuis in order to commission an instrument modelled on that which he used to discover the aberration of light, he must have comprehended that he may have to make public his work on the suspected nutation. At this stage after almost ten years of observation, there was little doubt in Bradley's mind. In accordance with his usual methods, he still sought to observe the phenomenon through a complete retrogression of the nodes before making his final conclusions.

Maupertuis sent a letter to Bradley in Oxford on 27 September 1737 NS after returning to Paris from Lapland, referring to the remarkable accuracy of Graham's zenith sector. He gave an account of the expedition explaining how the sector was transported at walking pace on a sledge, evidence of the care they had taken of Graham's handiwork. Maupertuis voiced his opinion that,

> notwithstanding the high idea we had of Mr. Graham's ability, we could not but with astonishment see, that taking the mean of the observations made by five observers, who agreed extremely well with each other, the arc of the limb did not differ but 1″ from what it ought have been according to its construction.[22]

It is at the end of this letter that the moment Bradley must have anticipated finally arrived. Maupertuis wrote,

> both from the experiments which we have made in the frigid zone, and from those which our academicians have sent us from the equator, that the gravity augments more towards the equator, than sir Isaac Newton has supposed in his Table. And all this is agreeable to the reflections which you have made upon the experiments of Mr. Campbell in Jamaica. But I have one favour to ask of you, sir, and hope you will not refuse it me; and that is, to tell me whether you have any observations actually made of the stars δ and α themselves, and whether we have made use of the proper correction for this aberration.[23]

The calculations had been performed by one of the leading mathematicians in France, the precocious Alexis Clairaut, so it was highly unlikely that the theory of annual aberration would have been misapplied. It was a courtesy request. But what Maupertuis could never have envisaged in Bradley's reply was news of yet another phenomenon and moreover one that supported the claims of the Lapland expedition, for the nutation was a consequence of the Earth's oblateness. Bradley had only observed the 'deviation'[24] through half of its retrogressive cycle. In a fragment of the letter addressed to Maupertuis,[25] Bradley wrote, 'And on this occasion, sir, I beg leave to do myself the honour to tell you what I have farther discovered, relating to some other changes that happen in the apparent situation of the fixed stars'.[26]

Bradley briefly described his observations, referring to the variable quantities of the annual precession of the equinoxes. Coming to his conclusion based as it was on some nine or ten years of observation of an infinitesimally small motion by the standards of the period, amounting to fractions of an arcsecond per year.

Compared to our age when measures of hundredths or thousandths of an arcsecond are routine, Bradley was making his measurements by the use of the eye and by touch as he adjusted his micrometer manually. By the standards of his age, his observations were almost unbelievably precise. Halley's scepticism may have seemed justifiable, but for the fact that he was surely aware of the integrity of his protégé's work. After describing the phenomenon Bradley revealed to Maupertuis, conscious at how his hypothesis supported their own objectives, that the

> seeming contradiction in the phenomena, with regard to the stars in those different situations, I now conceive to arise from the unequal action of the moon on the equatorial parts of the earth, which, varying on account of the different inclination of her orbit to the equator, will cause a small nutation in the earth's axis, as also an acceleration and retardation in the precession of the equinoxes. The whole quantity of this nutation, I find, amounts to about 9″ each way from the mean, the obliquity of the ecliptic being greatest when the moon's ascending node is in Aries, and least in Libra.[27]

Maupertuis accepted that all of the empirical evidences were coming together in support of the assertion that the Earth was shaped as an oblate spheroid even though, as the last chapter illustrates, many years of disputation yet remained.

Bradley's second great observational discovery, which was to be widely accepted as a nutation of the Earth's axis, had not been fully established, nor would it until he had observed the now highly predictable phenomenon through to a complete cycle or retrogression of the nodes of the Moon's orbit. His observations of the nutation began to fall away during the 1740s, concentrating on the motions of certain key stars, just as he had earlier when working on the aberration. He determined and verified the parameters of the nutation. The English climate sometimes rules out observation for several days or nights in succession. Whenever possible throughout 1727 and 1728 Bradley observed his selected stars assiduously, making up to thirty different precise stellar observations each day,[28] in combination with repeated observations of Jupiter's satellites.[29] During the following decade his observations settled into a steady routine. When he completed his observations of the nutation on 3 September 1747 he was only observing γ Draconis and β Draconis twice a year. This, after so many years, was quite sufficient to establish evidence to demonstrate the existence of a lunar-induced nutation of the terrestrial axis.

Between September and December 1747, Bradley worked on his account of his second major astrometrical discovery. Bradley's paper on the nutation of the Earth's axis was written as a letter addressed to his patron, George Parker 2nd Earl of Macclesfield.[30] Parker's father Thomas, the Lord Chancellor of England, one-time Regent of Great Britain, had been one of Bradley's electors when he acquired the Savilian Chair of Astronomy at Oxford. Bradley's patron was an excellent astronomer and chemist, with a first-class observatory and laboratory at his country seat at Shirburn Castle near Tame, about seventeen miles from Oxford. Bradley's paper on the nutation was read at two meetings of the Royal Society on 7 and 14 January 1748 after which Bradley was awarded the Copley Medal.[31] This is the highest scientific award of the Royal Society, here presented by Martin Foulkes PRS, the very person who had represented Bradley to his uncle, Archbishop Wake, who was the leading elector for the Savilian Chair in Astronomy at Oxford. By 1748 the application of Newtonian principles was so widely adopted that the long-awaited paper on the nutation came as additional evidence supporting the Newtonian contention that the Earth was an oblate spheroid. It also firmly established the validity of the universal law of gravitation as Cartesian natural philosophy became the preserve of a diminishing number of supporters.

Since 27 September 1737 NS, and the correspondence between Bradley and Maupertuis, a confirmation of the discovery of the nutation had been anticipated by many, within the European astronomical community. Another precocious member of the Lapland party, Pierre-Charles Le Monnier (1715–1799) who was admitted to the Royal Academy of Sciences at the age of twenty, was the first to publically apply the nutation in the reduction of his observations in a paper addressed to the Academy. It was opportunistic of Le Monnier when France was at war with Great Britain and Bradley was unable to respond to it. Fortunately this intrusion was largely ignored by the Academy. Le Monnier became familiar with the methods utilized by Bradley and was quick to capitalize on Bradley's work. According to Lacaille, Le Monnier was rather quick-tempered, which led to him being alienated from many of his astronomical contemporaries in France. His younger brother Louis-Guillaume (1717–1799) was more highly respected and worked alongside Lacaille and Cassini de Thury (1714–1784) extending the Paris meridian whilst working on both the geology and botany along the route.[32]

Nicolas-Louis de Lacaille wrote to Bradley in a letter posted from Paris on 28 July 1744 NS. Like Bradley he had been intended for the Church but rather than taking ordination as a priest, only took deacon's orders, so he was referred to as the Abbè. In 1739 he re-measured the French arc of the meridian, correcting what was increasingly recognized as the anomalous result of Jacques Cassini's in 1718. Lacaille's work on the Paris meridian took two years and no little measure of courage too. Lacaille was an outstanding astronomer who thoroughly admired Bradley's work, fully appreciating the skills and the insights of the Englishman. After informing Bradley that he was sending a book of ephemerides to him Lacaille wrote,

> And although I have not the honour of being known by you, I thought you would not refuse this small gift. It would be, for me, an opportunity of letting you know the singular esteem I have for your talents, and how much I rejoice in the reputation you have acquired among astronomers, who have no difficulty in recognising you as the first among them.[33]

This was a modest introduction by an astronomer who would prove to be the closest French response to the skills, hard work and integrity of Bradley. Over the years, they developed a strong mutual respect that transcended wars and international frontiers. Hardly had the correspondence between Lacaille and Bradley begun than one of the repetitive eighteenth-century wars between Great Britain and France intervened. The War of the Austrian Succession which began in 1740, involved Britain from 1744 to 1748 when it supported Maria Theresa's succession to the Habsburg monarchy. It began as a dynastic dispute and led to what one historian has called 'the first world war',[34] when the imperial and naval might of Spain, France, and Great Britain were set against each other in theatres of war ranging from Europe to the Americas, as well as India and the Indies. These are not our concern here, but it is important to note that after 1744 communications between French and British subjects became increasingly difficult, and were made no easier by the Jacobite rising of 1745–1746 that culminated in the last major battle fought on British soil at Culloden near Inverness. In such circumstances the flow of information between astronomers in France and England largely ceased.

On 2 May 1748, after the war had been concluded, Lacaille once more took up his pen and wrote a lengthy, rambling letter to Bradley, no doubt eager to regain contact with a man he undoubtedly admired. He was concerned whether Bradley had completed his work on the 'deviation' as Lacaille called it, the phenomenon Bradley was so certain was a nutation of the Earth's axis. He began,

> Sir,
>
> Having found an opportunity to get this letter to you, I take the liberty of writing to you in order to beg you at once to do me the kindness to indicate to

me what has been done in England towards the advancement of Astronomy[35] – should you have the leisure and opportunity to do so. Here in Paris we have been in a state of perfect ignorance on this matter for more than 3 years: we see neither books nor news of your country from where, usually, the most interesting news comes. I do not know whether it is the same for you in regard to us: in that case I would be most happy to please you in giving you a few lines on what comes to my mind concerning what has been happening here in Paris for some years. Although you may know it, perhaps, for the most part, I shall nevertheless tell you about it.[36]

Due to the lack of information between astronomers separated by war from 1744 to 1748, it is apparent that during hostilities very little scientific communication had taken place between two of the leading centres of research and scientific dissemination in Europe. It is apparent that the leaders of scientific opinion in Paris felt divorced from one of the principal foreign centres they looked to. Scientific activity appears to have gone into a relative decline in France for the duration of the war.[37] Lacaille complains that that since the return of the mathematicians[38] from Peru no account of their deliberations had been issued. He was even unsure about the measurement of the length of a degree of latitude at the equator. According to Cassini it was 56,746 toises.[39] Lacaille informed Bradley that research in the theory of the moon had led to Clairaut abandoning Newton's theory of gravity. Working on the three-body problem and applying the inverse square law led to Clairaut insisting that some modification to this law of the inverse ratio of the square of the distance was required. Lacaille also reported that Jean-Baptiste le Rond d'Alembert (1717–1783) and Clairaut had presented new methods of constructing tables of the moon very different from those of 'Mr. Newton'. He mentions other work undertaken by Clairaut, including the application of aberration to planets and comets. Work on the diameters of the Moon and the Sun revealed that they are not exactly equal though the Sun appears to have equal diameters across the poles and the equator.

All of these items and other news were passed eagerly to Bradley. Only twenty years earlier Voltaire could compare an 'English' physics with that officially recognized in France, and wrote disparagingly of the works of his countrymen[40] only to witness his work being condemned. Theories and practices in the two nations were now converging. There were divergences for sure, most particularly in the mathematical notation used. French mathematicians used Leibniz's more lucid version, while most English workers persisted in using Newton's notation which was less adaptable, giving Paris a head start over London, Oxford, and Cambridge. Drawing to the close of his lengthy epistle, Lacaille asked Bradley whether Halley's tables had been published. He referred to Bradley's work on the 'deviation' requesting whether he had established the theory of the variations in the obliquity of the ecliptic[41] 'which we wait from you'. Lacaille continued with his apologetic tone, though with candour, when referring to some of his fellow academicians writing,

> I would not have asked you directly if the war did not close to us all the means of communication with England; not if those who are able to have news from your country did not keep it to themselves in order to pass themselves off as marvellously knowledgeable. We have, unfortunately, some of this kind in our Academy.[42]

Amongst those Lacaille held in low regard was Pierre-Charles Le Monnier. Although Bradley gave permission to Maupertuis to use the zenith sector to make observations of the motions of the so-called fixed stars[43] it appears that Le Monnier began to take advantage of this permission to push his own claims concerning the importance of the role he played. Lacaille regarded this with contempt. Rigaud referred to the incident,

> In the letter to Maupertuis there was an invitation to use the sector, which Graham constructed for him, in making observations on the motions of the fixed stars. Le Monnier undertook it, and found them conformable to what Bradley's observations had determined. These results he sent to Greenwich; and if he had done nothing more, all would have been well; but, too eager in putting himself forward, he gave, in 1745, his own account of the doctrine of nutation,[44] and of the part which he had taken in settling the question. This was not handsome, after the freedom of communication which he had enjoyed, and Bradley, when he had made so fine a discovery, and had pursued the inquiry for so many years, ought to have been allowed in common courtesy (to say nothing of justice) the privilege of being the first to publish the details of it to the world. His name, however, was too high to be affected by such a circumstance, and he probably paid little attention to it.[45] He continued his observations as long as he thought right, and when they were completed he drew up the account of their results in a letter to the Earl of Macclesfield.[46]

With the publication of Bradley's paper on the nutation following two readings at the Royal Society on 7 and 14 January 1748 OS and the award of the society's

Copley Medal, Bradley's transformation of the science of eighteenth-century positional astronomy was almost complete, laying new foundations for the science of astrometry. There would be many more refinements in future by other astronomers, using technologies far beyond Bradley's imagination. The discovery and application of aberration and nutation, with the latter's interaction with precession, revolutionized the practice of the observational science of astrometry. Towards the end of the paper on the discovery of nutation Bradley summed up the consequences. He wrote,

> When the causes which affect the places of all the stars in general are known, such as the precession, aberration, and nutation, it may be of singular use to examine nicely the relative positions of particular stars; and especially of those of the greatest lustre, which it may be presumed lie nearest to us,[47] and may therefore be subject to more sensible changes, either from their own motion, or from that of our system.[48] And if at the same time that the brighter stars are compared with each other, we likewise determine the relative positions of some of the smallest that appear near them, whose places can be ascertained with sufficient exactness, we may perhaps be able to judge to what cause the change, if any be observable, is owing. The uncertainty that we are at present under, with respect to the degree of accuracy wherewith former astronomers could observe, makes us unable to determine several things relating to the subject I am now speaking of; but the improvements which have of late years been made, in the methods of taking the places of the heavenly bodies, are so great, that a few years may hereafter be sufficient to settle some points, which cannot now be settled by comparing even the earliest observations with those of the present age.[49]

Elsewhere in this account of his work, I have written that Bradley's discovery of the aberration of light was the greatest single advance in positional astronomy since the micrometer was first applied to the telescopic sight.[50] The discovery of nutation was an advance of equal importance when applied to the calculation of annual precession.[51] Bradley was conscious that his innovations made the comparison of contemporary positional observations with those prior to the discovery of the aberration of light inordinately difficult, unless the observations were accompanied by sufficient information to allow them to be reduced to agreed standards, which in most cases was not possible.

After the discovery of nutation, astronomers were enabled to determine 'the precession of the equinoxes for any given year' with confidence, a task that had eluded them from when Hipparchus first identified the phenomenon during the second century before the Common Era.[52] During the inception of heliocentric and heliostatic cosmologies between the work of Copernicus and Newton, most of the observed motions of the heavens had to be made with reference to the motions of the Earth itself. This transformation was accompanied by the rise of mechanical philosophies. Yet there has been too little attention paid to the causes of the emergence of the mechanical philosophy and its rise to dominance towards the end of the seventeenth century. Jim Bennett in his paper 'The Mechanic's Philosophy and the Mechanical Philosophy'[53] brings attention to the arbitrary distinctions made between scholars and craftsmen, intellectuals, and mechanics, without reference to what was happening on the ground, and how this imprinted itself on contemporary habits of thought or practice. My accounts of the work of James Bradley duly emphasize his dependence and interactions with the artisans and instrument makers that made his work possible. This common ground made the diminishing distinctions between scholars and mechanics meaningless. Both Bradley and Graham saw the other as equal partners and both shared insights into the other's work.

Although the aberration of light could have been revealed with the technology available to Bradley's predecessors, it was the confidence that he possessed in the skills of Graham and other craftsmen such as Jonathan Sisson, who laboured in his and other workshops that enabled him to achieve their objectives. In contrast to the social and political changes that took place in England, the status of artisans in France remained largely unconsidered and their labours unprotected. In France, the lack of a robust patent system inhibited the spread of new technologies. English craftsmen acquired a growing status with their innovations, protected in law by registered patents. It led not only to increases in innovation but also to the expansion of a scientific instrument business, concurrent with the expansion of commerce in general as the British economy grew.

In France, much intellectual effort was expended in support of Cartesian natural philosophy based on first principles. In England natural philosophy was led by observation and experiment, whereas in France it was engendered by arguments from first principles with observation and experiment taking a secondary role. The Cartesian approach to natural philosophy revealed how the intellect can be deceived without primary reference to observational or experimental evidence. Cartesians made observations and conducted experiments but largely to verify conclusions derived from first principles. Like Aristotle's metaphysical approach to natural knowledge, Descartes's was also equally based on casuistry. Jean (Johann) Bernoulli's attempted reconciliation of unobservable 'motions' of the vortices of Cartesian physics with Johannes Kepler's empirically derived laws of planetary motion, was an attempt at a harmonization with Newtonian experimental philosophy. The more open-ended

Newtonian analyses of natural processes was often suppressed in France,[54] placing restrictions on the dissemination of hard-won knowledge, just as surely as the work of artisans was held in abeyance by the support of outmoded social and economic norms. In France, several academicians refused to accept the conclusions of Bradley's paper on the aberration of light because they would not accept the notion of the progressive motion of light, contradicting as it did the Cartesian conclusion, based on first principles, that the velocity of light was instantaneous.

Bradley's discovery of the nutation of the Earth's axis, itself indirect evidence that the Earth is an oblate spheroid, was also opposed by several influential academicians due to their reliance on Cartesian natural philosophy. Bradley's work enabled astronomers to accurately anticipate the annual variations of the apparent precession of the equinoxes. It was the formidable precision and dependability of Bradley's work that made its mark with many of the astronomical community. Bradley's observations through the complete retrogression of the nodes of the lunar orbit were completed with the publication of a letter addressed to George Parker, the 2nd Earl of Macclesfield. Bradley's work was thorough and complete in every conceivable way. His paper on the nutation is full of observational tables, all them reduced, but the final table in the paper is entitled 'The Annual Precession of the Equinoctial Points' concluding a task that remained beyond the ability of astronomers for over 1900 years. I include the table in order that the clarity of the achievement may be acknowledged. Hitherto an estimate of about fifty to fifty-one arcseconds per annum had been sufficient, much as the observational limits of about five arcseconds had been, but no longer. The discoverer of the aberration of light had now advanced the observational expectations of the science of positional astronomy to degrees of accuracy and precision never before conceived of, let alone achieved.

The Annual Precession of the Equinoctial Points							
S.F.FPA	Sig.0	I	II	III	IV	V	Signs
0°	58.0″	57.0″	54.2″	50.3″	46.5″	43.7″	30°
5°	57.9″	56.6″	53.6″	49.7″	46.0″	43.4″	25°
10°	57.9″	56.2″	53.0″	49.0″	45.5″	43.2″	20°
15°	57.7″	55.7″	52.3″	48.4″	45.0″	43.0″	15°
20°	57.5″	55.2″	51.7″	47.7″	44.5″	42.8″	10°
25°	57.3″	54.7″	51.0″	47.1″	44.1″	42.8″	5°
30°	57.0″	54.2″	50.3″	46.5″	43.7″	42.7″	0°
Signs	XI	X	IX	VIII	VII	VI	S.F.FPA

S.F.FPA: Signs from the first point of Aries.

This table reveals the annual precession of the equinoctial points for any given year, for such is the consequence of Bradley's discovery of the nutation of the Earth's axis. It is divided into 'signs' and degrees, each sign of the zodiac being by convention 30°, so ten signs and 15° is equivalent to 315° from the 'first point of Aries' (FPA). The precession in any given year now ranges from 42.7″ to 58.0″ due to the combined action of nutation and precession. The nutation goes through a period of just over eighteen years and is superimposed on the motion of the precession, which retrogresses through a period of over 25,900 years.

Bradley observed the original objective star γ Draconis, originally chosen by Robert Hooke, because it transits close to the zenith at the latitude of London. This star was originally chosen by Hooke because its location enabled it to be observable through a long-focus suspended telescope, which enabled him to make precise observations free of the distracting effects of atmospheric refraction. No other star was observed by Bradley for as long or as many times than γ Draconis, so it is to this star that he refers to first in his nutation paper. He wrote,

As I made more observations of γ Draconis than of any other star, and it being likewise very near the zenith of Wansted, I will begin with the recital of some of them. The point upon the limb with which this star was compared was 38° 25′ from the north pole of the equator, according to the numbers of the arc of my sector. The first column in the following table shews the year and the day of the month when the observations were made; the next gives the number of seconds that the star was found to be south of 38° 25′; the third contains the alterations of the polar distance mean precession, at the rate of one degree in 71½ years, would cause in this star from the 27th day of March 1727 to the day on which the observation was taken; the fourth shews the aberrations of light; the fifth, the equations arising from the forementioned hypothesis [nutation]; and the sixth gives the mean distance of the star from the point with which it was compared, found by collecting the several numbers, according to their signs, in the third, fourth, and fifth columns, and applying them to the observed distances contained in the second.[55]

Bradley reveals that he made over 300 observations of γ Draconis, and taking the trouble of comparing all of them with the hypothesis, found that only eleven of them differed from it by as much as 2 arcseconds. Having compared many observations against the hypothesis, it is astonishing that so many of them were in such close agreement. Made in all seasons of the year as well as during the different positions of the Moon's nodes, it seemed sufficient proof of the validity, both of this hypothesis and that of the aberration of light. It is surely an indication of the profound transformations introduced by Bradley's discoveries and innovations, which laid the new foundations of contemporary positional astronomy during the first half of the eighteenth century.

The following table is derived solely from his observations of γ Draconis. After Bradley's explanations the following table of the observations of γ Draconis is simple to comprehend.

γ Draconis	South of 38°25'	Precession	Aberration	Nutation	Mean distance
1727 Sept. 3	70.5″	−0.4″	+19.2″	−8.9″	80.4″
1728 Mar. 18	108.7″	−0.8″	−19.0″	−8.6″	80.3″
Sept. 6	70.2″	−1.2″	+19.3″	−8.1″	80.2″
1729 Mar. 6	108.3″	−1.6″	−19.3″	−7.4″	80.0″
Sept. 8	69.4″	−2.1″	+19.3″	−6.4″	80.2″
1730 Sept. 8	68.0″	−2.9″	+19.3″	−3.9″	80.5″
1731 Sept. 8	66.0″	−3.8″	+19.3″	−1.0″	80.5″
1732 Sept. 6	64.3″	−4.6″	+19.3″	+2.0″	81.0″
1733 Aug. 29	60.8″	−5.4″	+19.3″	+4.8″	79.2″
1734 Aug. 11	62.3″	−6.2″	+19.3″	+6.9″	79.9″
1735 Sept. 10	60.0″	−7.1″	+19.3″	+7.9″	80.1″
1736 Sept. 9	59.3″	−8.0″	+19.3″	+9.0″	79.6″
1737 Sept. 6	60.8″	−8.8″	+19.3″	+8.5″	79.8″
1738 Sept. 13	62.0″	−9.6″	+19.3″	+7.0″	78.7″
1739 Sept. 2	66.6″	−10.5″	+19.2″	+4.7″	80.0″
1740 Sept. 5	70.8″	−11.3″	+19.3″	+1.9″	80.7″
1741 Sept. 2	75.4″	−12.1″	+19.2″	−1.1″	81.4″
1742 Sept. 5	76.7″	−12.9″	+19.3″	−4.0″	79.1″
1743 Sept. 2	81.6″	−13.7″	+19.1″	−6.4″	80.6″
1745 Sept. 3	86.3″	−15.4″	+19.2″	−8.9″	81.2″
1746 Sept. 17	86.5″	−16.2″	+19.2″	−8.7″	80.8″
1747 Sept. 2	86.1″	−17.0″	+19.2″	−7.6″	80.7″

This was one of several tables included in the paper on nutation. In spite of such accumulative evidence, a few figures in Paris still refused to accept the evidence revealed by Bradley's long series of observations of γ Draconis and several other stars. Those who accepted the evidence of the observations, but not the hypothesis, because it was claimed as evidence of the oblate figure of the Earth in contrast to the prolate figure of Cartesian theory, referred the motion due to 'nutation' as a 'deviation'. Quite what it was a deviation of was never fully explained. Even Lacaille referred to Bradley's observations as those of the deviation until he was fully satisfied of the reality of the nutation.

In this he was merely being cautious and referring to the deviation as that from the aberration of light. The English astronomer only revealed what he thought was evidence of a nutation to Maupertuis after observing the phenomenon for ten years. Bradley included similar tables for 35 Camelopardalis (HR 2123), α Cassiopeæ, τ Persei, α Persei, and η Ursæ Majoris, presenting evidence that it could only be a nutation directly connected with the retrogression of the nodes of the Moon's orbit around the Earth. As Bradley finally completed his observations at Wanstead with Graham's remarkable zenith sector, as robust as it was precise, he was turning to the necessity of re-equipping the Royal Observatory, having been appointed as the Astronomer Royal in succession to his mentor Edmond Halley. In 1748, the instruments at Greenwich were decidedly second-best to those at the disposal of the 2nd Earl of Macclesfield at Shirburn Castle in Oxfordshire. Bradley's paper was addressed to Lord Macclesfield and he makes reference to the quality of his patron's instruments relative to his own. Bradley wrote, 'I esteem the curious[56] apparatus at Shirburne Castle, and the observations there taken, as a most valuable criterion, whereby I may judge of the accuracy of those that are made at the Royal Observatory'.[57]

Having completed his work on the nutation, his main attention was now directed to his reform of the Royal Observatory. He needed new instruments and the old instruments to be repaired and realigned. He also required new buildings to house them. He reviewed his observational methods before he set himself on what may be regarded as his final outstanding achievement, the great series of observations made with the assistants that he trained in his own skills and methods. In doing so he transformed the work of the Royal Observatory.

Notes

1. Stephen Peter Rigaud, *Memoirs of Bradley*, p. lxxi. In Thucydides' *The Peloponnesian War*, the author described Themistocles, whose name means 'the glory of the law' and who was one of the generals who defeated the Persians at Marathon and whose naval policies were to have lasting impacts on Athenian foreign policy, as 'a man who exhibited the most indubitable signs of genius, indeed, in this particular he has a claim on our admiration quite extraordinary and unparalleled'. Rigaud's similar perception of Bradley's genius is clearly expressed in his *Memoirs of Bradley*.
2. This was in order to be able to observe the first magnitude star *Capella*.
3. James Bradley, 'A Letter to the Rt. Hon. George Earl of Macclesfield, Concerning an Apparent Motion Observed in Some of the Fixed Stars', in *Miscellaneous Works and Correspondence of James Bradley*, Oxford, 1832, pp. 17–41.

4. James Bradley, 'Zenith Observations at Wansted', in *Miscellaneous Works and Correspondence of James Bradley*, Oxford, 1832, pp. 201–284.

5. Ibid. p. 208. 'going out' refers to the object passing out of the field of vision of the telescope.

6. Ibid. p. 208.

7. James Bradley, 'A Letter to Dr. Edmond Halley, Astronomer Royal and etc. Giving an Account of a New Discovered Motion of the Fixed Stars', in *Miscellaneous Works and Correspondence of James Bradley*, p. 15.

8. The situation of the Bradley family during much of James' boyhood must indeed have been uncertain because of the lack of familial care, due to the gambling debts of the owner of the otherwise prosperous estate, of which Bradley's father William was a steward. This is why Jane Pound's younger brother James took over the responsibilities of ensuring her third son's education in preparation for a career in the Church of England.

9. Or else he would not have continued with his observations.

10. James Bradley, 'A Letter to the Rt. Hon. George Earl of Macclesfield, Concerning an Apparent Motion Observed in Some of the Fixed Stars', in *Miscellaneous Works and Correspondence of James Bradley*, Oxford, 1832, p. 18.

11. They were after all based entirely on the use of the senses even if aided by elementary optical aids.

12. James Bradley, 'A Letter to the Rt. Hon. George Earl of Macclesfield, Concerning an Apparent Motion Observed in Some of the Fixed Stars', in *Miscellaneous Works and Correspondence of James Bradley*, Oxford, 1832, p. 20.

13. Isaac Newton, *Philosophiæ Naturalis Principia Mathematica*, Book III, Prop 21, Theorem 17. 3rd edn. London, 1726. Trans. I. Bernard Cohen and Anne Whitman, London, 1999.

14. Ibid, Book III, Prop. 21, Theorem 17. 3rd edn. 1726.

15. James Bradley, 'A Letter to the Rt. Hon. George Earl of Macclesfield, Concerning an Apparent Motion Observed in Some of the Fixed Stars', in *Miscellaneous Works and Correspondence of James Bradley*, Oxford, 1832, p. 23.

16. Ibid. p. 23.

17. As it was he observed the phenomenon for a period of over twenty years.

18. George Graham was buried in Westminster Abbey next to his mentor Thomas Tompion.

19. Bradley regularly visited Wanstead to continue observing the nutation with the zenith sector that was in situ at Elizabeth Pound's house. He usually stayed at Matthew Wymondesold's house, almost certainly in the company of his aunt Elizabeth Pound, when visiting Wanstead, where he also commonly stayed during the Christmas vacation when he was able to make observations also of the Jovian and the Saturnian systems. In view of the arrangements that were probably negotiated with Mrs Elizabeth Jenkins, his visits to the town house were kept to a minimum, and when possible in socially acceptable hours. Bradley's other telescopes must have been housed in Matthew Wymondesold's handsome dwelling close by. Other instruments were kept in Oxford.

20. The Wymondesolds also appeared to own an estate in Dorset where Bradley may have spent holidays.

21. Estimated to be worth about £20,000. Bradley may have inherited £10,000 in 1740.

22. Ibid. p. 404.

23. Ibid. p. 406.

24. The term 'deviation' is how Lacaille described the motion until it was finally confirmed as a nutation.

25. James Bradley to Pierre Louis Moreau de Maupertuis, Oxford, OS,n *Miscellaneous Works and Correspondence of James Bradley,* Oxford, 1832. pp. 408–410.

26. Ibid. p. 408.

27. Ibid. p. 409.

28. Many of the brightest stars were also observed in bright daylight in order to observe them throughout the entire year, day and night.

29. For evidence of this refer to Appendix 1: A Chronology of the Life and Work of the Rev. James Bradley, DD, FRS, the third Astronomer Royal of England.

30. James Bradley, 'A Letter to the Rt. Hon. George Earl of Macclesfield, Concerning an Apparent Motion Observed in Some of The Fixed Stars'. In the *Miscellaneous Works and Correspondence of James Bradley*, Oxford, 1832, pp. 17–41.

31. The Copley Medal was instituted in 1731, two years after Bradley's paper on aberration.

32. Georges Cuvier, *Éloge historique de Lemonnier,* 7 October 1800.

33. Nicolas-Louis de Lacaille to James Bradley, 1744 July 28 NS. This is a translation from the original French found in the *Miscellaneous Works and Correspondence of James Bradley,* p. 430. The original text reads, 'Et quoique je n'aye pas l'honneur d'être connu de vous, j'aye cru que vous ne refuseries pas ce petit present. C'a été pour moi une occasion de vous faire connoître l'estime singuliere que j'ai de vous talents, et combine j'applaudis à la reputation que vous vous étes acquise parmi les astronomes qui ne font aucune difficulté de vous reconnoitre pour le premier'.

34. Micheal Clodfelter, *Warfare and Armed Conflicts: A Statistical Encyclopedia of Casualty and Other Figures, 1492–2015*, 4th edn, London, 2017.

35. This is very obviously a tangential reference to Bradley's work on the 'deviation', the nutation.

36. Nicolas-Louis de Lacaille to James Bradley, 2 May 1748 NS. Translation from the original French, in *Miscellaneous Works and Correspondence of James Bradley*, p. 438. The original text reads, ' Monsieur, Ayant trouvé une occasion sure de vous faire tenir cette lettre, j'ai pris la liberté de vous écrire, afin de vous prier instamment de ma faire la grace de me marquer ce qui s'est passé en Angleterre pour l'avancement de l'astronomie, au cas que vous en ayés le loisir et l'occasion par la suite. Nous sommes icy à Paris sur cet article dans une parfait ignorance depuis plus de trois ans: nous ne voyons ni livres ni nouvelles littéraires de votre pays, d'ou cependent les plus intéressantes ont coutume de venir. Je ne sçais s'il en est de meme de vous à notre égard: dans ce cas serois-je assés heureux de vous faire plaisir en vous disant en peu de mots ce qui me revient actuellement à la mémoire sur ce qui s'est passé icy à Paris depuis quelque années? Quoi que vous le sçachiés peut-être mieux et de meilleure part, je ne laisserai pas de vous le dire'.

37. The War of the Austrian Succession lasted from 1740 to 1748.

38. The term 'mathematician' had long been a term used for 'astronomer'. Astronomy was perceived as a branch of mathematics, as was geometry. Bradley's chair at Oxford was therefore regarded as a chair in applied mathematics. This traditional view of astronomy meant that the University of Oxford lacked an observatory until after the death of Bradley.

39. One toise is equal to 1.949 metres.

40. Voltaire, *Lettres Philosophique*, Paris, 1734.

41. The obliquity of the Earth's axis varies between 24.2° and 22.5° due to the gravitational effects of the other planets. It does not vary beyond these limits due mainly to the stabilizing effects of the Moon. It has been a subject of interest to astronomers for almost 2,000 years.

42. Nicolas-Louis de Lacaille to James Bradley, 1748 May 2 NS. Translation from the original French printed in *Miscellaneous Works and Correspondence of James Bradley*, p. 441. The original text reads, 'Je ne vous aurois pas importune directement, si la guerre ne nous fermoit tous les passages et toute communication avec l'Angleterre, ou si ceux qui peuvent avoir des nouvelles littéraires de ce pays-là, n'en faisoient pas de mystére, pour se faire passer dans l'occasion pour des homes merveilleux. Nous en avons malheureusement quelque-uns de cette humeur dans notre académie'.

43. By the 1740s, over twenty years after Halley had established what were later called 'the proper motions of the stars', the term 'fixed stars' was becoming a misnomer. Nevertheless, long usage maintained the habit. We still use the terms 'sunrise' and 'sunset', even though we recognize that it is a phenomenological consequence of the rotation of the Earth on its axis.

44. *Memoires de l'Académie Royale des Sciences*, 1745, p. 512.

45. It seems that Rigaud did not fully appreciate that communications between Paris and London were very much in abeyance at the time that Le Monnier made his own claims between 1744 and 1748.

46. Stephen Peter Rigaud, *Memoirs of Bradley*, p. lxv.

47. Here Bradley is assuming that the brightness of the stars is indicative of their distance from the Earth.

48. Bradley evidently appreciates that if the stars possess their own 'proper' motions then so does the solar system. This also insinuates that the solar system is not special or privileged, its motions being shared with those of other stars and their implied stellar systems.

49. James Bradley, 'A Letter to the Rt. Hon. George Earl of Macclesfield, Concerning an Apparent Motion Observed in Some of The Fixed Stars'. In the *Miscellaneous Works and Correspondence of James Bradley*, Oxford, 1832, p. 40.

50. In Chapter 5 of this work, 'And Yet It Moves'.

51. A major factor in the determination of the positions of celestial bodies reduced to any epoch.

52. Theon's daughter Hypatia continued the work of her father, but after her death to a mob in Alexandria, the science of astronomy practically ceased throughout the classical world.

53. J. A. Bennett, 'The Mechanic's Philosophy and the Mechanical Philosophy' *History of Science*, Vol. 24, No. 11, 1986, pp. 1–28.

54. As evidenced when Voltaire's popularization of Newtonian natural philosophy in his *Lettres Philosophique* was publically burned in Paris in 1734.

55. James Bradley, 'A Letter to the Rt. Hon. George Earl of Macclesfield, Concerning an Apparent Motion Observed in Some of The Fixed Stars', *Phil Trans,* No. 485, Vol. XLV, p. 1.

56. As before, the word 'curious' in this context means precise and accurate.

57. James Bradley, 'A Letter to the Rt. Hon. George Earl of Macclesfield, Concerning an Apparent Motion Observed in Some of The Fixed Stars', *Phil Trans,* No. 485, Vol. XLV, p. 1.

Boxwood model. Bradley devised this to illustrate the aberration of light and used it when giving lectures on the subject as lecturer in experimental philosophy at Oxford.

Zenith sector commissioned by Bradley and constructed by Graham in 1727. This is the instrument Bradley used to discover the aberration of light and the nutation of the Earth's axis. Nevil Maskelyne, the fifth Astronomer Royal 1765–1811, confirmed that it was accurate to 0.5 arcseconds.

Bradley commissioned John Bird to construct this eight-foot quadrant modelled on that of George Graham's 1725 commission from Halley. It was cast in brass and suspended in the New Observatory in 1750.

Commissioned by Bradley in 1749 this transit instrument was constructed by Jonathan Sisson. It was used as late as 1816.

A water-colour, c. 1770, of the Royal Observatory, Greenwich and its meridian buildings from the south-east.

The Meridian Building, the Royal Observatory, Greenwich, originally Bradley's New Observatory. At the far right is the assistant's quarters and to the left of this is Bradley's Transit Room. Further to the left is Airy's Transit Room constructed in the 19th century which became the Prime Meridian in 1884.

9

If Such a Man Could Have Enemies

Bradley's Appointment as the Third Astronomer Royal and his Reform of the Royal Observatory at Greenwich

'Astronomy is a science in which you are not able to touch anything you study'.

Allan Standage.

Throughout the 1730s, Bradley worked diligently on several projects, on the nutation, on isochronal pendulums, and on the development of the new observatory at George Parker, the 2nd Earl of Macclesfield's seat in Oxfordshire, as well as expounding his lectures in experimental philosophy and in astronomy at Oxford. He found time and energy to join his friend, the ailing Edmond Halley at Greenwich. The persistent fact that Halley never accepted the validity of Bradley's work on the nutation does not appear to have interfered with their friendship. Most of the work that Bradley performed for Halley was undertaken at Wanstead, but occasionally he assisted Halley in his observations at Greenwich.

It is remarkable that Halley in his sixty-fourth year when he was appointed Astronomer Royal, undertook a project that committed him to observing the Moon during an entire retrogression of the nodes of its orbit around the Earth, committing him to an observational programme lasting about twenty years. He sought to develop his lunar theory, enabling him to produce a set of lunar tables to aid mariners. Thomas Hearne, the Oxford University proctor, offered his opinion that Halley was 'somewhat lame'[1] but Halley had no doubt about his physical prowess when he was appointed in succession to John Flamsteed.[2] It was, as Rigaud remarked, that Halley 'entered upon a series of lunar observations which would require nearly twenty years for their completion. He was proud of his physical powers, so even at quite an advanced age he contemplated that he could realistically complete his programme'.[3]

Halley remained in fairly robust health until the early to middle 1730s when his right hand became progressively paralysed.[4] The Royal Observatory was beset by persistent instrumental problems. Margaret Flamsteed, blaming

The Life and Work of James Bradley. John Fisher, Oxford University Press. © John Fisher (2023).
DOI: 10.1093/oso/9780198884200.003.0010

Halley for contributing to her husband's downfall, understandably removed almost all of her late husband's astronomical instruments from the Royal Observatory after his death. When Halley took over the institution it was almost totally bereft of instruments. Halley had insufficient funds to replace them. The policy by which the Astronomer Royal was supposed to use his own instruments and resources was revealed to be impractical. The government made a grant of £500[5] in order to re-equip the observatory. Halley's first purchase was a 5-foot transit instrument in 1721. He housed this instrument in 'a little boarded shed between the study and the summer house'.[6] After tests, Halley determined that Flamsteed's brick meridian wall was subsiding, apparently because it was sited too close to the brow of the hill. At some time in 1723 or 1724, Halley had a new wall built on the same meridian but set back a few yards. On this he proposed to erect two identical 8-foot quadrants, one to observe the southern sky, the other the north.[7] George Graham constructed the first quadrant but Halley possessed insufficient funds for a second[8] or even for the large moveable quadrant he had also intended to purchase.[9]

The key to the success of Graham's 8-foot quadrant was its extraordinary rigidity. The quadrant was constructed by Jonathan Sisson under the direction of Graham, and in July 1725 Graham began dividing the quadrant in the Octagon Room in what is now known as as Flamsteed House.[10] Nevil Maskelyne, the fifth Astronomer Royal, later claimed it was divided with an accuracy and precision never before devised.[11] The quadrant had two sets of divisions and the read-offs from one could be checked against the other. The inner scale on the limb was divided according to the current convention into 90°. This was subdivided by vernier to 30″ and later in 1745 under Bradley's commission, with an application to 1″. The outer scale was divided into 96 equal parts and was achieved by continual bisection of the 60° arc laid off by inscribing an arc equal to the radius of the quadrant and which was numbered 64. Successive bisections then gave 32, 16, 8, 4, 2, 1, etc parts. One part was equal to 56.15″. A table was provided, converting the outer scale into degrees, minutes, and seconds. The outer scale reading could be directly compared with those of the inner scale.[12]

Although the quadrant performed well for several years, Halley began to experience problems from the early 1730s, largely because of the settling of the meridian wall he had constructed, to which the arc with its immense weight was attached. From 1734, the problems had become sufficiently pressing for Graham, Bradley, and John Machin to begin making visits to Greenwich in order to make adjustments to the quadrant. On 28 June 1734, Bradley wrote,

I examined Dr. Halley's Mural Arch and found that ye Center fallen towards ye South so much, that the wire of the Plummet, instead of bisecting the spot (that it was at first adjusting to) fall southward of it so that the north edge of ye wire touch ye south edge of ye spot; the difference of ye present situation of ye Arch from what it was when last adjusted (which I believe was about a year after it was first set up) amounting as far as I could estimate to about 20″ too small.[13]

Over a year later on 26 July 1735, Bradley, Machin, and Graham again felt obliged to repeat the trials and experiments on the 8-foot quadrant that they had performed the year before. The formerly determined variation of apparently twenty arcseconds, which could be allowed for in the reduction of Halley's observations with the quadrant, was inadequate for his purposes. The retaining wall was also slumping to the west as well as to the south. Bradley was not satisfied that the tests of 28 June 1734 were accurate or precise enough. In a very characteristic manner, he got Graham and Machin to rejoin him, this time to determine the exact variation since the the previous year.

Bradley rejected the notion that accuracy and precision guaranteed epistemological certitude. He held the view that empirical data was insufficient to ensure the validity of theory, that there was a disjunct between observations and experiments and the conjectures and hypotheses based on their results. In his mature work, Bradley developed the habit of recording every source of possible error, enabling later computers, astronomers, and experimental philosophers to apply different theoretical analyses when reducing these observations.[14] The connections between Bradley's nescient approach to data, and the careful recording of the perceived causes of error, in addition to his reiterative practice was fundamental to Bradley's later observational practice.[15] Bradley's concern with accuracy and precision was integral to his concern about the significance of error. It is a major reason why he was reticent to make unjustifiable claims. He always sought the completion of an experiment or a set of observations before he drew any conclusions. In the application of such reticence and caution, the science of positional astronomy was placed on firm practical and theoretical foundations. Bradley did not live to see the practical and observational benefits derived from his work, but this in itself would not have caused him undue concern. Bradley's entire approach to the acquisition of natural knowledge was conceived of in terms of community and the shared optimistic values of eighteenth-century philosophical and social life.

Edmond Halley died on 14 January 1742.[16] Although he had often expressed his wish to be succeeded by his protégé, even offering to stand down in Bradley's favour, this wish was always firmly refused. In spite of Halley's wishes, there was no guarantee that he would be succeeded by Bradley. Public offices in Hanoverian Britain were usually in the gift of holders of the major offices of state, particularly the prime minister. Early in 1742, the office of Astronomer Royal was in the power of the First Lord of the Treasury,[17] Sir Robert Walpole, a man who had used the award of senior offices as a means of advancing his interests and managing the affairs of state. The death of Halley coincided with a difficult juncture for the future interests of Bradley's astronomical career. It coincided with the moment when Walpole's peace policy, with the avoidance of wars, was coming under censure in Parliament.[18] It was against this volatile background that the actions of Bradley's patron George Parker and other figures, became important in the process of securing Bradley's succession to the office of Astronomer Royal. Even before Halley succumbed to his terminal illness, Macclesfield was making representations on his friend's behalf. On the day that Halley died Macclesfield wrote from his seat at Shirburn Castle to his former mathematics teacher Sir William Jones,[19] saying,

I have by this post written a long and pressing letter to the Chancellor[20] in behalf of Mr. Bradley, whom you know how rejoiced I should be to have it in my power to serve. But my way of thinking and voting is a great obstruction to having any interest at court, nor can I think of any other way I can recommend Mr. Bradley to this professorship, than by means of the Chancellor and Archbishop.[21] I do not know upon what footing my kinsman Mr. Justice Parker is with Lord Chief Justice Willes, it is certain the latter has great interest with Sir Robert, and it may be of advantage if cousin Parker would state the thing in a proper light to Willes, and represent to him how much it will for Sir Robert's discredit, that the only man in England fit to succeed Halley at Greenwich, confessed so by all who know anything of astronomy, should be put by, and the finest instrument in the universe[22] put into the hands of a person unable to make a proper use of it. I dare say my cousin's friendship for me will make him undertake this part, and endeavour to prevail upon Willes to interest himself for Mr. Bradley, if there be no impropriety in the thing.[23]

Macclesfield was not in favour in the highest circles of political life at this time. However, through representations by the 2nd Earl, Bradley enjoyed the support of many figures in English society who had the ear of Walpole and his closest allies. Amongst these was Sir William Jones who also had access to the Walpole Whig Sir Henry Pelham. Sir William wrote to Lord Macclesfield,

> it is not only my friendship for Mr. Bradley that makes me so ardently wish to see him possessed of the professorship, it is my concern for the honour of the nation with regard to science. For as our credit and reputation have hitherto not been inconsiderable amongst the astronomical part of the world, I should be extremely sorry we should forfeit it all at once by bestowing upon a man of inferior skill and abilities, the most honourable, though not the most lucrative, post in the profession,[24] (a post which was well filled by Dr. Halley and his predecessor), when at the same time we have amongst us a man known by all the foreign, as well as our own astronomers, not to be inferior to either of them, and one whom Sir Isaac Newton was pleased to call the best astronomer in Europe. And I dare assure your lordship, that if you are pleased to espouse Mr. Bradley's interest, you will have the satisfaction to find your recommendation of him approved and applauded universally by those who are versed in those studies, both home and abroad. As Mr. Bradley's abilities in astronomical learning are allowed and confessed by all, so his character in every respect is so well established and so unblemished, that I may defy the worst of his enemies (if so good and worthy man have any) to make even the lowest or most trifling objection to it. After all, it may be said, if Mr. Bradley's skill is so universally acknowledged, and his character so established, there is little danger of opposition, since no competitor can entertain the least hope of success against him.[25]

The admiration of Bradley as an astronomer was linked to his acknowledged skills. Yet as with his other offices, his accession to the post was always effected not just by his abilities but through the social and political patronage that in succession to his uncle James Pound, he had so diligently tended to.[26] This was strengthened after Bradley's marriage, when he and his wife Susannah acquired the friendship and support of Lady Catherine Pelham, the Ranger of Greenwich Park, the wife of the Prime Minister Sir Henry Pelham. It seems that she was impressed by Bradley. Apparently, he possessed an inborn ability to explain difficult ideas and concepts in a candid and patient manner: he possessed the 'common touch',[27] little doubt in part at least, due to his modest origins. Given his acute social skills, he must have impressed those that formed the social and political elites who were able to influence the elections and appointments that advanced his scientific career.

Walpole was defeated in a division of the House of Commons on 2 February 1742[28] and Parliament was adjourned; Bradley was duly appointed as the third Astronomer Royal the very next day.[29] It appears that one of the first acts that Walpole undertook after his defeat in the Commons was to appoint Bradley to the vacant office of Astronomer Royal. This was before he eventually resigned from office on 11 February 1742.[30] It appears that the King also expressed an interest. Bradley first came to the attention of George II when he was the Prince of Wales, and when Bradley's observational partner Molyneux was his private secretary. The King possessed an enquiring intellect and had availed himself of many advances in astronomy. He was familiar with Bradley's work on aberration. This interest was maintained when George II interceded to ensure that Bradley acquired sufficient resources to build the New Observatory, and equip it with the finest instruments in Europe. This was a privilege denied to Bradley's predecessors in post, John Flamsteed and Edmond Halley. Bradley was held in high regard and the University of Oxford took great pride in Bradley's associations with the University. This regard was clearly signalled on 22 February 1742 when Bradley was created Doctor of Divinity by the rare procedure of 'by diploma'. Thus, in a very short period Bradley had been appointed as the Astronomer Royal by the government and then Doctor of Divinity by the University of Oxford.

The difference in the practical observational skills of Halley and Bradley were made apparent as soon as Bradley took up the office of Astronomer Royal, his arrival at Greenwich being delayed by his teaching commitments in Oxford. Halley customarily used only a single vertical wire in the focus of the 8-foot quadrant's telescope, and also in that of the transit instrument. As he lacked an assistant, this made observation simultaneously in both instruments quite impossible, unless he used Bradley's 'eye and ear' method,[31] difficult in his infirmity. By way of contrast Bradley, utilized at least three, and often five, vertical wires enabling him to track the passage of a star in more than one instrument as it passed through the field of view. Following his appointment as Astronomer Royal, Bradley remained as Savilian Professor of Astronomy and Lecturer (soon to become Reader) in Experimental Philosophy at Oxford. Less than two years before his appointment at Greenwich, Bradley inherited about £10,000 from Elizabeth Pound, an equal sum being left to his cousin Miss Sarah Pound, who lived in Bradley's household in Greenwich after 1742. Unlike his forerunners, Bradley had sufficient resources to meet his requirements. He was intent on the appointment of an assistant. It would extend the observational continuity of the Royal Observatory, ensuring that all the programmes of observation be maintained, whilst allowing the continuation of his lucrative teaching programmes in Oxford.

Before Bradley could initiate any observational programmes, he needed to repair and rectify the instruments of the Royal Observatory. He was aware of the continuing problems associated with the 8-foot quadrant, designed and constructed for Halley by Graham.[32] The neglect of the Observatory's instruments during the final years of Halley's life, as he declined into inactivity, made them entirely unusable. From 5 July 1742, Bradley began to re-examine the state of the 8-foot quadrant. On the following day he wrote,

> I this day took off the Telescope and the centerwork having discovered the day before that the screws (which held the brass cilinder about which the steel plate turned) were all broke so that the Brass Cilinder itself moved round with the telescope which now was supported by means of the smaller part of the Brass center piece which turned in the hole of the square Brass Plate that is perseved[33] to the Ironwork. The Brass cilinder itself adhered so closely to the inside of the steel collar that I could not move it which must have been the occasion of the screws being broken the friction being greater than the strength of those screws.[34]

Bradley was unable to make many observations during the initial months of his incumbency. This was due to the demands of his teaching commitments at Oxford and also to the state of the Observatory and its instruments. Being in such a penurious state, his early work was concerned almost solely with examination, repair, and rectification. It must have been a vexing time for him. Patient, he was nevertheless a man of boundless ambition and energy. Bradley set himself the exacting task of investigating the state of the instruments. He discovered that he needed to pay close attention to the most fundamental requirements. Later Bradley wrote,

> When I came to reside at the observatory in June 1742 I made but few observations either with the mural Quadrant or Transit instrument, till I had made some alterations in both. For there being no good method of illuminating the wires of the Telescopes either of the Quadrant or Transit Instr.. Soon I found it necessary to make proper Apparatuses for that use. Dr. Halley seldom attempted to make Observations when the wires required to be illuminated, but when he found that to be necessary he usually placed a candle upon the south end of the wall on which it hangs, which shining upon the inside of the shutter (which was painted white) the light was thence reflected thro the

Object Glass upon the wires in the focus and by raising the shutter (which slided up and down) till the upper edge of it was in part before the Object Glass, more or less light was reflected into the Telescope.[35]

Bradley discovered these primitive arrangements impossible to reconcile with a practice that aspired to an extreme regard for precision and accuracy in a wide range of observational conditions. This underlines the realization that Halley was not as skilled an observer as Bradley. It suggests why Halley considered Bradley's observation and measurement of motions as small as half an arcsecond was impossible. In the terms of Halley's own practice, it certainly was. Bradley's observational practice had developed techniques that used refinements such as the use of a precise screw micrometer made available by Graham's consummate craftsmanship. Bradley perceived a virtue in precision, to him it seems to have been a moral issue involving personal integrity. Nevertheless, Bradley appeared to treat his predecessor's work with respect, despite its observational shortcomings. Although Halley had been increasingly infirm for several years prior to his demise, when Bradley took up the reins at Greenwich, he had to deal with the deterioration of years of neglect and the inadequacies of his predecessor's practice. As early as 2 September 1727, when Halley visited Wanstead to use the new zenith sector, constructed by Graham for Bradley, it was obvious astute that he was, he was nevertheless comparatively inadequate as an observational astronomer. Conversely, it reveals that Bradley's achievements were not just the product of his possession of some of the finest instruments of his age. Although it must have been frustrating for Bradley to accept Halley's negative opinions concerning his claims for the high precision of his observational work, he graciously accepted that his respected mentor possessed many other outstanding qualities, and simply accepted his friend's strictures.[36] He accepted that Halley lacked the demanding skills as an observer to fully appreciate that such precision was the product of an exacting methodology. This paid attention to the fine detail Halley overlooked, which Bradley regularly used in his observational practice.[37] Bradley appreciated that Halley had imbued him with much of his knowledge of Newton's *Principia,* that he was the person most responsible for getting Newton to compose the great work, supporting the financial cost for its publication out of his own pocket, because the Royal Society possessed insufficient funds at that time.[38]

By 3 September 1742, Bradley was already beginning to work in tandem with his first assistant, his 14-year-old nephew John Bradley (son of his eldest brother William).[39] All of Bradley's assistants were trained to the exacting standards he set for himself in all of his mature observational work. This began a tradition of rigorous observation and calculation at the Royal Observatory that continued after Bradley's demise. This was a crucially important aspect of Bradley's reform of the institution, which after the demise of his predecessors left it in disuse or chaos. At his passing on 13 July 1762, the Royal Observatory continued working unabated, an important legacy.

James Bradley's own observational skills had been developed under the mentorship of his uncle, and he in turn passed many of them onto John Bradley (assistant from 1742 to 1756), Charles Mason (assistant from 1756 to 1760), and Charles Green (assistant from 1760). Bradley's eyesight was probably better than average, much better we must assume, than the abused vision of our own age. Because of his well-practiced methodologies, those with average vision were capable, with training of becoming competent observers. It is Bradley's diligence to detail, as readers of this account of his work must now appreciate, together with the constant rectification of his instruments that distinguished his practice from almost all of his predecessors and contemporaries.[40] Typical of his approach to his work is his detailed preparation for the use of the transit instrument, and the 8-foot quadrant around 29 August 1742. In his notebook in August he wrote,

the line of collimation of the Transit Instrument was adjusted about July 24th, but it could not be properly directed till I had a clear view of the mark which Dr. Halley had made on the park wall; the sight of which was intercepted by the boughs of trees which had grown up since Dr. Halley had used this instrument:[41] these being cut away by the end of July 1742, I then set the instrument by his mark, which I at first supposed to be exactly in the meridian, but afterwards found that it lay 12″ or 15″ in the Azimuth to the west of the true meridian.

Between the first and 14th of August, several stars were observed by the Transit Instrument and the Quadrant and by comparing the times of their Transits the errors of the Place of the Mural Quadrant were found to be entered in the Table before August 14, 1742.

The Quadrant was not altered till Aug. 18 when it was rectified so that the center lay exactly over the beginning of ye Divisions or the 0 point, & then a Plummet hung from the notch at top at the center Plate was found to correspond well with the point on the limb below, to which it had been at first adjusted by Mr. Graham in 1726.[42]

Bradley continued making adjustments to this remarkable instrument and on 20 August he altered the plane of the quadrant 'to lie much truer than it did, as left by Dr. Halley, as appears by the Table of errors then found'.[43] He had already consigned the telescope that had been attached to the quadrant to Sisson so various alterations could be made to it.[44] On its return on 11 September 1742, Bradley adjusted the line of collimation, 'Mr Sisson having made for me a new apparatus for that purpose'.[45] Several additions were made by Sisson to the transit instrument, and the telescope was balanced so that it would stand in any inclination. The notes made by Bradley reveal just how little use Halley had made of the transit instrument, certainly in his final years. Bradley wrote, 'In Dr. Halley's time the eye end being the heaviest, he was obliged to support it at the proper altitude, upon which account observations could not be made so expeditiously as now'.[46]

This was an extraordinary revelation. It shows that the instrument was so out of balance that it was useless as a precision instrument. Halley had to support the instrument manually while he was observing with it. The lack of use of this instrument is confirmed by the notes made by Bradley, when he first began using the instrument. He explained how Sisson put in two other wires some fifteen minutes of arc from the middle wire to facilitate accurate observation of the transit of objects, particularly when clouds interfered with measurement as the object passed through the centre wire. Bradley spent much of the second half of 1742 on the rectification and improvement of the instruments of the Royal Observatory, a measure in part of his vocation and in part of the wretched state of the instruments when he took over the Royal Observatory.

Over the next year and a half, Bradley and his nephew continued using the transit instrument and the 8-foot quadrant, cross-checking the results obtained at Greenwich with those acquired at Shirburn Castle, when used by Bradley and the 2nd Earl of Macclesfield. The new Astronomer Royal was able to use the instruments at Shirburn Castle as if they were his own. These observations revealed the comparative inadequacy of the instruments of the Royal Observatory.[47] About 1,500 observations were made both at Shirburn and at Greenwich from 25 July to 31 December 1742 and beyond.[48] Through August 1742, Bradley made repeated observations of the first magnitude, circumpolar star *Capella,* above and below the pole at 12-hourly intervals. The error in the position of the mark was estimated. On 3 September 1742, the clock was fixed against the brick wall and its dial made to face north-west.[49] On 11 October 1742, Graham added a gridiron pendulum to the clock of the design first developed by John Harrison.[50] On 4 December 1742, there is an entry revealing Bradley's return from Oxford after presenting a course of lectures in experimental philosophy. From 16 October 1742, all of the observational entries at Greenwich are in John Bradley's hand.[51] The style of the observations is entirely that of his uncle. It reveals that John had acquired many of the transferable, tacit skills that his uncle had developed over three decades of precise and exacting observation.

In spite of the imperfect state of the observational instruments of the Royal Observatory, Bradley undertook programmes of observation. In the twenty years from 1742, Bradley was to transform the administration and performance of the Royal Observatory. He developed the science to levels of practical competence never before achieved. His work in 1743 offers insight into the new levels of industry achieved by Bradley and his young nephew. Not since John Flamsteed first acquired his arc in 1689 had the Royal Observatory witnessed such observational productivity.[52] The transit observations for this year alone amount to 177 folios.[53] On 8 August 1743, no less than 255 precise observations were made[54] together with accurate records of barometric pressure and atmospheric temperature. Several times over 200 observations a day were made during a continuous programme of sustained energy. This consistent programme of dedicated endurance was performed when Bradley encountered constant problems induced by the perennial necessity of having to adjust the collimation of the transit instrument. The quadrant observations for the same year come to 148 folios.[55] These observations involved even greater efforts than those displayed in the work with the transit instrument. The exact times of the passage of stars through the sights were recorded, as well as their declination. The arc was read off from the outer 96 divisions and then reduced to sexagesimal degrees, minutes, and seconds. On 8 August 1743, when Bradley and his nephew made 255 observations with the transit instrument, they also made 181 observations with the quadrant. The pages are ruled for 60 observations a page although Bradley regularly used some lines for notes about the collimation and other adjustments. If it is assumed that there are about 55 observations a page, it can be estimated that 18,000 rigorously precise observations were made in 1743. This assiduity was as characteristic of Bradley's practice as was his commitment to precision and accuracy.

Bradley constantly rectified his instruments in order to get them to perform to the rigorous standards that he expected of them. Such were the standards of the craftsmen who provided them. The voluminous notes he made are repetitive, making it is easy to overlook the fine adjustments leading to his satisfaction. There are several passages which record his increasing, but never full satisfaction, during 1744 and 1745. When testing the line of collimation of the quadrant he wrote,

The Instrument now (when I look Northward) points very near the same marks as it did when Mr. Graham & I adjusted the line of Collim in the year

1726 but as the Southern mark has been made near two inches more easterly than twas before ... and I judge from the present circumstances that 'tis the mark on the Park wall that has altered, for I find the marks that I made on a chimney in Greenwich do now exactly correspond with the same distant objects with which I formerly compared them.[56]

By 1744, Bradley was recording barometric and thermometric observations in his researches on atmospheric refraction, in order to improve the reduction of stellar observations. Always the impulse was to greater precision. On 1 July 1745, Bradley entered the following into his notebook,

> With a moveable mark on the Park wall I tried how near I would judge when it was truly bisected; and after several trials I found that the difference of the places of the mark did not exceed 4/10 of an inch which answers to 2½″. So from this experiment it may be concluded, that we can direct the Instruments to the same point always little more than one second in Azimuth.[57]

This scarcely met the requirements of the rigorous standards of Bradley's mature observational practice. At this time he was moving into the last two years of observation of the nutation at Wanstead, just five miles away, north of the River Thames.

When the transit observations are examined, the passage of the stars through the wires were usually recorded to the nearest second but where there was a perceptible interval Bradley commonly entered either + or − next to the data, signifying a variation of about ½ an arcsecond above or below.[58] Ten years after Bradley's demise, in 1772, Bradley's successor Nevil Maskelyne introduced a decimal notation, which became the standard method of recording small sub-second arcs.[59] Bradley's equipment lacked the refinements that a few decades would bring. His wires were thicker and his telescopes lacked achromatic lenses. The resolving power of his telescopes was considerably inferior to those that followed. For example in 1743 Bradley wrote, 'This day I put a shorter eyeglass into the transit telescope, its focus is 1½ inch, so the telescope magnifies now about forty times'.[60]

This compares unfavourably with the telescopes used by his successors, Maskelyne, John Pond, and George Biddell Airy. It is Maskelyne who first remarked that Bradley was the first observer who 'noting the proportional distance of the star from the wire at the two beats immediately preceding and following the transit across the wire',[61] improved the efficacy, accuracy and precision of his observations through the application of a standard observational technique he transmitted to his assistants. So much of Bradley's technique was acquired and refined through constant practice. Once developed, such skills could be passed onto assistants, who were trained to observe to the same exacting standards. When using Graham's 8-foot quadrant, Halley's maxim was 'secure your minute'.[62] With the divisions on the vernier for the outer arc of 96 divisions being equivalent to 13.18 arcseconds, Halley read off by estimate to about half this arc, Halley's observations were never more accurate than about seven arcseconds. Bradley was equally limited by these circumstances but on 18 July 1745, a new micrometer screw was fitted to the instrument. He wrote, 'the threads of which are much finer than those of the former, viz. 39¼ threads in one inch ... so that one revolution of it will alter the inclination of the telescope fifty-three seconds'.[63]

After the addition of this refinement, Bradley was able to read off angles as small as a second of arc. In the middle of August 1745, Bradley wrote that he

> fitted a shorter eyeglass into the quadrant telescope (viz. one of two inches focus, the old one being about one inch longer,) and likewise put a small horizontal wire, that is only 1/440th part of an inch in diameter, or subtending [an] angle of 5″ only; the former, that was of the same size with the perpendicular wires, seeming too gross for small stars.[64]

In spite of these adjustments and the constant rectification of his instruments, Bradley was displeased with the tools at his command. In comparison with the instruments he had access to at Shirburn Castle, he accepted that those of the Royal Observatory were lacking in the requirements necessary for the task he now set himself, a complete stellar survey of the sky observable from Greenwich. Bradley coordinated these early observational programmes with his nephew at Greenwich with those undertaken at Shirburn Castle in partnership with Lord Macclesfield. During 1746 Bradley discovered a distortion in the figure of the quadrant of up to 15¾ arcseconds. He wrote,

> in the whole part of the quadrant is occasioned by the cylinder [on which the telescope turned] being placed. ... about 1/191 part of an inch from the true centre of the arcs, in a line passing through it at 45° on the arch; for such a position of the cylinder will occasion such a defect in the quadrantal arc as was found in Sept. 1745.[65]

He repeated his examination of the quadrant during September 1747. At this time he was able to determine that the eccentricity did not amount to much more than 1/400th part of an inch, having also estimated that the length of the quadrant had gradually diminished. This testified to the changing figure of the instrument. The quadrant had been designed and constructed by Graham in order to preserve the rigidity of the framework. By the mid-1740s, after twenty years of use, this could no longer be relied upon. The sheer weight of the iron framework was distorting the figure of the quadrant. By 1745, Bradley was reconciled to the necessity of acquiring new instruments for the Royal Observatory. No matter how much he rectified and refined his instruments, they were simply not up to the demands required for a survey of the brighter stars observable from Greenwich.[66]

From early in 1747, as Bradley neared the end of his work on the nutation of the Earth's axis, he began planning the New Observatory at Greenwich. Bradley required an observatory that conformed to modern practice, as it should have been in 1675. It was an observatory designed to allow the observation of stars as they passed through a meridian at Greenwich without recourse to the use of temporary outhouses. The original observatory designed by Wren was already outmoded when it was built. Bradley's chief instruments were the quadrant, the transit instrument, the zenith sector, and the regulator. He began work on the acquisition of the new instruments and facilities he needed, in order to undertake the programme of systematic observation he had in mind. From 1749, he amended his teaching commitments at Oxford, from three courses of experimental philosophy a year to two. He was mindful of the task to which he was committed, a reform of the Royal Observatory and a complete survey of the sky observable from Greenwich. Much of the documentary evidence for these conclusions is missing.[67] Many of the expected documents from this crucial period are notable by their absence. Rigaud had no doubts of the reasons for the lack of documentation from this period, coinciding with the years that Thomas Hornsby worked on Bradley's observations. From 1749 Bradley's expanding commitments at Greenwich coincided with his major reforms of the Royal Observatory. He needed to build the New Observatory, as well as acquire the new instruments he needed to use in his great survey.

The series of registers recording the observations made by Bradley during the period of his incumbency as Astronomer Royal were originally donated to the University of Oxford by Bradley's executors in 1776. Any financial value envisaged by Bradley's executors fell by the wayside after the Board of Longitude abandoned their suit. The registers remained a persistent point of dispute until they were finally transferred to the safe keeping of the Royal Observatory in 1861.[68] The observations from 1742 until 1762 fill 14 registers.[69] In addition there were a series of copies in 'fair hand' which were commonly used as a basis of their later publication and reduction.[70] There are some observational records made at Greenwich still remaining in the possession of the Bodleian Library, including the observations made by the Bradleys at Greenwich from

1742 to 1743.[71] These extraneous records and registers are significant, coming after taking possession of the existing instruments at Greenwich. They compare the performance of the Royal Observatory's instruments with those at Shirburn Castle. A good case can be made for their transfer from the Bodleian Library to Cambridge University Library to be placed along with the other records of the Royal Observatory.[72]

I have assessed that the number of recorded observations made using the transits and quadrants by Bradley and his assistants at Greenwich from 1742 to 1762 is about 200,000 or more. The very first of what could be called Bradley's 'working registers' date from 1743.[73] It records transits of the major stars[74] using Halley's transit instrument, filling 177 folio pages. Bradley and Macclesfield were busily engaged cross-comparing the results of observations at Greenwich, with those made at what was then the finest observatory in England at Shirburn Castle.[75] A considerable number of entries at Greenwich are made in John Bradley's hand[76] just as the Shirburn registers[77] possess many entries in James Bradley's hand.[78] In this first register there are included calculations to discover the difference between the meridians at Oxford and Greenwich, and between Shirburn and Greenwich. Dating from this early period of Bradley's tenure at the Royal Observatory is a register of observations made with Graham's 8-foot quadrant.[79] The register is also interesting because it includes evidence that Bradley's eventual successor to the office of Astronomer Royal, Nathaniel Bliss, was working closely with him at Oxford.[80] It includes an account of Bliss's observations of Jupiter at Oxford on 21 May 1743.[81]

After a period of assessment, Bradley began a programme of observation leading to the inevitable conclusion that the instruments of the Royal Observatory were inadequate for the purpose they were designed for. The second transit register covers the period from 1744 to 1746.[82] The succeeding quadrant registers include a 'pot pouri' of miscellaneous items in addition to the record of observations from 1744 to 1747.[83] The most important of these items is a roughly written table of variations of the quadrant from 1743 to 1744 which exposes Bradley's continuing dissatisfaction with the instrument. The number of observations that the Bradleys made during these years meant they were forced to rectify the quadrant with a tedious repetitiveness that hindered them. In observing with an instrument subject to so many disparities, as well as coping with the changing position of the retaining wall, with motions east to west as well as north to south, observing was a tiresome task. There are several loose papers which refer to these problems, including a record of the collimation errors of the quadrant,[84] errors of the plane of the quadrant at diverse times from 1743 to 1745,[85] and various data relating to the adjustments made to the quadrant.[86]

There is documentary evidence, revealing that Bradley was not merely concerned with the state of the observational instruments of the observatory. There are loose papers on the errors and rates of his clocks, which also include part of a letter concerning the building of a new observatory. They are undated, but papers on the clocks range from 1743 to 1748.[87] It indicates that Bradley's early ambition was a rebuilding of the Royal Observatory. He resolved that a purpose-built new observatory was required, a building designed for modern observational practice. Having been instrumental in the construction and provision of the Shirburn Observatory from 1739 to 1740 on behalf of Lord Macclesfield,[88] Bradley surely considered a similar programme at the Royal Observatory early in his incumbency. The problems with the 8-foot quadrant must have increased his resolve. In the register containing the observations made with the Graham 8-foot quadrant from 1747 to 1753[89] is a useful summary inside the front cover, of the repairs and alterations made to the instrument from 1742 to 1753. It reveals the full extent of the problems that Bradley experienced before the Bird quadrant was constructed, divided, and mounted in 1750. The equivalent register for the observations made with the old transit instrument from 1746 to 1749,[90] contains a supplement on lunar eclipses from 1748 to 1751. Finally, there is a register from 1749 to 1750,[91] containing further transit observations from 1754 to 1755. It also contains a record of an earthquake shock recorded in Greenwich at 8.23 a.m. on 8 February 1750.

Although Bradley experienced frustrations using instruments deficient for his own exacting standards, his rigorous practice countered some of the inadequacies. The continuous rectifications helped to deal with the discrepancies experienced with the Graham 8-foot quadrant. As the summary of his repairs and alterations demonstrate,[92] he found the instrument almost impossible to rely upon. As John Bird (1709–1776) was to discover when he constructed a new quadrant just before 1750 modelled on George Graham's prototype of 1725, the rigidity of the frame was undermined by the immense weight of the ironwork.

The rise to prominence of John Bird is remarkable. A weaver by trade, he walked to London from County Durham in 1740. He was an excellent engraver, a skill he acquired to supplement his income. After employment in the workshops of George Graham and Jonathan Sisson, he began a small business of his own in the Strand. His work was remarkable for its quality. He came to the notice of James Bradley, who could no longer rely on George Graham, who was too elderly, and died in 1751. John Bird was set to work by Bradley, studying Graham's instrument, before constructing a new quadrant made entirely with brass. Its figure was less likely to be distorted by its own weight. It was divided by Bird to a degree of precision that Graham would have taken immense pride in. Bird's instruments were so highly regarded that he received important commissions from observatories all over Europe, beginning with the new Russian Imperial Observatory at Pulkovo near St. Petersburg.

Bradley's studies of the vibrations of seconds pendulums at Greenwich date from 1 September 1743 to 1 August 1745. His attentions returned to these inquiries during 1748 when he was advancing the future developments of the Royal Observatory. On 28 January 1749,[93] he began his final pendulum experiments writing up notes on the subject. It had been a topic that had focused his attention from 1719. From experiments he had conducted at Wanstead, he had discovered that the length of a seconds pendulum at the latitude of London was between 39.14 and 39.168 inches.[94] Rigaud mentions that nothing was known of the standards by which these results were obtained, or the temperatures under which they were conducted.

In the summer of 1748, Bradley represented the dire state of the instruments and facilities of the Royal Observatory to the Visitors. This was the very body that Flamsteed regarded as a group of hopelessly ill-informed and ill-disposed people and as the tools of Sir Isaac Newton. Now this same body was being used by Bradley towards the attainment of his own ends. We can be sure that this body of Visitors was better informed and in awe of Bradley's immense achievements, rewarded as he had been that very year with the Copley Medal. During August 1749, the Council of the Royal Society considered a petition to the Board of Admiralty, which had been prepared by Bradley.[95] Once it had been approved, it was signed by the President, Martin Foulkes, and the rest of the Council. Some alterations were made to the petition, so it was not presented to the Lords of the Admiralty until October. On 9 October 1749, the President informed the Council that he and Bradley 'had attended the lords of the admiralty, and that their lordships had promised their assistance in forwarding the petition.'[96]

There was a detailed schedule attached to the petition prepared by Bradley together with an account of the sum required which was estimated to be £1,000. It was granted by George II without delay.[97] Following the publication of his paper on the nutation of the Earth's axis and the award of the Copley Medal in January 1748, Bradley's influence was at its most pre-eminent. Bradley once more came to the notice of George II who admired his work. Bradley had already enjoyed his patronage when he was the Prince of Wales. In the twenty years or more since then, George II, or so it appears, had maintained a distant interest in the work of the man Newton had described as 'the finest astronomer in Europe'.

George II directed an order that the sum of £1,000 be paid to Bradley[98] in order to repair the old instruments and to provide new ones. Rigaud listed the details of the estimates in the *Memoirs of Bradley*. It included backdated payments of £85 16s 0d for instruments and repairs from December 1742 which had been paid for out of his own pocket.[99]

		£	s	d[100]
1742, Dec. 8	By Mr. Sisson's bill for altering the transit	19	8	0
1742, Aug. 10	By a pair of globes of Mr. Sennex	8	9	0
1742, Feb. 9	By Mr. Graham's bill for a pendulum	15	13	0
1744, Jan. 3	By Mr. Graham for a pendulum	10	0	0
1745, Oct. 1	By a micrometer for the mural quadrant, by Mr. Sisson	3	3	6
1746, Feb. 11	By an apparatus for trying the line of collimation, by Mr. Bird	2	12	6
1746, May 29	By an arch for the transit instrument, by Mr. Bird	8	13	0
1747, July 9	By a level by Mr. Bird	1	11	6
1748, Oct. 12	By alterations in the transit instrument, &c. by Mr. Bird	5	15	6
1743–1748	By alterations made in the quadrant and transit rooms by smiths and carpenters, their bill	10	0	0
	By glasses and a perspective of Mr. Mann	3	18	0
	By a chamber-alarum by Mr. Graham	3	3	0
	By a parallactic sector 12½ feet radius	45	0	0
	By a diurnal sector, with a telescope of 30 inches	35	0	0
	By an apparatus for frequently observing the variation and inclination, or dip of the magnetic needle	20	0	0
	By a 15 feet refracting telescope and micrometer	20	0	0
	By a clock per Mr. Graham	39	0	0
	By a brass mural quadrant by Mr. Bird	30	0	0
	By a moveable quadrant per ditto	20	0	0
	By a transit instrument ditto	73	13	6
	By a 20 feet refracting telescope ditto	7	10	0
	By a barometer per ditto	2	12	6
	By a thermometer per ditto	1	15	0
	By alterations to the old mural quadrant ditto	32	10	0
	By fees at the treasury	2	2	0
	By Mr. Short in part for a six feet reflecting telescope	100	0	0
		972	0	0
		28	0	0
		1000	0	0

The costs of the new buildings which were built from 1749 to 1750, to house the new and old quadrants as well as the new transit instrument, was the largest single outlay devoted to the Royal Observatory since it had been constructed in 1675. Even then the bricks used to build what is now called Flamsteed House were secondhand, salvaged from the demolition of Greenwich Castle. These new outlays were probably borne by the Board of Ordnance.[101] The £10 paid to smiths and carpenters must have been for work prior to the construction of the new observatory, probably on outhouses. The New Observatory is the building through which the internationally recognized Prime Meridian now passes. The present meridian dates from the nineteenth century, and marks the position of Airy's transit instrument. Bradley's meridian passes through the middle of the structure, 17 feet to the west of the present meridian. The position of Bradley's meridian is clearly marked on the exterior of the building. The maps of the British Ordnance Survey are based on the position of the meridian sited by Bradley. Shortly after the construction of the Transit House, or as it was commonly referred to at this time 'The New Observatory', Bradley wrote a memorandum which reveals further information about its construction. He wrote,

> In the year 1749 £1000 was given by his majesty, to be paid by the treasurer of the navy out of money arising from the old stores of the navy, (upon the representation of the lords of the admiralty, and principally upon Lord Anson's[102] recommendation) to buy some astronomical instruments for the use of the Royal Observatory; when it was proposed by Mr. Foulkes, Mr. Graham[103] and Mr. Robins, who were consulted with on that occasion, that in the catalogue of instruments to be purchased a parallactic sector should be inserted, as very useful for observing stars near the zenith; and the sector which I formerly hung up in Wansted, in 1727, (with which I afterwards discovered the laws of aberration of the fixed stars, as also the nutation of the earth's axis,) being judged by them worthy of a place at the Royal Observatory,[104] I removed it from Wansted in July 1749, and procured a new apparatus for suspending it, (made by Mr. Hearn, as the old one was;) and I likewise took care, while the rooms of the new observatory were building, that there might be made convenient places for hanging the sector, both in the new quadrant room and in the new transit room: my view (for providing for its suspension in either room was, to render it useful for settling the true zenith,) whereby errors of the line of collimation of the telescopes of the mural quadrants may be found with great ease and certainty.[105]

Rigaud states that the provision for two separate places of suspension was with the view to the advantage of reversing the instrument, the face of the sector being turned to the east when hung in the quadrant room, and to the west in that of the transit room.[106]

The clock placed at Graham's account was not made by him, for by now he was almost 75 and too old to work to the exacting degrees of precision required by Bradley. It was constructed by John Shelton under Graham's instructions.

Most notably, several of the instruments were constructed by John Bird. This was to be an association almost as important to Bradley as his former partnership with George Graham. Of all the instruments constructed by John Bird, the most celebrated was the new 8-foot quadrant, modelled on the quadrant designed and constructed for Halley by Graham in 1725. Bird constructed the frame out of brass so that the new instrument retained the strength and rigidity of the former, while avoiding many of the problems associated with internal distortion due to the weight of the frame of the instrument. The new quadrant had been ordered by Bradley in February 1749 [107] and was suspended on 16 February 1750.[108] In June, it was ready for observation, the first recorded observation being undertaken on 10 August 1750.[109] During Bradley's lifetime there was no measurable variation in the figure of the new quadrant,[110] although his own tests revealed that it was 2″ short of a complete quadrant which Bird verified, although he held that it was nearer 1½″ short of the full 90°.

John Bird stated that when he recognized that he was to be employed to construct a new quadrant for the Royal Observatory, he examined and thoroughly familiarized himself with the quadrant constructed for Halley by Graham in 1725.[111] This was written a few years after Bradley's demise with no fear of contradiction. Yet it was Bradley who chose Bird to undertake his commission in the light of Bird's earlier work and who had asked him to study Graham's instrument, before he gained the commission. Although Bird admired Graham's workmanship he early recognized that the alteration in its figure was due to several sections being forged in iron, which he contended was incapable of being forged as easily as brass into the form that was required. The new mural quadrant constructed by Bird to Graham's design, incorporated the same double division.[112]

In 1753 when Graham's quadrant was moved from where it had been situated since 1725 a new set of divisions was marked on it by Bird.[113] With these new marks, Bradley examined the arc on 12 August 1753 by means of his micrometer. He asserted,

> I measured with the screw of my micrometer the difference of the arcs (of 64/96) as set off by Mr. Graham originally, and by Mr. Bird when he put on a new set of divisions upon the old quadrant, and found that Mr. Graham's arc was less than Mr. Bird's by 8/40 divisions of my micrometer, which to a radius of 96 inches answers to 10.6″; so that the whole arc of 96 differs from a true quadrant 15.9″[114] which is the same difference that I formerly found by means of the level, &c.[115]

As well as the new quadrant, Bird constructed a new transit instrument which also reflected his immense skill and craftsmanship. It was mounted in the transit room, which defined the Greenwich meridian from 1750 until 1816.[116] The original specification reveals that it had an 8-foot focal length with an aperture of 2.7 inches, which was stopped down to 1.5 inches with an elliptical reflector. It possessed, in conformity with Bradley's earlier practice with the zenith sector at Wanstead, one horizontal and five vertical wires illuminated by a lamp attached to a nearby pier which could be reflected down a tube by the annular reflector in front of the object glass.[117] The transit instrument initially possessed a single eyepiece,[118] but the field of view could often become indistinct near the edge. To remedy this, a double eyepiece was applied to the transit in January 1753.

It was part of Bradley's specification for the new observatory that there should be a Newtonian reflector of about 6-feet in length. The estimate was that it would cost £115. Bradley paid £100 on account (10% of the entire grant from George II) to James Short for this purpose. It was mentioned in the minutes of the Council of the Royal Society (of which Bradley was a member) for 5 June1755.[119] It appears that there may have been some discussion about Short taking back a Gregorian reflector which he had provided to the Royal Observatory, while he was at work on the Newtonian that Bradley had ordered. Short agreed to this, also agreeing to deliver it in four months. When Bradley died in 1762 the money remained on account, so it appears that Short had not been able to fulfil the terms of his contract with the Royal Observatory. Short, by some accounts, was a difficult man to deal with. He made a fortune making the finest reflectors on the market until his death in 1768. Short was secretive about his methods. He destroyed his own craft instruments before he died, when John Bird was publishing his methods of constructing and dividing astronomical instruments. Short's response to his commission for the Royal Observatory must have been a disappointment to Bradley. He had pioneered the use of a viable reflector after it had been constructed by his friend John Hadley and his brothers in the early 1720s. All was not lost. Short had earlier given his services to the Royal Observatory with a letter sent to a 'Mr. Davall' of the Royal Society, on his methods of making large lenses. Rigaud writes in the *Memoirs of Bradley* that.

He delivered 'a full account' of it sealed up, to be preserved with the papers of the society,[120] and not to be opened till he gave leave for its publication. He appears to have made no communication on the subject during his lifetime, but after his death it was opened up by the council, and printed in the Phil.

Trans. (vol. LXIX, p. 507.) under the title of 'A method of working object glasses of refracting telescopes truly spherical'.[121]

The first mirrors that Short had made for his reflecting telescopes were made of glass. He described his method of producing a spherical figure to be such that in 1749 at Bradley's request he ground some object glasses for the new instruments at the Royal Observatory. Bradley considered them much superior to everything he compared them with, making much use of them in the following years. During the final decades of the eighteenth century telescope makers were able to capitalize on Short's methods.

In this account of the work of James Bradley, I have on occasion pointed to his reputation throughout Europe. His fame obviously rested on his discovery of the aberration of light or what Bradley commonly referred to as 'the laws of aberration'. Bradley's contribution was a step advance. When the position of stars could vary by angles as much as forty arcseconds a year[122] in latitude and longitude, the precise measurement of stellar positions, unless the time and date were clearly indicated, were made wholly obsolete. When Caroline Herschel attempted to reduce John Flamsteed's Greenwich observations to the current standards of the late eighteenth-century, she had to abandon the task. The first Astronomer Royal spent a lifetime of struggle producing a catalogue of 2,883 stars as precisely and accurately as his instruments allowed. Flamsteed's studies of the motions of *Polaris* revealed an annual motion of up to thirty arcseconds, that he precipitately forwarded as evidence of annual parallax, until his claim was demolished in a masterful paper by Jacques Cassini. Mari Williams writes,

The observations which Flamsteed believed demonstrated the existence of parallax had not been gathered expressly for that purpose; they were among the vast number of observations made by Flamsteed between the years 1689 and 1695 with his mural arc. He scrutinised those of *Polaris* after receiving a request from Newton for a new evaluation of the altitude of the north celestial pole which Newton required for his lunar theory. Once he had studied the data and concluded that they exhibited parallax, Flamsteed then decided to observe the Pole Star specifically with parallax in mind, but even these subsequent observations cannot be considered as an attempt to measure parallax in the way that Hooke's observations of γ Draconis were; Flamsteed's later observations of the Pole Star were made to test the theory based on his earlier ones.[123]

Flamsteed reveals a perceptual bias that led him 'to see what he wanted to see' and then make an unjustifiable claim for the observation of annual parallax. Flamsteed quickly retracted his claim in the face of Jacques Cassini's analysis, demolishing it on sound theoretical grounds. This sort of observational or theoretical bias was something Bradley fought against throughout his entire career, doubtlessly informed by the experiences of both Hooke and Flamsteed before him. Whether he acquired this sceptical approach from James Pound or whether it was a character trait is not possible to determine, but without doubt Bradley remained a careful, guarded observer and theoretician throughout his life. It was the key to his success. It led not merely to the discovery of aberration and nutation but laid the foundations of his final work, the distinguished series of observations he shared with his assistants at the Royal Observatory from 1750 to 1762. They transcended anything that had been achieved prior to his work and the work of his assistants. Until the work of Bessel, the unpublished, unreduced observations lay embryonic, without revealing the many treasures that the German astronomer disclosed. The reason for this is not only that these observations were so precise and accurate, but because every conceivable source of error was recorded. These included changes in the equipment, or variations in the seeing conditions, as well as changes in the ambient air temperatures, or the barometric pressure, and all sorts of situational changes. These were entered in the registers, so that anyone could reduce his observations to any degree of precision and accuracy. This was the measure of Friedrich Wilhelm Bessel's achievement in the *Fundamenta Astronomiæ* of 1818.

It is Bradley's sceptical practice that offers an insight into his friendship with Halley. His mentor's refusal to accept Bradley's widely admired claims to be able to discern motions as small as one arcsecond or less, reveals a similar sceptical mindset. This, Bradley accepted and understood.[124] Bradley recognized that Halley was incapable of making observations of such fine angles. He never pushed the point with his friend, sure as he was that he had discovered a new phenomenon after observing the nutation for several years. Even his relationship with Nicolas-Louis de Lacaille, whom he only knew through their mutual correspondence, was also marked by the Frenchman's due scepticism about the causes of the phenomenon. Bradley was sure the observed motion was a lunar-induced nutation. Lacaille consistently referred to it as a 'deviation' until he was sure that it was a nutation of the terrestrial axis. This due scepticism is a basis of so much modern science. Knowledge and scepticism go hand in hand.

Ever since the publication of the writings of Thomas Kuhn it has been commonly asserted that 'science' is simply one belief system amongst many others. Yet scientific knowledge is public knowledge. It is knowledge based on what can be observed or deduced. It is hard won, and sometimes it is hard lost too, when seemingly well established 'facts' must be relinquished in the face of overwhelming evidence to the contrary. Such 'facts' are not simples, they are complexes, based not just on observations and deductions, but often on the acceptance of presumptions and the interpretations applied to them. Even the most sceptical of us can accept beliefs that transcend our knowledge. We have to accept that the scepticism that so many hold about science itself, or at least what they believe science to be, is based on a misapprehension of what science claims, and what it doesn't. Science is not about 'proof': it is about knowledge determined by experiential evidence. It is knowledge that is open to challenge. Bradley, in his lectures on experimental philosophy at Oxford, claimed that natural philosophy was incapable of the same degrees of certainty possible in mathematics.

At the time that Bradley began work on his survey of all the brighter stars visible from Greenwich, his correspondent Lacaille was setting out to begin his own major work. Bradley made his first observations with the new 8-foot quadrant on 10 August 1750 OS, whilst on 21 October 1750 NS according to I. S. Glass, Lacaille began his journey from Paris to the Cape of Good Hope, to undertake his own survey of the southern stars.[125] He set sail from L'Orient on 21 November 1750 NS, finally arriving at the Cape on 20 April 1751 NS. From 6 August 1751 NS, Lacaille began his survey of the southern skies which lasted until 18 July 1752 NS. Following his measures of the Earth's radius, he finally left the Cape on 8 March 1753,[126] and after visiting Mauritius and Réunion to accurately measure their latitudes and longitudes, finally reached Paris on 28 June 1754.[127] Lacaille, like Bradley, was both highly intelligent and very personable. After arriving at the Cape, he immediately struck up a relationship of mutual regard with the Dutch governor Ryk Tulbagh. As with the British, the Dutch had recently been at war with the French, and the circumstances could so easily have led to suggestions that the Frenchman was an agent in the employ of the Kingdom of France. Fortunately, Tulbagh was a man of common sense, who possessed more than a passing interest in all matters scientific. The two men evidently impressed each other. Lacaille had already demonstrated his assiduity when working on the survey of the Paris meridian and his capacity for work had not at all diminished at the Cape. Glass brings to life the hard work and tedium that the life of an astronomer could involve during the eighteenth century. Lacaille reported to Grandjean de Fouchy in Paris that,

> The general catalogue of [the stars] between the South Pole [and] the Tropic of Capricorn is advancing with the season. I have done more than 3/5 of it and because I am always prepared to profit from clear nights, I am sure to have finished it by the end of June, assuming the weather is no worse than last year, and that my health always keeps up. It is very arduous work, which requires around 150 complete nights of observation, my head almost always tipped back, however, the more I advance, the more I am content to have undertaken to do it, because the furious southeaster prohibits any other kind of observation when it blows and because the sky is then of admirable clarity. I have seen double stars and the most singular nebulæ, all worth being studied. As soon as the work is finished I will put it all in order and I will send a copy before I leave; there will probably be no fewer than 10,000 stars.[128]

Whilst Lacaille was on his travels to the Cape and to various French possessions in the Indian Ocean, communications with Bradley and others were conducted through the offices of Joseph-Nicolas Delisle, who sought earnestly to improve communications throughout the scientific community.

Delisle had earlier met up with Halley in 1724. This led to him being elected an FRS during 1725. Shortly after this, from 1725 he was invited to the Russian Empire by Peter the Great, to create and run a school of astronomy, an important adjunct in Peter's modernization programme. Delisle's scientific rectitude led to differences with various figures associated with the *Atlas Rossicus* leading to his request to leave Russia in 1743. He finally left in 1747 and returned to Paris. Delisle had discussed with Halley the possibility of using the transits of Venus in 1761 and 1769 in order to organize an international effort to determine the mean distance from the Sun to the Earth, a distance which we now refer to as an *astronomical unit*, the baseline in the determination of stellar distances through the measurement of annual parallax. Delisle's efforts to open up communications between members of the scientific literati was hindered by a series of wars that often thwarted his aims. Although it was an open secret that Bradley was likely to complete his programme of observations of the nutation of the Earth's axis in 1747, his supporters in France were unable to ascertain whether he had completed his work, for Great Britain and France were at war with one another from 1744 to 1748.

As the evidence of Lacaille's lengthy letter to Bradley reveals, the war had almost completely closed down any constructive form of philosophical communication between the inhabitants of the warring states, certainly between London and Paris. In a century when war was regarded as an extension of politics, it genuinely surprises that communication between scientific workers was affected quite so radically, but apparently it was, as evidenced by Lacaille's letter of 1748 to Bradley. So with the Seven Years War fought from 1756 to 1763 when Great Britain was once again at war with France, we need not be surprised that useful communications between members of the scientific community in London and Paris all but ceased. Yet such was the respect with which James Bradley was held at the Royal Academy of Sciences in Paris, he was admitted as a full academician in 1761, the year before he died. By 1761 Bradley was ill and had more or less retired to live with his late wife's family in Chalford in Gloucestershire. Remarkably, in just a few months Tobias Mayer (b. 1723), Lacaille (b. 1713), and Bradley (b. 1692) all died, robbing the science of positional astronomy of all three of its leading exponents.

Notes

1. Stephen Peter Rigaud, *Memoirs of Bradley,* p. li. Edmond Halley was the Savilian Professor of Geometry at Oxford from 1704 to his demise in 1742, who on occasion took up residence in Oxford when he was detained from his duties in Greenwich from 1720. If Hearne spoke of Halley being 'somewhat lame', he must have been describing him in his later years when he began to suffer from the paralysis of one of his hands.
2. Ibid. xliv.
3. Ibid. xliv.
4. Ibid. xlv.
5. This was the equivalent of about £125,000 in today's monetary values.
6. Francis Baily, *An Account of the Rev. John Flamsteed, the First Astronomer Royal, Compiled from his Manuscripts and Other Authentic Documents, Never Before Published,* London, 1835, p. 343. Both Flamsteed and Halley were compelled to make observations from various outhouses, testimony that the design of the original observatory, now known as 'Flamsteed House', was both inadequate and outdated from the outset.
7. Derek Howse, *Greenwich Observatory: The Story of Britain's Oldest Scientific Institution: The Royal Observatory at Greenwich and Herstmonceux 1675–1975: Volume 3, The Buildings and Instruments,* Taylor and Francis, 1975, p. 6.
8. The Duke of Argyll, the Master General of Ordnance, refused a request for sufficient resources to purchase a second mural quadrant. Time and again the importance of the work of the Royal Observatory in the national interest appears to have escaped the notice of those who were in authority.
9. Ibid. p. 6.
10. George Graham was also engaged at this time with the design and construction of the zenith sector to the commission of Samuel Molyneux at Kew.

11. Nevil Maskelyne, 'Astronomical Observations 1766–1774', London 1776, Vol. 1, Section 1.

12. Derek Howse, *Greenwich Observatory: The Story of Britain's Oldest Scientific Institution: The Royal Observatory at Greenwich and Herstmonceux 1675–1975: Volume 3, The Buildings and Instruments,* Taylor and Francis, 1975. pp. 22–23.

13. James Bradley, Royal Greenwich Observatory Archives, Cambridge University Library, Document RGO 3/31, *Observations: Transits of major stars including adjustments to the telescope and observational method 1743,* fol. 1.

14. This is surely the reason why Bessel referred to Bradley as the 'incomparable'.

15. See Chapter 5, note 46, for discussion of Bradley's nescient approach.

16. Stephen Peter Rigaud, *Memoirs of Bradley,* p. xiv.

17. This is the official title of the office of 'Prime Minister' which is in a sense a courtesy title. The title of Prime Minister is first mentioned in official state documents early in the twentieth century.

18. There was a growing 'war party' that opposed Walpole's peace policy, evidently perceiving a financial interest in a war policy.

19. He is not to be confused with the later Sir William Jones, poet and noted orientalist, who was also also one of Sir Joseph Banks' correspondents.

20. The holder of the office at this time was Philip Yorke Earl of Hardwicke. He held the office from 21 February 1737 until 20 November 1756.

21. The Archbishop of Canterbury was Dr. John Potter, who held the office from 1737 to 1747.

22. I am sure here that Macclesfield is making reference to George Graham's 8-foot quadrant, made for Halley in 1725.

23. George Parker 2nd Earl of Macclesfield to Sir William Jones, 1741/2 January 14 OS, This is quoted in full in Stephen Peter Rigaud, *Memoirs of Bradley,* p.xlvi.

24. Here Sir William Jones is using the word 'profession' in the sense of a calling or a vocation. It does not indicate a position that is characterized by the receipt of a salary, although the post was associated with a small annual stipend of £100, barely enough to keep body and soul together.

25. Stephen Peter Rigaud, *Memoirs of Bradley,* pp. xlvii–xlviii.

26. The contrast with John Flamsteed's comparative lack of support is palpable.

27. James Bradley, Royal Greenwich Observatory Archives, Cambridge University Library, Document RGO 3/43 *Miscellaneous Observations and Calculations 1742–1757.* On a loose sheet of paper in a box.

28. A central plank of Walpole's foreign policy had been the maintenance of peace. Difficulties with Spain over the West Indies trade forced him to declare war when his hand was forced by a majority in Parliament. Walpole admitted his opposition to the war to his cabinet colleagues and his views eventually became known to the rank and file. The so-called 'War of Jenkins Ear' was indecisive and opposition within Parliament grew, though he succeeded in winning the 1741 general election. However, many Whig politicians and several independent members mistrusted his capacity to direct the war with the vigour considered necessary and his resignation was eventually forced over a minor issue. After his resignation, he was created 1st Earl of Orford by George II and was granted an annual pension of £4,000. An attempt to impeach him

over his ministry failed. In the final years of his life he attempted all in his power to have Carteret (who had followed him in the office of Secretary of State) dismissed. He sought the promotion of Henry Pelham, his protégé, who was the leader of the Walpole Whigs. Pelham became First Lord of the Treasury in 1743. Pelham dismissed Carteret (who was earlier created Earl Granville) in 1744 when the Secretary of State sought to increase Britain's involvement with the War of the Austrian Succession in conflict with France and Prussia. Both Walpole and Pelham remained close confidants of George II. Pelham's close relationship with the King was to have consequences when Bradley sought the reformation of the Royal Observatory.

29. Stephen Peter Rigaud, *Memoirs of Bradley,* p. xlviii. Citing the Supplement to the Biographical Dictionary, p. 56.

30. Ibid. xlviii. Rigaud wrote, 'It appears, therefore, that one of the first things that Walpole looked to, as soon as he had determined to retire, was to make use of the power which yet remained in his hands, to secure the office of astronomer royal for the man who of all others was the most deserving of it'.

31. One of the skills that Bradley developed and perfected was his 'eye and ear' method. Using his quadrant or his transit instrument he carefully determined the declination of a star (or any celestial object) whilst counting the audible vibrations of his clock or regulator. These would be recorded enabling him to determine the right ascension of the object whilst recording its declination. This and other skills were passed on to his assistants. This technique was aided by the use of multiple lines in north–south in addition to east and west at the focal point of the telescope.

32. Ibid. p. xlix. Rigaud suggests that Macclesfield's assertion that 'the finest instrument in the universe' was Graham's 8-foot quadrant. Macclesfield must have been aware of the state of that instrument for throughout the 1730s Bradley often attended to its faults and insufficiencies prior to his appointment as Astronomer Royal. He must also have been referring tangentially to the simple fact that Bradley of all people was the one person who possessed sufficient knowledge and skills to rectify the instrument.

33. This word is clearly written in the manuscript.

34. James Bradley, Royal Greenwich Observatory Archives, Cambridge University Library, RGO 3/31, *A Miscellaneous Collection of Handwritten Notes.* Doc. 1, fol. 1v.

35. Ibid. doc. 5. fol. 1r.

36. Bradley was a witness to Halley's Will and joint executor.

37. Bradley possessed two truly outstanding mentors. Halley imbued Bradley with the mathematical skills required to master Newton's *Principia,* and Pound inculcated the observational disciplines which were such an important part of his armoury of skills.

38. In 1686 the Royal Society largely emptied its coffers publishing Francis Willughby's *De Historia Piscium.* The society was unable to publish Isaac Newton's *Principia.* It was published in 1687 through the support of Edmond Halley, then the clerk of the society. He was recompensed with unsold copies of *De Historia Piscium.*

39. The appointment of his young nephew as his first assistant at Greenwich probably came after discussions with members of his family in Gloucestershire. The retention or destruction of Bradley's personal correspondence by his executors has deprived historians and biographers of any insights concerning John Bradley's appointment to this vital post.

40. I will at this point defend the abilities of John Flamsteed.

41. This surely indicates that some considerable time had elapsed since Halley had last used the transit instrument.

42. James Bradley, Royal Greenwich Observatory Archives, Cambridge University Library, RGO 3/31, *A Miscellaneous Collection of Handwritten Notes,* doc. 5, fol. 2r—2v.

43. Ibid. doc. 5, fol. 2r—2v.

44. Ibid. doc. 5, fol. 2r—2v.

45. Ibid. doc. 5, fol. 2r—2v.

46. James Bradley, 'The State of the Instruments at the Greenwich Observatory when Dr. Bradley became Astronomer Royal', in *Miscellaneous Works and Correspondence of James Bradley,* Oxford, 1832, p. 382.

47. Ibid. p. 382.

48. James Bradley, Bodleian Library, Oxford, Department of Western Manuscripts, Bradley MS31, *Miscellaneous Astronomical Observations at Greenwich with Two Transit Instruments from 25th July 1742 to January 1742/43. At folio 38 Lord Macclesfield's Similar Observations at Shirburne, September to December 1742 and March 1743.*

49. Stephen Peter Rigaud, *Memoirs of Bradley,* p. lii.

50. Ibid. p. lii.

51. James Bradley, Bodleian Library, Oxford, Department of Western Manuscripts, Bradley MS31, *Miscellaneous Astronomical Observations at Greenwich with Two Transit Instruments from 25th July 1742 to January 1742/43. At folio 38 Lord Macclesfield's Similar Observations at Shirburne, September to December 1742 and March 1743.* fol. 28r.

52. John Bradley remained as his uncle's assistant until 1756.

53. James Bradley, Royal Greenwich Observatory Archives, Cambridge University Library, RGO 3/1, *Observations: Transits of major stars including adjustments to the telescope and observational method, 1743.*

54. Ibid. fol. 108 r—108v.

55. James Bradley, Royal Greenwich Observatory Archives, Cambridge University Library, RGO 3/8, *Observations, using a Quadrant, of major stars; including adjustments to the telescope and observational method, 1743-1744.*

56. James Bradley, Royal Greenwich Observatory Archives, Cambridge University Library, RGO 3/31, *Miscellaneous Collection of Handwritten Notes,* doc. 6, fol. 1r.

57. Ibid. doc. 6, fol. 2v.

58. James Bradley, Royal Greenwich Observatory Archives, Cambridge University Library, RGO 3/8, *Observations, using a Quadrant, of major stars; including adjustments to the telescope and observational method, 1744 -1746.*

59. Stephen Peter Rigaud, *Memoirs of Bradley,* p. lv.

60. James Bradley, Royal Greenwich Observatory Archives, Cambridge University Library, RGO 3/1, *Observations: Transits of major stars including adjustments to the telescope and observational method, 1743.*

61. Nevil Maskelyne, 'Astronomical Observations, 1766–1774', Vol. III, Transit Observations, p. 339.

62. Stephen Peter Rigaud, *Memoirs of Bradley,* p. lv.

63. James Bradley, Royal Greenwich Observatory Archives, Cambridge University Library, RGO 3/2, *Observations, using a Quadrant, of major stars; including adjustments to the telescope and observational method, 1744 –1746.*

64. Stephen Peter Rigaud, *Memoirs of Bradley,* p. lv.

65. Ibid. p. lv.

66. Judging from the Catalogue of 3,222 stars included in the *Fundamenta Astronomiæ pro anno MDCCLV deducta ex obsivationibus viri incomparabilis James Bradley in specula astronomica Grenovicensi per annos 1750–1762 institutis,* Königsberg, 1818, Bradley observed stars down to the 8th magnitude.

67. The reasons for this are possibly connected to Thomas Hornsby's work on Bradley's Greenwich Observations from 1750 to 1755. Rigaud refers to the immense difficulties he had retrieving many of Bradley's papers from his predecessor as Savilian Professor of Astronomy. According to Rigaud, his predecessor had not treated Bradley's papers with the respect they deserved. Rigaud went through Hornsby's papers and gathered as many of Bradley's papers as he could locate, mindful that many had certainly been irretrievably lost. See Stephen Peter Rigaud, *Memoirs of Bradley,* p. xxxviii passim.

68. Bartholomew Price Papers, Pembroke College, Oxford, (PMB/S/14). Letters PMB/S/14 & 15. 1861.

69. James Bradley, Royal Greenwich Observatory Archives, Cambridge University Library, RGO 3/1 to 3/14.

70. Ibid. RGO 3/15 to 3/29. These are 'fair copies' of RGO 3/1 to 3/14.

71. James Bradley, Bodleian Library, Oxford, Bradley MS 32, *James Bradley's Observations at Greenwich with Two Transit Instruments from the 25th July, 1742 to January 1742/43.* At fol. 38r to fol. 43v, *Lord Macclesfield's Similar Observations at Shirburne, September to December 1742 and March 1743.* These latter observations are copies of the Shirburne registers.

72. James Bradley, Bodleian Library, Oxford, Bradley MS28∗ *Calculations concerning the Precession of the Equinoxes. Absolute and Relative Motion of Certain Fixed Stars compared with Flamsteed's Observations*; Bradley MS29 *Observations and Calculations by James Bradley of the Fixed Stars*; Bradley MS33 *Miscellaneous Astronomical Observations and Computations*; Bradley MS42 *Observations and calculations on refractions*; and Bradley MS43 *Miscellaneous Astronomical Tables by James Bradley. Tables of the Sun's declination. Mean precession in longitude.*

73. James Bradley, Royal Greenwich Observatory Archives, Cambridge University Library, RGO 3/1 *Observations: transits of major stars; including adjustments to the telescope and observational method, 1743.*

74. As revealed by F. W. Bessel's *Fundamenta Astronomiæ,* this includes stars of the 8th magnitude.

75. The observatory at Shirburne Castle was constructed under the direction of James Bradley and later served as a guide to the creation of the New Observatory at Greenwich, also under the direction of James Bradley.

76. James Bradley spent a large part of the second half of 1742 training his nephew in the skills required in order that he might be safely entrusted with the continuation of observations during the periods he was residing in Oxford. Until 1749, Bradley

maintained his teaching schedule, presenting three courses a year in experimental philosophy and his lectures in astronomy.

77. James Bradley, Bodleian Library, Oxford, Bradley MS 32, *James Bradley's Observations at Greenwich with Two Transit Instruments from the 25th July, 1742 to January 1742/43*. At fol. 38r to fol. 43v, *Lord Macclesfield's Similar Observations at Shirburne, September to December 1742 and March 1743*. These latter observations are copies of the Shirburne registers.

78. Stephen Peter Rigaud, *Memoirs of Bradley*, p. liv.

79. James Bradley, Royal Greenwich Observatory Archives, Cambridge University Library, RGO 3/8, *Observations, using a Quadrant, of major stars; including adjustments to the telescope and observational method, 1743–1744.*

80. Nathaniel Bliss was Savilian Professor of Geometry, from 1742 to 1764, in direct succession to Edmond Halley.

81. Bliss's observations were addressed to Bradley.

82. James Bradley, Royal Greenwich Observatory Archives, Cambridge University Library, RGO 3/2, *Observations: transits of major stars; including adjustments to the telescope and observational method, 1744–1746.*

83. Ibid. RGO 3/9 *Observations using a Quadrant, of major stars; including adjustments to the telescope and observational method, 1744–1747.*

84. Ibid. RGO 3/32, *Collimation errors of the Quadrant, 1742–1752*, doc. 12.

85. Ibid. RGO 3/32, *Errors of the planes of the Quadrant at various times from 1743 to 1745*, doc. 14.

86. Ibid. RGO 3/32, *Various data relating to adjustments taken on the Quadrant*, doc. 16.

87. Ibid. RGO 3/32, *Errors and rates of clocks, 1743–1748*, doc. 15.

88. It is very difficult to ascertain the exact dates when the Shirburne Observatory was constructed. Lord Macclesfield did not preserve his correspondence with Bradley and very little of the converse correspondence exists in Bradley's papers at Oxford or Greenwich.

89. James Bradley, Royal Greenwich Observatory Archives, Cambridge University Library, RGO 3/10, *Observations using a Quadrant, of major stars, including adjustments to the telescope and observational method 1747–1753.*

90. Ibid. RGO 3/3, *Observations: Transits of major stars; including adjustments to the telescope and observational method, 1746–1749.*

91. Ibid. RGO 3/4, *Observations: Transits of major stars; including adjustments to the telescope and observational method, 1749–1750.*

92. James Bradley, Royal Greenwich Observatory Archives, Cambridge University Library, RGO 3/10, *Observations using a Quadrant, of major stars, including adjustments to the telescope and observational method 1747–1753*. The inside of the front cover of this register details a record of repairs and alterations to the Graham quadrant.

93. James Bradley, 'Experiments to determine the Length of the Pendulum vibrating seconds at Greenwich' in the *Miscellaneous Works and Correspondence of James Bradley*, p. 385.

94. Stephen Peter Rigaud, *Memoirs of Bradley*, p. lvi.

95. Royal Society, *Minutes of the Royal Society*, 1749.

96. Stephen Peter Rigaud, *Memoirs of Bradley*, p. lxxiv.

97. Ibid. p. lxxiv.
98. Ibid. p. lxxiv.
99. This sum included the £3 13s 6d he paid Sisson for the micrometer for the Graham mural quadrant on 1 October 1745. It is recorded as a payment for £3 3s 6d but Rigaud states that 10 shillings be added to this sum in order to balance the accounts.
100. For those of my generation who were familiar with twelve pence a shilling and twenty shillings a pound the system was so useful for the multi-number of ways the pound could be divided. The decimalization of the British currency in 1971 eased calculations for many but the old pre-decimal currency still had its advantages. Actors at the Theatre Royal in Drury Lane early in the eighteenth century earned £10 per annum whilst supporting actors earned £6 13s 4d. This apparently incongruous amount is, however, ⅔ of £10, equal to £6.66666 recurring!
101. Nevil Maskelyne, 'Astronomical Observations, 1766–1774, Transit Observations', Vol II, p. 160.
102. George Anson had captured the Acapulco galleon near the Philippines on 20 May 1743 OS. The Neustra Señora de Covadonga was carrying 1,313,843 pieces of eight and 35,862 ounces of silver. Under the terms of prize money then in force, Anson took ⅜ of the prize money available from the Covadonga amounting to £91,000, which was roughly the equivalent of almost £23 million in present money values. This compares with the £719 he earned as a captain on the round the world voyage lasting almost four years. By any standards Anson had become a wealthy man from the proceeds of this voyage, even though it had sometimes bordered on the farcical and would according to the orders he sailed under have been deemed a failure.
103. This was probably Graham's last involvement in public affairs. He died on 20 November 1751.
104. This remarkable instrument now hangs on the west wall of the transit house at the Old Royal Observatory at Greenwich. It must qualify as one of the most remarkable instruments in the history of astronomy. Visitors tend to ignore it as they pass into the transit house.
105. James Bradley, Royal Greenwich Observatory Archives, Cambridge University Library, RGO 3/43, *Miscellaneous Observations,* doc. 2.
106. James Bradley, *Astronomical Observations Made at Greenwich from the Year MDCCL to the Year MDCCLXII. Vol I MDCCL to MDCCLV, Observations with Transit and Mural Quadrant from 1750 to 1755,* ed. Thomas Hornsby, Oxford, 1798. *Vol. II MDCCLVI to MDCCLXII. Observations with Transit and Mural Quadrant from 1756 to 1762,* ed. Abram Robertson, Oxford, 1805, Vol I, p. 91.
107. Derek Howse, *Greenwich Observatory:* Vol. 3, 'The Buildings and Instruments', p. 25.
108. John Bird, *The Method of Constructing Mural Quadrants,* London, 1768, pp. 23–24.
109. Derek Howse, *Greenwich Observatory:* Vol. 3, 'The Buildings and Instruments', p. 25.
110. Stephen Peter Rigaud, *Memoirs of Bradley,* p. lxxvii.
111. John Bird, *The Method of Constructing Mural Quadrants,* London, 1768, p. 1.
112. John Bird, *The Method of Dividing Astronomical Instruments,* London, 1767, p. 5.
113. James Bradley, *Astronomical Observations made at Greenwich from the year MDCCL to the year MDCCLXII. Vol I MDCCL to MDCCLV, Observations with Transit and Mural Quadrant from 1750 to 1755,* ed. Thomas Hornsby, Oxford, 1798. Vol. II

MDCCLVI to MDCCLXII, *Observations with Transit and Mural Quadrant from 1756 to 1762*, ed. Abram Robertson, Oxford, 1805, 'Zenith Distances to the Southward', Vol. I, ed. p. 57.

114. If 64 divisions out of the 96 in the full quadrant is 10.6″ short of 60°, then 96 divisions (64 + 32) equals 10.6″ + 5.3″, equal to 15.9″ short of a full quadrant of 90°.

115. James Bradley, 'Extracts from Dr. Maskelyne's Observations on the Latitude and Longitude of the Royal Observatory at Greenwich', in the *Miscellaneous Works and Correspondence of James Bradley*, p. 78.

116. Derek Howse, *Greenwich Observatory:* Vol. 3, 'The Buildings and Instruments', p. 34.

117. Ibid. p. 35.

118. James Bradley, *Astronomical Observations made at Greenwich from the year MDCCL to the year MDCCLXII. Vol I MDCCL to MDCCLV, Observations with Transit and Mural Quadrant from 1750 to 1755*, ed. Thomas Hornsby, Oxford, 1798. Vol. II MDCCLVI to MDCCLXII, *Observations with Transit and Mural Quadrant from 1756 to 1762*, ed. Abram Robertson, Oxford, 1805, Vol II, Preface, pp. iii–v.

119. Stephen Peter Rigaud, *Memoirs of Bradley*, p. lxxix.

120. James Short had been elected as an FRS in acknowledgement of his immense skill making reflecting telescopes.

121. Royal Society, *Journals of the Royal Society*,30 April 1752.

122. Across the *Orbis Magna*, the mean diameter of the Earth's orbit around the Sun.

123. M. E. W. Williams, *Attempts to Measure Annual Stellar Parallax: Hooke to Bessel*, PhD Thesis, University of London, 1981, p. 31.

124. Halley's observational practice belonged to an earlier generation.

125. I. S. Glass, Nicolas-Louis de la Caille: Astronomer and Geodesist, Oxford, 2013.

126. After the end of 1752, Old Style fell into disuse, so the designations OS and NS were no longer required. This followed the adoption of the Gregorian Calendar in Great Britain, Ireland, and the thirteen self-governing American colonies of the crown in North America, as well as all British colonies and possessions throughout the world.

127. I. S. Glass, *Nicolas—Louis de La Caille: Astronomer and Geodesist*, Oxford, 2013, p. 113.

128. Nicolas-Louis de Lacaille in Cape Town to Grandjean de Fouchy in Paris, 21 February 1752.

10

Observations Beyond Compare

The 'Incomparable' Observations of James Bradley and his assistants at Greenwich from 1750 to 1762 and their Reduction by F. W. Bessel from 1807 to 1818

> Astronomy has not only taught us that there are laws, but that from these laws there is no escape, that with them there is no possible compromise.
>
> Henri Poincare.

The observations made by Bradley and his assistants at the Royal Greenwich Observatory from 1750 to 1762[1] were the culmination of everything that Bradley had achieved. It was the consummation of an observational practice long in the making, constructed during the investigation of phenomena in which the imperatives of rigor, accuracy, and precision were pushed to the very limits of the available technology. Into these observations went almost forty years of accumulated knowledge, technique, and practice.[2] With John Bird's new instruments, Bradley achieved a greater degree of reliable precision and accuracy than with the instruments he inherited from Halley. This should not be inferred as a criticism of Graham's craftsmanship. The distortions in the great 8-ft quadrant constructed in 1725 were due almost entirely to the weight of the instrument which over a period of a quarter of a century became increasingly intrusive. All of Bradley's rectifications and calibrated instrumental errors were recorded. It was an essential aspect of Bradley's practice that he repeatedly rectified his instruments, often several times during a run of observations. It is due to such a rigorous application of his techniques that his observations could be reduced with such confidence and precision.[3]

By the time Bradley undertook his observations of all of the brighter stars[4] observable from Greenwich, he was systematically recording the barometric pressures and atmospheric temperatures. Bradley recorded the errors of his clocks, the collimation of his instruments, indeed every possible cause of bias or error as well as the 'seeing conditions' under which the observations were made. From 1750, Bradley and his three successive assistants were making

The Life and Work of James Bradley. John Fisher, Oxford University Press. © John Fisher (2023).
DOI: 10.1093/oso/9780198884200.003.0011

about 5,000 highly precise observations a year. If this does not impress, recall that the British climate does not allow year round uninterrupted observational conditions. A conservative estimate suggests that seeing conditions were only available for about 200 nights a year.[5] This would mean that about twenty-five precise observations were made each 'seeing night' together with all records of barometric pressure, indoor and outdoor air temperatures, with records of the 'seeing conditions'. This would be recorded along with aberration, nutation, precession, adjustments of the clocks, and so forth. Each and every observational calibration was accompanied by all of these constants and other factors, all of which would affect the recorded position of each of the twenty-five or more precise observations of every star observed on any particular night. Many of the 3,222 stars in Bradley's survey of the objects observable from Greenwich were recorded twenty to thirty times, half in declination, half in right ascension. Even insignificant stars of the eighth magnitude were observed about ten times over the decade. All of this recorded data was compiled by Friedrich Wilhelm Bessel and used to reduce the positions of each and every star to epoch 1 January 1755.

Over the 'long decade' from 1750 to 1762, Bradley's observational methods reached the limits of the technology available. John Bird's instruments were the last word in mid eighteenth-century precision. These habits were already well established by the time Bradley constructed Lord Macclesfield's Observatory at Shirburn Castle near Tame in Oxfordshire.[6] The observations made there by Bradley or Lord Macclesfield or by servants of his lordship's household, were all accompanied by barometric or thermometric measurements.[7] Reductions were applied for the associated effects of annual stellar aberration, for precession and for the nutation of the Earth's axis. No astronomer prior to Bradley was so well equipped either in terms of the instruments he used or in terms of his observational practice. None had approached him in the rigor of his methods.

It was a strict rigor that was carefully inculcated into his well-trained assistants, some of whom were to acquire an eminence of their own. An aspect of Bradley's rigorous practice was to analyse his methods before passing these skills on to his assistants, so they were capable of working to the same rigorous standards. It laid the foundations of a tradition of accuracy and precision at Greenwich, maintained long after Bradley's death, standards which were eventually emulated in every leading observatory in Europe and America. The importance of Bradley lies not merely in the value of his discoveries, but in the way that the techniques that he pioneered were adopted throughout much of the astronomical community. During his lifetime his methods were adopted by Nathaniel Bliss who occasionally worked alongside Bradley at Greenwich, briefly succeeding him as Astronomer Royal from 1762 to 1764. Bliss adopted his methods so that from 1760 until Bradley's death in 1762, he worked along-side Charles Green as they maintained the great series of observations in the name of Bradley. The registers containing the observations of Bliss and Green using the Bird transit instrument and the Bird 8-foot quadrant were contin-uous with Bradley's and by most criteria were almost indistinguishable in the terms of their rigor and exactitude.[8]

Although each of Bradley's assistants took part in the work of the Royal Observatory, they all achieved some eminence. John Bradley was employed by his uncle from February 1742 when he was fourteen years of age, act-ing as his uncle's assistant from 15 June 1742, when his uncle began work with the Graham quadrant commissioned by Halley, no doubt by his uncle's side. John remained in the service of the Royal Observatory until September 1756 when he was succeeded by Charles Mason. After leaving his post, John Bradley ventured on a 2-year voyage with Captain John Campbell, when var-ious instruments were tested and lunar observations were made at sea. It was during this 2-year voyage that Campbell used the prototype of the modern mariner's sextant, an improvement on Hadley's octant. In 1767, John Bradley was appointed Usher (second mathematical master) at the Royal Naval College in Portsmouth. In 1769, he was sent by the Board of Longitude to observe the transit of Venus at Lizard Point, the most southerly point in mainland Great Britain. An account of this expedition was included by Nevil Maskelyne in the 1771 *Nautical Almanac*.

At Greenwich, John Bradley was succeeded by Charles Mason who made his earliest recorded observation in October 1756, his last in November 1760. Like John Bradley, Mason was a Gloucestershire man, from Wear the same parish that James Bradley's wife Susannah came from. It is probable that Mason was appointed through the strength of these local connections. He was a good mathematician, having received his education from Robert Stratford in nearby Sapperton. He was a competent successor to the Astronomer Royal's nephew. It was Mason who assessed the accuracy of Tobias Mayer's solar and lunar tables, which had been submitted to the Board of Longitude. He left Greenwich on Bradley's recommendation, when he was selected by the Royal Society to travel with his friend Jeremiah Dixon, a reliable astronomer from County Durham, to observe the transit of Venus in 1761 in Sumatra. Originally it had been intended that Mason travel to St Helena as Maskelyne's assistant, but this role was taken by Robert Waddington. Mason's role was suitably upgraded on Bradley's advice, so that he was provisioned to observe the transit of Venus at Bencoolen in Sumatra on 6 June 1761 though they left port late, in January 1761.

Even though British and French astronomers corresponded with each other, it was a setback for the advance of scientific research that the transit of Venus coincided with another conflict between Great Britain and France, better known in Britain and its American colonies as the Seven Years War, that lasted from 1756 to 1763. After the frigate *HMS Sea Horse* left Portsmouth with Mason and Dixon as passengers on 12 January 1761, she was attacked by a more powerful French frigate, all masts being damaged, with eleven men killed and thirty-seven wounded. The ship returned to port for repairs. At the Royal Society's insistence the ship was set once again for Bencoolen, even though it had been reported that the station in Sumatra had fallen to French forces. On 27 April 1761, *HMS Sea Horse* reached the Cape of Good Hope with insufficient time to reach the intended destination and to set up an observatory. With the added possible inconvenience of being hounded by French privateers along a busy trade route from the Cape to the straits between Java and Sumatra, it was a sensible decision to stay at the Cape. Charles Mason set up an observatory in Cape Town (ten years after Lacaille's important sojourn), so they would be fully prepared and ready for the transit on the 6th of June.[9]

Charles Mason and Jeremiah Dixon successfully observed the entire transit of Venus in good seeing conditions, which is more than could be said of those that beset Nevil Maskelyne and Robert Waddington at St Helena. Observations there were hindered by cloud and windy conditions. After the transit, Mason joined Maskelyne at St Helena in order to make tidal and gravitational measurements before returning to England on 7 April 1762. On his return to Greenwich, Mason resumed work on lunar tables and a stellar catalogue[10] which was published in the 1773 *Nautical Almanac*. Late in 1762, Mason and Dixon left for the American colonies where they were enjoined between 1763 and 1767 to survey the boundaries between Pennsylvania, Maryland, and Delaware, a line that acquired infamy as the boundary between the slave-owning and non-slave owning states of the American Union. Returning to England in 1768, Mason was requested by the Royal Society to travel to Ireland to observe the transit of Venus due on 3 June 1769. It was Mason who selected the distinctive cone-shaped Schiehallion in the Scottish Grampians as the mountain most suitable for Maskelyne's determination to measure the density of the Earth, an extension of the gravitational studies they had undertaken at St Helena in 1761. Mason died in Philadelphia on 25 October 1786, three years after Great Britain finally recognized the independence of the nascent United States of America.

At Greenwich, Mason was succeeded in 1760 as Bradley's assistant by Charles Green (1735–1771), the son of a prosperous farmer near Swinton in Yorkshire. He was educated at a school in Denmark Street in Soho, where he became assistant to his older brother Rev. John Green. Having studied astronomy, he was to join Bradley as his third assistant during 1760 just as ill health forced the Astronomer Royal to give up teaching experimental philosophy at Oxford, and before he was forced to retire from the Council of the Royal Society. Green worked alongside Nathaniel Bliss during the final eighteen months of Bradley's life, as they continued with the observations in his name, both men having been trained to the Astronomer Royal's standards. On 6 June 1761, Bradley was too ill to work. The transit of Venus was observed at Greenwich instead by Bliss assisted by Green.[11] Bliss used a reflecting telescope of 2-feet focal length to which was fitted Dollond's micrometer. Both the telescope and micrometer were made by Short. The telescope had been fitted for Charles Mason's eyes. Other observers were given insufficient time to make the fine adjustments required. Green was the observer of the transit with the reflector. Bliss made use of a fifteen-foot refractor[12] using a screw micrometer in the form of those made by George Graham. Under cloudy conditions, much of the early period of the transit was missed. There was little likelihood of conditions improving. Thomas Hornsby observed the transit at Shirburn Castle where observing conditions were better. The observations at Greenwich were constantly thwarted by scudding clouds.

When Bradley died on 13 July 1762, Green continued as his successor's assistant, Bliss being appointed as the fourth Astronomer Royal. In 1763 the Board of Longitude asked Green to accompany Maskelyne to Barbados to make observations in connection with the sea trials of John Harrison's chronometer H4. He returned to Greenwich in July 1764, shortly before Bliss died on 2 September 1764. At the unexpected death of Bliss only just over two years after Bradley's, the Board of Longitude purchased many of his observations from his widow Elizabeth, they being considered useful to the resolution of the longitude problem. Bradley's executors may subsequently have believed that they held a financial interest in the third Astronomer Royal's more extensive observations. As late as 13 June 1765, Bradley's executors appeared ready to relinquish Bradley's important observational registers, but by 10 June 1767 the Peach family had changed their minds, and the settlement made by the Board of Longitude with Elizabeth Bliss may have influenced this decision. The Board initiated a legal suit against Bradley's executors. On 28 November 1772, Samuel Peach was ready to settle in order to be relieved of the lawsuit by the Board, but even then he demanded a gratuity. On 2 March 1776, the Earl of Sandwich[13] approached the Board with a suggestion that Bradley's observations be given up to the University of Oxford. Bradley's Greenwich registers were in fact surrendered to Lord North as the Chancellor of Oxford University who enjoined the Clarendon Press to print and publish the observations as soon as possible. At this, the Board of Longitude, seeing that the observations were to be published, gave up their law suit.

In 1861 the registers were still located at the Bodleian Library, and it took a period of careful negotiation before Bradley's registers were finally sent to the Royal Observatory to be placed alongside the registers of the other Astronomers Royal. George Biddell Airy's opinions of Bradley's executors during this affair leave no doubts about where his sympathies lay, He wrote, 'Of the conduct of Mr. Peach in this transaction my opinion is too strong to admit of expression here'.[14]

This was surely as strong an opinion as could be expressed by a Christian gentleman in Victorian England. Later in the same letter Airy elicits his view that, 'Dr. Hornsby's printed book is not an exact counterpoint of the original'.

So that Thomas Hornsby, who had possession of Bradley's registers to 1797, only transcribing the observations made between 1750 and 1755 in a period of twenty-one years, was unable to produce a faithful copy of the original. That he found the task both irksome and quite unimportant there can be little doubt, but his conduct in the handling of Bradley's observations was almost as reprehensible as that of Samuel Peach.

Nevil Maskelyne was appointed as the fifth Astronomer Royal on 8 February 1765. Charles Green continued as an assistant until he made his final entry in the registers on 15 March 1765, remarkably the day before Maskelyne took up residence. It is open to conjecture that relations between Maskelyne and Green were not as they ought to be, following their voyage testing Harrison's chronometer H4. Abram Robertson, editor of the second volume of Bradley's Greenwich Observations, states that in fact Green officially left his post on Lady Day, 25 March 1765. Green accompanied James Cook as the expedition's astronomer for the 1769 transit of Venus expedition to Tahiti. Cook took umbrage because Green as the expedition's astronomer was paid twice what he as the ship's master received for the voyage. On the return voyage after exploring the eastern seaboard of Australia or 'Terra Australis Incognita', HMS Endeavour had to be refitted in port at Batavia. Although being the main Dutch stronghold in the East Indies, the port was rife with many tropical diseases of the kind I briefly discussed in Chapter 2. Charles Green died of dysentery in Batavia on 29 January 1771. All of Bradley's assistants were trained by him and all worked in conformity with his exacting standards of observation. Certainly

Bessel thought that Bradley's assistants were as rigorous, accurate, and precise as Bradley was himself, even though he clearly identified differences in their observational habits which he referred to as the 'personal equation'.

Researching the Bradley archives in Oxford, Cambridge, and elsewhere, various gaps and omissions have come to light. The most notable omission is the lack of almost any personal correspondence. Yet the most extraordinary gap is an almost complete lack of correspondence between Bradley and his assistants, when he was detained in Oxford. There is a letter addressed to Bradley from his nephew dated 9 April 1752 concerning a portrait of Tycho Brahe which had been sent to Oxford by John Flamsteed, as well as observations made with John Bird's new transit instrument reduced to 1 January 1751. In his letter John Bradley wrote,

> I have sent the particulars of the observations, and under each difference, the mean difference by the number of observations against it; beginning with the difference between Aldebaran and Capella, then Capella and Rigel, and so on to α Cygni, and then between α Cygni and Aldebaran; and I make the sum of differences to be 360° 0′21.1″.[15]

Characteristically, James Bradley used his more finely honed skills to recalculate the mean differences, and amended them to a much more acceptable 360° 0′ 4.3″.[16] This, however, could hardly be referred to as 'personal' correspondence. He was after all the Astronomer Royal's nephew, and given that John was effectively in charge of the Royal Observatory for months at a time, I simply cannot accept that he barely corresponded with his uncle when he was in Oxford, and never in close or familial terms.

If the correspondence between Bradley and his assistants is apparently nonexistent, it seems judicious to assume that they were among a host of papers that may have been mislaid or lost by Thomas Hornsby. This of course is conjectural, yet it is difficult to accept that Bradley's executors withheld such papers during their legal dispute with the Board of Longitude. The most surprising aspect of the dispute over the ownership of Bradley's Greenwich observations, made in the service of the nation's interests, is that after the disputes that followed the demise of John Flamsteed, this contentious issue had never been settled by a legal ruling or legislative action before Bradley's death. It appears that all of Bradley's personal papers were deliberately withheld by his executors in the belief, no doubt, that they were private. The papers dealing with his correspondence with his assistants would, I am conjecturing, have held little but the minutiæ of the day-to-day affairs of the Royal Observatory.

The assumption to be drawn is that much of the correspondence may have intruded into the personal life of Bradley and his family.[17] It must be regretted that there is such an omission from the Bradley archive. It is not possible to establish how James Bradley trained and communicated with his assistants. This is evident from the final registers of the great series of observations from 1750 to 1762 when observations made by Nathaniel Bliss and Charles Green with the transit instrument[18] and the new quadrant[19] were recorded in conformity with those attributed to Bradley. In Bradley's miscellaneous papers at the Bodleian Library, there are some observations made by Charles Green of an eclipse of the Moon on 8 May 1762.[20] Bradley was then living as an invalid in Chalford in Gloucestershire. He had just two months to live.

The real measure of the great series of over 60,000 precise, accurate, and *dependable* observations was lost for over half a century due to the obduracy of his executors, who possessed no concept of their true value beyond some financial return. The Peach family, or more specifically Samuel Peach, acting on behalf of Bradley's daughter Susannah, claimed a pecunary interest in Bradley's observations. The Board of Longitude thought differently, in spite of the payment made for Bliss's observations to his widow. The Board evidently perceived that the observations made by James Bradley at the Royal Observatory, using instruments provided out of the public purse were public property.[21] The Board initiated a law suit in order to make a claim to Bradley's registers. This was the beginning of a lengthy process denying access to Bradley's Greenwich observations, only to be resolved with Friedrich Wilhelm Bessel's reduction in 1818 of the observations made by Bradley and his assistants from 1750 to 1762. He produced an unprecedentedly reliable catalogue of 3,222 stars down to the eighth magnitude[22] reduced to epoch 1 January 1755.[23] The methods employed by Bradley were inherited and perfected by his successors. Bessel's achievement in the publication of the *Fundamenta Astronomiæ*[24] was equal in measure to that of Bradley himself. The *Encyclopædia Britannica* (11th edition) article on 'Astronomy', in the section on precise calculations and observations states,

> A major aspect of 19th century astronomy was the move toward greater precision both in methods of calculation and in quantitative methods of observation. Here the natural successor to Bradley was Friedrich Wilhelm Bessel who reduced Bradley's enormous collection of star positions for aberration and nutation and in 1818 published the results in a new star catalog of unprecedented accuracy, the *Fundamenta Astronomiæ* (Foundations of Astronomy).

No better demonstration of improved methods could be wished for than the near simultaneous measurements of stellar parallaxes by Friedrich Georg Wilhelm von Struve of the star Vega in 1837, by Bessel of the star 61 Cygni in 1838, and by Scottish astronomer Thomas Henderson of the triple star Alpha Centauri in 1838.[25]

The author of this article, Hugh Chisholm, does not appear to have any real understanding of the significance of Bessel's achievement or of the observations made at Greenwich from 1750 to 1762. Although Bradley's registers comprised 931 folio pages, these remained at the Bodleian Library until transferred to the Royal Observatory during 1861 to be placed alongside the registers of the other Astronomers Royal. They were published in two unreduced volumes in 1798 and 1805.[26] Complaints were made by interested parties to the Clarendon Press over the fate of Bradley's observations. After twenty years a volume was published. The observations were not reduced but transcribed from the registers, and as Airy's letter reveals, not always accurately. It seems extraordinary that such a resource was in the hands of Bradley's successor at Oxford for over twenty years, only for Hornsby[27] to produce a copy of the original. There was a failure to recognize what a valuable resource these observations were. As if to underline the point, he only managed to complete the first five or six years of Bradley's work. It is acknowledged that Hornsby suffered from ill health at various periods, but there were no attempts to relieve him of his duties. No doubt this was because he was widely respected and because there was no sense of urgency by the Syndics of the Clarendon Press. The second volume edited by Abram Robertson, covering the years from 1756, was produced much more urgently and published only seven years later in conformity with the first volume. The role played by Heinrich Olbers (1758–1840) is overlooked yet his intercession was both critical and timely.

Olbers first notice of the work of Friedrich Wilhelm Bessel in 1804, when the young mathematician sent a remarkable treatise containing his orbital calculations of Halley's Comet, using data provided from the observations of Thomas Harriot and Nathaniel Torporley in 1607.[28] Whilst studying medicine at Göttingen, Olbers devised the first satisfactory way of calculating cometary orbits. He was the obvious person for Bessel to send his treatise to, especially as both men lived in Bremen where Olbers had his medical practice. Olbers converted the top storey of his house into an observatory and though an amateur he was, like Sir William Herschel in England, an astronomer of the first order. Olbers, a comet-hunter, was well known at this time for his identification of two of the first four known asteroids, discovering *Pallas* on 28 March 1802 and *Vesta* on 29 March 1807. He was so impressed with the maturity and quality of Bessel's work that he arranged to have it published. At some time between 1805 and 1807 Olbers purchased both volumes of the Clarendon edition of James Bradley's Greenwich observations[29] before passing them on to Bessel, believing he was capable of reducing this immense series of observations. This led to the publication of the *Fundamenta Astronomiæ* in 1818.

The final decade of Bradley's productive life began with the acquisition of John Bird's new mural quadrant on 16 February 1750.[30] On 10 August 1750, the new quadrant was suspended to face north in order to study *Polaris* and other circumpolar stars. The old quadrant remained in its original position until 1753 to continue observing the stars south of the Royal Observatory. Whilst Bradley's main programme of research over the next few years was his survey of the brighter stars[31] observable from Greenwich, he remained a busy academic. His correspondence with his French copains was maintained, even though his closest international correspondent was now busily engaged on his expedition to the Cape of Good Hope. Writing from Greenwich to Delisle in Paris on 12 October 1750 OS Bradley wrote,

> Sir
>
> Dr. Mortimer having informed me that you are preparing to publish a collection of all the observations you can procure of the eclipses of Jupiter's satellites, and that you desired I would communicate to you those that had been made at Wansted, either by Mr. Pound or myself, I have enclosed a copy of those that have been taken there, and hope they will not arrive too late to be inserted in their proper place.
>
> The time in which all our observations (except the first seven) are registered, being mean time, it may be proper to take notice, that the equation of time which we made use of is the same as may be collected from Dr. Halley's Tables.
>
> The greatest part of these observations were made with the 15f. refracting telescope; a few that are marked with an asterisk were made with the Hugenian glass of 123f. focus; and those which have the letter R annexed were taken with a very good 5f. reflecting telescope of the Newtonian construction, which was made by the late Mr. John Hadley, and presented by him to the Royal Society; of which some account has formerly been given in the Philosophical Transactions, No. 376 and 378.
>
> Besides the observations made at Wansted, I have added some that were made at Oxford, either by myself or my worthy colleague Mr. Bliss, which are

denoted by the letter B. Oxford lies 5′ 0″ in time westward; and Wansted 0′ 8″ eastward of Greenwich.

> I am, sir, with great respect,
> Your most obedient humble servant,
> J. B.

Observations at Wansted	170
at Oxford	30
In all, about	200.[32]

From 25 October 1750 OS with the old quadrant *in situ* observing the stars to the south of the Royal Observatory, Bradley began using the new quadrant to make observations of the stars that passed under the pole, a situation that continued until the 24 July 1753 NS. During 1750, Bradley received letters from St Petersburg inviting him to be a correspondent of the Imperial Academy. Bradley superintended the 8-foot mural quadrant being constructed by John Bird, similar to the one he made for the Royal Observatory which was now being made for the Russian Imperial Observatory. Bird's transit instrument at Greenwich located the telescope in the middle of the axis and possessed five wires, Bradley himself determined their exact separations. The transit wires were silver and measured 1/750th of an inch in diameter. Bradley's working practice reveals a level of preparation as exacting as his observations. From 1750, all observations made at Greenwich were accompanied by readings from a barometer and two thermometers, one inside and one outside the New Observatory. The quadrants proved to be highly reliable instruments. In 1818, Bessel discovered that the mean error from 300 sightings of five selected stars to be only 1.45 arcseconds.

From 25 October 1750 OS Bradley continued through to 24 July 1753 NS using the new quadrant facing north, when he determined as exactly as he was able, the latitude and longitude of Greenwich. Nevil Maskelyne later wrote,

> Dr. Bradley having been furnished by the Government in the year 1750 with a brass mural quadrant of eight feet radius, constructed by that excellent artist Mr. John Bird, an instrument far superior to any before used in the practice of astronomy, assiduously observed the pole star and other stars lying to the north of the zenith with it for upwards of three years, and then removed it to the opposite side of the wall, making it change place with the iron quadrant of the same radius constructed by Mr. Graham, likewise an excellent instrument, though inferior to this, and commenced a regular series of observations

of the sun, planets, and fixed stars, which have been ever since continued in the same manner. Moreover, the temperature of the air, shewn by the barometer and thermometer, is affixed to each observation; and the zenith point of the quadrant settled from time to time by the help of a zenith sector of 12½ feet radius, turned alternately contrary ways, the same with which Dr. Bradley had before made his two useful and admirable discoveries of the aberration of light and the nutation of the earth's axis.[33]

A major issue intruded on Bradley's ambitions at this moment. This was the introduction on 25 February 1750/51 OS of Lord Chesterfield's Parliamentary Bill for the Reform of the Calendar. For many years educated opinion had sought a reform of the calendar. Conformity with most continental calendars was one objective now that Great Britain was eleven days out of step with much of Europe where the Gregorian had long replaced the Julian calendar. In England, the New Year officially began on Lady Day on 25 March, so that, for instance, the date of 25 February 1751 would be written as 1750/51. Subsequently, the 'Calendar (New Style) Act' was dated 1750 even though it was 1751. It was an Act of the Parliament of Great Britain. Parliament held that the Julian calendar then in use and the start of the year on 25 March was,

> attended with divers inconveniences, not only as it differs from the usage of neighbouring nations, but also from the legal method of computation in Scotland, and from the common usage throughout the whole kingdom, and thereby frequent mistakes are occasioned in the dates of deeds and other writings, and disputes arise therefrom.

Although Scotland used the Julian Calendar, the new year began on 1 January from the year 1600 which added to the general confusion. Although many of the educated members of society saw the advantages of calendrical reform, a majority of the population was set against it. In some quarters the Gregorian calendar was seen as a papist plot and its introduction was regarded with deep-seated suspicion. Such a Bill could not have been laid before Parliament before the Jacobite rising of 1745/46. After the Act had been passed, there were riots in various parts of the country including London, with calls of 'give us back our eleven days!' During the transition from Old Style to New Style with the adoption of the Gregorian Calendar, the legal year was of only 282 days running from 25 March to 31 December 1751. The year 1752 began on 1 January and the Gregorian calendar was adopted when the calendar was advanced by eleven days with 2 September followed by 14 September, so this was also a short year of 355 days. Little wonder that the ill-informed and credulous could be led to believe that they had been robbed of eleven days of their lives.

The calculations and confirmations supporting this Act of Parliament were made by James Bradley in his official role as the Astronomer Royal. When several years later, Bradley entered his painful terminal illness and was cared for by the Peach family, various locals in Chalford and Minchinhampton ascribed his painful death as a divine punishment for robbing people of eleven days of their lives. The cause of science has often been beset by ignorance, superstition, and gullibility. One long-standing consequence of this reform has been the peculiar date that marks the beginning of the 'financial year' in the United Kingdom. The beginning of the year for matters financial, used to be Lady Day (25 March) but when the calendar was advanced eleven days it meant that accounts had to be rendered by the beginning of the new financial year on 6 April, even though the calendrical year begins 1 January. This peculiarity remains.

The second, though minor, intrusion into his affairs as the Astronomer Royal came when the vicar of Greenwich the Rev. R. Skerret died in May 1751.[34] Bradley's income as the Astronomer Royal was only £100 per annum, the income set for John Flamsteed in 1675. Henry Pelham, the First Lord of the Treasury (the official title of the Prime Minister), offered the living to Bradley in order to augment the meagre income that came with the most honoured post in astronomy in England. Bradley declined the offer. According to Rigaud, Bradley placed so much store in the 'care of souls' that he believed it demanded more than he felt able to deliver. Rigaud surprisingly forgot that if Bradley accepted the living, the conditions of the Savilian Chair in Astronomy at Oxford would have demanded that he relinquish his position. It may also have jeopardized his readership in experimental philosophy, his most lucrative source of regular income. After his refusal, Pelham procured a warrant of £250 per annum to be paid quarterly (£62 10s a quarter) in addition to his bursary of £100 per annum for his Greenwich post. It was given him under the Privy Seal on 15 February 1752 to be reckoned from the previous Christmas.[35] The instrument stated that the King was pleased to grant this pension, 'in consideration of the great skill and knowledge of our trusted and well-beloved James Bradley, doctor in divinity, in the several branches of astronomy, and the other parts of mathematics which have proved so useful to the trade and navigation of the kingdom.'[36]

From 25 October 1750 OS using the new mural quadrant, Bradley began his survey of the stars north of the Royal Observatory. Bradley began his studies of the position of *Polaris* and that of the celestial North Pole, giving the latitude of Greenwich. From the 6 August 1751 NS, Lacaille began his survey of the stars of the southern hemisphere. By 18 July 1752, NS the French astronomer had completed his survey of the southern stars, having observed and located 10,000 stars and 42 southern nebulae.[37] Lacaille gave names to fourteen new southern constellations. Bradley continued with his work which was described by Bessel as 'vir incomparabilis' an assertion that was resented by Jean-Baptiste-Joseph Delambre.[38] He writes, 'Qui sait d'ailleurs si toute ces observations sont de Bradley? Nous n'avons pas ce doute en lisant celles de La Caille; il n'avait point d'assistant'. Which I translate as, 'Who knows if all these observations are from Bradley? We do not have this doubt by reading that of Lacaille; he had no assistant'.

Rigaud stoutly defended Bradley against this heedless intrusion by a source he holds in contempt. He strongly asserts,

> If the meridian observations were to be carried on simultaneously at the transit and the quadrant, it became necessary for the astronomer Royal to employ an assistant. But the assistants were trained by himself, and acted under his control; he was answerable for the performance of their duties, and he therefore had a right to connect their labours with his own.[39]

We possess the testimony of Charles Mason, his second assistant from 1756 to 1760, when he felt impelled to write to Thomas Hornsby on 27 July 1774, saying,

> If your good-nature will excuse my freedom, I will beg leave to observe to you that I have seen one of the greatest of men lost. Some poor faithful hand might have been found, that, drove by necessity, would gladly for about £30 a year have reduced Dr. Bradley's observations one year under another, and placed them in form proper to have appeared in the annals of fame forever.[40]

As one of Bradley's assistants, an astronomer of worth trained by Bradley, there is no doubt about Mason's admiration or his loyalty to a man who advanced his position in society. This letter was written when Bradley's observations were still held in the possession of Samuel Peach,[41] denying access to them by the Board of Longitude or any other body.

Lacaille was an ardent admirer of Bradley and would have resented Delambre using his name as a stick to beat Bradley with. Delambre was unaware of the awe with which Bradley was held throughout the astronomical community and throughout polite society. In 1751, Dr Robert Lowth, Professor of Poetry at Oxford University and later Bishop of London, gave an elegant speech in the Sheldonian Theatre in the presence of and in honour of James Bradley. On 25 January 1752, Bradley was elected a member of the Council of the Royal Society. He remained on the council until 1760, when ill health led to his resignation. During the whole of this period, the President of the Royal Society was his patron and friend George Parker, the 2nd Earl of Macclesfield.

On 22 July 1752, Bradley wrote to Professor Grischow, Secretary of the Russian Imperial Academy of Sciences, to inform him that John Bird had completed the construction of the 8-foot brass mural quadrant (an exact copy of the instrument made for Bradley at Greenwich). It was intended for the Imperial Observatory at Pulkovo near St Petersburg. He informed Grischow that he had minutely examined the instrument and found it satisfactory.[42] Bradley advised Grischow that because of the great weight and bulk of the quadrant, it should not be moved any more than was necessary and he further advised him that it should not be erected anywhere before its intended location. The precision of Bird's instruments were acquiring a European-wide reputation, partly because of their connection with Bradley. It is a sad reflection that although Bird had a steady order book for his scientific instruments from all over Europe, when he died in 1776 he was living in very modest circumstances indeed. His instruments were after all without compare, so he really should have made a better living than he was able to acquire.

On 22 August 1752 OS, Bradley replied to Delisle apologizing for the delay in the reception of the earlier letter sent from Paris in July, explaining that everything had been delayed because Delisle had addressed his letter to London instead of Greenwich. It was a misapprehension amongst some Parisian correspondents that Greenwich was in London, rather than several miles downriver from the English capital. Delisle had requested that Bradley observe a selection of comparison stars being used by Lacaille at the Cape. On 30 November 1752, NS Delisle wrote to Bradley. This was apparently received on 13 January 1753 NS. It was included in the *Memoires de l'Academie Royale des Sciences*. The observations that were sent to Delisle were printed in the *Memoires*.[43] At this time France and Great Britain were either at war or in an armed peace, imperial rivalry constantly leading to conflict. The development of the natural sciences transcended these conflicts.

A regular correspondent of Bradley who should be mentioned is Dr John Bevis. During the summer of 1752, Bevis published Halley's Tables, including those of Jupiter's Satellites, a resource denied to the astronomical community. It was Bevis who revealed that most of the observations and calculations were made by the young James Bradley.[44] Bevis a physician, natural philosopher, and astronomer, regularly consulted with Bradley and possessed a fine observatory at his home in Stoke Newington,[45] which was then one of several villages to the immediate north of the City of London. It was Bevis who first reported the discovery of the Crab Nebula in 1731.[46] During the same summer of 1752, Guillaume de Saint-Jacques de Silvabelle wrote to the Royal Society. His paper resolved the nutational ellipse. It was translated by Bevis and included in the forty-eighth volume of the *Philosophical Transactions*.[47]

The Greenwich programme of observation continued with John Bradley undertaking manifold observations when Dr Bradley was in Oxford, giving his lectures in astronomy and experimental philosophy. Throughout 1752, John Bradley tested the new transit instrument to measure the differences in right ascension between various first magnitude stars.[48] James Bradley put the new mural quadrant through its paces, using it to observe circumpolar stars. The old mural quadrant constructed by George Graham for Halley in 1725 was laid flat so John Bird could put new divisions on the instrument. The structure was reinforced and strengthened to counter the weight of the instrument which had gradually distorted the frame. With these improvements, the Graham 8-foot quadrant was suspended from 1753 where Bird's new 8-foot quadrant had been suspended from August 1750. From now on the Graham mural quadrant was used to observe the stars to the north of the Royal Observatory. Bird's mural quadrant was taken to the west side of the quadrant room so that now the most accurate astronomical instrument ever constructed was suspended in the position formerly occupied by Graham's instrument to observe the stars to the south of the Greenwich Observatory.

During 1754, Joseph Nicholas Delisle through whom Bradley had corresponded with Lacaille whilst he was engaged on his various studies at Cape Town, took up the post in charge of the Russian Imperial Observatory at St Petersburg. It was apt that Delisle of all people was taking over the use of John Bird's new quadrant. At this moment Bradley, already a corresponding member was elected and admitted a full member of the Russian Imperial Academy of Sciences. This was for Bradley's role in the commission and supervision of John Bird's 8-foot mural quadrant for the observatory at Pulkovo. In the meantime, Lacaille had returned to Paris, taking up the post he'd held since 1739 as the professor of mathematics at the Collège Mazarin on the roof of which he had an observatory.

Although Lacaille was received like a conquering hero at the Royal Academy of Sciences on 3 July 1754, he maintained a low profile, embarrassed by his reception, preferring the company of his friends. Lacaille's integrity was more than equal to Bradley's for he cared little for money.[49] Lacaille's recent biographer I. S. Glass refers to an incident on his return from the Cape. Glass writes,

> There are several stories concerning his utter disdain for financial matters. When returning to France he was offered a large sum of money to smuggle in some goods, by hiding them with his equipment, for which he had been granted customs immunity. He refused indignantly both 'as an ecclesiastic and as an honest man'.[50]

When Bradley was at Greenwich during the winter of 1754/55, he had new counterweights fitted to the new transit instrument. This was completed on 4 January 1755, the instrument now being so finely balanced that it behaved as if it weighed just 3 lb. This greatly increased the facility and ease by which the transit instrument could be used. On 1 December 1755, John Cleveland,[51] Secretary to the Admiralty, wrote to Bradley concerning Tobias Mayer's Lunar and Solar Tables which Bradley was comparing with the observations made at the Royal Observatory. On 10 February 1756, Bradley replied to Cleveland. In his letter Bradley wrote,

> In obedience to their lordships' commands I have examined the same (the theory and tables of the moon's motions ... relating to finding the longitude at sea), and carefully compared several observations that have been made (during the last five years) at the Royal Observatory at Greenwich, with the places of the Moon computed by the said tables: in more than 230 comparisons which I have already made, I did not find any difference so great as 1'½ between the observed longitude of the moon and that which I computed by the tables.
>
> The method of finding the longitude of a ship at sea by the moon hath often been proposed, but the defects of the lunar tables hath hitherto rendered it so very imperfect and precarious, that few persons have attempted to put it into practice; but those defects being now in great measure removed, it may well deserve the attention of any lords commissioners of the admiralty (as also of the board of longitude) to reconsider what other obstacles yet remain, and what trials and experiments may be proper to be made on shipboard, in order to enable them to judge whether observations for this purpose can be taken at sea with the desired accuracy.[52]

The scientific rectitude displayed by Bradley is worthy of comment. A recent popular account has represented 'the astronomers' as being opposed to the work of John Harrison.[53] I have represented that John Harrison had been financially supported by George Graham. The friendship between Bradley and Graham, Bradley's experiments with isochronic pendulums, his paper on the longitudes of New York and Lisbon, together with the personal characteristics of the man, gives credence to the view that he would have been both familiar and supportive of Harrison's work. This reasoning is underlined by the way Bradley supported the remarkable skills of John Bird, a weaver who had walked from County Durham to London in 1740. Bird's skills as an engraver were developed before he arrived in the capital. Bradley gave him his first major commission, the new 8-foot quadrant for the Royal Observatory, showing his great confidence in Bird's capabilities. Likewise the work of Tobias Mayer based at Göttingen led to the practical application of lunar theory, enabling mariners to utilize observations to determine the longitude at sea using a mariner's sextant. Yet Bradley remained reticent about such astronomical methods until trials as strictly applied as those to test John Harrison's H4 chronometer had been applied. Bradley stressed that it should be assured, what trials and experiments may be proper to be made on shipboard, in order to enable them to judge whether observations for this purpose can be taken at sea with the desired accuracy'.[54]

Mayer died prematurely[55] in 1762, only a few months before Bradley, due to being overworked through the intensity of the development of his Lunar and Solar Tables. His widow was given a share of the Longitude Prize to protect her from penury.

By the mid-1750s, astronomers' attention was turning to two future events. The first was two transits of Venus in 1761 and 1769, in order to calculate the length of what is now termed the *astronomical unit*, the mean distance of the Earth from the Sun. More immediately, attention was focused on the expected return of the comet identified by Edmond Halley, that would be referred to as 'Halley's Comet'. Bradley began to receive correspondence from various gentlemen throughout England of whom the most outstanding was the correspondence he entered into with Thomas Barker, the nephew of William Whiston, one of Newton's acolytes. Bradley recognized the quality of Barker's correspondence. Bradley wrote to him replying to his letter of 17 December 1754.[56] Bradley wrote,

The inconvenience arising from the shortness of Dr. Halley's general table of the Parabola, is what every one must sometimes find that makes much use of it; and I am glad to hear that you have so effectually removed it by constructing one that is so much longer, and which, in other respects likewise, as you observe, may be more convenient for practice. Such general tables being wholly designed for the computations in which they are to be used, the fuller they are the more effectively they answer the purpose for which they are framed; though therefore the difference of 5' in the angle from the perihelion may seem less than is needful in some parts; yet a little increase of the length of the table need not, I think, be an objection to the making those differences run uniform throughout the whole Table, from the beginning to the end of it.[57]

Bradley was sufficiently impressed with Barker's work to think it worthy of publication. This was followed through, and in 1757 Barker published the very timely *An Account of the Discoveries concerning Comets, with the Way to find their Orbits, and some Improvements in constructing and calculating their Places.* He 'provided a handy table for determining parabolic trajectories and orbits.'[58] Barker was the first modern astronomer to note the observed discrepancy between the ancient description of *Sirius* as a red star, and its categorization as a fiercely blue-white colour in all modern observations. Barker is a figure largely forgotten. Aside from Whiston, his illustrious relative, his sister Anne married the naturalist Gilbert White whose *Natural History and Antiquities of Selborne* must grace the library of anyone with an interest in natural history. Barker represented a strong current of contemporary life amongst gentlemanly natural historians and philosophers. Barker's greatest contribution to natural science was in the field of meteorology. His meteorological records are outstanding in the manner in which they were made. He recorded barometric pressure, temperature, clouds, wind, and rainfall precisely. In September 1749, he recorded, given his description of the phenomenon, a tornado. A year later Barker wrote to Bradley. After a discussion of his methods of determining the orbits of comets following Newton's method of calculating orbits from three observations, he declared the development of his methods following instruction from his uncle George Whiston. His letter ends rather sadly, for it reveals how so many scientific workers in the eighteenth century were lone enthusiasts, amateurs with little collegial support. He concluded his later letter, 'I beg pardon, sir, for troubling you with this long, intricate letter, but having no friend in this country conversant in this problem, I hoped you would not take my intruding on you ill.'[59]

The exchange of letters between Bradley and Barker suggested an easy method of finding Halley's Comet on its expected return in 1758.

It was during 1755 that James Bradley's assistant John Bradley became a father; his son also John Bradley was to work in the store-keeper's office at the Royal Naval Yard in Portsmouth before he died at the premature age of thirty-three in 1788. His second son succeeded him as Usher (second mathematician) at the Royal Naval Academy at Portsmouth, whilst his youngest son William Bradley was a naval officer and cartographer who surveyed the area around Sydney Harbour in New South Wales. A local feature near Sydney is named after him. No doubt Bradley's nephew must have been forced to think of advancing his prospects now that he had a family to support, leading him to resign his office as assistant to his uncle in September 1756 when he was succeeded by Charles Mason. John Bradley went to sea with his friend Captain John Campbell[60] on a two-year voyage when a prototype of the mariner's sextant was used.

From the turn of the eighteenth century, English scientific instrument makers, with London the centre of the trade, created an upward trend in the provision of high-precision scientific instruments. This was in response to an expanding market protected by a vibrant patent system and fed by a growing interest in natural knowledge, including astronomy. In comparison, few instrument makers in France found it easy to develop their skills in a market that was risky and often unprotected. The French Lapland expedition used a zenith sector commissioned and constructed in George Graham's workshops in London. It was modelled on that used by Bradley in his discovery of the aberration of light and later the nutation of the Earth's axis.

There was, however, a notable French instrument maker who must be mentioned. Claude Langlois received the accolade around 1740, being appointed *ingénieur en instruments de mathématiques* by the Royal Academy of Sciences in Paris, being given quarters within the Louvre. He provided academicians with the instruments they required, including the 1.9 metre[61] zenith sector used by Lacaille under the direction of Cassini de Thury, who did the Abbé a disservice in claiming Lacaille's work as his own in the survey along the meridian, formerly surveyed by his father Jacques Cassini and his uncle Jacques Maraldi. This had been claimed as empirical support for the Cartesian conclusion that the Earth was a prolate spheroid. This assertion had been disputed within and without the Academy. This zenith sector was provided by the Academy for Lacaille for his use when he left for the Cape of Good Hope. Lacaille also used a quadrant of 98cm and a sextant of 1.9 metres.[62]

Bradley's correspondence with two excellent French astronomers Lacaille and Delisle, who were as committed to rigor, precision, and accuracy as he, was hindered by the successive wars between their countries. Lacaille returned to Paris from Cape Town and the Indian Ocean on 28 June 1754. He had formerly worked as Jacques Cassini's assistant at the Paris Observatory. Although Lacaille had issues with Jacques Cassini's son, he was treated with kindness and respect by Jacques. Lacaille's work on the length of a degree of longitude demonstrated the inaccuracy of Jacques Cassini's observations. In spite of this, it appears the two men maintained their friendship.[63] Cassini's conclusions had been in conformity with the expectations of Cartesian theory. It was a pointed example of observers seeing what they wanted to see.[64]

The complementary work undertaken by Bradley and Lacaille was supported by the equally talented German astronomer Tobias Mayer. Lacaille's southern sojourn was to lead to the best catalogue of the southern stars until the nineteenth century. Bradley's observations of the 3,222 stars down to the eighth magnitude observable from Greenwich, with their reductions by Bessel, were augmented by Giuseppe Piazzi's positional data from his late eighteenth-century catalogue of 7,646 stars. Piazzi's most remarkable stellar achievement was his demonstration that most stars were in motion relative to the solar system. In comparison with the stellar observations made by Bradley and Lacaille, the complementary work of Mayer was equally demanding and laborious in the determination of viable lunar and solar theories. The combined effort of the two stellar astronomers and the work on lunar theory by Mayer was to make the astronomical method of determining the longitude at sea a real possibility. Later in the eighteenth century, British capital ships[65] would carry several chronometers modelled on Harrison's H4, whilst the chronometers would be regularly verified by calculations based on 'the method of lunars'.

There was an exchange of letters in the final years of Bradley's life with the Rev. Dr. Charles Walmsley, a Benedictine monk with a European wide reputation as a mathematician. It was through the offices of Bradley that Walmsley had a paper on the precession of the equinoxes published in the *Philosophical Transactions*.[66] Bradley received a short letter of thanks from Walmsley. He wrote,

Reverend Sir,

I thought so much was proper to be said by way of preface, in case you judged the papers worth presenting to the Royal Society, of which I have the honour to be a member; but if you judge otherwise, you are master to suppress them, as also to reform whatever you meet therein that is erroneous. I shall further observe in this note, which is private to yourself, that I have chosen to write these papers in Latin rather than English, as theories of this nature have generally been delivered in that language. As I should be glad to know whether they be worth any attention or not, I must desire of you the favour to let me know whether they be worth any attention or not, I must desire of you the favour to let me know what is your opinion of them. The liberty I have taken I hope will not be offensive, though coming from a person who is a stranger to you. I have lived for some years in Paris, though native of England, and left that city about a year ago to pass some time at Rome. As it is a little uncertain where I shall be at the time you write, you'll please to direct your letter as follows: à M. Walmesley vis à vis la fontaine des Carmelites ruë St. Jacques, à Paris, and it will be immediately forwarded to me. I am glad to have this occasion of paying due respect to your distinguished merit, and am, with the highest esteem, reverend sir,

Your most obedient humble servant,
C. Walmesley.

Bradley's reputation throughout Europe was so well established that he must have received much unwanted correspondence. Some correspondents were welcome, some of whom like Gael Morris, John Bevis, and John Machin were valued as friends for many years. In his final years he welcomed correspondence with figures of scientific worth such as James Barker and Charles Walmesley. These figures will not be found in many surveys of eighteenth-century science, but they inhabited the same republic of letters within which Bradley himself lived. During the final years of his life, it was the exchange of letters with Walmesley that revealed a growing friendship based on mutual respect. On 29 June 1756 Walmesley wrote a letter to Bradley from Rome. He stated that the previous December he had written to Bradley but received no reply. He assumed it had been intercepted or misdirected. He stated that his correspondence could be directed to his address in Paris. This letter found its way to Bradley in Greenwich.

The storm clouds of war were again making correspondence between France and England difficult. These difficulties saw Walmesley remove himself to the city of Bath. Walmesley was the Procurator General of the English Benedictine Congregation and had been Prior of St Edmund's Priory in Paris. He had graduated Doctor of Divinity from the Sorbonne. Although he was to become an important figure in the Roman Catholic Church in England, his international scientific reputation rested on his abilities as an astronomer and mathematician. During the period after Bradley's wife Susannah had died, when his own physical powers were in decline, he evidently enjoyed the younger man's company. They met more than once and in a letter from Walmesley to Bradley dated 21 October 1758, he wrote,

Reverend Sir,

From the kind reception you gave me when I waited upon you last year at Greenwich, and the regard you seemed to have for the theory I had sent you some time before on the precession of the equinoxes, &c. I have been induced to address to you the paper[67] here enclosed, presuming this liberty will not be offensive, as you are the best judge of all performances relating to astronomy. I have, indeed, been hindered of late by bad health from attending much from these studies; but I hope that what is here offered, though imperfect, will not be entirely unacceptable. If you please to favour me with a line, you may direct to me at the Belltree house in Bath.[68]

In return, on 1 December 1758 Bradley wrote an equally short reply. He wrote,

Sir,

Some time ago I received the enclosed papers, in a letter from the author, desiring me to transmit them to you. But not being able to learn where you then resided, I was unwilling to venture to send them as directed for you at Bath, because I was uncertain whether you were returned from the north, whither you told me you intended to go soon after I had the pleasure of seeing you here. I hope they will now come safe to your hands, as your curious[69] dissertations lately did to mine; which I shall lay before the Royal Society, being highly sensible of the honour you do me by transmitting to the public through my hands such elegant theories on subjects so well deserving the attention of all lovers of astronomical and natural knowledge.[70]

The correspondence continued with the exchange of various letters up to 11 February 1761, when Bradley began to display the signs of his impending terminal illness, after resigning from his position as reader in experimental philosophy and shortly before he moved his residence to Chalford, where he was cared for by the Peach family. The subject of what was probably the final exchange of letters between the two men was the gravitational influences and possible resonances that Venus and the Earth had upon each other, and the way this influenced their orbital motions. This was a timely subject, for Venus would shortly transit the Sun on 6 June 1761.[71] The passage of the planet was closely observed in locations all over the world.

Three days after the transit of Venus, Thomas Barker wrote to James Bradley reporting his observations of the phenomenon. Barker was only able to observe the latter stages of the transit, apparently in the company of a party of friends, some of whom were able observe the passage of Venus whilst others, probably those with little or no experience of making astronomical observations, were unable to see anything of the phenomenon. It is easy to conclude that the entire occasion was treated as much as a social occasion by many of

Barker's neighbours. Barker used a refractor with a focal length of fifteen feet, similar to that regularly used by Bradley, who recognizing his failing health at least gave serious thought to the oncoming transit of Venus in the years prior to the phenomenon. He gave considerably more attention to it than the return of Halley's Comet. He did observe the return of the comet, but perhaps he regarded it as having little importance. The issues at stake had already been well established.

Referring to these earlier observations on 30 April 1759 at Greenwich, Bradley began observations of the expected comet of 1758/59 which had been predicted by his friend and mentor Edmond Halley. The comet proceeded in a highly elliptical orbit with a period of about seventy-six years, eight years less than the period of the planet Uranus which was discovered in 1781 by William Herschel. It is surprising that the seventy-six-year elliptical orbit of 'Halley's Comet' didn't lead to widespread speculation that such an orbit had been governed by an unknown planet or planets beyond Saturn. It had already been recognized that some short period comets had periods determined by the gravitational influences of Jupiter and Saturn. Such questions were forming in the minds of some astronomers. Bradley's intensive studies of Jupiter and the four Galilean satellites revealed the various gravitational influences between these bodies. Bradley began observing 'Halley's Comet' on 30 April 1759. It had been located by other astronomers on the continent including Jérôme Lalande, a one-time law student who under the influence of Delisle had taken up the pursuit of astronomy and had assisted Lacaille when he was making observations at the Cape, by making comparable observations in Berlin. Bradley was ignorant of Lalande's observations as once again Great Britain and France were in a state of war making easy communication between astronomical collaborators and copains difficult.

Bradley made observations of Halley's Comet to the end of May 1759 and determined its orbit. The characteristic industriousness of his observational practice was in decline, evidence of his failing powers with the onset of his terminal illness.[72] A month later on 21 June 1759, while Great Britain and France were again at war, the issue of a certificate confirming James Bradley's acceptance[73] as an academician of the the Royal Academy of Sciences was celebrated in Paris.[74] During the summer, Bradley was sent the certificate signed by La Condamine, Clairaut, and d'Ortous de Mairan which had been read before the Academy.

On 31 January 1761, Bradley attended his final meeting of the Council of the Royal Society. He had been a member of the council for most of the previous decade with his patron and friend George Parker, 2nd Earl of Macclesfield the

President of the Royal Society. Bradley's drive and strength was diminishing, leading to his resignation from the readership in experimental philosophy at Oxford. His work on the survey of the stars visible from Greenwich down to the eighth magnitude was becoming difficult to attend to. After Charles Mason was appointed by the Royal Society on Bradley's personal recommendation, to observe the transit of Venus in the Indies, he appointed Charles Green to replace him towards the end of 1760. Bradley's illness was now so marked that he was unable undertake many observations, which meant Green undertook an increasing share of the observations alongside Bradley's kinsman[75] Nathaniel Bliss. Then Bliss in his turn was afflicted by illness before he died in 1764, again with Green undertaking a large share of the observations at the Royal Observatory. Bradley's enculturation of a rigorous tradition of high precision and accuracy was maintained by his successors. At Bradley's demise, the Royal Observatory was Europe's leading centre of astronomical observation and calculation. The leadership displayed by Bradley meant that the Royal Observatory was the leader in a transformation in the astronomical sciences. Even without the evidence of the extensive series of observations made from 1750 to 1762, to be reduced by Bessel, it was recognized that Bradley had led an advance in the practice of astronomical observation from an inexact art into precise science.

It is worth remarking that in spite of his terminal illness and his 'retirement' before his demise on 13 July 1762, the work of the Royal Observatory continued without interruption. Some may consider that Bradley made himself redundant but this would be a fundamental misunderstanding. Both of his predecessors Flamsteed and Halley had in their very different ways left the Royal Observatory in chaos. Bradley commenced training assistants, when his teaching commitments at Oxford made such an appointment a necessity. Between 1749 and 1750, he built a 'New Observatory', later better known as the 'Transit House' through which the Prime Meridian passes. It possessed living quarters for the assistant. It is apparent that without Bradley's organizational skills, the situation inherited by Nathaniel Bliss and Nevil Maskelyne might have been similar to Bradley's in 1742.

Yet some authors, including Francis Baily, who was Flamsteed's first major biographer, have indicated something they considered a deficiency in Bradley's work, his failure to reduce the observations made from 1750 to 1762. Bradley was incorporating far more constants in his reductions than Flamsteed ever conceived of including annual aberration and nutation. British astronomers used Bradley's equations, reducing observations in differing atmospheric conditions. Bradley was a pioneer in recording barometric and thermometric readings in preparation for the reduction of his observations. His critic argued that because Flamsteed reduced his observations, his achievement was greater than Bradley's. I am not going to deny Flamsteed's abilities, many of which were enacted in trying circumstances. To decry those of Bradley because he hadn't reduced the observations at Greenwich seems preposterous. It was his drive regarding ever greater rigor when recording observations that obstructed the reduction of his observations. He was sceptical about his methods of reducing atmospheric factors such as barometric pressure, atmospheric temperature, as well as aberration, nutation, and precise measures of precession. Flamsteed was untroubled by such factors in the reduction of his observations.

More importantly, than reduced observations, Bradley left dependable and precise data including sources of error. His successors could apply any methods of reducing them to any required degree of accuracy and precision. This was first achieved by Bessel with the publication of the *Fundamenta Astronomiæ* in 1818 which included the fully reduced catalogue of all 3,222 stars. The so-called 'failure' to reduce all of his observations was part of Bradley's perception of what 'science' was. His approach was open and cooperative. Flamsteed's was more defensive and acquisitional. It is tragic that Flamsteed could so easily have made the discovery that would have established his name and enabled him to reduce his own observations with much greater dependability. If he had identified the phenomenon, which after Bradley's discovery was called the 'aberration of light', the subsequent reduction of his observations would have been considerably enhanced. Flamsteed was so fixated on the identification of annual parallax, that he had no mind to seeking a clarification of the causes of the apparent motions of *Polaris*. Once it had been demonstrated by Jacques Cassini that it could not be due to parallax, he failed to determine the causes of the observed motions of the pole star, losing interest. Did Flamsteed really believe that the regular motions of *Polaris* could simply be reduced to observational error?

Margaret Flamsteed's steely determination to establish her late husband's reputation by seeing through the publication of the *Historia Coelestis Britannica,* the authorized version of his observations in 1725, was admirable. The stellar atlas, the *Atlas Coelestis,* published in 1729 established her late husband's reputation. Without Margaret Flamsteed's interventions much of her husband's work might have been overlooked almost as surely as Bradley's nearly was. She was a woman of strong character and courage.[76] Bradley was able to make use of the catalogue and its 2,833 stellar positions. It undoubtedly saved much time in the location of many of the stars that Bradley and his assistants observed from 1750 to 1762. Bradley surely admired the work of his predecessor in the office of Astronomer Royal, for he must have appreciated the dedication and industry of Flamsteed as only an active observational astronomer could. When later in the eighteenth century, another remarkable woman, Caroline Herschel, sister of William, attempted to reduce Flamsteed's great *magnum opus* she was often incapable of reducing them by applying aberration, nutation, precession, and the other factors at play in determining stellar positions at the end of the century.

In 1832, Stephen Peter Rigaud published his *Memoir of Bradley* included with his edition of the *Miscellaneous Works and Correspondence of James Bradley*, a single volume of his major papers and correspondence. Rigaud's work was invaluable for the assiduous way he preserved so much of the Bradley archive. Without Rigaud's work, later writers might have been unable to write about Bradley. My own account of Bradley's work is not equal to the measure of the man. I leave it to future historians and scholars to reveal more about Bradley's work. Yet before I hand this account on to others, I must present the equal genius of the young man who revealed the significance and value of the work of the 'incomparable' observations of James Bradley and his assistants, in the publication of the *Fundamenta Astronomiæ* in 1818. A majority of the people who have a passing interest in astronomy hold the belief that it was Copernicus or Galileo who revealed that the Earth was in motion around the Sun. *But saying that something is so is not the same as showing it is so.* Bradley 'revealed' that the Earth was in motion around the Sun and not vice versa. The evidence of Bradley's discovery of the aberration of light shifted the opinions of all but the most conservative of Roman Catholics and the last lingering defenders of Cartesian orthodoxy, such was the value of the empirical evidences he presented.

Notes

1. James Bradley, Royal Greenwich Observatory Archives, Cambridge University Library, RGO 3, and the registers designated as RGO 5, RGO 6, and RGO 7 for observations made with the new Transit Instrument, and RGO 11, RGO 13, and RGO 14 for the observations made with the new 8-foot mural quadrant.
2. It is likely that James Bradley was first introduced to the science of astronomy in or about the year 1711 when he first lived with his uncle, the outstanding astronomer James Pound. The earliest recorded independent observations date from 1715. These reveal a degree of skill and adeptness that had evidently been acquired well before these results had been recorded.

3. Later Bessel was able to distinguish the observational patterns of Bradley's technique from those of his various assistants, and recognized small but perceptible variations which he identified as the 'personal equation'. See Simon Schaffer, 'Astronomers Mark Time: Discipline and the Personal Equation', *Science in Context*, Vol. 2, No. 1, 1988, pp. 115–145.

4. By 'brighter' I refer to the catalogue produced from Bradley's observations by Bessel in the *Fundamenta Astronomiæ*. In this catalogue, Bessel includes stars down to magnitude 8.5, two and a half or three magnitudes fainter than those observable by the naked eye in good seeing conditions.

5. Even this is possibly an over-estimate. It may be nearer 180 days a year. Half the year observations would be very difficult or impossible.

6. This appears to have been during 1739.

7. These readings were made in order to allow for possible variations in the refractive index of the atmosphere, a possible cause of error when making observations where accuracy and precision were so essential.

8. I am not suggesting that Graham's quadrant wasn't also in continuous use.

9. Here I discern the conservative habits of his mentor Bradley, to prepare for the transit at the Cape in preference to the chancy voyage over the regular trade routes to Sumatra, when he would have less time to set up his observatory. The observations made by Mason and Dixon were made in good seeing conditions.

10. John Mason, *A Catalogue of the Places of 387 fixed stars, in Right Ascension, Declination, Longitude and Latitude; Adapted to the Beginning of the Year 1760: with their Magnitudes and Annual Variations in Right Ascension and Declination. Calculated from the Late Dr. Bradley's Observations*, 1773.

11. Nathaniel Bliss, 'Observations on the transit of Venus over the sun, on the 6th of June 1761: In a letter to the Right Honourable George Earl of Macclesfield, President of the Royal Society, from the Reverend Nathaniel Bliss, M. A. Savilian Professor of Geometry in the University of Oxford, and F. R. S', *Philosophical Transactions*, Vol. 52, pp. 173–177.

12. This refers to the focal length of the instrument.

13. This was John Montagu, the 4th Earl of Sandwich.

14. George Biddell Airy to Bartholomew Price, 3 April 1861.

15. John Bradley to James Bradley, 9 April 1752 OS, Bodleian Library, Oxford, Bradley MS 45 *Correspondence to James Bradley and his assistants*. This letter accompanied a letter written in French addressed to James Bradley from the Russian astronomer A. N. Grischow in St. Petersburg, dated 2 April 1752 NS. See also James Bradley, *Miscellaneous Works and Correspondence of James Bradley*, pp. 466–470.

16. On the letter from John Bradley to James Bradley dated 9 April 1752 James Bradley crossed out the 21.1″ and substituted 4.3″ after making his own assessments of the incorporated data

17. Such as the correspondence between James and John Bradley.

18. James Bradley, Royal Greenwich Observatory Archives, Cambridge University Library, RGO 3/7, *Observations: Transits of major stars: Including adjustments to the telescope and observational method, 1758–1762*.

19. James Bradley, Royal Greenwich Observatory Archives, Cambridge University Library, RGO 3/13 and RGO 3/14, *Observations, using the new quadrant, of major stars, including adjustments to the telescope and observational method 1758–1762 and 1762–1765*.

20. James Bradley, Bodleian Library, Oxford, Bradley MS 32, *Miscellaneous Astronomical Observations by James Bradley*, fol. 72r.

21. Though quite how this could be reconciled with the payment made to Nathaniel Bliss's widow is difficult to see.

22. The evaluation of stellar magnitude during the eighteenth century was subjective.

23. Bessel's reductions also included reference to later observations made by Piazzi using more refined instruments.

24. Friedrich Wilhelm Bessel, *Fundamenta Astronomiæ pro anno MDCCLV deducta ex observationibus viri incomparabilis James Bradley in specula astronomica Grenovicensi per annos 1750–1762 institutis*. Königsberg, 1818.

25. Hugh Chisholm, 'Friedrich Wilhelm Bessel', in *Encyclopædia Britannica*, 11th edn, Cambridge, 1911 .

26. James Bradley, *Astronomical Observations made at Greenwich from the year MDCCL to the year MDCCLXII*. Vol I, MDCCL to MDCCLV, *Observations with Transit and Mural Quadrant from 1750 to 1755*, ed. Thomas Hornsby, Oxford, 1798. Vol. II, MDC-CLVI to MDCCLXII, *Observations with Transit and Mural Quadrant from 1756 to 1762*, ed. Abram Robertson, Oxford, 1805.

27. Thomas Hornsby was the Savilian Professor of Astronomy from 1763 in succession to Bradley to 1810 when he was succeeded by Abram Robertson. Hornsby's leading achievement was in his part in the planning and construction of the Radcliffe Observatory.

28. Hugh Chisholm, 'Friedrich Wilhelm Bessel' in the *Encyclopædia Britannica*, 11th edn, Cambridge, 1911.

29. James Bradley, *Astronomical Observations made at Greenwich from the year MDCCL to the year MDCCLXII*. Vol I, MDCCL to MDCCLV, *Observations with Transit and Mural Quadrant from 1750 to 1755*, ed. Thomas Hornsby, Oxford, 1798. Vol. II, MDC-CLVI to MDCCLXII, *Observations with Transit and Mural Quadrant from 1756 to 1762*, ed. Abram Robertson, Oxford, 1805.

30. Stephen Peter Rigaud, *Memoirs of Bradley*, p. lxxvii.

31. Of stars down to magnitude 8.5.

32. James Bradley to Joseph-Nicolas Delisle, 12 October 1750 OS. *Miscellaneous Works and Correspondence of James Bradley*, Oxford, 1832, pp. 462–463.

33. Nevil Maskelyne, 'Extracts from Dr. Maskelyne's Observations on the Latitude and Longitude of the Royal Observatory at Greenwich', *Philosophical Transactions*, Vol. LXXVII, p. 154

34. Stephen Peter Rigaud, *Memoirs of Bradley*, p. viii.

35. It is unfortunate that some sources believe that Bradley and his successors commanded an annual salary of £350 per annum. The annual bursary remained at £100 pa and in addition an annual pension of £250 paid quarterly that remained the Astronomer Royal's income up to and including Nevil Maskelyne.

36. Ibid. p. xci.
37. A friend asked me how it was that Lacaille was able to locate the position of 10,000 southern stars in a year whilst Bradley was only able to locate the position of *only* 3,222 stars in 12 years. It was when I revealed the catalogue in the *Fundamenta Astronomiæ* that gives the reduced stellar positions and how many observations were made of each and every star, often 20 or 30, leading to ever greater precision, that he clearly comprehended the difference between the two surveys. Even so, Lacaille's workload is quite stupendous.
38. Jean Baptiste Joseph Delambre, *Histoire de 'astronomie au dix-huitième siècle.* Ed. Claude-Louis Mathieu, Paris, 1827. Delambre died in 1822.
39. Stephen Peter Rigaud, *Memoirs of Bradley,* p. xcii.
40. Charles Mason to Thomas Hornsby, amend to 27 July 1774. This letter is extensively quoted by Rigaud in his *Memoirs of Bradley,* p. lxxxix, p. xcii, and passim. It is of interest that Mason wrote to Hornsby and not to Nevil Maskelyne, the Astronomer Royal. This was two years before Thomas Hornsby had access to Bradley's observations at Greenwich.
41. This Samuel Peach should not be confused with his son, the Rev. Samuel Peach who married Bradley's daughter Susannah. He gained custody of Bradley's papers from his father after which he and Susannah gave them up during 1776 to Lord North in his capacity as Chancellor of the University of Oxford. Lord North gave them to the University with strict instructions that Bradley's Greenwich observations be printed and published. The task was given to Thomas Hornsby, Savilian Professor of Astronomy.
42. If James Bradley had examined John Bird's handiwork as closely as he surely did, given that his reputation depended on it, it is certain that this quadrant was one of the two finest astronomical instruments in the world.
43. *Memoires de l'Academie Royale des Sciences,* 1752, pp. 426–434.
44. Stephen Peter Rigaud, *Memoirs of Bradley,* p. v.
45. I was informed of this by Dr Allan Chapman for which I am grateful.
46. Better known to astronomers as M1. The Chinese astronomers reported the supernova that led to the formation of the Crab Nebula in 1054 CE.
47. Guillaume de Saint-Jacques de Silvabelle, *Philosophical Transactions,* Vol. XLVIII, p. 325.
48. These observations were included in the quoted letter addressed from John Bradley to his uncle in Oxford.
49. Neither man was ever motivated by money issues though in Bradley's case he became wealthy through an inheritance and earned a substantial living through his Oxford lectures in experimental philosophy.
50. I. S. Glass, *Nicolas-Louis De La Caille: Astronomer and Geodesist,* Oxford, 2013. p. 119.
51. Rigaud has his name as Cleveland, but several other sources have it as Clevland.
52. James Bradley to John Cleveland secretary to the Admiralty, 10 February 1756. In James Bradley, *Miscellaneous Works and Correspondence of James Bradley,* pp. 84–85.
53. Dava Sobel, *Longitude: The True Story of a Lone Genius who Solved the Greatest Scientific Problem of his Time,* New York, 1995.
54. James Bradley to John Cleveland secretary to the Admiralty, 10 February, 1756. In James Bradley, *Miscellaneous Works and Correspondence of James Bradley,* pp. 84–85.

55. He was just 39 years of age.

56. *Philosophical Transactions,* Vol. XLIX, p. 347.

57. James Bradley to Thomas Barker, 1755, *Miscellaneous Works and Correspondence of James Bradley,* pp. 488–489.

58. Thomas Hockey, *Biographical Encyclopedia of Astronomers,* 2009.

59. Thomas Barker to James Bradley, 8 December 1755, *Miscellaneous Works and Correspondence of James Bradley,* p. 492.

60. John Campbell (1720–1790) was on board the HMS Centurion, Anson's flagship that captured the Acapulco Galleon near the Philippines on a circumnavigation of the world.

61. The introduction of the metric system came after the French Revolution. These dimensions are approximate estimates.

62. Lacaille's instruments were, of course, measured in toises, pieds, and lignes.

63. It is easy to be reminded of the relationship between Halley and Bradley.

64. Bradley's observational practice was inclusive of a determined attempt to avoid 'seeing what he wanted to see'. He was inordinately aware of the effects of bias in precisional observation. Refer to the immense care taken by Bradley and Graham in the pendulum experiment conducted in Jamaica. Bradley was discomforted when the results were so close in agreement with prediction.

65. And American ships after the War of Independence.

66. *Philosophical Transactions,* Vol. XLIX, p. 700.

67. *Philosophical Transactions,* Vol. L, p. 809.

68. Charles Walmesley to James Bradley, 21 October 1758 in *Miscellaneous Works and Correspondence of James Bradley,* p. 498.

69. The Oxford English Dictionary offers various meanings for this word, including the common meaning of being inquisitive. The common meaning when used in scientific circles in England during the eighteenth century bears a greater dependence on the original Old French 'curios', from the Latin 'curiosus', meaning 'careful'. Curious in this sense dates from the early eighteenth century in England. By way of example, Bradley referred to his friend Graham (both being members of the Royal Society) as 'our curious member'. He refers of course to the accuracy and precision of his instruments a reflection of his immense care. The same root leads to the title of 'curator' as someone who takes care of precious or important items.

70. James Bradley to Charles Walmesley, 1 December 1758, in *Miscellaneous Works and Correspondence of James Bradley,* p. 498.

71. If the two planets interacted gravitationally with each other then observations during the transits of 1761 and 1769 would require reductions including calculations of their gravitational interactions.

72. It is reported in various sources that he died from 'a chronic abdominal inflammation'. Given that various locals knew well that he was in great pain during his final months, being attributed as a punishment from God for taking away eleven days of their lives, we can assume that it was a progressive illness that took his life.

73. Bradley's acceptance of the honour suggests that some degree of communication survived the inconveniences imposed by the war that had started in 1756 and would only be concluded in 1763, the year after Bradley's demise.

74. Stephen Peter Rigaud, *Memoirs of Bradley,* p. lxx.
75. In various documents Nathaniel Bliss is described as James Bradley's kinsman. I have taken this to mean that like Bradley he was a Gloucestershire man.
76. Margaret Flamsteed (née Cooke) (1670–1730) was the daughter of a London-based lawyer. Well educated and fully literate and numerate, she sometimes assisted her husband John Flamsteed's observations. She was married to her husband when he was 46 and she 22 in 1692. She published the *Historia Coelestis Britannica,* the official catalogue of her husband's observations in 1725 and the *Atlas Coelestis* in 1729. She died when she was 60 in 1730.

Bradley's tomb, Holy Trinity, Minchinhampton, Gloucestershire.

The plaque was removed from Bradley's tomb and placed, for safety, in the nave of Holy Trinity, Minchinhampton.

Bradley's tomb inscription translated from the Latin by Rachel Chapman.

TO THE EARLY INSTRUCTION AND
FATHERLY CARE OF THE
REV. JAMES POUND, RECTOR OF WANSTEAD,
ARE DUE IN A GREAT MEASURE
THE ASTRONOMICAL DISCOVERIES
OF HIS NEPHEW THE REV. JAMES BRADLEY,
SAVILIAN PROFESSOR OF ASTRONOMY,
OXFORD 1721–1762.
ASTRONOMER ROYAL 1742–1762.
THIS STONE WAS PLACED IN 1910 BY SOME
FELLOWS OF THE ROYAL ASTRONOMICAL SOCIETY
AND OTHERS, TO MARK AND PRESERVE HIS GRAVE.

This memorial stone, at St Mary the Virgin, Wanstead, London, was added to
the grave of James Pound by some fellows of the Royal Astronomical Society
and others in 1910 in recognition of Pound's 'early instruction and fatherly
care' of his nephew James Bradley.

This commemorative plaque was placed on The Corner House, Wanstead, by the London Borough of Redbridge in March 2000 close to the original location of Bradley's zenith sector.

11

Fundamenta Astronomiæ

Friedrich Wilhelm Bessel's Reductions of James Bradley's Observations at the Royal Greenwich Observatory from 1750 to 1762, leading to the Revolution in Multi-Factoral Reductions of Astronomical Data.

> It constitutes a milestone in the history of astronomical observations, for until then positions of stars could not be given with comparable accuracy: through Bessel's work, Bradley's observations were made to mark the beginning of modern astrometry.
>
> From a catalogue of Blackwell's Rare Books.

Bradley's reservations about his empirically based ad hoc approach to the reduction of his barometric and thermometric records, leaves it open to discussion whether he ever really intended to reduce them. By the mid-1750s, he may no longer have possessed the strength to undertake such a task. The time and effort involved in the reduction of 60,000+ observations is immense. It took Friedrich Wilhelm Bessel, a highly motivated young man, well over ten years to complete the task up to 1818. After completing this immense task, Bessel urged that reduction should always accompany observation, and in the *Fundamenta Astronomiæ* he revealed how this could be achieved. Bradley was thwarted by the onset of the painful abdominal complaint that brought his life to an end. He may have favoured recording all of his data[1] sufficiently to allow future mathematicians and computers to reduce it proficiently, with a much improved theory of reductions. Bradley now incorporated many more variables, all of which had to be reduced separately. The reduction of his earlier Wanstead observations was surely time-consuming and these did not include records of barometric pressure or atmospheric temperature. Observing stars at the zenith, he did not have to factorize for atmospheric refraction. Robert Grant summarizes the problems faced by astronomers in the reduction of observations. He wrote,

> With respect to refraction, its magnitude in any particular instance depends upon the *altitude* of the object above the horizon, and the state of the *barometer* and *thermometer* at the time of observation. It is obvious,

The Life and Work of James Bradley. John Fisher, Oxford University Press. © John Fisher (2023).
DOI: 10.1093/oso/9780198884200.003.0012

therefore, that it cannot be conveniently combined in calculation with precession, aberration or nutation, the respective values of which depend on
the *position* of the object in the celestial sphere and on the *time* of observation. An objection of a similar nature occurs, when the question relates to
the combination of the geometric parallax with any other correction.[2]

This didn't even touch on other factors such as proper motion or the orbital
motion of stars in binary systems. What really concerned Bradley at this time
was his dissatisfaction with the ad hoc state of his theory of refraction. He was
appreciative of the increasing difficulties posed for the reduction of his observations, not merely because of the increasing complexities, but also because
all of the various factors demanded quite different mathematical procedures
to resolve.

The Greenwich observations included data of the barometric pressure at
the time of every observation and also the internal and external thermometric readings for every observation as well as all the corrections made to the
clocks. The Wanstead observations were made close to the zenith where the
problems associated with atmospheric refraction were minimal and the use
of the clock was unnecessary for observations of the declination of the stars.
The Greenwich observations were the culmination of Bradley's observational
practice including reiterative observations of many stars at different altitudes
and in contrasting conditions. Bradley was confident he had left sufficient data
for precise reduction by someone who was qualified in the science of multi-
factorial reduction, including atmospheric refractions. Bradley believed his
own rather tedious ad hoc approach was out of balance from the rest of his
precise approaches to observation and reduction. As I stressed in Chapter 10,
Bradley apprehended that scientific research was a cooperative activity as
befitted a man who worked with well-trained assistants. Not only did he think
that it involved cooperation across national frontiers, he also trusted that
it involved cooperation through time. Improvements in theory and practice
together with the accurate recording of data, would allow his observations to
be reduced with ever more precise degrees of certainty. This was in conformity
with a 'Bradleian' open-ended approach to scientific research as revealed in his
lectures in experimental philosophy at Oxford. Bradley discerned the importance of error and recorded all the perceived sources in the observations he
and his assistants made. To Bradley the processes of dealing with error were
key to the reduction of observations.

After Bradley's death his work on refractions was published by Nevil Maskelyne in his *British Mariner's Guide*[3]. Later it appeared in the *Philosophical*

Transactions[4] Bradley's formula gives 57″ tang. (2.D—3r) b/29.6 × 400/t + 360 = r. The coefficient of r wasn't settled until after a multiplicity of practical trials. Rigaud reveals that there is a paper written in John Bradley's hand. He wrote, 'From the mean zenith distance subtract 4 times the refraction to find the apparent refraction by the tangent'[5]

As this was written in John Bradley's hand, it had to have been written before the end of 1756, possibly in 1753. This was after the time when the new mural quadrant was suspended in order to observe the stars to the south rather than to the north of the Royal Observatory. As already discussed, the old quadrant originally made for Halley in 1725 was then suspended in order to observe the stars north of the Royal Observatory. The words recorded by John Bradley were probably to lead to results that were found to be too large. It is probable that the next step in the resolution of a satisfactory reduction of the observations of atmospheric refraction was entered by James Bradley as a memorandum in the Greenwich Observation Book. Bradley recorded, 'The refractions in all altitudes above 5″ are nearly as the tangents of the apparent distances, lessened by about three times the refraction; in summer 3½″ in winter 2½″'[6]

Rigaud in the *Memoirs*[7] stated that it simply isn't possible to determine when Bradley wrote this memorandum. I include the account written by Rigaud for the light it throws on the problems that Bradley wrestled with over the years of his great series of observations at Greenwich. Rigaud's account continues,

in the beginning of 1753, he certainly had not satisfied himself on the subject, but in writing to Lacaille[8] about 1757, he speaks with more confidence when he refers[9] to the same fundamental quantities. Below 5° indeed of altitude, there are still great difficulties in all attempts to reduce the refraction to precise rules; but there is another remarkable circumstance, for Bradley evidently considered at first that the coefficient might be variable which was to be applied to the quantity of refraction subtracted from the apparent zenith distances. It is possible, however, that this might have been occasioned by an attempt to introduce the correction for temperature into this part of his formula, which he afterwards applied to the whole quantity.

The mean refraction at 45° being nearly 1′, Bradley took his constant of 57″ from 45°3′; because at that apparent altitude the tangent of (Z. D.—3 r) should be considered as unity, and it then remained for him to settle the corrections which must be applied to the changes in the state of the atmosphere.[10]

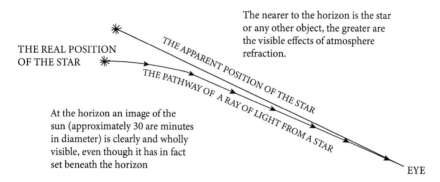

The nearer to the horizon is the star or any other object, the greater are the visible effects of atmosphere refraction.

THE REAL POSITION OF THE STAR

THE APPARENT POSITION OF THE STAR

THE PATHWAY OF A RAY OF LIGHT FROM A STAR

At the horizon an image of the sun (approximately 30 are minutes in diameter) is clearly and wholly visible, even though it has in fact set beneath the horizon

EYE

The Visible Effects of Atmospheric Refraction

Rigaud suggests that Bradley came to the opinion that the refraction was proportional to the height of the mercury in the barometer, having no reason to alter his estimate of this ratio. In the determination of the coefficient for heat or cold according to Rigaud, he found no evidence that Bradley performed any original experiments on the subject.[11] He appears to have estimated that the air contracted 1 part in 450 by a variation of temperature that was equivalent to a single degree on the Fahrenheit scale. He later introduced such modifications that would best accommodate the quantity of elasticity to the effects he detected in his own observations.

The manner by which Bradley's observational practices had developed may be inferred from Charles Mason's papers alongside his correspondence with Thomas Hornsby in Oxford. In one of these items dated 27 July 1774,[12] twelve years after Bradley's death and two years before Bradley's papers were surrendered to Lord North at the end of an entirely unnecessary law suit when the public was denied all access to Bradley's works. Mason wrote,

in computing the refraction, Dr. Bradley made use of the mean of the thermometer within and without, (:. Taking half the sum :) and then the general rule followed was, as 350+ the thermometer: the barometer : : the constant number 77″ to the refraction at 45° 3′:: tang. zenith dist. lessened by three times the refraction: the refraction required.[13] This rule was, I believe, partly deduced from observations,[14] and partly from Mr. Simpson's (late of Woolwich) theory. But I very well remember Dr. Bradley's paying Mr. Simpson a visit in the year 59 or 60, however it was after the lunar computations were nearly gone through, that the Dr. at his return told me they had had some discourse on the subject of refraction, and that Mr. Simpson rather supposed that the thermometer in the open air only should be used. But I believe (though am not sure) the Dr. continued his computations as before.[15]

Mason made his own notes on the rule stating that it was supposed to hold good to 75° from the zenith, adding, 'Dr. Bradley took no notice of any inequalities, besides those of refraction and parallax, in reducing the sun's true zenith distances about the time of the equinoxes, that I can remember: but I have not one operation by me, though I made many hundreds'.[16]

I will conclude this brief examination of Bradley's evolving difficulties in the reduction of refractions, in attempting to understand the causes of his dissatisfaction. His astronomical career, from his earliest observations and through his Wanstead-based discoveries of the aberration of light and the nutation of the Earth's axis, including his early problems with the instruments at his command in Greenwich, when he took over the Royal Observatory was concerned mostly with the achievement of ever greater accuracy and precision. This final problem of establishing a practical means of reducing refraction was never satisfactorily resolved in Bradley's ever-questing mind. He avoided fractional quantities more than was fully justified, so that his coefficients were the subject of subsequent and near constant revision. The comparative simplicity of his formula, when combined with its near approximation to the observations he made of the inferred refractions, made its wider use inevitable in Great Britain and its American colonies. Although Bradley was one of the pioneers of the regular recording of barometric and thermometric readings, he did not record them quite as regularly as his successors. This is in the nature of a lot of scientific development.

Later advances in record making of scientific data came with its mechanization, just as some of the major advances in the science of positional astronomy came with the application of photography and spectroscopy. Bradley's achievements mark a pinnacle in the development of the classical science of astrometry, when the skills of observation were reducible to human sensory abilities, using such basic instruments as the quadrant, the transit instrument, the zenith sector, and the regulator. Bradley's capacity to reduce his observations at Greenwich to the degrees of accuracy and precision he sought was diminished due to illness and the effects of aging. He believed that his observations would be reduced by his immediate successors and assistants. He could not have conceived in his wildest imaginings that his trusted executors, especially Samuel Peach,[17] would have denied public access to his observations. In his possession, Bradley's precious observations lost their immediate value. The cause of science was hindered by the stubborn resistance of a man of wealth who was incapable of apprehending their scientific value, only offering to release the observations on the receipt of a gratuity.

When Bradley's papers and correspondence were in the possession of his executors there is a strong suspicion they were also carefully sorted. There is no documentary evidence tying Bradley to Elizabeth Pound, his uncle's widow. They lived together from 1725 to 1732 at Wanstead and to 1737 at Oxford, before she returned to Wanstead during her last illness. Bradley did not marry until 1744, four years after her passing. Bradley was 52 when he married Susannah Peach. Even if the relationship between Bradley and his aunt was, by the social standards of the time quite acceptable, the Peach family evidently concluded differently. Elizabeth was a wealthy woman who left half of her substantial estate to Bradley, and this may have led them to place a very different interpretation over the entire matter. The Bradley archive was shorn of everything that was in any way related to his personal life, all of it being removed and/or destroyed. One item I discovered relating to Bradley's private life was a short note from a young niece, thanking him for the gift he sent her on her birthday. The only reason this survived is because he used the back of the letter in order to make calculations for the orbit of the comet of 1737. It is so unfortunate that none of Bradley's personal papers are available.

When Bradley's 'public' papers were surrendered to Lord North as the Chancellor of Oxford University, the Greenwich registers were passed on to the Clarendon Press. He gave instructions to print and publish Bradley's Greenwich observations as soon as possible, all of which still needed to be reduced. Long before there was any indication that they were ever to be published, Charles Mason, Bradley's former assistant produced a catalogue of 387 stars, *all reduced*, published along with the 1773 *Nautical Almanack*. Mason admired Bradley and was concerned that the law suit issued by the Board of Longitude which was contested by Bradley's executors had diminished the standing of the third Astronomer Royal's name, once unrivalled throughout Europe. Mason's catalogue was also included in the first volume of the Greenwich observations edited by Thomas Hornsby. Bradley's observations were published in an unreduced form in 1798 some twenty-two years after Lord North had given his instructions to print and publish them as soon as possible. Hornsby had no inkling of the riches contained in Bradley's observations. He was primarily a daytime observer and may not have appreciated Bradley's stellar observations. When Rigaud prepared the *Memoirs of Bradley*, he was shocked at the negligent manner with which Hornsby had treated Bradley's papers. They had been carelessly mixed up with his own or simply mislaid altogether. It took Hornsby an unpardonable number of years to produce a catalogue of 'unreduced' observations made between 1750 and 1755. He evidently found the task irksome and of little importance. Abram Robertson completed the task with more urgency, with a catalogue of unreduced observations in conformity with Hornsby's work, of observations made between 1756 and 1762 in a period of less than seven years, being published in 1805.

Friedrich Wilhelm Bessel was the second son of a civil servant who was apprenticed to an import–export company in Bremen. The reliance of the business on maritime trade led Bessel to develop his mathematical skills, to resolve problems in navigation. He had a growing interest in astronomy and the resolution of the problem of determining longitude at sea through astrometrical means. The precocious Bessel developed an elegant refinement of the orbital calculations for Halley's Comet in 1804, bringing him to the attention of Heinrich Olbers a leading astronomical practitioner. Like Bessel he lived in Bremen. Olbers once devised a method of calculating cometary orbits in 1779 when studying medicine at Göttingen. Although a physician by profession, he had equipped the upper part of his house as an astronomical observatory. In 1807, two years after the second volume of Bradley's Greenwich observations, edited by Abram Robertson, was published, Olbers handed both volumes of Bradley's observations to Bessel. He was aware that the young mathematician possessed outstanding abilities and was capable of reducing Bradley's observations. At this time Bessel had taken up a post on Olbers' recommendation as an assistant to Johann Schröter at nearby Lilienthal Observatory.

In January 1810, Bessel's abilities had come to the attention of Friedrich Wilhelm III of Prussia who appointed him director of the newly instituted Königsberg Observatory. For several years, Bessel oversaw the construction and development of the observatory. Bessel's alternative long-term project was the reduction of the substantial series of observations made by Bradley and his assistants at Greenwich from 1750 to 1762, reducing them to epoch 1 January 1755. I have extolled the skills of James Bradley as an observer and a perceptive natural philosopher. To understand the achievement of the man who brought Bradley's Greenwich observations to life, reducing over 60,000 precise and accurate observations, it is essential to comprehend the problems associated with data reduction. Bessel produced a catalogue of 3,222 stars down to magnitude 8.5 of unprecedented rigor and reliability, combating the problems of data reduction as they had developed by the beginning of the nineteenth century. Bessel developed procedures of multi-factoral reduction allowing astronomers to deal with the problems created by increasing precision.

It is not necessary to describe the development of instruments of angular measurement during the period between the demise of Bradley in 1762 and the publication of the *Fundamenta Astronomiæ* in 1818; as it is covered comprehensively in Allan Chapman's excellent history of the development of angular measurement from 1500 to 1850.[18] A major development after the death of Bradley was the replacement of mural quadrants with full circles. In this chapter, I will present a narrative of the road that led to the reduction of Bradley's observations. He took his observations to the limits of what was conceivable using the instruments at his command from 1750 to 1762. After his demise the measurement of smaller angles became routine. These new instruments exposed the urgent problems associated with data reduction, earlier brought to light by Bradley. Increasing observational precision revealed the near impossibility of making standardized and reliable reductions.

The opening decades of the nineteenth century witnessed a growing interest in the detection of annual parallax. It was no longer necessary in order to give support to the heliocentric hypothesis, earlier confirmed by Bradley's discovery of annual aberration, but rather to determine the true scale of the stellar universe. With increasing precision in the measurement of multiple constants required in the reduction of observations came confusion. Bradley was aware of the significance of error, understanding that no matter how precise and accurate his observations were the reduction of data always involved estimates and approximations. The problems associated with the reduction of observations had become ever more pressing. From 1807 when Olbers handed Bradley's observations to Bessel until 1818 when he published the *Fundamenta Astronomiæ,* he not only reduced them, he also transformed the science of data reduction. Bessel's work opened the way for astronomers to measure the annual parallax of the stars, though some problems still remained. This was not easily achieved. The complexity due to increasing precision combined with the increasing number of constants made reduction and the detection of annual parallax ever more fraught with difficulty.

By the close of the eighteenth century, the annual value of the displacement of the first point of Aries,[19] was calculated by Jerome Lalande to be about 50¼″. But whatever value was placed on it, few astronomers ever considered whether this value was correct, or whether its sources of error had been considered, or its parameters had been fully determined. Mari Williams writes,

> It was not only Lalande who made assumptions of this kind during the closing years of the eighteenth century, and it was in such assumptions that many of the problems of positional astronomy lay. There was no concept of [the] degree of accuracy as far as astronomical constants were concerned, and the difficulties of isolating one phenomenon from all others were far from being recognised, and were certainly therefore not being coped with. The tendency was to calculate a value for a constant and from that point onwards

to assume that this was its correct value. Astronomers naturally dealt with one phenomenon at a time, but they did so by assuming at each stage that the phenomenon they were studying was the only one affecting the observations under consideration. As a result there was little agreement between astronomers over the values to ascribe to astronomical constants; it is hardly surprising, therefore, that the reduced data of one did not agree with that of any others to any great degree of accuracy.[20]

The different values quoted by astronomers for precession was amongst the least problematic of all the various constants cited. It was more problematic when other constants were quoted, such as those for the aberration of light, nutation, and refraction. At the end of the eighteenth century, there remained no general agreement over the accepted value of any of these important constants. With increasing precision, astronomers met with differing values for the various constants, prohibiting agreement in the reduction of observations. It led to a rapid diversification of the values acquired before reducing observations. Given such heterogeneity in the various constants and in the results of reduction, the ability to acquire any agreed values for the observation of annual parallax, by way of example, were vanishingly small. There were, for instance, two rival sets of refraction tables. These were the Greenwich Tables, compiled during the 1750s by Bradley, and the French Tables, originally compiled by Lacaille, which were generally favoured on the continent. These were modified by Laplace after many observations by Piazzi and Delambre. Bradley's method was discussed earlier in this chapter, but for ease of comprehension I will present Bradley's formula again.

$$r = a/29.6 \tan (z - 3r) \times 400/350 + h \times 57''$$

Where a = atmospheric pressure in inches of mercury.
       z = observed zenith distance.
       r = 57'' tan z.
       57'' = mean refraction at 45° latitude

In contrast to this, the French formula gives:-

$$R = \frac{a\,(1 \pm y)\tan\theta}{(1+.00375x)\,(1 + x/5 + 12)} \quad + \quad \frac{1/2\,a^2\,(1 \pm y)}{(1+.00375x)\,(1 + x/5 + 12)} \quad \times \frac{(1 - \cos^2\theta)\tan\theta}{\cos^2\theta}$$

$$\frac{-a(1 \pm y)}{(1+.00375x)\,(1 + x/5 + 12)} \quad \times \frac{.00125254\tan\theta}{\cos^2\theta} \quad - \quad \frac{a \times .00375x \times .0125254\tan}{\cos^2\theta}$$

Where θ = apparent zenith distance.

.76 (1 + y) = atmospheric pressure in meters of mercury.

x = temperature in °C.

a = 60″.616, a constant determined by a multiplicity of observations by Piazzi and Delambre.

Mari Williams offered an opinion that it was tempting to use Bradley's formula. I cannot disagree. Digital computers can resolve such equations in nanoseconds. Yet when they could only be solved through repetitious manual computations, they led to unavoidable errors in calculation. The supposition that the more complex French formula offered more accurate results is rather debatable. Bradley's formula was based on reiterative empirical observation. Laplace was a physicist of genius, but the French formula was based on theoretical assumptions that still required the empirical interventions of Piazzi and Delambre.

In addition to the differences in the formulae their application led to significant dissimilarities in the reductions attributed to refraction. John Brinckley, the first Astronomer Royal of Ireland, demonstrated this in an 1818 paper on the determination of annual parallax, at how much a difference in atmospheric temperature could affect the apparent displacement of a star, as computed with the two different formulae. I refer again to Mari Williams's account of the history of attempts to measure annual parallax. She writes,

From Bradley's formula it can be seen that to obtain the displacement due to refraction at a temperature different from 50°F the multiplying factor was

$$400/350 + h;$$

according to Brinckley, the equivalent multiplier used by the French was

$$500/450 + h;$$

the difference between the two therefore being

$$h - 50/2000$$

Brinckley then applied this difference to his data for the star Procyon the mean refraction (that is its refraction at 50°F and normal pressure, 29.6 inches of mercury) of which he gave as 58″.[21]

A difference of 58″, almost an entire minute, between the two formulæ is not satisfactory. There were several factors affecting the apparent displacement of a star, all of which required different means of reducing them including the aberration of light, the nutation of the Earth's axis, the precession of the equinoxes, and its proper motion. In the pursuit of determining the annual parallax of a star all of these processes needed to be reduced differently. There were also adjustments to clocks and instruments. The increasing levels of precision and accuracy created ever more complexity to the problems of the reduction of observations.

Bradley used his instruments to their limits. The greater precision of the instruments available to astronomers following him, made by highly skilled instrument makers such as Jesse Ramsden and Edward Troughton, allowed observers to refine their observations with greater clarity and reliability. It was commonplace to observe the apparent locations of stars to within a tenth of an arcsecond. It was essential that an agreed standard of reduction of all this data was achieved. The possibility of this was revealed by Bessel working on Bradley's Greenwich observations. In reducing this vast store of data and producing a stellar catalogue of 3,222 stars, Bessel transformed the methods available to astronomers in the reduction of their own observations. These methods were revealed in 1818 with the publication of the *Fundamenta Astronomiæ*.

This important publication is concerned with the best methods of handling increasing amounts of data, such that the position allocated to each celestial object was as precise, accurate, and reliable as was feasible, inclusive of the parameters of possible error. Bessel applied attention to the current problems affecting positional astronomy. Consequently, the *Fundamenta Astronomiæ* deals extensively with instrumental errors, clock errors, and the determination of the latitude of the observatory, as well as refraction, aberration, nutation, precession, proper motion, and annual parallax. The reduction of a single observation was achieved through the execution of several distinct processes of calculation. The immense amount of labour required to reduce even a single observation led to many being accumulated without being reduced at all.[22] The various sections of the *Fundamenta Astronomiæ* deal with different reductive procedures. A complete analysis of this work would require an entire book to do it justice. My interest in Bessel's great work concerns his reduction of the Greenwich observations. Bessel produced an immensely accurate and reliable stellar catalogue of stars visible from the Royal Observatory down approximately to the eighth magnitude.[23] It converted Bradley's 60,000 plus observations into the earliest dependable, fully reduced stellar catalogue.[24]

I discussed the problems Bradley experienced on the reliable reduction of atmospheric refraction in Chapter 10. Besides other considerations such as his mental and physical decline, the problems related to atmospheric reduction may have been the major factor inhibiting his reduction of the Greenwich observations during his time in office. In the *Fundamenta Astronomiæ* Bessel began the section on parallax with an attempt to determine the parallax of the bright stars *Sirius, Vega,* and *Procyon* from Bradley's measurements of zenith distance. It was a task without any real prospect of success. Bessel was also dissatisfied, because Bradley's collimation errors were too uncertain. Mari Williams writes, 'when it came to the consideration of so small an angle as parallax Bessel preferred to work in right ascension, believing the clock errors to be calculable to a sufficient degree of accuracy'.[25]

The stars on which Bessel concentrated his attention were the four on which he had written in one of his earliest papers on the subject of annual parallax in 1809. These were *Sirius, Vega, Procyon,* and *Altair.* Mari Williams adds,

> He had no new observations to present, but it was not for developments in practical astronomy that the book was so important. It was after all a detailed account and reduction of Bradley's observations, not of any new data, so Bessel was bound to work with what was available.[26]

In *Sectio IX* of the *Fundamenta Astronomiæ* Bessel simplified the formula for the difference in parallax in right ascension between two stars lying about 12 hours apart. This he had derived from Olbers's formula that Bessel then continued with in order to formulate the difference in parallax that he needed. Ingenious though Bessel's work was in manipulating trigonometrical formulæ, of far greater significance was his treatment of errors, particularly his introduction of the notion that errors existed *within* the constants used for reduction. This was a profound insight. Mari Williams reveals the value of Bessel's work in his reduction of Bradley's great series of observations. She writes,

> Bessel suggested that these constants could only be known within certain limits but that the limits could be determined from observations. Prior to Bessel's work specific values for constants were deduced from observation and then applied to subsequent reductions; Bessel however calculated constants to precisely determinable errors and his subsequent reduction was carried out in terms of these errors. This was a feature of the whole of the *Fundamenta,* and the impact of such a technique on astronomical calculation in general

could only be appreciated fully from a consideration of the book as a whole. However, from Bessel's treatment of parallax, which he calculated in terms of the error in the constant of aberration, we can see how the method works and perhaps appreciate its potential when applied to the larger issues of the reduction of astronomical data.[27]

After Bessel derived the formula for the quantity he was seeking he then calculated the expression for each of the pairs considered. He derived the required expression for each of the pairs of stars, reaching the final formula in terms of the declination of each star and the longitude of the Sun at the moment of the observation.

For the difference in parallax in right ascension between two stars lying nearly 180° or 12 hours in right ascension apart, he obtained the formula:

$$-u \left[\cos \lambda \cos \alpha + \sin \lambda \sin \alpha \sin \omega \right]$$

where $u = \pi \sec \delta + \pi^1 \sec \delta^1$

$\lambda = \frac{1}{2} (\odot + \odot^1)$
$\alpha = \frac{1}{2} (\alpha + \alpha^1)$
$\omega =$ obliquity

and $(\alpha, \delta,), (\alpha^1, \delta^1,)$ are the equatorial coordinates of the two stars

$\odot$ the longitude of the Sun when the first star was observed,
$\odot^1$ the longitude of the Sun when the second star was observed,
π the parallax of the first star,
π^1 the parallax of the second star.

This was followed by specific equations for the pairs of stars involved. By way of example Bessel wrote the difference for *Procyon* and *Altair* as

$$+0.9874 \left[\pi \sec \delta + \pi^1 \sec \delta^1 \right] \cos \left(\Theta + \Theta^1/2 - 21° \, 26' \right)$$

Bessel now proceeded with the introduction of a term to allow for the effect of aberration to which he assigns the value of $20''.255 + \Delta$ for the constant of aberration[28]. The introduction of this innovation was of primary importance, for Bessel was challenging the whole notion that a constant such as that for the aberration of light had to possess a *specific* value. With the proliferation of different values for an increasing number of constants Bessel appreciated that these values might be inaccurate and he assessed the amount by which they were inaccurate. By introducing the term Δ Bessel was assigning the parameters within which the constant must lie. He wrote that the difference in right ascension between *Procyon* and *Altair* resulting from aberration was

$$+ 1.9900 \, \Delta \sin \left(\Theta + \Theta^1/2 - 21° \, 26' \right)^{29}$$

In *Sectio* IX of the *Fundamenta* there is a table entitled 'The differences in right ascension between α Canis Minoris and α Aquilæ = 12h 12′ +'[30] This is a table reducing all of the observations made by Bradley and his assistants at Greenwich, from 17 September 1750 to 5 October 1761, a series of over 180 observations reduced to epoch 1 January 1755.[31] Bessel obtained a value of one arcsecond for the sum of the parallaxes of these two stars but he did not claim that he had detected traces of parallax. This was because there was still a great amount of uncertainty concerning the value of the constant of aberration which ruled out the possibility of making any positive claims for parallax. From the observations of *Procyon* (α Canis Minoris) and *Altair* (α Aquilæ), Bessel acquired a low value for Δ but this was not repeated when he considered other observations made by Bradley. When the observations of *Sirius* and *Vega* were reduced, the value of Δ he acquired was $0''.6247$ with a probable error of $0''.1417$. When he dealt with the observations made of *Polaris* where he was seeking the single parallax of the star, Bessel calculated $0''.5001$ for Δ with a probable error of $0''.0928$. It was obvious that if the uncertainty of the constant of aberration could amount to half an arcsecond, this was not in any sense a well supported claim for the observation of annual parallax. I hope I have clarified that the real significance of Bessel's work in the *Fundamenta* 'lay in the attention he paid to the errors'.[32] In this he was fully in agreement with Bradley's own outlook and methodology. In the *Fundamenta* what impresses is the way Bessel calculates errors at succeeding stages of his work. Carrying the errors through the rest of his calculations meant that the final errors were more convincing than anything previously derived because throughout they had been calculated as rigorously as the data had allowed.

What was so important in the *Fundamenta* was that Bessel's methodology could be applied to any set of observations or any number of constants. He now offered a system of analysis that allowed a standard reduction of multiple constants into a feasible task. This great work, ostensibly undertaken in order to reduce Bradley's extensive series of observations, now opened the door to advances in the science of positional astronomy. The *Fundamenta* developed a high standard, making obsolete all prior methods of reducing observations. Just as Bradley's discovery of the aberration of light made obsolete all prior positional observations, then Bessel's innovations in the treatment of errors formed a new step change in the science of positional astronomy. The importance of Bradley's large series of observations in the development of the science of positional astronomy was in making available a consistent bank of data of unprecedented rigor. By refusing to reduce the observations made at Greenwich from 1750 to 1762 and preferring to leave a bank of unprecedented data Bradley left a golden legacy. That Bradley assumed that the task of reduction would be undertaken by future staff at the Royal Observatory in no way nullifies his achievement. The Greenwich observations took observational standards to new levels of rigor at the time that they were recorded. In the hands of Bessel, an astronomer whose concerns with error were the equal of Bradley's so were his reductions. Bessel's work opened the way to new levels of reliability that brought the work of Bradley to its fulfilment.

The significance of the work of Bessel is often misunderstood or misrepresented. It is so often recognized as a major advance, but the reasons given are invariably lacking in any real understanding. Some even suggest that the beginnings of precise astrometry date back to the mid eighteenth century. In her work on nineteenth-century astronomy, Agnes Clerke wrote that, 'The eminent value of the [*Fundamenta Astronomiæ*] consisted in this, that by providing a mass of entirely reliable information as to the state of the heavens at the epoch 1755, it threw back the beginning of *exact* astronomy almost half a century'.

Similarly, Walter Fricke wrote in the *Dictionary of Scientific Biography*, where he claimed that the *Fundamenta* 'constitutes a milestone in the history of astronomical observations, for until then positions of stars could not be given with comparable accuracy: through Bessel's work Bradley's observations were made to mark the beginning of modern astronomy'.

This claim is probably justified, but it does not convey the enduring quality of Bessel's work, for it merely emphasizes the significance of the results of his endeavours.[33]

The outstanding aspect of Bessel's achievement lay elsewhere, in the way he dealt with observations, manipulating them into forms that could prove to be useful in any number of ways. Bradley's observations, after many delays caused by legal contests sustained by his executors, the procrastination of his first editor, and the complacency of the publishers, were finally published in 1798 and 1805. Without being reduced, little could be made of them. The lengthy delays in the final publication of the Greenwich observations meant that few grasped the enormous amount of *reliable information* contained in them. When Olbers handed the young Bessel the two recently published volumes of Bradley's Greenwich observations, he must have perceived that placed in the right hands, the promise offered by the massive collection of 60,000 + observations would yield a mine of information. Olbers must have recognized that Bradley and his assistants had recorded every possible factor determining the causes of possible error, so they could be reduced reliably. After a decade or more of calculation and reduction, Bessel transcended Olbers's expectations. Bessel's enormous amount of work on Bradley's observations opened the way for him to develop his extraordinary mathematical insights into the problems of the reduction of observations. His persistence and industry matched that of Bradley himself, for it was an undertaking of repetitious tedium that demanded immense dedication and determination. Such reiteration often reveals new insights.

Bessel's achievement far transcends the production of a dependable catalogue of 3,222 stars, valuable though this was. The methods he developed to produce the catalogue were more important than the catalogue itself. Like Bradley before him, Bessel paid great attention to the assessment of error, and the final values given to the positions of stars in the stellar catalogue were taken to be reliable and convincing. Bessel was grateful that his data was provided by an astronomer concerned as he was with the recording of all conceivable forms of error. The observations made between 1750 and 1762 at Greenwich reduced to 1 January 1755 were an undeveloped resource without equal during the earlier decades of the eighteenth century.

The first revolution in astrometry came with the application of the micrometer to the telescopic sight. The second was in the application of Bradley's fundamental discoveries of aberration and nutation. But the third revolution came with the *Fundamenta*, allowing the manipulation and development of observations, enabling astronomers to reduce their own observations in any required yet standardized form. The ninth section of the *Fundamenta* is concerned with the problem of annual parallax. Like many other features of this publication, it is the culmination of many years of work on Bradley's observational data. What originally began as a quest to measure annual parallax, an angle expected to be well within the capability of the instrumentation and techniques applied by seventeenth-century astronomers, proved to be more difficult than they expected. The discoveries of annual aberration and terrestrial nutation resolved other constants such as precession, exposing the difficulties imposed by recognition of the importance of observational error. Bradley wrestled with these problems in his attempts to resolve the difficulties associated with atmospheric refraction. It was a quandary that resisted any convincing attempt to reduce the Greenwich observations to Bradley's satisfaction. More than any other factor, this inhibited the dependable reduction of the observations made at Greenwich. Bessel first recognized that Bradley had left all of the data he required to resolve the problems that defeated Bradley during his final years. Bradley was defeated by age and illness but Bessel was young and ambitious. In the *Fundamenta Astronomiæ*, Bessel opened the way to the resolution of the potential problems associated with the errors imposed by increasing precision. He enabled astronomers to incorporate many constants, including the known limits of errors for each constant in turn, within his calculations.

Bessel's methodologies, because of their various superior advantages, supplanted the many different ad hoc methods applied by his predecessors and contemporaries. The complete reduction of a star's position taking into account such factors as precession, nutation, and aberration, whether by right ascension or in declination is developed in terms of four products, each one of which is composed of two factors, one of which is dependent on the *place* of the star and the other on the *time* of the observation. Let me quote Robert Grant at this point, for he writes with immense clarity that,

> The factors which depend on the time of observation are the same both for right ascension and declination but those depending on the place of the star are different. Hence the reduction of an observation... demands the computation of twelve constants, eight of which depend on the *place* of the star, and four on the *time* of observation. Now, when the mean place of a star is once determined, the constants of the first-mentioned group may be computed by an easy process, and it is manifest that when this is once accomplished, the results will serve equally well for all observations of the star. On the other hand the four constants which depend on the time of observation, will vary for every day of the year, and will even be different on the same day of each successive year. Hence these constants, unlike the former, will require to be calculated for each day of observation. They possess this peculiar advantage, however, that when once calculated for any day of the year, they will apply equally well to any star in the heavens on that day.[34]

It was an immense task reducing Bradley's stellar observations. This was a long-term programme of work that Bessel constantly fell back on. When he was not fully pre-occupied with the oversight of the construction and administration of the new Königsberg Observatory over a period of eleven years, he gave it utmost attention. I must therefore conclude with Bessel's reduction of Bradley's Greenwich observations. In 1830, Bessel published the *Tabulæ Regiomontanæ*. In the 'Report to the Twenty-second Annual General Meeting' of the Royal

Astronomical Society[35] the immense gratitude of the astronomical community was expressed for the consequences of Bessel's work. The report states

> It had long been a subject of regret that the immense magazine of facts contained in the Annals of the Royal Observatory from the time of BRADLEY'S appointment, downwards, till a very recent epoch, should remain in a great degree unavailable for astronomical use. Our illustrious associate Bessel, in his *Fundamenta Astronomiæ*, corrections to the solar tables, and finally by his *Tabulæ Regiomontanæ*, rendered this vast labyrinth permeable, and extracted and exhibited in a finished shape much of its valuable contents.[36]

It is not my intention to examine Bessel's acquisition of his Fraunhofer heliometer and his work on the *Tabulæ Regiomontanæ*, which eventually helped him to determine the annual parallax of 61 Cygni. The subject of my tale is of course James Bradley, for his work was the culmination of a tradition beginning in the 1660s, which laid the foundations of the work undertaken by astronomers up to the early 1800s. With the introduction of new technologies as well as new techniques, the practice of positional astronomy was set onto a path that loosened its links with the classical age, epitomized by men like Bradley and Lacaille. This was an age that reached its climax in the work of these men and others, with its reliance on a heuristic located in the use of basic instruments such as the mural quadrant, the transit instrument, the zenith sector, and the regulator. This age of observational skills was to be succeeded by ever more sophisticated instrumentation and technologies. Astronomy during the nineteenth century was characterized by the importance of stellar catalogues, and it was the work of Bessel based on the data provided by Bradley that laid the foundations.

Bessel, through the intercession of Olbers, rescued Bradley's final works from oblivion. It was the publication of Bradley's observations at Greenwich, reduced by Bessel to epoch 1 January 1755 and published in 1818 that finally brought attention to the remarkable work undertaken at the Royal Observatory by the third Astronomer Royal and his assistants. It focused attention, if only momentarily, on an astronomer who had sunk into near obscurity. It led to the invaluable work of Stephen Peter Rigaud, who in 1832 published his *Memoirs of Bradley* as an attachment to his compilation of Bradley's more important cogent works and correspondence in the *Miscellaneous Works and Correspondence of the Rev. James Bradley*. Anyone seeking to use it to write a biography of Bradley will, as I have insinuated, be disappointed. Some authors have expressed their own disappointment and cannot believe that such an important figure in the development of the science of astrometry has left such a paucity of material. Bradley's executors decided to 'censor' Bradley's archive. To a degree this is quite understandable if we accept that the Peach family were socially conservative. For whatever reasons open to conjecture, they successfully prevented any hope of a coherent biography. Any account of Bradley necessarily relies on the surviving manuscripts, his scientific papers, his lecture notes, and his scientific correspondence. Bradley's daughter Susannah was related by birth and marriage to the Peach family. They may have believed that in protecting Bradley's imputed reputation they were protecting hers. But what you may ask, could possible sully a life so dedicated to the pursuit of knowledge and whose character was considered to be so blameless by all who knew him? What could the Peach family wish to hide? Bradley's wife Susannah was married to him in 1744 just four years after the death of Elizabeth Pound. Bradley was fifty-two years of age when he finally married. In an age when late marriage was common in the middling sort, this in itself could be no cause of concern, but his uncle's widow lived with Bradley from 1725 until 1737 when she contracted her final illness and returned to her brother Matthew's household in Wanstead. Nowhere is there any explicit suggestion that Bradley's relationship with his aunt was anything but socially acceptable in polite society, but it undoubtedly delayed Bradley's marriage. It was a close relationship, for Elizabeth left half of her substantial fortune to Bradley when she died in 1740.

The truth of the matter possibly arises from the fact that the Peach family having no personal experience of Elizabeth or of the Wymondesold family in Wanstead, drew their own conclusions, perhaps after reading some of his surviving correspondence. They may have placed the worst possible interpretation of the relationship between Elizabeth and Bradley. Perhaps their correspondence offered negative interpretations. Until he married he often spent his vacations with the Wymondesold family, including holidays at their estate in Dorset. Much is of course conjectural, but the lack of any personal correspondence suggests that the Peach family simply withheld any such material. I say 'withheld' but in view of its absence over a period of over 250 years it can only be assumed that such material was destroyed. This at least may explain why Bradley's correspondence is so lacking in almost any personal material, when from other sources we can be sure that he was a kindly man who would have written to many of his friends for most of his life. Unfortunately his personal correspondence is visible by its absence.[37]

So far as historians are concerned the Peach family ill-served the memory of James Bradley in more ways than one. Rigaud does not refer to the lack of Bradley's personal correspondence though it must have weighed heavily on his mind. Rigaud had problems of his own, for sizable portions of Bradley's remaining archive was either missing or lost due to the apparent negligence of the first editor of his Greenwich observations Thomas Hornsby. Many of Bradley's papers were randomly interfiled with his own. One obvious and serious omission is the lists of the attendees enrolled on his courses on experimental philosophy from 1729 to 1745. Given that they exist from 1746 to 1760 I find it inconceivable that such lists were never compiled until 1746. More than likely they were mislaid or destroyed by Hornsby possibly carelessly given Rigaud's stated difficulties in disentangling Hornsby's archive from that of Bradley's.

Historians should regard Stephen Peter Rigaud with nothing short of gratitude, for so much of the remaining Bradley archive might have been lost to future use without his intervention. Rigaud reclaimed several of Bradley's papers which had already been dispersed. Without Rigaud's timely interjection much of the Bradley archive would surely have been permanently lost. I have little doubt that various miscellaneous papers relating to Bradley will from time to time reappear, though not always being identified as such. Rigaud tutored at Exeter College, Oxford and from 1805 he deputized for the ailing Hornsby, who like Bradley his immediate predecessor was Savilian Professor of Astronomy and reader in experimental philosophy. In 1810, Abram Robertson editor of the second volume of Bradley's observations was elected to the Savilian Chair in Astronomy in succession to Hornsby. Rigaud was elected to the chair in geometry in Robertson's place. Rigaud was elected an FRS in 1805 and was widely respected as a scholar as well as an astronomer assisting his grandfather at the King's Observatory at Kew.[38] When Robertson died in 1827, Rigaud was transferred to the Savilian Chair in Astronomy, continuing as the reader in experimental philosophy. No one was better placed to research the works and correspondence of Bradley. He acquired Bradley's two posts at Oxford and earlier worked at the King's Observatory, only a stone's throw from where Bradley joined Samuel Molyneux to begin work with the new 'parallax instrument'. He possessed an outstanding library of scientific and mathematical writings from the seventeenth and eighteenth centuries. As a member of the Board of Longitude he was criticized by Flamsteed's champion Francis Baily as 'one of the learned professors who seldom attend'. Most of the words used to describe Bradley's work in works of reference are invariably quoted from the *Miscellaneous Works and Correspondence of James Bradley,* which has long been the standard source and will long remain so.

Given the critical importance of the work of James Bradley and his pivotal role in the development of the science of positional astronomy, the question that comes to mind is why did Bradley become the invisible man? One of the great attractions to many historians when dealing with the likes of men such as Hooke, Newton, and Flamsteed is the ready availability of so much of their correspondence. In addition, the great antipathies expressed towards each other and many others are grist to the mill for those researching the development of the sciences up to and after the turn of the eighteenth century. By way of contrast, the personal correspondence of Pound and Bradley is lacking, giving the impression that they were indolent with regards to their correspondence, which I firmly contend could not have been the case. Consider Pound's busy life. Consider the figures he was associated with in the rise of the Whig ascendancy in English public life. These include Thomas Parker, the regent of Great Britain between the death of Queen Anne and the accession to the throne by King George I, later the Lord Chancellor and the 1st Earl of Macclesfield. In addition there were men such as Benjamin Hoadly, the most controversial churchman of the early eighteenth century, the man who ordained Bradley as a priest in the Church of England. I could continue. I simply cannot believe that a man like Pound, a man I have described as a 'networker' to use an anachronistic but nevertheless accurate assessment of Pound's capacities, left no 'personal' correspondence. We simply have insufficient knowledge of Elizabeth Pound's character to assess how she may have dealt with her late husband's many effects. Pound wrote effusively during his many travels and it is due to his correspondence that our knowledge of the Pulo Condore massacre may be ascertained. Some letters survive to be sure, but where is the general day-to-day correspondence of his busy life? The result is that this interesting and pivotal figure has been relegated to the footnotes of history.

As for that of Bradley where are we to begin? I regard this volume as the start of a new conversation. If I were younger and subsequently more energetic, I would trawl my way through the writings and correspondence of all those that Pound and Bradley were associated with, in an attempt to reconstruct the ins and outs of their lives. Yet the truth of the matter is stark. Whether it was the Peach family or others, the fact remains that it is nigh on impossible to contend with the lack of Bradley's personal correspondence. In a life that regularly shuttled between Oxford, Greenwich, Wanstead, and London as well as visits to friends and relatives in Gloucestershire and the Cotswolds, it is simply inconceivable that he would not have written to all and sundry in a busy personal and working life. From his observation books and his working materials I have been able to construct the outlines of a working chronology of Bradley's life, which clearly reveals how busy his life really was. Some sources have attempted to explain the paucity of Bradley's correspondence by the suggestion that he found writing time-consuming. But any astronomer based in the British Isles will be the first to inform you that more than a mere handful of successive nights of skies clear enough to make observations are few and far between. Bradley would have experienced plenty of times when observations would have been impossible, time enough for the voluminous calculations forming part of his archive and time enough to tend to his correspondence with friends and relatives, a correspondence evident by its complete absence. To suggest that he found writing time-consuming flies in the face of the evidence of his patient character. In view of his noted civility, this is as absurd as the suggestion that Elizabeth Pound was his housekeeper. As a historian I sadly have to accept that for whatever reasons, Bradley's personal life, for the moment at least, remains hidden, persistently remaining out of our grasp.

Notes

1. And that of his assistants.
2. Robert Grant, *History of Physical Astronomy,* pp. 343–344.
3. Nevil Maskelyne, *British Mariner's Guide,* 1763, p. 120.
4. *Philosophical Transactions,* Vol. LIV, p. 265 and also Vol. LXXVII p. 157.
5. John Bradley in Stephen Peter Rigaud, *Memoirs of Bradley,* p. lxxxvii.
6. James Bradley in Stephen Peter Rigaud, *Memoirs of Bradley,* p. lxxxvii.
7. Stephen Peter Rigaud, *Memoirs of Bradley,* p. lxxxvii.
8. Rigaud uses the alternative form of the name, La Caille.
9. *Philosophical Transactions,* Vol. LIV, p. 265 and also Vol. LXXVII p. 157.
10. Ibid. p. LXXVII, p. 157.
11. Given the imputed loss of important documents from Bradley's archive, this does not mean that such experiments were not made.
12. Charles Mason to Thomas Hornsby 27 July 1774, in the *Memoirs of Bradley,* p. lxxxix.
13. Rigaud remarks that this is very nearly the same as was given by Maskelyne in the *Philosophical Transactions,* Vol. LXXVII, pp. 156–157.
14. As I have stated elsewhere Bradley's refraction parameters were based primarily on empirical data.
15. Charles Mason to Thomas Hornsby 27 July 1774, in the *Memoirs of Bradley,* p. lxxxix.
16. Ibid. p. lxxxix.
17. Not to be confused with his son, the Reverend Samuel Peach, who married Bradley's daughter Susannah and gave up Bradley's papers to Lord North in 1776. His father believed that he possessed an interest in them and had resisted a law suit by the Board of Longitude. Whoever's interest he thought he was serving, it certainly wasn't that of James Bradley or indeed his daughter-in-law. He served only to undermine Bradley's scientific legacy. His conduct in this affair was remarked upon by George Biddell Airy when transferring Bradley's registers from the Bodleian Library to the Royal Observatory in 1861. Airy wrote, 'Of the conduct of Mr. Peach in this transaction my opinion

is too strong to admit of expression here'. Airy simply did not believe that Peach (Snr) was a gentleman. During the law suit, Peach offered to give up Bradley's papers but only after receiving a gratuity.

18. Allan Chapman, *Dividing the Circle: The Development of Critical Measurement in Astronomy 1500–1850*, Chichester, 1990. 2nd edn. 1995.

19. The first point of Aries is the point where the rising node of the ecliptic intersects the plane of the Earth's equator.

20. M. E. W. Williams, *Attempts to Measure Annual Stellar Parallax: Hooke to Bessel*, PhD Thesis, London University, 1981. p. 98.

21. Ibid. pp. 103–104.

22. This was the problem already being experienced by Bradley in the 1750s.

23. At this time the determination of stellar magnitudes was largely subjective.

24. The *Gaia* satellite launched in 2013 is locating the positions, velocities, and directions of up to 1.5 million stars an hour which surely makes even this catalogue redundant.

25. Ibid. p. 143.

26. Ibid. p. 143.

27. Ibid. p. 145.

28. Ibid. p. 146.

29. Ibid. p. 146.

30. In the Latin of the text it reads 'Adscensionum rectarum differentiæ in stellis α Canis minoris et α Aquilae = 12h 12′ +'.

31. Friedrich Wilhelm Bessel, *Fundamenta Astronomiæ*, pp. 114–116.

32. M. E. W. Williams, *Attempts to Measure Annual Stellar Parallax: Hooke to Bessel*, PhD Thesis, London University, 1981, p. 148.

33. Ibid. p. 149.

34. Robert Grant, *History of Physical Astronomy*, London, 1852, p. 344.

35. Royal Astronomical Society, *Memoirs of the Royal Astronomical Society*, Vol. 12–13, p. 469.

36. Ibid. p. 469.

37. I write as an occasional researcher seeking days when I was free of the commitments I had with full-time occupations. I have never had sufficient time or energy to examine the archives of most of Bradley's correspondents, an area of future research that may reveal the shortcomings of my own work. In my latter years physical disability made travel difficult and time-consuming.

38. Robert Henry Scott, 'The History of the Kew Observatory' by Robert Henry Scott, MA, FRS, Secretary to the Meteorological Council. Received and read 18 June 1885, in *Proceedings of the Royal Society of London,* Vol. XXXIX. From 19 November 1885, to 17 December 1885. p. 42. Scott writes,'This was known as the "King's Observatory," and in a paper by the late Major-General Gibbes Rigaud, it is further styled 'the King's Observatory at Kew'.

Conclusion

The Man Who Moved the World

The work of James Bradley during the half century from 1711 to 1761 covered a period of major transition in the development of astrometrical observation and calibration. It witnessed advances in instrumental sophistication and observational technique leading to the culmination of the 'classical' science of positional astronomy, when the only tools available to positional astronomers were observational instruments reliant on the use of the human senses and pendulum clocks. Following the death of Bradley, instrumentation was enhanced and yet many of the observational techniques that Bradley and his contemporaries pioneered were still being used. Two more radical transformations were yet to come. The first was in the methodical reduction of observations, an advance led by Friedrich Wilhelm Bessel. The second advance completely revolutionized the science of astrometry, coming with the invention and development of new technologies such as spectroscopy and photography. Astrometrical observation utilizing such advances was increasingly mechanized, relying less on the manual and tacit skills utilized by Bradley and his co-workers and more on the adoption and development of these new technologies.

Science and its methodologies generally move on to become publically ascertainable forms of knowledge, and as such is largely a cooperative human activity. Although there is often competition for priority at the forefront of a lot of research, scientific knowledge is usually public and shared knowledge. It can be tested and challenged against new evidence or old evidence reinterpreted. Although science may often be complex, scientific change stems from positing simple direct questions, leading to positive or null results. Scientific knowledge is often hard won. It is usually gained slowly with many false leads. Even Newton acknowledged that he had to 'stand on the shoulders of giants' for the *Principia* was not born in proud isolation. It was a high point in the acquisition of hard-won knowledge. Bradley also repeatedly confessed his dependence on predecessors, such as John Flamsteed or Edmond Halley, and contemporaries, such as George Graham or Jonathan Sissons (1690–1747) who constructed many of the instruments that made his work possible. Bradley clearly perceived that he was a member of a community made up of gentlemen

The Life and Work of James Bradley. John Fisher, Oxford University Press. © John Fisher (2023).
DOI: 10.1093/oso/9780198884200.003.0013

(and a few gentlewomen) of like mind, transcending national boundaries. Scientific laws and principles did not end at international frontiers and natural knowledge was often shared even by those who worked in the service of the state. Bradley conceived of this community transcending generations too. He in his turn believed he was passing on knowledge that others would capitalize on in the future. I have absolutely no doubt that when Bradley died, he was sure that his observations at Greenwich would be processed and reduced by the future staff of the Royal Observatory. This presumption was prevented by actions of the executors of his will who possessed no comprehension of the scientific value of these observations beyond their asking price.

Although Bradley's discoveries have been absorbed into that immense body of knowledge that is still utilized universally, it remains something of a scandal that we have to accept, as one of his assistants Charles Mason once feared, that 'one of the greatest of men is lost'. Bradley's work culminated in the immense achievement of over 60,000 of what were then the most dependable positional stellar observations ever made. Every observation was accompanied by every item of information required in order to reduce each and every one to any required precision and/or accuracy. Much of the blame for the dispute between Bradley's executors and the Board of Longitude has to lie in the failure of the Board, the Visitors of the Royal Observatory, and the Royal Society. They left unresolved the ownership of any observations made at the Royal Observatory using instruments provided out of the public purse. The payment made by the Board to the widow of Nathaniel Bliss for the observations made by her husband at Greenwich from 1762 to 1764, after the demise of Bradley, may have created an important precedent. It may have led Bradley's executors to the belief that Bradley's daughter Susannah possessed a pecuniary interest in Bradley's many observations made at Greenwich. It was a tragic circumstance that Bradley's executors saw nothing in his work beyond its financial value. The observations made at the Royal Observatory were never legally established as public property. This failure can justifiably be partially attributed to Bradley himself. He was a member of the Board of Longitude as well as serving for the best part of a decade as a member of the Council of the Royal Society. Yet it was a problem that should have been resolved four decades earlier following the death of John Flamsteed. Being a 'reasonable man', I am sure James Bradley never conceived for a moment that his executors would have denied access to his fellow astronomers of his observational legacy.

After fourteen years, when the observations remained out of the public sphere, the registers were surrendered to Lord North in his capacity as Chancellor of the University of Oxford. Bradley's observations must have been regarded by the few who were aware of them as a historical collection of little worth. Coming in the year when Lord North, as the First Lord of the Treasury, had more pressing matters to deal with, including the American Declaration of Independence, Bradley's papers and registers were given little thought. I am fully convinced that Thomas Hornsby, who was Bradley's successor at Oxford, regarded the publication of his predecessor's Greenwich observations as a troublesome task of no value. This may explain why he did not treat the Bradley archive with the respect it deserved. We should at least be grateful that he published the observations made from 1750 to 1755 before Abram Robertson completed the task in 1805, half a century or more after they were made. In spite of the concern expressed by some figures including Charles Mason, formerly Bradley's assistant from 1756 to 1760, the Syndics of the Clarendon Press do not appear to have regarded the task of publishing Bradley's observations with any degree of urgency at all, in spite of Lord North's injunction to publish them as soon as possible.

There had been a legacy of opposition to Bradley's lectures at Oxford and this influence in certain conservative quarters together with a degree of indifference may possibly have stemmed from this former hostile heuristic. This must remain highly conjectural. The influence of Bradley's lectures amongst the intelligentsia was extensive for much of the middle of the eighteenth century. The radical political and social consequences of much of the new knowledge helped to give rise to new movements, including the Declaration of Independence of the American colonies, a circumstance that reactionary powers like France and Spain took full advantage of, supporting the colonies against Great Britain. Ironically these actions by the forces of repression helped to support the rising tide of radical and revolutionary movements sweeping aside such absolutist forces within a generation.

Those concerned about the influences of Bradley's lectures in experimental philosophy and the challenges they posed to the established curricula within the university, expressed concerns that had been mooted since the new knowledge was believed to pose a threat to all forms of established moral instruction. A present common perception that the establishment of the new astronomy was simply an argument about the arrangement of the planets, fails to understand the wider implications. What was commonly perceived to be at stake was the moral foundations of Christian society. Men such as James Bradley were capable of reconciling their confessions of the faith with the new forms of knowledge based on observation and experiment. In polite society in England during much of the eighteenth century, all forms of religious and political 'enthusiasm' were resisted. After the seventeenth century, 'common sense' and 'reason' came much more to the fore, at least within the optimistic and expansive commitments of Whig society.

In spite of Bradley's parents' inferred wishes that under the direction of his maternal uncle James Pound he enter into the service of the Church of England as an ordained priest, his wider education opened up another possibility, acquiring an entirely different 'heavenly' vocation. There was a possibility that like Flamsteed and his uncle and many other Anglican divines, Bradley might have continued life as a country cleric making astronomical observations during his leisure hours. This was a common occurrence in England during the eighteenth century. John Hadley the designer of the six-foot reflecting telescope made in 1721 was the vicar of Barnet, a village north-west of London. Hadley is now commemorated by the Hadley Rille, explored during the Apollo 15 expedition to the Moon. Nearby is located Mons Bradley, the highest point in the Lunar Apennines on the Earth-facing side of the Moon. Rigaud believed that Bradley was well set for a very successful career in the Church of England, evidenced by his native intelligence together with the patrons he had acquired through his uncle's connections. This opinion was supported by Allan Chapman and he suggests that in turning his back on a clerical vocation, as wished for by his parents, he was very likely turning his back on a very lucrative career. The highest paid bishoprics in England were amongst the most remunerative public offices in the land.

Serendipity sometimes plays an important role in the lives of men and women. Being in the right place at the right time, Bradley's life was transformed from that of a country cleric to that of an Oxford academic following the death possibly from food poisoning of John Keill, an acolyte of Newton's. After Pound refused the offer of the Oxford chair, attention was redirected to his nephew Bradley, particularly as a small but highly influential party based largely in the fellowship of the Royal Society appreciated his burgeoning skills. Newton referred to him as 'the finest astronomer in Europe'. This was probably an overestimate of Bradley's knowledge and skills at this moment, but it reveals that he was highly regarded, especially in the light of the complete breakdown in communications between the President of the Royal Society and the Astronomer Royal John Flamsteed. Bradley's career as an astronomer was indicative of his immense capabilities, yet it must never be lost sight of that his success was also dependent on the associations he developed with various important patrons, who often had access to the highest in the land.

The opening chapter of this account revealed a few of the difficulties experienced by Flamsteed. He found it difficult to acquire supporters, alienating some who may have given him succour during difficult times. In seeking to defend his reputation, he appears to have lost sight of the simple fact that the cultivation of supportive relationships, especially within the Royal Society, would have enhanced the reputation he so jealously defended. Jim Bennett's characterization of Flamsteed as morose also contrasts sharply with that so consistently displayed by Bradley. Neither Bradley nor his uncle were easily offended. Flamsteed was ready to take offence at almost every slight he thought was directed towards him. Flamsteed was his own worst enemy, in spite of his remarkable skills as an astronomer. It was less that he was regarded as an 'outsider' that brought him down, it was the simple notion that he perceived of himself as an outsider. This, as much as any other factor fettered his career as the first Astronomer Royal. In the coffee-house culture within which so much discussion took place following formal meetings of the Royal Society, Flamsteed felt ill at ease. Halley was relaxed and felt able to enter into easy discourse with his fellow virtuosi. Could this be one of the reasons why the defensive Flamsteed took such issue with Halley?

Flamsteed's voluminous writings and correspondence give full witness to his dealings with his fellow men. Such papers detailing Bradley's life and dealings are obvious by their absence. More than one person has expressed the view that Bradley had an aversion to writing letters because they took him away from his observations. That a man of such obvious conviviality and ease in society, in every sense a man of the world, did not enter into correspondence with many of his acquaintance is simply unbelievable. This supposed aversion to correspondence apparently 'explained' the paucity of Bradley's archive. The seemingly intimate relationship between Bradley and Elizabeth may offer some explanation as to why much of Bradley's correspondence was denied public access after his death. The Peach family who were his executors may have believed they were acting in Susannah Bradley's interest in retaining these personal effects. Given the shameful way that Bradley's executors retained Bradley's observational registers during the lawsuit with the Board of Longitude, I incline to the view that Bradley's papers were minutely examined in order to deny public access to any details of his familial or personal life.

I am sure Samuel Peach had few scruples about destroying Bradley's personal papers. Some may interject and express the opinion that historians have no right to the minutiæ of a person's life and they would be fully justified. I hold the view that his executors did the memory of Bradley's name a great disservice, for by destroying or withholding all of his more personal correspondence and effects, they impeded a fuller appreciation of the life and work of a remarkable figure in the history of science.

Bradley's registers and working documents were retained by the Peach family until 1776 when they were given over via Lord North to the possession of Thomas Hornsby, who was Bradley's successor as the Savilian Professor of Astronomy, a chair he held until his death in 1810. He possessed an excellent reputation at Oxford, derived from his lectures and was partly responsible for the construction of the Radcliffe Observatory, although he appears to have restricted his observations to the daylight hours, which may have contributed to an underestimate of the importance of Bradley's 'stellar' observations. One thing is more certain and this is evidenced by the desultory progress he made in producing an edited publication of Bradley's observations. He evidently thought the task was without value. If Rigaud's evaluation is justified, Hornsby was careless regarding Bradley's surviving archive and as I have intimated, many of Bradley's papers were carelessly mixed in with his own or even displaced completely. It was an unfortunate combination of the actions of the Peach family and those inferred of Hornsby that almost led to Bradley becoming an 'invisible man'. If Rigaud had not gone to the trouble of gathering Bradley's miscellaneous papers and correspondence, then 'one of the greatest of men' would indeed have been lost. In one sense we should thank Hornsby, for slow though his progress may have been, the delays in the publication of his edition of Bradley's observations from 1750 to 1755, when published alongside Robertson's edition of the observations made from 1756 to 1762, were eventually to finish up in the hands of the young Bessel. His reductions of Bradley's observations took as long to accomplish as it took Bradley and his assistants to make them. Bessel argued assertively that reduction should accompany observation. If Bessel had not reduced Bradley's observations, included in the *Fundamenta Astronomiæ,* then Bradley would have been even more obscure. Rigaud may not have felt compelled to publish Bradley's miscellaneous observations and what could best be called his surviving working correspondence.

Bradley was fortunate in his acquisition of patrons largely through his uncle's connections. Some of these figures moved in the highest echelons of British political society. Many of these connections were acquired through common membership of the Royal Society. Men such as Thomas Parker and Benjamin Hoadly formed friendships through their common fellowship of the Royal Society and as fellow Whigs. In a divided society where even the names of the political parties were terms of mutual abuse, it required great social skills to transcend these divisions. What is remarkable about Bradley's progress as an astronomer of note is the way he seemed to attract support from all quarters. When he gave his inaugural lecture after his election to the Savilian Chair in Astronomy, his performance was criticized by Thomas Hearne. As a prominent non-juror Hearne had little or no reason to admire a man elected by a London Whig power base. Yet a few years later when Bradley sought to become the Keeper of the Ashmolean Museum, the building where he gave his lectures on experimental philosophy, his only elector was the self-same Thomas Hearne, who now supported a man he formerly regarded as being a creature of all that he opposed. That Bradley could turn a man as difficult as Hearne says much for his personal skills. Bradley counted Tories such as Halley amongst his close friends. This might suggest that the divisions in English society were not quite as deep as sometimes characterized. Bradley appears to have possessed great personal charm, which helped to advance his interests. Bradley was deferential in his dealings with his fellow men, so unlike the defensive Flamsteed.

The importance of James Pound in the advancement of Bradley's prospects is inestimable. It is not merely that he supported his nephew's education and prepared him for the cloth, he also introduced him to some of the movers and shakers of English society, as well as the leaders of English science. Equally importantly, Pound's leadership, so obviously displayed during his years in the Indies, instructed the young man in the ways of the world. Whilst he may have shared some of his maternal uncle's familial characteristics, Bradley learned very quickly where his interests lay. His native intelligence counted for much in the advancement of his career, but many highly intelligent men and women have been passed over throughout history. Bradley was fortunate to be in the right place and at the right time. With many in the Royal Society also being figures of importance in the political life of the kingdom, Bradley soon gained friends in high places. Flamsteed was a man from Denby in Derbyshire who unfortunately possessed only one real patron, Sir Jonas Moore. After his death in 1679 only four years after Flamsteed became the King's Observator, it never seemed to trouble him that he needed other forms of support in a violent and unstable society. He perceived of himself as a court-based astronomer like his great hero Tycho Brahe. New forms of patronage were rising and men like Pound, and later Bradley, openly accepted and utilized it for their own purposes and ends.

Bradley's astronomical career probably began in 1711 when he moved from Sherborne, a small village in Gloucestershire, to Wanstead in Essex about eight years before Flamsteed's demise. The two astronomers inhabited entirely different worlds. Flamsteed began his career in an age of courtly rivalry whilst Bradley entered an entirely different milieu in which the power of the court was greatly diminished in comparison to the new forces of secular society, in which new financial and political powers held sway. Bradley's success was gained against the background of a confident and expansive society with outlooks empowering a commercial revolution. It enhanced new forms of material and human exploitation by companies such as the East India Company. In a society where maritime and naval interests were even more urgent than they were when Flamsteed's career began, the work of Bradley and his assistants at Greenwich acquired greater importance. The scandalous fate of the Greenwich observations after Bradley's death in 1762 must have caused concern in some quarters. How ironic that it took a young German astronomer to reveal the riches that had lain hidden for over half a century.

When I attended a conference of the Society for the History of Astronomy held at the Institute of Astronomy at Cambridge, I remember the complete lack of knowledge of Bradley when I was politely asked about my own interests. He was the first person to demonstrate that the Earth was a planet in orbit around the Sun when he discovered the aberration of light. The almost complete ignorance of Bradley might have been understandable in a general setting, but coming from a small group of people who were members of a society dedicated to the history of astronomy it came as a shock to me. It was this incident that finally impelled me to attempt to write an account of the work of the third Astronomer Royal. Admittedly, the same group of attendees also had difficulty even recognizing who Flamsteed was. I accepted that I had met a group of people whose interests were fixed on nineteenth and twentieth-century astronomy. The group did know about Halley, but their knowledge of him was largely restricted to the comet that bears his name. They knew nothing about Halley's discovery of the proper motion of the stars, or his catalogue of the southern stars which he compiled from observations made at St Helena.

Although my studies of the life and work of James Bradley began during the mid-1980s, I have published very little. I have given talks to astronomical societies and occasionally to more general audiences, but most of my time has been devoted to research during the infrequent periods when I was free to travel. There is no substitution for reading and transcribing the manuscripts that have survived the various depredations of Bradley's archive. The work of Rigaud in the preservation of the surviving archive is something historians of astronomy should be thankful for and I am grateful for his timely intercession. I share his vexation at Hornsby's apparent negligence, in spite of the good reputation he had acquired amongst his contemporaries. Like Bradley's first biographer, I attempted to give Hornsby the benefit of the doubt, but the length of time he retained the third Astronomer Royal's papers, and the seemingly careless manner in which he dealt with the Bradley archive, elicits justifiable anger. In an archive already partially suppressed by his executors, the loss of even the most mundane documents is to be regretted. Close attention to what remains is a rewarding experience. Both Bessel and Rigaud were surprised by the riches that a close study of Bradley's experiments or observations revealed. Thirty-five years of study and ample time to reflect have revealed Bradley as a scientist of the highest order. He was an investigator of industrious energy and incisive intelligence who refused to come to easy answers. He refused to fall into the traps that had for so long misled investigators, to confuse their observations of the phenomenon later identified as the aberration of light as evidence for annual parallax. Bradley possessed a clarity of mind that had the capacity to access the slightest clue and then follow its consequences to its ultimate conclusions.

It was Bradley's mastery of the geometry of parallax that enabled him to recognize immediately that the motion of γ Draconis observed in December 1725 at Kew could not possibly be reduced to that of annual parallax. The long-held belief in some quarters that Bradley discovered the aberration of light at Kew is an example of the misapprehension of so much of his work. It leads to the underestimation of what was involved in establishing the phenomenon. What he observed at Kew was an anomaly. To Bradley's clear mind it was evidential that whatever else the observation was it was not the one motion that so many astronomers of that period appeared bent on observing, annual parallax. Others had observed the phenomenon and also recognized that it could not be annual parallax, but only Bradley sought to resolve the causes of such an unexpected observation. The discovery of the aberration of light took three years of continuous investigation involving the observation of up to seventy different stars. The belief that Bradley discovered aberration at Kew following a handful of observations has led to a common interpretation that it was a 'fortunate' consequence following his 'failure' to discover annual parallax. More than a few secondary and tertiary sources have reported the discovery of the aberration of light in these terms. In front of Kew Palace within the grounds of the Royal Botanical Gardens are the remains of a monument to commemorate the discovery of the aberration of light at Kew. It has long reinforced the false apprehension that the discovery of the aberration of light was a fortuitous circumstance and not the result of a persistent investigation brought to fruition at Bradley's aunt's town house in Wanstead. Today there is a small green plaque to be found very close to the location of the zenith sector in Wanstead with which Bradley made his momentous series of observations. He was able to determine 'the law of a universal phenomenon' and then from his repeated observations of the satellites of Jupiter construct a convincing hypothesis that explained the observations. This settled once and for all that the Earth really was a planet in motion around the Sun. It fully vindicated Galileo 95 years after he was found guilty of the vehement suspicion of heresy by the Holy Office of the Inquisition, and 112 years after Copernicus's *De revolutionibus* was placed on the Index of Prohibited Books. Bradley's discovery created a stir within the Papal States and elsewhere as well as in Cartesian France. Bradley usually sought to avoid unnecessary controversy for most of his working life. He was deferential in his outlook. More seriously, however, he loathed all forms of hypothesizing based on first principles. This was strongly demonstrated in his lectures at Oxford, which begin with a polemic directed against all attempts of basing natural knowledge on first principles, whether those of Aristotle or Descartes.

If Bradley's discovery of the aberration of light is an example of a mind unconstrained by lazy expectations, his exposure of the phenomenon of the nutation of the Earth's axis is even more remarkable. In his paper detailing his discovery of nutation, Bradley is fulsome in his regard for George Graham the man largely responsible for the instrument he used in his observations of a phenomenon so infinitesimally small by eighteenth-century standards. Remarkably his mentor Edmond Halley never accepted the validity of his protégé's observations of nutation. Even so, Bradley's friendship with Halley never wavered, for both men were sceptical in their scientific outlook and practice. Bradley was an executor of Halley's will. It is something to be deeply regretted that most of the Halley archive was lost during a German bombing raid during the Second World War that destroyed much of the Guildhall in London.

The work on the nutation at Wanstead (from 19 August 1727 to 3 September 1747) was of such precision that his astronomical contemporaries throughout Europe and America regarded Bradley with admiration. That such a miniscule phenomenon should have been observed for so long and with such tenacity was admirable enough, but it also supported the validity of the arguments in Newton's *Principia* beyond reasonable doubt. After the reading of the paper on the nutation at the Royal Society on 7 and 14 January 1748, Bradley was awarded the highest scientific award of the society, the Copley Medal, instituted in 1731, three years after his discovery of aberration. A consequence of the discovery is that it was possible for astronomers to predict the exact amount of motion due to the phenomenon of precession for any given year. It is remiss that such a figure as James Bradley has been largely overlooked in the place where he made his two major scientific discoveries. There has been some passing interest in 'the astronomers of Wanstead' usually with reference to the location of the aerial telescope used by Pound and Bradley, which some mistakenly believe was on George Green rather than close to the former location of the parsonage near the Red Bridge.

Although Bradley has been variously described as 'the founder of observational astronomy' or even as 'the founder of modern science' I find both epithets meaningless. Observational astronomy is as old as humanity and as for 'modern science' what can possibly qualify for such a descriptor? If understood that Isaac Newton opened a portal into a more open-ended approach to the acquisition of natural knowledge, then it can be suggested that James Bradley was one of the earliest scientific workers who entered through it. In attempting to resist a tendency to categorize Bradley as a 'Newtonian', it was the ever perceptive Rob Iliffe who suggested to me that I call him a 'Bradleian' and ever since I have adopted his sage advice. There is something distinctive about Bradley's work that suggests that this descriptor is quite apposite. What can be meant by this term that makes his work both distinctive and yet indicative of a new way of doing science? I say 'doing' quite deliberately. Bradley was opposed to armchair speculation. In his own period this meant condemning the approaches to natural philosophy represented by the work of the Cartesians. A mind-matter dualism was the fundamental foundation of Cartesian physics. There is no distinction between objects and the space in which they move. It is all matter. The phenomenon of light was conceived as a pulse transmitted 'instantaneously' through the cosmos. Light was by *definition* transmitted infinitely. It was due to this and similar forms of reasoning that some Cartesians rejected Bradley's hypothesis. The phenomenon of aberration with its association with the progressive or the finite velocity of light, based on years of observation of the motions of the satellites of Jupiter, was rejected by some influential Cartesians including Jacques Cassini.

Bradley had spent many years observing the motions of the satellites of Jupiter and it was evident that the timings of the occultations of Jupiter's satellites by the planet varied according to where Jupiter and the Earth were in their respective orbits, revealing that the velocity of light must be finite signified by an equation of time in his ephemerides of the motions of the satellites. The aberration of light independently reveals a very accurate calculation of the velocity of light. Estimates of the value of the velocity of light based on Bradley's determination of the constant of aberration produces a conclusion that is within 1% of the modern value. In opposition to the methods of the acquisition of natural knowledge practised by the Cartesians, Bradley believed that scientific knowledge was only to be determined by observation and experiment and not from the acquisition by reasoning from some metaphysical principle. It is ironic that Descartes's attempts to replace Aristotelian physics failed precisely because he also created a metaphysical 'system'. Bradley's erstwhile partner Samuel Molyneux believed that he had 'completely destroyed the Newtonian system'. Aside from his deluded apprehension, his failure also revealed something that Bradley understood very well. There is no 'Newtonian system'. It is an open-ended contingent approach to the acquisition of knowledge based on experiments and observations. When Molyneux made his premature approach to Newton it was reported by Conduitt that the elderly President of the Royal Society merely responded with the words that 'there is no arguing with facts or experiments'. There is every reason to believe that Conduitt wished to present Newton in the best possible light, but it was completely characteristic of the means by which almost all modern scientific knowledge is acquired.

This volume cannot possibly be the last word on a remarkable figure in the history of science. It cannot be more than an attempt to open up a new conversation about the origins of modern empirical science and its presumptions. One of the more remarkable factors that any serious student of the work of the third Astronomer Royal will become aware of is Bradley's nescience. By this I mean to infer that he had difficulties establishing knowledge claims. Studying his work suggests that he believed there was a disjunct between the results of observation and experiment and the theories and hypotheses derived from them. A classic example of this is the complementary observations made by Bradley and Manfredi that led to diametrically opposed conclusions. Bradley revealed that the 'new discovered motion of the fixed stars' was evidence of the motion of the Earth. He left it open to the 'Anti-Copernicans' to explain the phenomenon in geocentric terms. Manfredi wrote a treatise on 'aberrations', showing that none of the observed motions of the stars had anything to do with annual parallax. This Bradley entirely agreed with. Bradley identified a universal motion leading to the conclusion that the Earth moved. Manfredi identified many separate motions fully compatible with those observed by Bradley, leading him to the conclusion that the Earth did not move.

Bradley, more than most, apprehended the real difficulties in the establishment of new knowledge claims. This may explain his generous response to Halley's scepticism concerning observations of angles that he considered beyond the possibility of contemporary technology. Bradley was always modest in his claims and left it open to further observation or experiment. His discovery of the nutation of the Earth's axis challenged all of the expectations of the age concerning what was and what was not possible with the contemporary means at the disposal of astronomers. Halley should not be criticized too severely for his non-acceptance of his protégé's claims to observe motions as small as half an arcsecond or even less.

Bradley's observations were made possible by a number of remarkable craftsmen based in London which was becoming the leading centre of precision instrument making in Europe. In Great Britain and its colonies there was a commercial revolution powered by the growing demands of a free market that was financed increasingly by the Atlantic tripartite slave trade. Unlike most of the other countries in Europe where the court was usually the main driver of demand, the Dutch Republic being an obvious exception, the English markets were driven by the demands of new classes in society. The scientific instrument trade expanded because of the demands of many gentlemen seeking to make their own observations and experiments, reporting their results in journals including the *Philosophical Transactions*. Bradley inhabited that expanding community made up very largely of men and alas of too few women, who were perceived to be of the 'middling sort'. By the standards of modern life they were 'amateurs': men who followed their studies purely for their own entertainment or enlightenment. In this Bradley in no way differed from most of his fellow scientific workers. This did not mean that the leading practitioners of what we now call science were anything but exacting or serious. Like scientists of the modern age they recognized that their pursuits demanded great insight and imagination, and yet also demanded persistence and hard work alongside an adamant refusal to accept easy answers. The work of Bradley and many of his contemporaries reveals that scientific labour was as demanding during the eighteenth century as it appears during our own age. We cannot call them 'professionals' but in their investigations they laid the foundations of the modern world. James Bradley and his fellow experimental and natural philosophers moved the world in more ways than one.

A Chronology of the Life and Work of James Bradley (1692–1762), DD, FRS, Third Astronomer Royal of England

Compiled by John Fisher PhD, FRAS

Selected events before the beginning of Bradley's astronomical career in 1711

All page numbers in square brackets refer to page numbers in James Bradley's *Miscellaneous Works and Correspondence* compiled by Stephen Peter Rigaud in 1832. The *Memoirs of Bradley* written by Rigaud was published with the *Miscellaneous Works* and the pages are signified by lower case Roman numerals.

1608	**October 2** Hans Lippershey attempted to claim a patent for a 'kijker' (a looker) or a primitive telescope. This was refused as it was already available in markets throughout the Netherlands.
1610	Galileo Galilei published his *Siderius Nuncius* in Venice.
1611	Johannes Kepler published his *Dioptrice*.
1633	Galileo Galilei found guilty of the 'vehement suspicion of heresy' by the Holy Office of the Inquisition in Rome for teaching the motion of the earth contrary to the teaching of the Church Fathers.
1639	**November 24** Jeremiah Horrocks observed the transit of Venus. He and William Crabtree were the only known observers of this rare event. His *Venus in sole visa* was published two decades later by Johannes Hevelius in Danzig (Gdansk).
1640	**December 28** William Crabtree corresponded with William Gasgoigne informing him of Jeremiah Horrock's admiration of his micrometric device.
1641	**January 3** Jeremiah Horrocks died at Toxteth Park near Liverpool, he was just 22.
1644	**July 16** William Gascoigne perished at the Battle of Marston Moor, an engagement of the English Civil War. Gascoigne held a commission in King Charles I's army. Crabtree was a merchant supportive of the Parliamentary cause. **July 19** William Crabtree wrote his Will. He was interred on 1 August 1644.
1656	Christiaan Huygens described a clock using a pendulum which controlled the timepiece by the stability of its oscillations, greatly increasing the accuracy and dependability of the clock.
1666	Robert Hooke proposed a plan to observe the parallax of the fixed stars. This was diverted by the consequences of the Great Fire of London when much of the City was razed to the ground between 2 and 6 September.

1667   **January 9**   Henry Oldenburg reported to the Royal Society, Adrien Auzout's use of a micrometer. Christopher Wren and Robert Hooke reported that they were already in possession of such a device.

March   Robert Hooke appointed as the City Surveyor, staking out and certifying about 3,000 foundations over the next five years.

1669   **July 6**   Robert Hooke began his observations of γ Draconis (Eltanin) in order to observe annual parallax, chosen because it passed very close to the zenith at the latitude of the City of London, avoiding most of the major problems associated with atmospheric refraction.

November 4   John Flamsteed sent his calculations of an eclipse of the Sun and five occultations of stars by the Moon to Henry Oldenburg at the Royal Society.

1674   In one of his Cutlerian Lectures, Hooke revealed the results of his 'Attempt' to observe the parallax of γ Draconis and made his claim that it revealed an annual parallax of 27 or 30 arc-seconds. All this on the basis of just four observations made between July and October 1669.

1675   George Graham, Bradley's major instrument maker, was born in Cumberland.
John Flamsteed was appointed as the first Astronomer Royal by Charles II.

1676   G. D. Cassini and Ole Römer first proposed the finite velocity of light to explain observed variations in the expected timings of eclipses of the satellites of Jupiter. Cassini abandoned his claim, recognizing that it contradicted both the Cartesian theory of the instantaneous velocity of light, and more importantly for a devout Roman Catholic, the motion of the Earth. His assistant Ole Römer established his claim, for as a Danish Lutheran he had no inhibitions about defying Cartesian theory or Catholic doctrine.

1678   William Bradley (b.1657), third son of Lewis and Eleanor Bradley, married Jane Pound (b.1656) of Bishops Canning, near Devizes in Wiltshire.

1679   **August 27**   Sir Jonas Moore, Ordnance General, John Flamsteed's first and only patron died.

1685   The Revocation of the Edict of Nantes by Louis XIV. Many Huguenots were forced to flee to England, including the forebears of Stephen Peter Rigaud, who was to become Bradley's first biographer.

1690s   John Wallis discussed Galileo's proposition of observing pairs of stars to determine annual parallax and Francis Roberts suggested that the Earth's orbit might not offer a large enough base to determine annual parallax.

1690   **June 17**   John Flamsteed (the Astronomer Royal) established the length of the sidereal day at 23hr 55min 50sec. (The modern accepted value is 23hr 56min 4sec).

1692   Christiaan Huygens presents a glass with a focal length of 123ft to the Royal Society.

October 3   James Bradley baptised on this date, suggesting that he was probably born late September 1692. He was the third son of William and Jane Bradley. (The third son of a third son.)

1693   Ole Römer discovered a motion he was sure was not annual parallax. He was unable to ascribe a cause but it may have been what James Bradley later identified as the aberration of light.

G. D. Cassini published tables of the motions of Jupiter's four Galilean satellites. James Pound modified these tables following his own observations and it was these amended tables that formed the basis of Bradley's observations of the Jovian system, leading to his discovery of the gravitational resonances of satellites I, II, and III (Io, Europa, and Ganymede).

1694   **June 16**   John Flamsteed now measured the length of the sidereal day at 23hr 56min 15sec. (The modern value is 23hr 56min 4sec.)

1697	James Pound received his medical degree (BM) and a licence to practice medicine.
1698	John Flamsteed corresponded with John Wallis claiming to have discovered the annual parallax of *Polaris*.
1699	John Flamsteed's claim of a parallax of 40 arc-seconds for *Polaris,* which was countered by Jacques Cassini on sound theoretical grounds, was later identified by James Bradley as an example of annual aberration, even incorporating the data in his discovery paper as further empirical evidence in support of his claim.
	James Pound was employed by the English East India Company as a chaplain to the company's community at Madras in India.
	November 30 James Pound elected a Fellow of the Royal Society, but admittance was deferred until 13 July 1713.
1700	James Bradley began his education at Westwood's Grammar School at Northleach about three miles from his home village of Sherborne.
1702	**November** John Flamsteed corresponded with Christopher Wren retracting his claim for an annual parallax of 30 arc-seconds, suggesting that he would return to his observations. He never revisited them.
1705	**March 3** The total destruction of the East India Company's settlement on the island of Pulo Condore, near the Mekong delta, during an insurrection by native employed troops, known as Macassars, from near the locality, possibly Borneo or Celebes (Sulawesi). James Pound was one of only fifteen who escaped in a sloop (eleven Europeans and four Black servants or slaves), reaching Malacca in Malaya well over a week later after an extremely hazardous voyage in hostile waters.
	May 2 James Pound wrote from Batavia to the Managers of the East India Company informing them of the massacre at the Company's former station at Pulo Condore.
	July 7 James Pound wrote from Batavia to the Astronomer Royal, John Flamsteed, informing him that he had lost all of his observations and the quadrant so kindly lent by him. Pound's reputation with the Astronomer Royal never really recovered from this.
1706	**July** James Pound returned to London after being recalled by the Honourable Company.
1707	**July** James Pound appointed Rector of Wanstead by Sir Richard Child, third son of Sir Josiah Child, former Governor of the Honourable Company.
1708	**June 23** John Dutton issued an indenture of assignment taking over the management of the Sherborne Estate from his father Sir Ralph Dutton Bt. Heavy gambling debts were threatening the future probity of the wealthy estate. The gross mismanagement of the estate by Sir Ralph meant that steward, William Bradley, James Bradley's father, felt unable to guarantee his third son's education. He gave over responsibility for this to his younger brother-in-law James Pound, to prepare him for a career in the Church of England.
1709	John Bird born in County Durham; he became Bradley's second notable instrument maker. A weaver by trade he also supplemented his income by becoming a highly proficient engraver. He died in 1776.
1709/10	**February 4** A further indenture of assignment was issued by John Dutton issuing a sum of £5,000 to his father Sir Ralph and a life annuity of £400 per annum. It indemnified his mother and sisters of all further claims due to any debts accumulated by Sir Ralph.
	February 14 James Pound married Sarah, widow of Edward Farmer. His only son died immediately after birth, late in 1710.
1710	John Desaguliers began teaching experimental philosophy at Oxford.

The Life and Works of James Bradley, following his move from Sherborne in Gloucestershire to Wanstead in Essex

1711   **March 15** Bradley had continued at Westwood's Grammar School at Northleach in Gloucestershire before he went up to Balliol College at Oxford. Bradley took up residence with his maternal uncle James Pound, who was the Rector of Wanstead, Essex. This was probably the start of Bradley's astronomical career as assistant to his uncle, one of the best observers in England.

1712   **November 4** Samuel Molyneux elected Fellow of the Royal Society along with Jean Bernoulli, the Swiss mathematician.

1713   John Desaguliers after teaching experimental philosophy moved from Oxford to London.
        July 13 James Pound elected a Fellow of the Royal Society in 1699, admitted from this date.
        September 16 Sarah Pound gave birth to daughter Sarah, who later died unmarried at cousin Bradley's residence at Greenwich 19 October 1747. James Bradley was then the third Astronomer Royal.

1714   Jacques Cassini began observing *Sirius* in order to determine the annual parallax of the star.
        August 1 Queen Anne died. Thomas Parker appointed as the Regent of Great Britain until the accession of George I on 18 September 1714.
        September 18 George I takes the thrones of Great Britain and Ireland.
        October 15 James Bradley graduated with the degree of BA. We have no way of knowing whether he ever attended Desaguliers' lectures.

1715   George Graham was researching the expansion of different metals in different conditions. John Harrison had devised his 'grid-iron' pendulum rod of brass and steel, to counter the effects of expansion due to increasing atmospheric temperature. It would increase the accuracy of pendulum clocks used as regulators in observatories.
        June Sarah Pound died. We do not know the cause of her death. Could it have been associated with pregnancy or childbirth? The couple lost a son soon after birth late in 1710.
        Bradley's **earliest recorded observations**. There was an emphasis on observations of Jupiter and its four Galilean satellites. He corrected his uncle's tables, which were based on those made by G. D. Cassini in 1693.
        There were many improvements including the remarkable arrangement by which all the equations were made additional. These were communicated to Edmond Halley who printed them in *Philosphical Transactions* (*Phil. Trans.* Vol. 30, p. 1023).
        July 14 Pound observed an occultation of a star by Jupiter when he and Bradley were timing Jupiter's occultations of the four Galilean satellites.
        October 21 James Bradley's **earliest clearest recorded observation.** The entry reveals a consummate practitioner, so it was not in any way his first observation, which was probably made during 1711.
        October 29 Bradley measured Jupiter's diameter and RA + Dec + 5 other observations. [p. 340]. These observations reveal his burgeoning abilities.
        October 30 Pound observed an eclipse of the Moon.
        November 3 Observed Jupiter's limb + five other observations. [p. 340].
        November 30 Observed difference in Dec. of Saturn and γ Virginis and several other observations. [p. 340].
        December 4 Observed difference in Dec. of Saturn and γ Virginis and six other observations. [p. 340].

1716 Bradley's papers include observations by which Pound determined the relative positions of the stars in '*sinistro calcaneo Persei*' referred to in Newton's *Principia*. The distances from each other were measured in September 1716. [*Principia*].

Edmond Halley pointed out the usefulness of the transits of Venus in 1761 and 1769 to determine the distance between the Earth and the Sun (the astronomical unit, AU), the basis of all parallactic calculations determining the distance of nearby stars up to 50 or 100 light years (15 to 30 parsecs).

March 6 Halley and Pound observed an aurora in Wanstead.

March 22 Bradley reports observing the same aurora tending to an area between *Castor* and *Pollux* 25° from the zenith.

August 12 Measured Jupiter's distance from *Propus*. [p. 340].

August 14 Measured diff. of RA between Jupiter and *Propus*. [p. 341].

August 19 Centre of Jupiter, same declination of an unidentified star. [p. 341].

August 24 Measured Jupiter's diameter + five other observations. [p. 341].

August 26 Measured Jupiter's centre from two stars. Also its distance from Venus. [p. 341].

September 6 Halley requests Pound and Bradley to make observations on his behalf.

September 10 Four observations of Venus. [p. 341].

September 12 Three observations of Venus. [p. 341].

November 29 Position of Jupiter relative to *Propus*. [p. 341].

November 30 Position of Jupiter relative to *Propus*. [p. 341].

December 1 Position of Jupiter relative to *Propus*. [p. 341].

December 4 Position of Jupiter relative to *Propus*. [p. 341].

December 5 Position of Jupiter relative to *Propus*. [p. 342].

December 6 Position of Jupiter relative to *Propus*. [p. 342].

December 7 Position of Jupiter relative to *Propus*. [p. 342].

December 18 Diameter of the Sun, several observations. [p. 342].

1717 **January 17** Observed Jupiter's 3rd satellite (Ganymede) disappear behind primary. [p. 380].

March 11 Bradley contracted smallpox shortly before this date. He was cared for by his uncle, James Pound, who possessed a degree in medicine and a licence to practice. Under daily care by a nurse. [*Memoirs*, p. ii]. [Pound's Accounts Book].

March 21 Dr Benjamin Hoadly, Bishop of Bangor, gave a sermon before King George I of Great Britain and Ireland on *The Nature of the Kingdom of Christ*. His text was John 18:36, 'My kingdom is not of this world'. It gave rise to the so-called 'Bangorian Controversy', leading to the suspension of the Convocation of the Church of England for over a century. Hoadly was a close friend of James Pound, and later ordained Bradley into the priesthood of the Church of England when he was the Bishop of Hereford.

June 21 Bradley graduated MA. [*Memoirs*, p. ii].

September 18 James Pound paid for tin and brass work to hold the Huygens 123ft glass in place.

September 27 On or close to this date James Bradley was twenty-five and old enough to take Holy Orders. He appears to have delayed this due to his astronomical work and studies.

December 5 Observations recorded by Halley. [*Phil. Trans.* Vol. 30, p. 354] [p. 380].

1718 **January 15** Observed and located the RA and declination of Jupiter and the RA, Declination of η Cancri. [p. 342].

February 18 Pound and Bradley observed an eclipse of the Sun. [p. 342].

March 11 Pound's observations of the position of Saturn concerning κ Virginis and other stars in Virgo. [p. 342].

March 12 Bradley repeats the observations made by Pound the previous night, of α and δ Virginis plus the two stars of γ Virginis. [p. 342].

March 25 Pound's writing, Bradley observed α Geminorum (*Castor*) in a parallel through β Geminorum (*Pollux*) leaving κ to westward—observations also of g, l Geminorum. [p. 346].

April 10 Bradley determined the position of Mars of various stars. [p. 342].

April 15 Measured the position of Mars relative to τ Tauri. [p. 342].

April 25 Pound purchased an eyeglass for use in conjunction with Huygens' 123ft glass. Cost £1 2s 6d. [*Memoirs* p. ix] [Pound's Accounts Book].

April 30 Bradley observed 1st satellite of Jupiter (Io). [p. 342].

May 12 Observed 3rd satellite of Jupiter (Ganymede) [p. 342].

May 13 Entries in Pound's Accounts Book reveal costs for the erection of the Maypole once located at Charing Cross. Probably relocated close to the parsonage in order to suspend Huygens' 123ft glass.

May 16 The Maypole erected for use with the Huygens 123ft glass.

May 23 Observed 1st satellite of Jupiter (Io) [p. 342].

August 29 Pound and Bradley observed an eclipse of the Moon. See entry for October 23.

September 4 Observations of lunar occultations in Taurus. Observed the Moon. Observed conjunction of Venus and Jupiter. [pp. 342, 343].

September 5 Observations of Venus and Jupiter. [p. 343].

September 7 Further observations of Venus and Jupiter determining positions of each planet. [p. 343].

September 17 Observed position of Jupiter related to *Regulus* and other stars. [p. 343].

October 17 Observed distance between Venus and θ Virginis. Observed 2nd satellite (Europa) emerge from Jupiter and further observations of the 1st, 2nd, 4th Jovian satellites (Io, Europa, Callisto) plus observations of Saturn and its satellites. [p. 343].

October 23 James Pound presented several astronomical observations to the Royal Society, made with James Bradley, of an eclipse of the Moon. [*Phil. Trans.* Vol. 30, p. 855]. At this meeting of the Royal Society Edmond Halley proposed James Bradley as a person well qualified to be a member of the Society. Referred to the next Council meeting 6 November.

November 6 James Bradley elected a Fellow of the Royal Society with Sir Isaac Newton in the chair.

December 19 Jupiter's 4th satellite (Callisto) emerged from behind the primary. [p. 344].

December 27 Bradley's immense skill in the management of long-focus aerial telescopes is demonstrated by his precise measurement of the angular diameter of Venus across the 'horns' using a telescope with a focal length of 212¼ feet. This at dusk, with the planet just above the horizon.

1719 **January to May** Bradley spent much effort observing the motions of Jupiter and its satellites on behalf of his mentor Edmond Halley.

January 2 Observed the position of Jupiter and ρ Leonis + Jupiter's 4th satellite (Callisto) [p. 344].

January 16 Observed Mars and ω Scorpii. [p. 344].

January 21 Jupiter's 2nd satellite (Europa) emerged. [p. 344].

January 28 RA and declination of Jupiter. [p. 344].

February 8 Observed Jupiter's 1st and 4th satellites (Io, Callisto) in conjunction. [p. 344].

February 16 Observed 4th satellite (Callisto) in contact with Jupiter's east limb – several observations of the 4th satellite. [p. 344].

March 5 Observations of Jupiter. Shadow of the 4th satellite (Callisto) on Jupiter. 2nd satellite (Europa) observed emerging from primary. Black and bright spots observed. [p. 344].

March 6 Continued observations of the black and bright spots on Jupiter's surface. [p. 344].

March 9 Observed the bright spot on a belt on Jupiter. Third satellite (Ganymede) emerged from the primary. Observations of the 1st satellite (Io). [p. 344]. Observed 1st satellite touch Jupiter's west limb in apogee. [p. 345]. Observed the 1st satellite emerge out of Jupiter's shadow. [p. 345].

March 10 Observed Jupiter's 1st satellite (Io) several times, 2nd satellite (Europa). [p. 344]. There is a note about versed sines of parallax of Earth's orb. [p. 345] Observed 1st satellite touch Jupiter's east limb in perigee. Other observations of the 1st satellite. Observed shadow of the 1st satellite touch the east edge of Jupiter. Determined the mean motion of the 1st satellite + various notes. [p. 345].

March 11 Observed 1st satellite (Io) emerging from behind Jupiter. [p. 345].

March 12 Observed black spot in middle belt of Jupiter. Observed 2nd satellite (Europa). [p. 345].

March 13 Observed 4th satellite (Callisto) emerging from Jupiter. Good seeing conditions. [p. 345].

March 16 Observed 3rd satellite (Ganymede) emerge from Jupiter. [p. 345].

March 17 Several observations of the 4th satellite (Callisto). The 1st satellite (Io) touched Jupiter's east limb in perigee. [p. 345].

March 18 Observed the 1st satellite (Io) emerging from Jupiter. Good observation. [p. 345].

March 27 Observed 1st satellite (Io) emerge from Jupiter. Observed shadow of the 3rd satellite (Ganymede) touch Jupiter's west limb. [p. 345].

March 30 Using the 123 ft glass donated by Huygens to the Royal Society, Bradley observes the 4th satellite (Callisto) emerge from the shadow of Jupiter. This instrument was used because of its excellent light grasp. Observed near contact of the 2nd satellite (Europa). Bradley used both the Huygens' glass and a 15ft refractor only a minute apart – an indication of Bradley's immense skills with his instruments. All in good seeing conditions. [pp. 345, 346].

Bradley also observed the double star of α Geminorum (*Castor*) nearly parallel to line κ and σ Geminorum to β Geminorum (*Pollux*). [p. 346].

Bradley began observations of *Castor* using Wallis's method to attempt observation of annual parallax. [p. 346]. It failed because the use of a double star in this case was undermined because the star was, in fact, a binary with the stars gravitationally connected in a system.

April 6 Observed 2nd satellite (Europa) emerge from Jupiter. [p. 346].

April 7 Observed 4th satellite (Callisto) – seen over Jupiter's disc appearing very black (as on 16 February 16). Long note concerning the 4th satellite. Measured the diameter of Saturn. Measured the diameter of Saturn's rings. [p. 346].

April 9 Observed the 1st satellite (Io) touch the east limb of Jupiter and various observations of the 1st satellite (Io). [p. 346]. Measured the diameter of Jupiter. Measured the diameter of Saturn and of Saturn's rings. [p. 347].

April 10   Observed the 1st satellite (Io) touch Jupiter's west limb in apogee. The 3rd satellite (Ganymede) touch Jupiter's east limb – described as dark. The 1st and 3rd satellites coming off Jupiter's disc. The shadow of the 3rd satellite coming onto Jupiter. [p. 347].

April 11   Measured the diameter of Saturn. Measured the diameter of Saturn's rings. Observed the 4th satellite of Jupiter (Callisto) reach greatest elongation.

April 14   Observed the 3rd satellite of Jupiter (Ganymede) emerge from the primary. [p. 347].

April 21   Observed the 3rd satellite of Jupiter (Ganymede) both immerse and emerge from the primary. [p. 347].

April 22   Observed the shadow of the 2nd satellite (Europa) touch Jupiter's west limb, then poor seeing conditions. μ Libræ preceded Saturn. Measured the difference in RA. [p. 347].

Bradley's observations not always distinguishable from Pound's. [*Memoirs* p.iv].

April 23   Measured the diameter of Saturn and its rings. [p. 347].

April 27   Measured the 4th satellite of Saturn (Titan) from Saturn's centre. [p. 347] There were five satellites known at the time Bradley observed the satellites of Saturn. These were I Tethys, II Dione, III Rhea, IV Titan, V Iapetus. Titan was discovered by Huygens whilst the other four were discovered by G. D. Cassini.

May 6   Observed 4th satellite of Jupiter (Callisto) reach greatest elongation. [p. 347].

May 7   Measured the greatest and the least diameters of Jupiter. Observed the 3rd satellite (Ganymede) and measured angular distance from both the nearest and furthest limbs of Jupiter. Observed transit of the 3rd satellite over the disc of Jupiter. Measured diameter of Saturn, of Saturn's rings, and of diameter of the inside edge of the rings. Observed the 1st, 2nd, and 5th satellites of Saturn: Tethys, Dione, and Iapetus. [p. 348].

May 8   Observed the 2nd satellite of Jupiter (Europa) emerge from the primary. [p. 348].

May 16   Observed *Regulus* precede Jupiter in RA. [p. 348].

Halley uses Bradley's observations of Jupiter and its satellites. [*Phil. Trans.* Vol. 31, p. 114]. Uses Cassini's methods.

May 18   Observed the 1st satellite (Io) touch Jupiter's east limb. [p. 348].

May 24   Bradley was ordained a deacon by Robinson, Bishop of London. Testimonials signed by Pound as the Rector of Wanstead, Chishall the Vicar of Walthamstow, Chisenhall the Vicar of Barking. [*Memoirs* p. vi].

May 27   Observed the 4th satellite (Callisto) touch Jupiter's east limb. Note on diameter of Jupiter. [p. 348].

May 28   Observed the 4th satellite of Saturn, (Titan) the rings, and the planet. Measured the diameter of Saturn, the rings, and the inner ring. [p. 348].

May 29   Observed the 4th satellite of Saturn (Titan) from near and far ansae. Notes. [p. 348].

June 5   Observed the 4th satellite of Jupiter (Callisto) emerge from the primary while near the horizon and in dull conditions. [p. 349].

June 6   Observed the 4th satellite of Saturn (Titan). Measured distance from rings and planet. Measured diameter of the rings. Notes on the 4th satellite of Jupiter (Callisto). Notes on the 1st and 4th satellites of Jupiter (Io, Callisto) and the diameter of Jupiter. Notes on the mean diameter of Jupiter. Notes on the diameter of Saturn's rings. [p. 349]. Notes on the diameter of the rings. [p. 350]. Notes on the greatest elongation of the 5th satellite (Iapetus) from Saturn. [p. 350].

July 2 A paper of James Bradley's was read at the Royal Society on the irregularities observed in the motions of Jupiter's satellites in the observations of Pound and Bradley. This appears identical to remarks inserted in Halley's Tables. [p. 81].

July 13 Free gift of 50 guineas (£52 10s) received by Pound from Newton. [*Memoirs* p. ii].

July 25 Dr Benjamin Hoadly, Bishop of Hereford, presented James Bradley to priestly orders. [*Memoirs* p. vi].

July 27 Bradley was instituted to the living at Bridstow near Ross on Wye. [*Memoirs* p.vi]. Rigaud asserts that Bradley 'was becoming known and esteemed amongst the scientific community'. [*Memoirs* p.vi].

December 31 John Flamsteed, the first Astronomer Royal, died. He would be succeeded by his hated protagonist Edmond Halley. [*Memoirs* p.xliv].

During this year of 1719, Bradley and Pound deduced the solar parallax to between 9″ and 12″. (The accepted modern value is 8.794″.)

1720 **January** Lord Chancellor Thomas Parker presented the rectory of Burstow in Surrey to James Pound, a living made vacant with the death of John Flamsteed. This in addition to the rectory of Wanstead made Pound a prosperous man. [*Memoirs* p.ii].

At the beginning of 1720 Bradley acquired the part living of Llanddewi felfry in the hundred of Narberth and the County of Pembroke. This was procured by his friend and patron Samuel Molyneux who as the Prince of Wales' private secretary was able to acquire this part-living for his friend.

There is a common misconception in Wanstead that Bradley was a curate at Wanstead. Bradley with livings in Herefordshire and Pembrokeshire was never a curate. Bradley still regularly visited London and Wanstead and this may have led to a belief that he was a curate. [*Memoirs* p.vi].

Edmond Halley succeeds John Flamsteed as the second Astronomer Royal. Flamsteed's executors remove all of his astronomical instruments from the Royal Observatory, leaving Halley with an almost empty building.

Halley replies to Jacques Cassini's paper on the annual aberration of *Sirius* in the *Phil. Trans.* Vol. 31, pp. 1–5.

February 2 The House of Commons accepted the proposals of the South Sea Company for paying off the national debt.

April 28 Newton gives Pound another gift of 50 guineas. [*Memoirs* p.iii].

April 28 Bradley's observations are notable by their absence. This was his first complete year as a cleric. The lack of access to astronomical instruments in Herefordshire in part explained this, though he may have joined his uncle James Pound during his visits to Wanstead. There is no record of any observations.

June 2 The day that the South Sea Bubble burst. It led to the ruin of many in society.

1721 **January 12** There is a note about the time the new contrived micrometer has been completed. Sure evidence that Bradley visited Wanstead.

April 6 There is no notice of when the Huygenian glass was placed in Pound's possession but on this date the Royal Society engaged him to make use of it. Bradley had previously made use of it.

August 31 The Savilian chair in astronomy at Oxford was vacated by the sudden and unexpected death of Dr John Keill, possibly from food poisoning. [*Memoirs* p.iv].

September 1 Lord Chancellor Thomas Parker Fellow of the Royal Society was a friend and patron of James Pound and one of the electors of the chair. [*Memoirs* p.vi] Pound was regarded as 'the fittest man in Europe' as recorded in Archbishop Wake's papers. [*Memoirs* p.vii].

September 1 Pound declined the offer of the Savilian chair. He had two lucrative church preferments and was planning to get married. One of the conditions of the two Savilian chairs was the relinquishment of all church livings.

September 1 Much concern amongst the Oxford based largely Tory interest who largely favoured the Rev. J. Whiteside FRS of ChristChurch, who stood against Keill when the Scottish acolyte of Newton was elected.

September 2 The various expenses paid to secure Bradley's election as Savilian Professor of Astronomy included in Pound's Accounts Book. It reveals the financial costs borne in the acceptance of patronage. [Pound's Accounts Book].

September 4 Martin Foulkes wrote to Archbishop Wake in support of James Bradley's canditure for the Savilian Chair in Astronomy. Wake was Foulkes uncle and the chief of the electors for the post.

October 14 Observations of Mars. Extensively over several hours. Measured difference in RA between Mars and a star. Bradley was residing at Wanstead. [p. 350].

October 15 Further extensive observations of Mars and the measurement of the difference in RA between the planet and the same star. [p. 350].

October 20 Observed and measured the RA of the reference star and observed Mars for several hours. [p. 350].

October 25 Observed another small star and measured its RA before making extensive observations of Mars. [pp. 351, 352, 353].

Bradley observed Mars in opposition. [p. 350]. Multiple observations to improve the Sun's parallax by 1.2″.

October 31 Bradley was elected to the Savilian chair in astronomy by a London based Whig interest including Archbishop Wake and Lord Chancellor Parker. He was accompanied to the House of Lords by Pound.

Rigaud remarks that Bradley was turning his back on a possibly lucrative career in the Church of England. But as Rigaud also remarks, astronomy had become an all consuming passion for Bradley. [*Memoirs* p. viii].

November16 Observed conjunction of Jupiter and Venus. [p. 343].

December 18 Bradley admitted to the office of Savilian Professor of Astronomy at Oxford University.

Halley acquired his 5ft transit instrument at Greenwich. [p. 382].

1722   James Pound married a wealthy heiress Elizabeth Wymondesold, the sister of the financier Matthew Wymondesold, an astute speculator in South Sea Company stock, which he sold just before the financial crash in 1720. [*Memoirs* p. ii].

George Graham made observations on the diurnal variation in his magnetic studies. [*Memoirs* p. lxxvi] [*Phil. Trans.* Vol. 33, p. 96].

George Parker (the future 2nd Earl of Macclesfield and future President of the Royal Society) was elected Fellow of the Royal Society. He was to become Bradley's most important patron.

April 26 Bradley gave his inaugural lecture as Savilian Professor of Astronomy. According to Hearne, who was a supporter of Whiteside, it did not add to his reputation. Rigaud suggests that Hearne was 'no judge of the subject'. [*Memoirs* p. ix].

May 21 James Pound visited Samuel Molyneux at his residence in Kew. [*Memoirs* p. xviii].

June 11 Observed the 2nd satellite of Jupiter (Europa) emerge from the planet using both the 123ft Huygens glass and the 15ft tube – seen distinctly. [p. 354].

June 14 Observed the shadow of the 3rd satellite (Ganymede) over Jupiter. Also the 4th satellite (Callisto) in conjunction. Observed the 1st satellite (Io) emerge from behind the planet using the 123ft glass and Hadley's 6ft Newtonian reflector at the same time. Observed exactly. Observed the 3rd satellite (Ganymede) in contact with Jupiter's limb. [p. 354].

June 19 Observed the greatest diameter of Saturn's rings. [p. 354].

July 7 Observed the 1st satellite of Jupiter (Io) emerge from the planet in the 6ft Newtonian reflector and 18sec later in the 15ft tube. [p. 354].

July 12 Observed the difference in RA between Jupiter and κ Libræ observed using a chronometer. Also determined the declination. [p. 354].

July 23 Observed the 1st satellite (Io) emerge from Jupiter with the 15ft tube and concurrently using the 6ft Newtonian reflector. [p. 354].

August 21 Observed the difference in both the RA and the declination of Jupiter and β Scorpii using the 15ft tube. [p. 354].

October 1 Observed the line through the double star α Geminorum (*Castor*) parallel to the line through β Geminorum (*Pollux*) and κ Geminorum. Bradley was attempting to use Wallis's method of deriving annual parallax. [p. 354]. It failed for reasons given above. The star is a binary.

October 7 Observed the difference in RA between Jupiter and ω Scorpii. [p. 354].

November 27 Samuel Molyneux communicated with James Pound and possibly James Bradley on his observations of a solar eclipse. [p. 391].

December 27 Observed angular distance between the horns of Venus using 212¼ft glass. [p. 354].

1723 John Hadley presented a reflecting telescope to the Royal Society. [*Phil. Trans.* Vol. 32, p. 303] [*Memoirs* p. ix].

Pound and Bradley compared the reflector's efficacy with Huygens' 123ft focal length refractor. Pound asserts that although the focal length of the object-metal was not quite 5½ feet, it had the same magnifying power as the glass. [*Memoirs* p. ix] [*Phil. Trans.* Vol. 32, p. 382]. It was only inferior in terms of its 'light-grasp'. Bradley applied himself to the construction of mirrors. [Smith's *Optics*, Section 782]. [*Memoirs* p. ix]. Bradley corresponded with Dr Smith. [p. 401]. Bradley too busy to return to it. Bradley's skills were furthermore confined to the development of his outstanding observational skills.

April 14 Observed the distance of Jupiter from two stars in the 'eye' of Sagittarius. [p. 355].

April 16 Observed the 2nd satellite of Jupiter (Europa) disappear. Used the 6ft Newtonian reflector. Air hazy. Jupiter low but the observation accurate to within a minute (in time). [p. 354].

April 23 Observed the 1st satellite of Jupiter (Io) using the 6ft reflector. Also observed the 2nd satellite (Europa). [p. 355].

June 1 Observed the 1st and 4th satellites (Io, Callisto) of Jupiter disappear. [p. 355]. Observed the 2nd, 3rd, and 4th satellites of Saturn (Dione, Rhea, Titan) with the 6ft Newtonian reflector. Moon bright. [p. 355].

July 7 Observed the 2nd satellite of Jupiter (Europa) emerge. [p. 355].

September 2 Observed the 2nd satellite of Jupiter (Europa) emerge. Used the 15ft tube. [p. 355].

October 9 Halley observed small comet at Greenwich and informed Bradley in Wanstead. [p. 46].

October 10 Bradley observed the small comet mentioned by Halley. [p. 46].

October 11   Pound travelled to Richmond, a residence of Samuel Molyneux. [Pound's Accounts Book] [*Memoirs* p. xviii].

October 12   Bradley and Pound observed the recently discovered comet. [p. 46].

October 13   Bradley and Pound observed the comet. [p. 46].

October 14   Bradley observed the comet. [p. 46].

October 15   Observed the comet. [p. 46].

October 21   Observed the 1st satellite of Jupiter (Io) emerge. Used the 15ft tube. [p. 355]. Observed the small comet first seen by Halley. [p. 46].

October 22   Bradley observed the comet. [p. 46].

October 24   Observed the comet. [p. 46].

October 29   Observed a transit of Mercury. Made several observations. measured the diameter of Mercury. [p. 355]. Pound's and Bradley's notes on the transit of Mercury referred to by Halley. [*Phil. Trans.* Vol. 33, p. 229] [p. 355]. Also observed the comet. [p. 46].

October 30   Observed the 4th satellite of Jupiter (Callisto) disappear. Used the 15ft tube. Twilight strong. [p. 355]. Observed the comet. [p. 46].

November 5   Bradley observed the small comet. [p. 46].

November 8   Observed the small comet. [p. 46].

November 14   Observed the comet. [p. 46]. Bradley then returned to Oxford. [*Memoirs* p. x].

November 20   Pound observed the comet with both George Graham and Edmond Halley in attendance. [p. 46].

November 28   James Bradley delivered a lecture *De Cometa* at Oxford detailing his observations of the comet in Wanstead. [*Memoirs* p. x].

December 3   Whilst visiting friends or relatives in Cirencester, Bradley observes the comet which was faint and dull. [p. 46].

December 7   Now returned to Wanstead Bradley observed the comet for the final time. See Table. [p. 46]. Bradley both combined his observations and calculated the elements of the comet's path. These calculations together with the original observations form Bradley's first contribution to the *Phil. Trans.* [p. 42]. [*Memoirs* p.x]. It contains the comet's parabolic path calculated using Newton's method.

1724   **March 19**   Observed the planet Jupiter at the same declination as 19 Capricornii in the British Catalogue, preceding the star in RA by 1′ 6½″ by the clock. [p. 355]. This would have referred to the edition of Flamsteed's observations edited by Halley but without Flamsteed's permission.

June 25   Observed Saturn and measured distance from a small star using the 15ft tube, and distance from the star using the 123ft glass. [p. 355]. With notes. [p. 356].

July 24   Observed Jupiter's 4th satellite (Callisto) emerge using the Newtonian reflector. [p. 356].

September 7   Observed Jupiter's 1st satellite (Io) emerge using the reflector. [p. 356].

September 27   Observed Jupiter's 2nd satellite (Europa) as it began to emerge using the reflector. [p. 356].

November 9   James Pound had business in London.

November 16   James Pound died suddenly and unexpectedly intestate (his marriage to Elizabeth, his second wife in 1722, nullified all prior Wills). His last entry in his Accounts Book was on 9 November for expenses for 'journey to town' [*Memoirs* p. xi]. His widow administered his effects. Bradley was left bereft of all resources beyond his modest income from the Savilian chair. [*Memoirs* p. xi]. The possession of the parsonage was soon relinquished.

Bradley remained part of Elizabeth's household and with the exception of the 123ft glass, which was returned to the Royal Society, Bradley retained all of Pound's astronomical instruments.

1725   Margaret Flamsteed published the *Historia Coelestis Britannica*, the authorized catalogue of her husband's observations at Greenwich. Edmond Halley took possession of George Graham's 8ft quadrant during this year.

Early Samuel Molyneux commissioned George Graham to construct a 'parallax instrument' to repeat Robert Hooke's observations in 1669 during his attempt to observe the annual parallax of a star. [*Memoirs* p. xiii]. It was constructed 'on the same principles' as Hooke's instrument of 1669. [p. 1].

July 24 Observed 1st satellite of Jupiter (Io) disappear using the Hadley 6ft reflector. [p. 356].

July 28 NS The immersion of satellite I (Io) observed at Wanstead was at 12hr 50min 00sec. At Lisbon this immersion was observed at 12hr 12min 26sec apparent time.

August 4 Using the Hadley reflector Bradley observed the immersion of Jupiter's satellite I (Io) some 45sec after the time calculated by his tables.

August 18 Observed 1st satellite of Jupiter (Io) disappear using an 8ft reflector at Kew. [p. 356].

August 27 Observed 2nd satellite of Jupiter (Europa) emerge from the limb of Jupiter in reflector. Notes. [p. 356].

August 29 The immersion of satellite I (Io) was 1min 55sec earlier than the time in Bradley's tables at 12hr 48min 45sec apparent time.

September 16 Observed 4th satellite of Jupiter (Callisto) disappear in Hadley's 6ft reflector. Later emerge in the reflector. Notes. [p. 356].

September 17 Observed 1st satellite of Jupiter (Io) emerge in Hadley's 6ft reflector. [p. 356]. Notes.

October 3 Observed 1st satellite of Jupiter (Io) emerge in the reflector. [p. 357].

October 5 Observed 1st and 2nd satellites of Jupiter (Io and Europa) in the reflector. [p. 357].

October 10 Observed a total eclipse of the Moon. Cloudy – hindered until after emersion from total darkness. Observed the enlightened part in the 15ft reflector + other observations. [p. 357].

October 14 NS After making observations of emersion of satellite I (Io) at Wanstead and Lisbon on this date and also on October 16 NS it led to an apparent difference of 36m 20s.

October 16 NS The observations of the emersion of satellite I (Io) at Wanstead and Lisbon revealed an apparent difference of 36m 20s.

October 23 Observed the 2nd satellite of Jupiter (Europa) emerge from the planet in the 15ft tube. Observed through thin cloud and saw it ½ minute late. [p. 357].

November 26 The new 'parallax instrument' (zenith sector) commissioned by Samuel Molyneux and made under the direction of George Graham was suspended in the White House at Kew with James Bradley in attendance. [p. 99]. [*Memoirs* p. xv].

November 27 A chair or couch for the use of the zenith sector was set up for the observer. [p. 100]. Molyneux observed a star brought to the cross-hairs. [p. 116]. Back at Wanstead Bradley observed the 1st satellite of Jupiter (Io) emerge using the 15ft tube in hazy conditions.

November 28 Molyneux put back the movable clock at Kew. [p. 117].

November 29 Molyneux returned to his town house in London 'for the winter'. [p. 118].

December 2   Came from London with Graham in order to make adjustments to the sector. [pp. 118–119]. Observations made with the zenith sector at Kew. [p. 104]. Cloudy conditions.

December 3   A clock was fixed up for the sector. [p. 105]. It was a rainy, windy night. The sector not affected. Graham returned to London. Molyneux made various trial observations. Molyneux makes a careful note of the declination of the object star γ Draconis in its passage through its transit. [*Memoirs* p. xvi].

December 4   Molyneux made various tests of the clock, glass and plumbline, etc. [p. 121].

December 5   Adjusting the Kew sector. [p. 105]. Molyneux worked on the sector with his servant (possibly Giles). [p. 123]. Molyneux made a further check of the position of γ Draconis as it passed through his meridian. [*Memoirs* p. xvi].

December 10   Work setting up the sector was interrupted by a storm passing through Kew. [p. 107].

December 11   Cross-hairs lit up – made for the sector by George Graham. [p. 107] Graham adjusts the bob, etc. [p. 126]. Molyneux once again checked the passage of γ Draconis as it passed through his meridian. [*Memoirs* p. xvi].

December 12   A rainy night, so Graham examined the zenith sector and the glass. [p. 108]. Molyneux observed a star with Graham. Then he left for London. [p. 128]. Molyneux once again checked the passage of the object star through his meridian to check the declination. There still did not appear to be any motion – nor was any motion expected. [*Memoirs* p.xvi].

December 15   At 22h.15′ (10.15am) Bradley determined the Sun's azimuth was 26° 8′ 10″ [*Memoirs* p. xxiv].

December 17   The sector was set up for observations with Bradley in attendance. [p. 109]. Molyneux observed Sun's transit and several stars with Bradley. [p. 129]. Bradley repeated observation of the passage of the object star γ Draconis as it passed through the Kew meridian. He perceived that the star was passing a little more southerly (even though no such motion was expected). [p. 2] [*Memoirs* p. xvi]. Bradley carefully examined the instrument to determine whether the observed motion could be explained by systemic or instrumental error. [pp. 109, 110] [*Memoirs* p. xvi].

December 18   Bradley observes several telescopic stars. [p. 110]. Molyneux and Bradley continued observing several stars. [p. 131].

December 19   A dark and cloudy day. Bradley observed Cap. Persei. [p. 132].

December 20   A day of adjustments to the sector. [p. 133].

December 21   Bradley examined and adjusted the telescope by the plumbline. [p. 114]. Bradley observed Cap. Draconis. There is a reproduction of Bradley's note. [p. 134].

December 24   Cloudy morning. Observations made of Cap. Persei. [p. 135].

December 25   Another cloudy morning. Observations of Cap. Persei. [p. 135].

December 27   Hazy night – vapours. Observed Cap Persei. [p. 135].

December 28   Adjustments made to the Index. [p. 136].

December 29   Molyneux made five trials or tests of the sector. [p. 137].

December 30   Weather cloudy, calm evening. Plumbline steady in 2min. [p. 137].

1726   **Early**   Twelve copies of the *Principia* privately printed on large thick paper bound with gilt leaves in red morocco to a pattern used in the Harleian Library were printed. One was presented by Sir Isaac Newton to James Bradley in person. [*Memoirs* p. xi]. Bradley communicated his earliest major paper to the Royal Society. It was a remarkable achievement determining the longitudes of Lisbon and New York from eclipses of Jupiter's 1st satellite (Io) [p. 58] [*Memoirs* p.xi]. The calculations were made from observations that were not simultaneous. [*Memoirs* p. xi].

Bradley elected a member of the Council of the Royal Society. [*Memoirs* p. lxxi].

January The partners began observing 35 Camelopardalis as a control star, some 12hr in RA difference from the object star γ Draconis.

January 3 The zenith sector was carefully examined by George Graham. Index 5.7–17h 48′ 42″ Cap. Draconis. Much snow. Mr Graham observed the star well bisected. Cap. Persei. [p. 139].

January 5 Bradley observes both the 1st and the 3rd satellites of Jupiter (Io & Ganymede) emerge from the planet with the 15ft tube. [p. 357].

January 7 35 Camelopardalis located as a control star at Kew. [*Memoirs* p. xix]. Graham made observations of Cap. Persei. Hard frost since the 3rd of January. Further observations were made 3, 12, 20 February. Its motion took it 5 arc seconds north, between 3 January and 13 February the motion of the object star γ Draconis was 9.1 arcseconds. By the end of February the hypothesis of an annual nutation led to a null result.

Bradley observed an occultation of Mars by the Moon at Wanstead. Observed for over 2hr. [p. 357].

January 9 Samuel Molyneux went to town. [p. 142].

January 15 Bradley observed Mars precede ξ Cancri. Determined the differences in RA and declination. [p. 357].

Molyneux observed Cap. Persei. Looked for 'Telescopica in Auriga'. This was Molyneux's term for Bradley's 'Anti-Draco' which was later identified as 35 Camelopardalis. [p. 142]. This star is now designated as HR2123.

February 2 Molyneux working with his servant Giles. Notes. [p. 143].

February 3 The position of 'Anti-Draco' Bradley's term for 35 Camelopardalis (only later *clearly identified* as such decades later) was checked and recorded. [*Memoirs* p. xix].

Molyneux observed 'Telescopica in Auriga'. There is a comment by Molyneux about his servant (unidentified) allowing the clock to stop. [p. 144].

February 4 Library clock (at Kew) 2′ 28″ slow. [p. 145].

February 12 The position of 35 Camelopardalis was clearly recorded. [*Memoirs* p. xix]. Giles found that the library clock was 1′ 16″ slow. It gains 10″ a day. [p. 145].

February 13 Molyneux observed Cap. Draconis, also Cap. Persei and 'Telescopica in Auriga'. [p. 146].

February 20 Once again the position of 35 Camelopardalis was located. [*Memoirs* p. xix]. γ Draconis was 9.1″ to the south of its initial position. [*Memoirs* p. xix]. [p. 148]. The hypothesis of an annual nutation had to be abandoned before the end of February 1726. Molyneux's approach could not have been later than some time in mid February. His actions betray an eagerness to discomfort Newton.

March The motion southward of γ Draconis initially detected in December 1725 ceased at this time when it was 20″ south. [*Memoirs* p. xviii].

March 5 Molyneux wrote that 'Mr Bradley here'. The sidereal clock 1′ fast. [p. 149]. The southward motion of γ Draconis came to a halt from this time, by April 1 the star was discerned to be moving northwards.

March 6 Bradley rose early. He adjusted the instrument at 5.30am. Observed γ Draconis. [p. 149].

March 19 Molyneux and his servant Giles determined that the clock was 3′ 37″ too fast. [p. 149].

March 20 Molyneux observed the control star 'Telescopica in Auriga' (35 Camelopardalis) etc. Now designated as HR2123. The star is now located in Auriga. [p. 150].

March 21   Molyneux rose early and observed γ Draconis. [p. 151].

March 22   Graham made various observations at Kew. [p. 152].

March 26   Bradley and John Hadley observed 35 Camelopardalis and γ Draconis. [p. 153].

March 27   Observed 35 Camelopardalis. [p. 153].

April   The object star γ Draconis was now definitely moving northwards. [*Memoirs* p. xviii].

April 2   Molyneux and Bradley assisted by servant Giles make observations with the zenith sector. [p. 154].

April 6   Molyneux observed γ Draconis. [p. 154].

April 7   Molyneux observed γ Draconis. [p. 155].

April 12   Molyneux made various observations with the zenith sector at Kew. [p. 155].

April 14   Molyneux observed γ Draconis etc. at Kew. [p. 156].

April 18   Molyneux observed γ Draconis. [p. 157].

April 22   Molyneux observed γ Draconis. [p. 157].

April 23   Observations and notes. Graham observed a small star. Observations of Herculis. Observations of γ Draconis. Notes. [p. 159] [Kew Observation Book K14].

April 24   Observed γ Draconis and other stars. Notes. [p. 161]. [Kew Observation Book K14].

April 29   Molyneux came to Kew with William Whiston. Observed γ Draconis. Notes. [p. 162]. [Kew Observation Book K14].

May 5   Tests were made of the zenith sector. [p. 162]. [Kew Observation Book K14].

May 11   Molyneux observed γ Draconis. [p. 163]. [Kew Observation Book K14].

May 12   Halley reported that the money advanced by the Treasury had been expended in the acquisition of the 8ft quadrant and other instruments. The RGO still required another 8ft quadrant but did not possess the resources to acquire one. [*Memoirs* p. lxxiii].

May 14/15   Observed γ Draconis. [p. 164] [Kew Observation Book K14].

May 23/24   Observed γ Draconis. [p. 164] [Kew Observation Book K14].

May 24   Observed γ Draconis. [p. 164] [Kew Observation Book K14].

May 25   Observed γ Draconis. [p. 164] [Kew Observation Book K14].

May 27   Observed γ Draconis. [p. 164] [Kew Observation Book K14].

May 29   Observed γ Draconis. [p. 164] [Kew Observation Book K14].

May 30   Plumbline broken. Notes about the plumbline. [pp. 165–166].

May 31   Graham came from London to Kew to attach a new thicker plumbline. High winds. Observation made of γ Draconis. [p. 167].

June   A report by the Royal Society led to Mr Conduitt commissioning the Master-General of Ordnance. Even so the Duke of Argyle granted Halley no more resources, who was therefore unable to purchase the needed second 8ft quadrant. [*Memoirs* p. lxxiii].

June   γ Draconis has now returned to the position it was first observed in December 1725, though now moving northwards. [*Memoirs* p. xix].

June 1   Molyneux observed γ Draconis. [p. 167].

June 6   Molyneux observed γ Draconis. [p. 167].

June 7   Molyneux observed γ Draconis. [p. 168].

June 8   Bradley came down to Kew with a heavy plummet. Observed γ Draconis. [p. 168].

June 9   Molyneux observed γ Draconis. [p. 169].

June 10 Molyneux observed γ Draconis. Hot day. [p. 169].

June 11 Molyneux observed γ Draconis. [p. 170].

June 12 Molyneux observed γ Draconis. [p. 170].

June 13 Molyneux observed γ Draconis. [p. 170].

June 14 Molyneux observed γ Draconis. [p. 171].

June 16 Molyneux observed γ Draconis. [p. 171].

June 19 Molyneux observed γ Draconis. [p. 171].

June 25 Molyneux observed γ Draconis. [p. 172].

June 27 Molyneux observed γ Draconis. [p. 172].

June 29 Molyneux observed γ Draconis. [p. 172].

June 30 Molyneux observed γ Draconis. [p. 172].

July Bradley and Graham adjusted the 8ft quadrant at Greenwich on behalf of Halley.

July 1 Molyneux observed γ Draconis. [p. 173].

July 2 Molyneux observed γ Draconis. Notes. [p. 173].

July 3 Molyneux observed γ Draconis. Note about Giles. Note by James Bradley. [p. 174].

July 6 Bradley and Molyneux observed γ Draconis. [p. 174].

July 8 Molyneux observed γ Draconis. [p. 174].

July 9 Molyneux observed γ Draconis. [p. 174].

July 10 Molyneux observed a star. [p. 175].

July 12 Molyneux observed a star. [p. 175].

July 15 Molyneux observed a star. [p. 175].

July 18 Molyneux observed a star. [p. 175].

August 5 The 1st satellite of Jupiter (Io) disappeared in Hadley's 6ft reflector. At Wanstead. Conditions serene – very good. [p. 358].

August 5 Molyneux observed γ Draconis at Kew. [p. 176].

August 9 Molyneux observed γ Draconis. [p. 176].

August 14 The 1st satellite of Jupiter (Io) disappeared in Hadley's 6ft reflector. Very good conditions. [p. 358].

August 15 The 3rd satellite of Jupiter (Ganymede) emerged using 14ft tube at Greenwich. Good conditions. [p. 358].

August 17 Molyneux observed γ Draconis. [p. 176].

August 22 Observed the 3rd satellite of Jupiter (Ganymede) disappear and emerge using Hadley's 6ft reflector. Clear and distinct. [p. 358].

August 25 Molyneux and Bradley observed γ Draconis. [p. 176].

August 26 Molyneux observed γ Draconis. [p. 177].

August 28 Observed 1st satellite of Jupiter (Io) disappear from the planet using the 6ft reflector in very good seeing conditions. Also observed the 2nd satellite (Europa) disappear. [p. 358].

Molyneux observed γ Draconis. [p. 177].

August 29 Observed the 3rd satellite of Jupiter (Ganymede) disappear using the 6ft reflector in the twilight. [p. 358].

Molyneux observed γ Draconis. [p. 177].

September The object star now 39″ north of its most southerly position in March. It begins to move south again. [*Memoirs* p. xix].

September 1 Bradley and Graham went to Greenwich to adjust the meridian telescope for Halley.

September 2 Jupiter preceded σ Piscium by 50.5″ in time and by 12′ 40″ in RA. Observed in the 15ft tube at Wanstead. [p. 358].
Molyneux observed γ Draconis. [p. 178].

September 5 Molyneux observed γ Draconis. [p. 178].

September 7 Molyneux observed γ Draconis. [p. 178].

September 8 Molyneux observed γ Draconis. [p. 178].

September 9 Molyneux observed γ Draconis. [p. 178].

September 11 Observed the 2nd satellite of Jupiter (Europa) disappear though the actual moment of immersion not seen due to clouds. [p. 358]. Molyneux observed γ Draconis. [p. 179].

September 12 Jupiter preceded star by 1.25″ in time or 20″ in RA. [p. 358].

September 13 Observed Jupiter precede the same star observed the previous night by 26.5″ in time and 3′ 47″. [p. 358]. Observed 1st satellite (Io) disappear from planet. Seen distinctly in the 6ft reflector. [p. 358].

September 14 Observed an Eclipse of the Sun at Wanstead using the 15ft tube. Measured the diameter of the Sun and corrected his earlier estimates. Bradley compared this with the data in Halley's Tables and other observations of the Sun's diameter. Timed the beginning of the first contact of the Moon's and Sun's limbs at the beginning of the eclipse. Description and timings of the eclipse. Observed for 45min before clouds and trees prevented any further observations. [p. 359]. Molyneux observed γ Draconis. [p. 179].

September 15 Bradley observed the 1st satellite of Jupiter (Io) disappear using the 6ft reflector. Jupiter was low in the sky though seeing conditions were good. [p. 359]. Molyneux observed γ Draconis. [p. 179].

September 18 Molyneux observed γ Draconis. [p. 179].

September 25 Molyneux observed γ Draconis. [p. 180].

September 27 Molyneux observed γ Draconis. [p. 180].

September 28 Bradley visited Kew and observed both 35 Camelopardalis and γ Draconis. [p. 180].

September 30 Bradley observed the Sun's altitude and diameter. Note about the clock. [p. 359].

October 1 Bradley observed and calibrated the Sun's diameter at Wanstead. [pp. 359–360].
Molyneux observed γ Draconis. [p. 181].

October 7 A short notice or paper on the effects of refraction was written by Bradley. [*Memoirs* p. xix].

October 8 Observed Jupiter's 1st satellite (Io) emerge in the 6ft reflector. Only fairly accurate. [p. 360].

October 11 Molyneux observed γ Draconis. [p. 181]. From this date the object star was discerned to be moving southwards again.

October 17 Molyneux observed γ Draconis. [p. 181].

October 26 Molyneux observed γ Draconis. [p. 181].

November 13 The sidereal clock at Kew was wound up. [p. 181].

November 20 Molyneux observed γ Draconis. [p. 182].

November 23 Bradley observed Jupiter's 1st satellite (Io) emerge using the 15ft tube. Bradley judged that the true emergence began 25″ or 30″ sooner. [p. 360].

November 25 Bradley observed Jupiter's 1st satellite (Io) emerge from the planet using the 15ft tube. Good conditions – some flying clouds. Observed 2nd satellite (Europa). Clouds. [p. 360].

November 26 Bradley at Kew who observed 35 Camelopardalis but due to clouds was unable to observe the passage of γ Draconis as it passed through the meridian. [p. 182].

November 27 Bradley remained in Kew for the whole weekend and observed α Persei.

December Bradley and Graham rectified the new quadrant made for Halley at Greenwich.

December 2 Memorandum at the end of the first year's observations of γ Draconis. The existence of the motion had, in the minds of the partners, been established. [*Memoirs* p. xxii].

Observed the 1st satellite of Jupiter (Io) emerge from the planet exactly in the 15ft tube. Also observed the emersion of the 2nd satellite (Europa) in the 15ft tube. Jupiter appeared distinct. [p. 360].

December 13 Bradley wrote a memorandum concerning the clock which had stopped. [p. 183].

December 15 Observed Jupiter's 3rd satellite (Ganymede) emerge from the planet in the 15ft tube. Very good results. [p. 360].

December 17 Bradley travelled to Kew, but no observations were made. [p. 183]. It was due to several instances like this that Bradley resolved to acquire his own sector.

December 25 Observed the 1st satellite (Io) emerge in the 15ft tube. Observation was very exact. [p. 360].

Molyneux observed α Persei. [p. 183].

December 26 Molyneux observed γ Draconis and compared with the position the previous year of 1725. [p. 183]. Molyneux observed 35 Camelopardalis. [p. 184].

December 31 Molyneux observed 35 Camelopardalis. [p. 184].

1727   **Early** Bradley began making plans for a new zenith sector of his own based in Wanstead. This was motivated by the need to observe as many stars as possible in order to ascertain whether the motion observed in γ Draconis was visible in other stars. [p. 194]. [*Memoirs* p. xxiii].

January 1 Bradley observed the 1st satellite of Jupiter (Io) very exactly in the 15ft tube. [p. 360]. Bradley also made nine measures of the Sun's diameter. [p. 360].

January 3 Bradley observed the 2nd satellite of Jupiter (Europa) emerge in the 15ft tube. Conditions were clear but low towards the horizon. [p. 361].

January 10 Bradley observed 1st satellite of Jupiter (Io) emerge. [p. 361].

January 15 Molyneux observed 35 Camelopardalis at Kew. [p. 184].

January 29 Molyneux observed 35 Camelopardalis or 'Telescopica in Ursæ Majoris'. [p. 184].

February 5 Molyneux observed 35 Camelopardalis. Unable to observe γ Draconis due to clouds. Molyneux wrote a memo to speak to Mr Graham. [p. 185].

February 9 Bradley observed Jupiter's 1st satellite (Io) emerge in the 15ft tube. Seeing very good. [p. 361].

February 11 Bradley at Kew – observed 35 Camelopardalis. [p. 186].

February 16 Nebula 4° south of *Sirius* was observed by Bradley. Notes about other stars. [p. 361].

February 19 Molyneux observed γ Draconis. [p. 186].

February 26 Bradley at Kew and observed 'Telescopica in Ursæ Majoris'. [p. 186].

March 11   Bradley wrote an entry on the azimuth of the chimneys with respect to which the new zenith sector was to be erected. At this time, Bradley's means were very limited. In 1724, his professorship brought £138 5s 9d. It is not known if he had any other source of income. He may have been supported by Elizabeth Pound or conversely George Graham, a generous man, may have deferred payment.

March 28   Bradley wrote a memorandum respecting the instrument at Wanstead. [p. 194].

April 15   Bradley measured the Sun's diameter in the 15ft tube. [p. 361].

May 25   Bradley cleaned the 'parallax instrument' (zenith sector). Tests. [p. 186].

May 28   Examined the instrument and observed γ Draconis. Clock was 6' slow. [p. 187].

June 6   Molyneux observed γ Draconis. [p. 187].

June 7   Molyneux altered apparatus and observed γ Draconis. [p. 188].

June 9   Molyneux observed γ Draconis. [p. 188].

June 12   Molyneux observed γ Draconis. [p. 188].

July 2 Bradley   observed Jupiter's 1st satellite (Io) disappear in the 15ft tube. Good. [p. 361].

July 19   Molyneux observed γ Draconis. [p. 189].

July 20   Molyneux observed γ Draconis. [p. 189].

July 31   Bradley at Kew. Observed γ Draconis. [p. 189].

August   Bradley selected 70 out of over 200 observable stars passing within 6¼° of the zenith. [*Memoirs* p. xxviii].

August 1   Bradley observed γ Draconis at Kew. [p. 190].

August 2   Samuel Molyneux was appointed as one of the Lords of the Admiralty to be addressed henceforth as the Honourable. [*Memoirs* p. xxviii].

August 4   Bradley observed Jupiter's 2nd satellite (Europa) disappear in the 15ft tube. Very good. [p. 361].

August 6   Molyneux observed γ Draconis. [p. 191].

August 9   Bradley's paper on the longitudes of Lisbon and New York determined from London and Wanstead was published. [*Phil. Trans.* No. 394, Vol. 34, p. 85].

August 10   Observed Jupiter's 1st satellite (Io) disappear in the 15ft tube. [p. 361].

August 19   Note by James Bradley stating 'My instrument set up' at Wanstead accompanied by Molyneux and Graham. [p. 191]. Bradley observed γ Draconis at Wanstead. The very first observation with the Wanstead zenith sector. [p. 203]. There is no record of Molyneux visiting Wanstead between 19 August 1727 and his death on 13 April 1728. [*Memoirs* p. xxix]. Bradley was occasionally accompanied by various visitors including Lord C. Cavendish [p. 237], Dr Benjamin Hoadly (probably the son of Bishop Hoadly) [pp. 254, 257], Edmond Halley [pp. 208, 248].

August 20   Bradley observed γ, β Draconis with the Wanstead sector. [p. 203].

August 21   Bradley observed γ, β Draconis with the Wanstead sector. [p. 203].

August 22   Molyneux observed γ Draconis at Kew. [p. 191]. Bradley observed *Capella*, β, γ Draconis, χ, θ, 1π Cygni, ξ Cephi, β Cassiopeia, φ, θ, τ, γ, α, δ Persei with the Wanstead sector. [p. 204].

August 23   Bradley observed *Capella*, β, γ Draconis with the Wanstead sector. There is a note about future entries. [p. 205].

August 25   Bradley observed *Capella*, β, γ Draconis, β, π Cygni, ξ Cephei, 5 Lacerti. [p. 206].

August 26   Bradley observed *Capella*. Notes. [p. 206].

August 27   Bradley observed γ Draconis. [p. 207].

August 30   Bradley observed β, γ, d, 138 Draconis, χ, θ, 20, 1f, 1π Cygni, 3, 7 Lacerti, 3 Andromedæ, τ, β, λ, α, π, θ Cassiopeæ, φ, 4 Persei before the lantern broke. [p. 207].

August 31   Bradley observed *Capella*, γ Draconis. [p. 207].

September 1   Bradley was joined by Edmond Halley at Wanstead. Due to clouds unable to observe β, γ or 138 Draconis. Then Halley observed θ, 2o Cygni. [p. 208].

September 2   Halley observed *Capella* – see note. Then Bradley observed β, γ Draconis. [p. 208].

September 3   Bradley observed an eclipse of the Sun (observed only in part). Observed when the eclipse began for about 33min before clouds intervened to end the observations. [p. 361]. Bradley observed β, γ Draconis. [p. 209].

September 3   Molyneux observed γ Draconis at Kew. [p. 191].

September 4   Bradley observed [mani?]. In 15ft tube at Wanstead. [p. 361]. Bradley observed *Capella*, β, γ Draconis, 2o, 3ω Cygni, 3 Andromedæ, τ, β, λ, α Cassiopeii. [p. 209].

September 5   Bradley observed *Capella*.at Wanstead. [p. 209]. Molyneux observed γ Draconis at Kew. [p. 191].

September 7   Bradley observed *Capella*, β, γ Draconis at Wanstead. [p. 210].

September 8   Bradley observed *Capella*, σ, β, λ, α, 2υ, θ, 34 Cassiopeii, φ Persei at Wanstead. Then clouds. [p. 210].

September 9   Bradley observed *Capella*, 18 Camelopardalis, γ, d, c, 136 Draconis, χ, ι, θ, ψ, 2o, 1f, 1π Cygni, μ, ξ Cephei, 3, 9 Lacerti, 3, 8 Andromedæ, τ, β, λ, α, θ Cassiopeii, ξ Andromedæ, φ Persei. [p. 211].

September 10   Bradley observed *Capella* then clouds. [p. 211].

September 11   Bradley adjusted the plumbline and index of the zenith sector at Wanstead. [p. 211]. Bradley observed θ, 2o, 3ω, 1f, 1π Cygni, ζ Cephei, 3, 9 Lacerti, β, λ, α, 2υ, θ Cassiopeii, ξ Andromedæ, φ, 4, h Persei, 65 Andromedæ, θ, τ Persei. [p. 212]. He began observing 35 Camelopardalis at Wanstead. [*Memoirs* p. xxx]. Bradley began work on his 'sine hypothesis' [p. 6] [*Memoirs* p. xxx].

September 12   Bradley observed *Capella*. [p. 212].

September 14   Bradley observed *Capella*, 18, 27, 35 Camelopardalis, δ, 46 Aurigæ, η Ursæ Majoris, γ, c, 136 Draconis, χ, ι, ψ, 2o, 3ω, 1f Cygni, ζ Cephei, 3, 9 Lacertæ. [p. 213].

September 15   Bradley observed β, γ, ε Ursæ Majoris, Note about adjustment, β, γ, d, τ, 136 Draconis, χ, ι, θ, ψ, 2o, 3ω, 1f, g, 1π Cygni, ζ Cephei, 7 Andromedæ, τ, σ, β, λ, α, 2υ, θ Cassiopiæ, φ Persei and then clouds. [p. 214].

September 16   Bradley observed β, λ, α, π, 2υ, θ Cassiopeæ, ζ Andromedæ, φ, 4, h, θ, τ, γ Persei, 65 Andromedæ, then clouds. [p. 215].

September 19   Bradley observed γ, d, c, 136 Draconis, χ, ι, θ Cygni, τ, σ, β, λ, α, 2υ, θ Cassiopeæ, ζ Andromedæ, φ, 4 Persei. [p. 215].

September 20   Bradley rotated the zenith sector. The instrument was entirely reversible.

September 22   Bradley observed γ Ursæ Majoris. [p. 216].

September 23   Bradley observed β, γ, d, c, 136 Draconis, χ, θ, ψ, 2o, 3ω, 1f, g, 1π Cygni, μ, ζ Cephei, *Capella*, 18, 35 Camelopardalis, δ, 46 Aurigæ, 13 Lyncis. [p. 216].

September 24   Bradley observed γ, ε, ζ Ursæ Majoris, β, γ Draconis. [p. 216].

September 25   Bradley observed η Ursæ Majoris, γ Draconis. [p. 216].

September 27   Bradley observed γ, c, 136 Draconis, χ, θ, ψ, 2o, 3ω, 1f, g, 1π Cygni, ζ Cephei, τ, β Cassiopeii, *Capella*, 18, 35 Camelopardalis, δ, 46 Aurigæ, 13 Lyncis. [p. 217].

Molyneux observed γ Draconis at Kew. [p. 192].

September 29   Bradley observed γ Draconis, χ Cygni, β Cassiopeii. [p. 217].

October 1   Bradley cleaned the tube etc. [p. 192].

October 8   Bradley observed γ Draconis, χ, θ, ψ, 2o, 3ω, 1π Cygni, ζ Cephei, τ, β, α Cassiopeii. [p. 217].

October 10   Bradley observed χ, θ, ψ, 2o, 3ω, 1π Cygni, ζ Cephei. [p. 217].

October 12   Bradley observed α, 2υ Cassiopeii. [p. 218].

October 13   Bradley observed ζ, η Ursæ Majoris, γ Draconis, χ, θ, ψ, 2o, 3ω, 1π Cygni, ζ Cephei, τ, β, α, θ Cassiopeii, ζ Andromedæ, φ, θ, τ, γ, α, δ Persei, 9 Aurigæ, *Capella*, 18, 35 Camelopardalis. [p. 218].

October 16   Bradley observed γ, ε, ζ, η Ursæ Majoris. [p. 219].

October 17   Bradley observed ε, ζ, η Ursæ Majoris, β Draconis. [p. 219].

October 20   Bradley observed Jupiter's 1st satellite (Io) disappear in the 15ft tube. However, the Earth's Moon was very bright and near Jupiter. [p. 362].
Bradley also observed ε, ζ, η Ursæ Majoris, γ Draconis, χ, θ, ψ, 2o Cygni, ζ Cephei, β, α Cassiopeii, ζ Andromedæ, φ, θ, τ, γ, α, δ Persei, *Capella*, 18, 35 Camelopardalis, 46 Aurigæ. [p. 219].

October 21   Bradley observed γ Draconis, χ, θ, ψ, 2o Cygni. [p. 220].

October 22   Bradley observed β, γ, ε Ursæ Majoris. [p. 220].

October 23   Bradley observed 1π Cygni, ζ Cephei, 7 Andromedæ, τ, σ, β, α Cassiopeii. [p. 220].

October 25   Bradley observed the Sun pass his meridian in Wanstead.

October 26   Bradley observed ζ, η, 1f, g, 1π Cygni, μ, ζ Cephei, β, α Cassiopeii. [p. 220].

October 27   Bradley observed 27, 35 Lyncis, ι, f, θ, φ, 36, β, ψ, γ, ε, ζ, η Ursæ Majoris. [p. 221]. Bradley made a note. 'I went to Oxford'.

November 12   Bradley observed λ, α, θ Cassiopeii, ζ Andromedæ, φ Persei. [p. 221].

November 13   Bradley observed ε, ζ, η Ursæ Majoris, Note. [p. 221].

November 14   Bradley observed ζ, ε, η Ursæ Majoris, β, γ Draconis, ζ Cephei, 3 Lacertæ, τ, β, λ, α, majoris2υ, θ Cassiopeii, ζ Andromedæ, τ, ν, α, δ Persei, *Capella*. [p. 221].

November 15   Bradley observed ε, ζ, η Ursæ Majoris. [p. 222].

November 16   Bradley observed 9 Aurigæ, *Capella*. [p. 222].

November 17   Bradley observed β, ψ, γ, ε, ζ, η Ursæ Majoris, β, γ Draconis. [p. 222].

November 22   Bradley observed τ, β, λ, α Cassiopeii, ζ Andromedæ, φ, θ, τ, γ, α, δ Persei, 9 Aurigæ, *Capella*, 18, 35 Camelopardalis, 46 Aurigæ, 13 Lyncis, ι, f, θ, β, ψ Ursæ Majoris. [p. 222].

November 23   Bradley observed γ Draconis. [p. 223].

December   The Moon's ascending node was near the beginning of the first point of Aries. [p. 23].Subsequently, the orbit was as much inclined to the equator as it could be. [*Memoirs* p. lxiv]. The stars near the equinoctial colure changed their declination 1½″ or 2″ a year more than if precession was 50″ per annum. [*Memoirs* p. lxiv].

December 5   Bradley observed Jupiter's 1st satellite (Io) emerge in the 6ft reflector. [p. 362].
Bradley also observed β, γ Draconis in bright conditions. ζ Cephei, τ, β, λ, α, π, θ Cassiopeii, ζ Andromedæ, φ Persei, *Capella*. [p. 223].

December 6   Bradley observed ψ, γ, ε, ζ, η, β Ursæ Majoris, β, γ Draconis, 7Andromedæ, τ, β Cassiopeii. [p. 223].

December 9   Bradley observed *Capella*. [p. 224].

December 12   Bradley observed ζ Cephei, 3 Lacertæ, τ, β, λ, α, π, 2υ, θ Cassiopeii, ζ Andromedæ, φ, θ, τ, γ, α, δ Persei, *Capella*, 35 Camelopardalis, 46 Aurigæ. [p. 224].

December 13   Bradley observed β, ψ, γ, ε, ζ, η Ursæ Majoris, β Draconis, β, λ, α Cassiopeii, 18, 35, Camelopardalis, 46 Aurigæ. [p. 224].

December 14   Bradley observed β, ψ, γ, ε, ζ, η Ursæ Majoris, β, λ, α, π Cassiopeii. [p. 225].

December 15   Cold and snowy. Bradley observed γ Draconis. [p. 225].

December 21   Bradley observed Jupiter's 1st and 2nd satellites (Io, Europa) emerge in the 6ft reflector. [p. 362].

Bradley also observed τ, β, λ, α, π, θ Cassiopeii, ζ Andromedæ, φ, τ, γ, α Persei, *Capella*, 18, 35 Camelopardalis. [p. 225].

December 27   Bradley observed γ Draconis. [p. 225].

December 28   Bradley observed β, γ Draconis. [p. 226].

December 29   Molyneux observed γ Draconis. This was the last ever observation made by Molyneux before his death 13 April 1728. The instrument fell into disrepair and Molyneux had insufficient time after his appointment as one of the Lords of the Admiralty together with his parliamentary affairs at Westminster and Dublin. [*Memoirs* p. xviii].

Bradley observed β, ψ, χ, γ Ursæ Majoris, 3 Canes Venatici, ε Ursæ Majoris, 21 Canes Venatici, ζ, η Ursæ Majoris, β, γ Draconis, λ, α, π, 2υ, θ Cassiopeii, ζ Andromedæ, φ, θ, τ, γ, α, δ Persei. [p. 226].

1728   **January 3**

Bradley observed β, λ, α, π, 2υ, θ Cassiopeii, ζ, Andromedæ, φ, θ, τ, γ Persei. [p. 226].

January 6/7   Bradley observed ι, f, θ, β, ψ, χ, γ, ε Ursæ Majoris, 3 Canes Venatici, then cloudy. [p. 227].

January 12   Bradley observed *Capella*, δ, 46 Aurigæ, 18, 35 Camelopardalis, 13 Lyncis. [p. 227].

January 16   Bradley observed β, γ Draconis, β, α, θ Cassiopeii, ζ Andromedæ, φ, θ, τ, γ, α, δ Persei, 9 Aurigæ, *Capella*, 18 Camelopardalis, δ Aurigæ, 35 Camelopardalis, 46 Aurigæ, 13 Lyncis, ι, f, θ, β, ψ, χ, δ Ursæ Majoris. [p. 227].

January 20   Bradley observed β, α Cassiopeii, ζ Andromedæ, φ, θ, τ, γ Persei. [p. 228].

January 23   Bradley observed β, γ Draconis, 46 Aurigæ, ι, f, ζ, η Ursæ Majoris. [p. 228].

January 24   Bradley observed γ Draconis, 9 Aurigæ, *Capella*, 18 Camelopardalis (Note), δ Aurigæ, 35 Camelopardalis, 46 Aurigæ, 13 Lyncis, 21 Canes Venatici, ε, ζ, η Ursæ Majoris. [p. 228].

January 25   Bradley observed β, γ Draconis, ε, ζ, η Ursæ Majoris. [p. 229].

January 26   Bradley observed 9 Aurigæ, *Capella*, η Ursæ Majoris. [p. 229].

January 27   Bradley observed *Capella*, 18, 35 Camelopardalis, δ, 46 Aurigæ, 13 Lyncis, ι, f, θ, β, ψ, χ, γ Ursæ Majoris. Fluttering all night. [p. 229].

February 1   Molyneux's zenith sector at Kew was out of order at this time. There is no notice of the instrument ever being repaired. Molyneux was busy with his duties at the Admiralty and in Parliament. [p. 186].

February 2   Bradley observed β, α Cassiopeii, θ, τ, γ, α, δ Persei, 9 Aurigæ, *Capella*, 18, 35 Camelopardalis, δ Aurigæ. [p. 229].

February 3   Bradley observed β, α Cassiopeii (θ Cassiopeii was too faint for accurate obs.). θ, τ, γ, α, δ Persei, 9, 46, δ Aurigæ, *Capella*, 18, 35 Camelopardalis. [p. 229].

February 4   θ, τ, γ, α, δ Persei, 9, δ, 46 Aurigæ, *Capella*, 18, 35 Camelopardalis. [p. 230].

February 5   Bradley observed the 1st satellite of Jupiter (Io) emerge from the planet in the 15ft tube. Clear. [p. 362]. Bradley also observed β, α Cassiopeii, θ, τ, γ, α, δ Persei, 9, δ, 46 Aurigæ, *Capella,* 18, 35 Camelopardalis. [p. 230].

February 10   Bradley observed β, γ Draconis. [p. 231].

February 14   Bradley observed δ Persei, *Capella,* [p. 231].

February 16   Bradley observed β Draconis, ι Herculis, γ Draconis, γ, α, δ Persei. [p. 231].

February 17   Bradley observed β, ψ, χ, γ, ε, ζ, η Ursæ Majoris, 3, 21 Canes Venatici. [p. 231].

February 18   Bradley observed the 3rd satellite of Jupiter (Ganymede) emerge in the 15ft tube. [p. 362]. Bradley also observed *Capella,* 9, δ, 46 Aurigæ, 18, 35 Camelopardalis, 13 Lyncis. [p. 231].

February 21   Bradley observed α, γ Persei, *Capella,* 9, δ, 46 Aurigæ, 18, 35 Camelopardalis, β, ψ Ursæ Majoris. [p. 231].

March   Bradley elected to the Council of the Royal Society for the third time. [*Memoirs,* p. lxxi].

March 1   Bradley observed 35 Camelopardalis, 46 Aurigæ. [p. 232].

March 2   Bradley observed *Capella,* δ, 46 Aurigæ, 35 Camelopardalis, 13 Lyncis, ι, f, θ Ursæ Majoris. [p. 232].

March 5   Bradley observed θ, β, ψ Ursæ Majoris. [p. 232].

March 6   Bradley observed *Capella.* [p. 232].

March 7   Bradley observed β, γ Draconis, ι Herculis, *Capella.* [p. 232].

March 8   Bradley observed *Capella.* [p. 232].

March 12   Bradley observed α Cassiopeii, *Capella,* β, ψ Ursæ Majoris. [p. 232].

March 13   Bradley observed f, θ, β, ψ Ursæ Majoris. [p. 232].

March 15   Bradley observed f Ursæ Majoris. [p. 232].

March 17   Bradley observed β, ψ, χ, γ Ursæ Majoris. [p. 232].

March 18   Bradley observed β, γ Draconis, ι Herculis, *Capella.* [p. 232].

March 19   Bradley observed *Capella.* [p. 232].

March 20   Bradley observed β, γ Draconis, ι Herculis. Bradley put up another wire. [p. 232].

March 21   Bradley observed β, α Cassiopeii, *Capella,* 21 Canes Venatici, ε, ζ, η Ursæ Majoris. [p. 232]. Bradley's observation of η Ursæ Majoris was recorded on the same batch of paper as used for his paper on the new discovered motion (aberration of light). This suggests that Bradley was developing his ideas early during the year. [*Memoirs* p. xxiv].

March 22   Bradley observed Jupiter's 1st satellite (Io) in 15ft tube. [p. 362]. Bradley also observed β, γ Draconis, ι Herculis, *Capella,* γ Ursæ Majoris. [p. 233].

March 23   Bradley observed β, α Cassiopeii, α Persei. [p. 233].

March 24   Bradley observed β, γ Draconis, ι Herculis, β, ψ, δ Ursæ Majoris. [p. 233].

April 2   Bradley observed f, θ, χ, γ, ι Ursæ Majoris. [p. 233].

April 6   Bradley observed 21 Canes Venatici, ε, ζ Ursæ Majoris, ι Herculis, β, ζ, γ, d, c Draconis. [p. 233].

April 7   Bradley observed β, α Cassiopeii, α Persei, 21 Canes Venatici, ε, ζ, η Ursæ Majoris, ι Herculis. [p. 234].

April 10   Bradley observed η Ursæ Majoris. [p. 234].

April 13 Molyneux died following his collapse in the House of Commons   a few days earlier. It is not known what happened to Molyneux's sector. This is to be regretted as it was the prototype of all such instruments. From this date all observations are Bradley's own.

April 15   Bradley observed ζ, η Ursæ Majoris. [p. 234].

April 16   Bradley observed *Capella*, 21 Canes Venatici, β, ψ, χ, γ, ε, ζ, η Ursæ Majoris. β, ζ, γ Draconis. [p. 234].

April 17   Bradley observed ι, θ, β, ψ, χ, γ, ε, ζ, η Ursæ Majoris. [p. 234].

April 18   Bradley observed β, α Cassiopeii, β, ψ, ε, ζ, η Ursæ Majoris. [p. 234].

April 23   Bradley observed *Capella*. [p. 235].

April 24   Bradley observed ε Ursæ Majoris. [p. 235].

April 25   Bradley observed α Persei. [p. 235].

April 26   Bradley observed β, ψ, γ Ursæ Majoris. [p. 235].

April 28   Bradley observed *Capella*. [p. 235].

May 5   Bradley observed ε, ζ, η Ursæ Majoris, 21 Canes Venatici, ι Herculis, β, ζ, γ Draconis. [p. 235].

May 6   Bradley observed ι Herculis, β, ζ, γ Draconis. [p. 235].

May 7   Bradley observed *Capella*, β, ψ, χ Ursæ Majoris, ι Herculis, β, ζ, γ Draconis. [p. 235].

May 8   Bradley observed β, α Cassiopeii, α Persei, *Capella*. [p. 235].

May 10   Bradley observed *Capella*, ι Herculis, β, ζ, γ Draconis. [p. 235].

May 11   Bradley observed 21 Canes Venatici, ε, ζ, η Ursæ Majoris. [p. 236]. Bradley then wrote 'Then I went to Oxford'.

June 4   Bradley observed ι Herculis, β, ζ, γ Draconis. [p. 236].

June 5   Bradley observed *Capella*, γ, ε, ζ, η Ursæ Majoris, ι Herculis, ζ, γ Draconis. [p. 236].

June 7   Bradley observed β, α Cassiopeii, *Capella*. [p. 236].

June 8   Bradley observed β Cassiopeii, β, γ, ε, ζ, η Ursæ Majoris. [p. 236].

June 9   Bradley observed ε Ursæ Majoris. [p. 236].

June 12   Bradley observed *Capella*. [p. 236].

June 15   Bradley observed γ, ε, ζ, η Ursæ Majoris, ι Herculis, ζ, γ Draconis. [p. 236].

June 17   Bradley observed β, α Cassiopeii, α Persei, *Capella*. [p. 236].

June 20   Bradley delivered the Huygenian glass and the furniture of the telescope to the Royal Society.

June 21   Bradley observed η Ursæ Majoris, ι Herculis, β, ζ, γ Draconis. [p. 237].

June 22   Bradley observed α Persei, *Capella*, γ, ε, ζ, η Ursæ Majoris, ι Herculis, β, ζ, γ Draconis. [p. 237].

June 23   Bradley observed β, α, Cassiopeii, α Persei, *Capella*. [p. 237].

June 24   Bradley observed β, α Cassiopeii, α Persei [plumbwire broke] then observed *Capella*, β, γ, ε, ζ, η Ursæ Majoris. [p. 237].

June 25   Bradley observed *Capella*, ε, η Ursæ Majoris, β, γ Draconis. [p. 237].

June 28   Bradley observed β, α Cassiopeii, γ, α Persei, ζ, η Ursæ Majoris. [p. 238].

June 29   Bradley observed η Ursæ Majoris. [p. 238].

June 30   Bradley observed ι Herculis, ζ, γ Draconis. [p. 238].

July 2   Bradley observed ζ, η Ursæ Majoris, ι Herculis, β, ζ, γ Draconis. [p. 238].

July 3   Bradley observed γ Ursæ Majoris, ι Herculis, β, ζ, γ Draconis. [p. 238].

July 4   Bradley observed *Capella*, ι Herculis, β, ζ, γ Draconis. [p. 238].

July 5   Bradley observed β, α Cassiopeii, α Persei, *Capella*, ε Ursæ Majoris, β, ζ, γ Draconis. [p. 238].

July 6   Bradley observed *Capella*, ε Ursæ Majoris. [p. 238].

July 12   Bradley observed *Capella*. [p. 238].

July 13   Bradley observed *Capella*, ι Herculis, β, ζ, γ Draconis. [p. 238].

July 17   Bradley observed η Ursæ Majoris. [p. 238].

July 18   Bradley observed ι Herculis, β, ζ, γ Draconis. [p. 239].

July 19   Bradley observed α Persei, *Capella*. [p. 239].

July 20   Bradley observed γ, α, δ Persei, *Capella*. [p. 239].

July 22   Bradley observed α Persei. [p. 239].

July 26   Bradley observed ι Herculis, β, ζ, γ Draconis. [p. 239].

July 27   Bradley observed β, α Cassiopeii, γ, α, δ Persei, *Capella*. [p. 239].

July 28   Bradley observed ι Herculis, β, ζ, γ Draconis. [p. 239].

July 29   Bradley observed β, α Cassiopeii, τ, α, δ Persei, *Capella*. [p. 239].

July 31   Bradley observed β, α Cassiopeii. [p. 239].

Summer   This is the supposed time of Thomson's 1812 account of Bradley's boating party on the River Thames published in his *History of the Royal Society*. [p. 346] [*Memoirs* p. xxxi]. There is no documentary evidence to support this account.

During the period when Bradley was seeking the causes of the phenomenon he was observing in many stars, he knew from his observations of the satellites of Jupiter that the velocity of light was finite, the light from Jupiter taking time to reach the Earth. This notion was earlier proposed by Galileo, even before G D Cassini and Romer proposed it on the basis of their observations. Cassini soon relinquished his claims for philosophical and/or religious reasons.

Maupertuis visited London. Like Voltaire who lived in London from 1726 to 1729 he was made aware of the differences between the mathematical sciences in England and France.

August 1   Bradley observed ι Herculis, β, ζ, γ Draconis. [p. 239].

August 2   Bradley observed ε, ζ, η Ursæ Majoris, ι Herculi, β, ζ, γ Draconis. [p. 239].

August 3   Bradley observed τ, γ, α, δ Persei, *Capella*, η Ursæ Majoris. [p. 240].

August 5   Bradley observed β, ε, ζ, η Ursæ Majoris, ι Herculis, β, ζ, γ Draconis. [p. 240].

August 7   Bradley observed τ, α, δ Persei, *Capella*. [p. 240].

August 8   Bradley observed τ, α, δ Persei, *Capella*. [p. 240].

August 10   Bradley observed ι Herculi, β, ζ, γ Draconis. [p. 240].

August 12   Bradley observed Jupiter's 1st satellite (Io) disappear behind Jupiter in 15ft tube. [p. 362]. Bradley also observed ζ, η Ursæ Majoris, [p. 240].

August 13   Bradley observed τ, γ, α, δ Persei, *Capella*. [p. 241].

August 14   Bradley observed γ Draconis. [p. 241].

August 15   Bradley observed ι Herculi, β, ζ, γ Draconis. [p. 241].

August 17   Bradley observed ζ, γ Draconis. [p. 241].

August 18   Bradley observed ι Herculii, β, ζ, γ Draconis. [p. 241].

August 19   Bradley observed τ, γ, α, δ Persei, *Capella*. [p. 241].

August 21   Bradley observed β Draconis, ι Herculis, [p. 241].

August 22   Bradley observed *Capella*, ζ, γ Draconis. [p. 241].

August 23   Bradley observed ι Herculis, β, γ Draconis. [p. 241].

August 24   Bradley observed β, γ, ε, ζ, η Ursæ Majoris. [p. 241].

August 25   Bradley observed ι Herculis, β, ζ, γ Draconis. [p. 241].

August 26   Bradley observed *Capella*. [p. 242].

August 27 Bradley observed *Capella*. [p. 242].

August 29 Bradley observed β, γ Draconis. [p. 242].

August 31 Bradley observed *Capella*. [p. 242]. On or about this date Bradley became sure that the residuals of his observations of the 'new discovered motion of the fixed stars' betrayed evidence of another natural phenomenon.

September 1 Bradley observed γ Draconis. [p. 242].

September 3 Bradley removed the wire of the plumb line because the observations appeared to differ from the previous year. After tests it made no difference. [p. 242]. Bradley observed ζ Ursæ Majoris, ι Herculis, β, γ Draconis. [p. 242].

September 4 Bradley observed γ Draconis. [p. 242].

September 6 Bradley observed ε Ursæ Majoris, β, γ Draconis. [p. 242].

September 8 Bradley observed η Ursæ Majoris. [p. 242].

September 9 Bradley observed τ, γ, α, δ Persei, β, γ Draconis. [p. 242].

September 10 Bradley again took off the wire – rectified – just to check out any suspicions – Referring to his actions on September 3. [p. 242]. Bradley observed β, γ Draconis. [p. 243].

September 11 Bradley observed β, γ, δ Ursæ Majoris, β, γ Draconis. [p. 243].

September 12 Bradley observed *Capella*. [p. 243].

September 13 Bradley observed *Capella,* δ Aurigæ, 18, 35 Camelopardalis, β, γ, ε Ursæ Majoris, β, α Cassiopeii. [p. 243].

September 15 Bradley observed β, γ Draconis. [p. 243].

September 16 Bradley observed *Capella,* β, γ Draconis. [p. 243].

September 17 Bradley observed β Ursæ Majoris, α Cassiopeii. [p. 243].

September 18 Bradley observed 9 Aurigæ, *Capella,* δ Aurigæ, 18, 35 Camelopardalis, β, α, Cassiopeii. [p. 243].

September 19 Bradley observed α Cassiopeii. [p. 244].

September 20 Bradley observed β, α, η Ursæ Majoris, β, α Cassiopeii. [p. 244].

September 21 Bradley observed γ, ζ, η Ursæ Majoris. [p. 244].

September 22 Bradley observed *Capella,* 18 Camelopardalis. [p. 244].

September 26 Bradley observed *Capella,* δ, 46 Aurigæ, 18, 35 Camelopardalis. [p. 244].

September 27 Bradley observed γ Draconis. [p. 244].

September 28 Bradley observed ζ, η Ursæ Majoris, β, γ Draconis. [p. 244].

September 30 Bradley observed *Capella,* 18 Camelopardalis. [p. 244].

September to October Rigaud suggests that this was the period when Bradley reduced his zenith sector observations. This, even though he was intensely busy with his observations at Wanstead and his teaching duties at Oxford.

October 1 Bradley observed *Capella,* δ Aurigæ, 18, 35 Camelopardalis. [p. 244].

October 9 Bradley observed ε, ζ, η Ursæ Majoris. [p. 244].

October 11 Bradley observed γ, ε, ζ, η Ursæ Majoris, β Draconis. [p. 244].

October 12 Bradley observed β Ursæ Majoris. [p. 245].

October 13 Bradley observed β Ursæ Majoris. [p. 245].

October 15 Bradley observed Jupiter's 1st satellite (Io) disappear behind Jupiter in 15ft tube. [p. 362]. Bradley also observed ψ, 2o, 3ω Cygni, β, α Cassiopeii. [p. 245].

October 16 Bradley observed γ Draconis. [p. 245].

October 18 Bradley observed γ Draconis, χ, ι, θ, 2o, 3ω, Cygni, β, α Cassiopeii, τ, γ, α, δ Persei, [p. 245].

October 19 Bradley observed ψ, 2o, 3ω Cygni, β Cassiopeii. [p. 245].

October 20 to 23 A great fire destroyed a quarter of the city of Copenhagen, including half of the medieval city. It destroyed most of Ole Rômer's observations and his instruments. One fragment survived, suggesting he made observations of *Sirius* and *Vega* in order to determine annual parallax.

October 22 Bradley observed γ Ursæ Majoris, β, γ Draconis. [p. 245].

October 29 Bradley observed ε, ζ, η Ursæ Majoris. [p. 245].

October 31 Bradley observed τ, β, α Cassiopeii, τ, γ, α Persei. [p. 245].

November 1 to 15 With the exception of a fortnight when Bradley was detained in Oxford, he never relaxed his observations during the last four months of 1728. [*Memoirs* xxxiii]. The velocity of light was determined from Bradley's work on aberration. He determined that light took 8min 12sec from the Sun to the Earth, comparing well with the modern figure of 8min 19sec.

November 14 Bradley discussed the new discovered motion – reported in minutes on this date.

November 16 Bradley wrote 'I came from Oxford'. [p. 245]. Bradley observed 9 Aurigæ, *Capella*. [p. 245].

November 17 Bradley observed ε, ζ, η Ursæ Majoris, β, λ, α Cassiopeii. [p. 246].

November 19 Bradley observed η Ursæ Majoris, β, γ Draconis, τ, γ, α, δ Persei, 9 Aurigæ, *Capella*. [p. 245].

November 22 Bradley observed β Draconis, λ, α Cassiopeii, τ, γ α, δ Persei. [p. 246].

November 23 Bradley observed β, ψ, χ, γ, ε, ζ, η Ursæ Majoris, β, γ Draconis, β, α Cassiopeii, τ, γ, α, δ Persei. [p. 246].

November 24 Bradley observed β, γ, ε, ζ, η Ursæ Majoris. [p. 246].

November 28 Bradley observed β, γ Draconis. [p. 247].

November 29 Bradley observed γ Draconis, β, λ, α Cassiopeii. [p. 247].

November 30 Bradley observed β, λ, α Cassiopeii. [p. 247].

December Bradley determined that the maximum variation for annual aberration was 40″ or 41″. [p. 12] [*Memoirs* p. xxxv].

December 1 Bradley observed β, λ, α Cassiopeii, τ, γ, α, δ Persei, *Capella*. [p. 247].

December 4 Bradley observed β, λ, α Cassiopeii. [p. 247].

December 7 Bradley observed β, ψ, γ, ε Ursæ Majoris. [p. 247].

December 9 Bradley observed β, λ, α Cassiopeii, γ, α, δ Persei, 9 Aurigæ, *Capella*. [p. 247].

December 11 Bradley observed ε, ζ, η Ursæ Majoris, *Alcor*, β, α Cassiopeii, τ, γ, α, δ Persei. [p. 247].

December 13 Bradley observed β, λ, α Cassiopeii, γ, α, δ Persei, 9 Aurigæ, *Capella*. [p. 248].

December 14 Bradley observed β Ursæ Majoris. [p. 248].

December 15 Bradley observed α Cassiopeii. [p. 248].

December 16 Bradley observed β, γ, ε, η Ursæ Majoris, β, α Cassiopeii, *Capella*. [p. 248].

December 17 Bradley observed γ Draconis. [p. 248].

December 19 Bradley observed β, γ Draconis. [p. 248].

December 20 Bradley observed γ Draconis. [p. 248].

December 21 Bradley observed γ Draconis, β, λ, α Cassiopeii, τ, γ, α Persei. [p. 248].

December 23 Bradley observed γ, α Persei, *Capella*. [p. 248].

December 24 Bradley observed *Capella*, δ Aurigæ, 35 Camelopardalis. [p. 248].

December 25 Bradley observed β Ursæ Majoris. [p. 248].

1729   Margaret Flamsteed published the *Atlas Coelestis* based on her husband's observations at Greenwich.

January 3   Bradley observed τ, γ, α Persei. [p. 249].

January 4   Bradley observed γ Ursæ Majoris. [p. 249].

January 9   Bradley's paper on the 'new discovered motion of the fixed stars' was read at a meeting of the Royal Society. [*Memoirs* p. xii]. Rigaud mentions a copy dated 1st January. This is different from the third unpublished version dated 5 January transcribed and edited by John Fisher for the MSc dissertation completed in 1994. The reading of the paper was continued and completed on 16 January.

January 14   Bradley observed τ, γ, α, δ Persei. [p. 249].

January 16   Bradley observed ε, ζ, η Ursæ Majoris. [p. 249]. Bradley's paper on the aberration of light was read to completion at the Royal Society. [*Memoirs* p. xii]. Bradley continued observing with his zenith sector because he was aware that the residuals in his observations possibly signified the existence another phenomenon, possibly the nutation mentioned by Newton.

January 19   Bradley observed *Capella*. [p. 249].

January 20   Bradley observed ε, ζ, η Ursæ Majoris, *Alcor*, β Draconis. [p. 249].

January 21   Bradley observed ε, ζ, η Ursæ Majoris. [p. 249].

January 22   Bradley observed ε, ζ, η Ursæ Majoris, *Alcor*, β Cassiopeii. [p. 249].

January 24   Bradley observed Jupiter's 1st satellite (Io) emerge from the planet using Hadley's 6ft reflector. [p. 362]. Hearne composes a memorandum [No.124, p. 121] revealing that J. Whiteside has sold the apparatus for more than £400. This was corrected during 1730. [*Memoirs* p. xxxviii]. Bradley was to use this apparatus teaching his courses on experimental philosophy from 1729 to 1760.

January 25   Bradley observed ε, ζ, η Ursæ Majoris, *Alcor*, β, γ Draconis, α Cassiopeii, τ, γ, δ Persei, *Capella*. [p. 249].

January 27   Bradley observed ε, ζ, η Ursæ Majoris, *Alcor*. [p. 249]. Bradley also observed Jupiter's 3rd satellite (Ganymede) emerge from the planet using Hadley's 6ft reflector. [p. 362].

January 28   Bradley observed β, γ Draconis. [p. 250].

February 1   Bradley observed τ, α, δ Persei, *Capella*. [p. 250].

February 5   Bradley observed β, γ Draconis. [p. 250].

February 6   Bradley observed β, γ Draconis. [p. 250].

February 7   Bradley observed γ Draconis, *Capella*. [p. 250].

February 12   Bradley observed *Capella*. [p. 250].

February 13   Bradley observed γ Draconis. [p. 250].

February 14   Bradley observed β, γ Draconis, *Capella*. [p. 250].

February 15   Bradley observed *Capella*. [p. 250].

February 21   Bradley observed β, γ Draconis. [p. 250].

February 24   Bradley observed *Capella*. [p. 250].

February 25   Bradley observed ι Herculi, β, γ Draconis, α Cassiopeii, γ, α, δ Persei, *Capella*, δ Aurigæ, 35 Camelopardalis. [p. 250].

February 26   Bradley observed ι Herculi, β, γ Draconis, β, α Cassiopeii, γ, α, δ Persei, *Capella*, 35 Camelopardalis, 46 Aurigæ. [p. 250].

February 27   Bradley observed β, γ Draconis, *Capella*. [p. 251].

March 1   Bradley observed γ Draconis. [p. 251].

March 5   Bradley observed γ Draconis. [p. 251].

March 6   Bradley observed ι Herculi, β, ζ, γ Draconis. [p. 251].

March 9   Bradley observed *Capella*, δ Aurigæ, β, γ, ε Ursæ Majoris. [p. 251].

March 10   Bradley observed *Capella*. [p. 251].

March 17   Bradley observed ι Herculi, β, ζ, γ Draconis, β, α Cassiopeii, *Capella*. [p. 251].

March 18   Bradley observed ι Herculi, β, ζ, γ Draconis, β, α Cassiopeii. [p. 251].

March 20   Bradley observed ι Herculi, β, ζ, γ Draconis. [p. 251].

March 24   Bradley observed β Draconis. [p. 251].

March 25   Bradley observed *Capella*. [p. 251].

April 4   Bradley observed ε, ζ, η Ursæ Majoris, *Alcor*. [p. 252].

April 5   Bradley observed ε, ζ, η Ursæ Majoris, *Alcor*. [p. 252].

April 8   Bradley observed ζ, η Ursæ Majoris. [p. 252].

April 9   Bradley observed β, α Ursæ Majoris. [p. 252].

April 12   Bradley observed ζ, η Ursæ Majoris, *Alcor*. [p. 252].

April 13   Bradley observed ζ, η Ursæ Majoris. *Alcor*. [p. 252].

April 16   Bradley observed γ, ζ, η Ursæ Majoris, ι Herculis, β, ζ, γ Draconis. [p. 252].

April 18   Bradley observed *Capella*. [p. 252].

April 21   Bradley left for Oxford returning to Wanstead on May 10. [p. 252].

May 11   Bradley observed β, γ, ε, ζ, η Ursæ Majoris, *Alcor*. [p. 252].

May 13   Bradley observed γ Ursæ Majoris. [p. 252].

May 14   Bradley observed *Capella*. [p. 252].

May 17   Bradley observed ζ Ursæ Majoris, *Alcor*. [p. 252].

May 26   Bradley observed *Capella*. [p. 252].

May 29   Bradley observed β, γ Draconis. [p. 253].

May 30   Bradley observed β, γ, ε, ζ, η Ursæ Majoris, *Alcor*. [p. 253].

May 31   Bradley observed *Capella*, β, η Ursæ Majoris, β, γ Draconis. [p. 253].

June 2   Bradley observed β, γ Draconis. [p. 253].

June 3   Bradley observed β, α Cassiopeii, *Capella*. [p. 253].

June 4   Bradley observed β, α Cassiopeii, *Capella*, β, γ, ε, ζ, η Ursæ Majoris. [p. 253].

June 5   Bradley observed β, α Cassiopeii, ι Herculis, β, ζ, γ Draconis. [p. 253].

June 6   Bradley observed β, α Cassiopeii, *Capella*, γ, η Ursæ Majoris, γ Draconis. [p. 253].

June 7   Bradley observed β, α Cassiopeii, *Capella*, γ, ε, ζ, η Ursæ Majoris, ι Herculis, β, ζ, γ Draconis. [p. 253].

June 8   Bradley observed β, α Cassiopeii. [p. 254].

June 9   Bradley observed γ Draconis. [p. 254].

June 10   Bradley observed β, γ Ursæ Majoris. [p. 254].

June 11   Bradley observed β, α Cassiopeii. [p. 254].

June 18   Bradley observed ε Ursæ Majoris. [p. 254].

June 20   Bradley observed *Capella*, γ Draconis. [p. 254].

June 23   Bradley observed α Persei, *Capella*, β, γ, ε, ζ, η Ursæ Majoris. [p. 254].

June 24   Bradley observed β, α Cassiopeii, α Persei, *Capella*, γ Draconis. [p. 254].

June 25   Bradley observed ζ, η Ursæ Majoris. [p. 254].

June 26   Bradley joined by Dr Benjamin Hoadly (son of the bishop). Bradley observed *Capella*. Hoadly observed ζ Ursæ Majoris. [p. 254].

June 27   Bradley observed γ, α Persei. [p. 254].

June 28   Bradley observed ε Ursæ Majoris. [p. 254].

June 29   Bradley observed γ, α Persei, *Capella*, ζ, η Ursæ Majoris. [p. 254].

June 30   Bradley observed γ, α Persei, *Capella*, δ, ε, ζ, η Ursæ Majoris. I Herculis, β, ζ Draconis. [p. 255].

July 1   Bradley observed α Persei, *Capella*, β, γ Draconis. [p. 255].

July 4   Bradley observed ε, ζ Ursæ Majoris. [p. 255].

July 5   Bradley observed ε, ζ, η Ursæ Majoris. Bradley notes that when he observed ζ he was also able to see the small star. [p. 255].

July 10   Bradley observed ι Herculis, β, ζ, γ Draconis. [p. 255].

July 12   Bradley observed γ, α, δ Persei, *Capella*. [p. 255].

July 14   Bradley observed *Capella*. [p. 255].

July 18   Bradley observed ε, ζ Ursæ Majoris. [p. 255].

July 21   Bradley observed *Capella*, ε, ζ, η Ursæ Majoris. [p. 255].

July 27   Bradley observed ι Herculi, β, ζ, γ Draconis. [p. 255].

July 28   Bradley observed a total eclipse of the Moon at Wanstead. He observed the phenomenon for about 3hrs 40mins. [pp. 363, 364]. With the zenith sector he also observed ι Herculi, β, ζ, γ Draconis. [p. 256].

July 29   Bradley observed ι Herculi, β, ζ, γ Draconis. [p. 256].

July 30   Bradley observed ι Herculi, β Draconis. [p. 256].

August 5   Bradley observed ι Herculi, β, ζ, γ Draconis. [p. 256].

August 6   Bradley observed ι Herculi, β, ζ, γ Draconis. [p. 256].

August 7   Bradley observed γ, α Persei, *Capella*. [p. 256].

August 10   Bradley observed ι Herculi, β, ζ, γ Draconis. [p. 256].

August 11   Bradley observed *Capella*. [p. 256].

August 12   Bradley observed ι Herculi, β, γ Draconis. [p. 256].

August 13   Bradley observed *Capella*, ι Herculi, β, ζ, γ Draconis. [p. 256].

August 15   Bradley observed τ, γ, α, δ Persei. [p. 256].

August 17   Bradley observed ι Herculi, β, ζ, γ Draconis. [p. 256].

August 19   Bradley observed τ, γ Persei, *Capella*, ι Herculi, ζ, γ Draconis. [p. 257].

August 20   Bradley observed τ, γ, α Persei. [p. 257].

August 21   Bradley observed τ, γ, α Persei, ε Ursæ Majoris. [p. 257].

August 22   Bradley observed *Capella*. [p. 257].

August 23   Dr Benjamin Hoadly observed β, γ Draconis. [p. 257].

August 24   Bradley observed Jupiter's 1st satellite (Io) disappear behind the planet in the 15ft tube. [p. 364]. Bradley also observed *Capella*. [p. 257].

August 25   Bradley observed β, γ Draconis. [p. 257].

August 28   Bradley observed ι Herculi, β, ζ, γ Draconis. [p. 257].

August 29   Bradley observed ι Herculi, β, ζ, γ Draconis. [p. 257].

August 30   Bradley observed Jupiter's 2nd satellite (Europa) disappear behind the planet in the 15ft tube. [p. 364]. Using the zenith sector the wire broke. It had been used since 10 September 1728. Bradley then observed d, c, Draconis, χ, ι Cygni. [p. 258].

August 31   Bradley observed *Capella*, ζ, γ, c Draconis, χ, ι Cycni. [p. 258].

September 1   Bradley observed *Capella*. [p. 258].

September 2   Bradley observed *Capella*, β, γ Draconis. [p. 258].

September 3   Bradley observed *Capella*. [p. 258].

September 8   Bradley observed β, γ Draconis. [p. 258].

September 9   Bradley observed *Capella*, β, γ Draconis. [p. 258].

September 10   Bradley observed η Ursæ Majoris. [p. 258].

September 13   Bradley observed β, γ Draconis. [p. 258].

September 18   Bradley observed γ Draconis. [p. 258].

September 19 Bradley observed *Capella*, δ Aurigæ, 18, 35 Camelopardalis. [p. 258].

September 20 Bradley observed δ Aurigæ, 35 Camelopardalis. [p. 258].

September 22 Bradley observed η Ursæ Majoris. Bradley notes windy conditions. [p. 258].

September 24 Bradley observed δ Aurigæ, 35 Camelopardalis. [p. 258].

September 26 Bradley observed β, γ Draconis. [p. 258].

September 27 Bradley observed α Cassiopeii. [p. 258].

September 29 Bradley observed *Capella*, δ Aurigæ, 18, 35 Camelopardalis. [p. 258].

October 2 Bradley observed Jupiter's 1st satellite (Io) disappear behind the planet in the 15ft tube. [p. 364].

October 4 Bradley observed β, γ Draconis. [p. 258].

October 7 Bradley observed η Ursæ Majoris, β, γ Draconis. [p. 259].

October 8 Bradley observed *Capella*, δ Aurigæ, 18, 35 Camelopardalis. [p. 259].

October 10 Bradley observed β Draconis, β Cassiopeii. [p. 259].

October 12 Eustachio Manfredi addressed a letter to Sir Thomas Dereham 'approving of the theory of Mr. Bradley' explaining the new discovered motion. [Manfredi Oct. 12].

October 13 Bradley observed ε, ζ, η Ursæ Majoris. [p. 259].

October 14 Bradley observed ε, η Ursæ Majoris, β Draconis, β, α Cassiopeii. [p. 259].

October 15 Bradley observed β, α Cassiopeii. [p. 259].

October 17 Bradley observed β, α Cassiopeii. [p. 259].

October 18 Bradley observed β, γ Draconis. [p. 259].

October 20 Bradley observed *Capella*. [p. 259].

October 24 Bradley observed ε, η Ursæ Majoris, γ Draconis. [p. 259].

October 27 Bradley travelled to Oxford. [p. 259].

November 11 Bradley returned to Wanstead. [p. 259].

November 21 Bradley observed γ Draconis. [p. 259].

November 23 Bradley observed β, α Cassiopeii. [p. 259].

November 25 Bradley observed *Capella*. [p. 259].

November 26 Bradley observed β, γ Draconis. [p. 259].

November 28 Bradley observed γ Draconis, β, λ, α Cassiopeii, τ, γ, α Persei. [p. 259].

November 30 Bradley observed γ Draconis. [p. 260].

December About this period Bradley composed his paper entitled 'Demonstration of the Rules Relating to the Apparent Motion upon account of the Motion of Light'. [p. 287].

December 2 Bradley observed λ, α Cassiopeii, τ, γ, α Persei, *Capella*. [p. 260].

December 3 Bradley observed β, γ Draconis, β, λ, α Cassiopeii, ζ Andromedæ, τ, γ, α Persei. [p. 260].

December 5 Bradley observed λ, α Cassiopeii. [p. 260].

December 6 Bradley observed β, γ Draconis. [p. 260].

December 8 Bradley observed Jupiter's 3rd satellite (Ganymede) disappear behind the planet using the 6ft Hadley reflector. Jupiter was distinct. [p. 364].

December 10 Bradley observed γ Draconis, β, λ, α Cassiopeii, ζ Andromedæ, τ, γ, α, δ Persei, *Capella*, δ, 46 Aurigæ, 18, 35 Camelopardalis. [p. 260].

December 11 Bradley observed β, λ, α Cassiopeii, ζ Andromedæ. [p. 260].

December 12 Bradley observed Jupiter's 1st satellite (Io) disappear behind the planet using the 6ft Hadley reflector. [p. 364].

December 15 Bradley observed γ Draconis. [p. 260].

December 18 Bradley went to Oxford. [p. 260]. Next observation made 1730 January 31.

1730 **January** Frederic, Prince of Wales, took a long lease on Samuel Molyneux's house on Kew Green. Eventually it became a residence of King George III. [*Memoirs* xv].

January 29 Sir Thomas Dereham wrote to the Royal Society from Italy mentioning Manfredi's Letter of 12 October 1729 (NS).

January 31 Bradley returned to Wanstead before January 31. Bradley observed τ, γ, α, δ Persei, *Capella*. [p. 261].

February 13 Hearne corrects the sum Whiteside sold his apparatus to Bradley from £400 to £170. Whiteside was the Chaplain of Christ Church and Keeper of the Ashmolean Museum. [*Memoirs* p. xxxviii].

February 21 Bradley observed Jupiter's 1st satellite (Io) emerge using Hadley's 6ft reflector. [p. 364]. Bradley also observed *Capella* in the zenith sector. [p. 261].

February 24 Bradley observed *Capella*. [p. 261].

March 3 Bradley observed *Capella*, δ Aurigæ, 35 Camelopardalis. [p. 261].

March 10 Bradley observed ι Herculi, β, ζ, γ Draconis. [p. 261].

March 16 Bradley observed ι Herculi, β, ζ, γ Draconis. [p. 261].

March 18 Bradley observed ι Herculi, β, ζ, γ Draconis. [p. 261].

April 3 Bradley went to Oxford.

May 7 A letter was read to the Royal Society from Mr. Atwell dated 14 April 1730 in Rome, describing the manner the new doctrine of the new discovered motion was received in Italy – seemingly not ready to receive it, stating that Manfredi of Bologna was writing an answer to it. (This appears to go against the correspondence of Dereham.) Atwell states that he sought vindication of Bradley's hypotheses through Cardinal Doria.

May 23 Bradley returned to Wanstead from Oxford.

May 25 Bradley observed *Capella*. [p. 261].

May 26 Bradley observed β Ursæ Majoris. [p. 261].

May 27 Bradley observed *Capella*. [p. 261].

June 6 Bradley observed *Capella*, ι Herculi, β, ζ, γ Draconis. [p. 261].

June 7 Bradley observed β, α Cassiopeii, ι Herculi, β, ζ, γ Draconis. [p. 261].

June 11 Bradley observed β, α Cassiopeii. [p. 261].

June 17 Bradley observed ε Ursæ Majoris. [p. 261].

June 19 Bradley observed β, α Cassiopeii, *Capella*. [p. 262].

June 21 Bradley observed γ Ursæ Majoris. [p. 262].

July 14 Bradley observed *Capella*. [p. 262].

July 15 Bradley observed α, δ Persei, *Capella*. [p. 262].

July 17 Bradley observed δ Persei, *Capella*, ζ, η Ursæ Majoris. [p. 262].

July 19 Bradley observed η Ursæ Majoris. [p. 262].

July 24 Bradley observed ε, ζ, η Ursæ Majoris. [p. 262].

July 25 Bradley observed *Capella*. [p. 262].

August 5 Bradley observed *Capella*, ζ, η Ursæ Majoris. [p. 262].

August 6 Bradley observed *Capella*. [p. 262].

August 9 Bradley observed *Capella*. [p. 262].

August 12 Bradley observed *Capella*, η Ursæ Majoris. [p. 262].

August 13 Bradley observed *Capella*. [p. 262].

August 16 Bradley observed *Capella*. [p. 262].

August 17   Bradley observed γ Draconis. [p. 262].

August 18   Bradley observed *Capella*, ε, η Ursæ Majoris, ι Herculi, β, ζ, γ Draconis. [p. 262].

August 20   Bradley observed *Capella*. [p. 262].

August 23   Bradley observed ε, ζ Ursæ Majoris, ι Herculi, β, ζ, γ Draconis. [p. 262].

August 24   Bradley observed *Capella*, ι Herculi, β, ζ, γ Draconis. Bradley made various tests. [p. 263].

August 25   Bradley observed Saturn with the Hadley 6ft reflector. The ring reappeared. The 1st and 2nd satellites (Tethys, Dione) seen to the west of the planet. The Huygenian satellite (Titan) the 4th also observed to the west of the planet in line of the ansae. [p. 364]. Bradley also observed *Capella*, β, ζ, γ Draconis. [p. 263].

August 26   Bradley observed *Capella*, ε, ζ, η Ursæ Majoris, ι Herculi, β, γ Draconis. [p. 263].

August 29   Bradley observed *Capella*. [p. 263].

August 31   Bradley observed ιHerculi, β, ζ, γ Draconis. [p. 263].

September 1   Bradley observed γ Draconis. [p. 263].

September 3   Bradley observed *Capella*. [p. 263].

September 7   Bradley observed γ Draconis. [p. 263].

September 8   Bradley observed *Capella,* γ Draconis. [p. 263].

September 9   Bradley observed γ Draconis. [p. 263].

September 16   Bradley observed γ Draconis. [p. 263].

September 18   Bradley observed ζ, η Ursæ Majoris, β, γ Draconis. [p. 263].

September 20   Bradley observed β, γ Draconis. [p. 263].

September 22   Bradley observed β, γ Draconis. [p. 264].

September 25   Bradley left for Oxford, returning to Wanstead November 27.

November 27   Bradley returned to Wanstead.

December 9   Bradley observed Jupiter's 4th satellite (Callisto) disappear in the 15ft tube. Jupiter was bright and distinct. [p. 364]. Bradley also observed γ Draconis, β, α Cassiopeii. [p. 264].

December 11   Bradley observed β, α Cassiopeii. [p. 264].

December 12   Bradley observed β, γ Draconis, β, λ, α Cassiopeii. [p. 264].

December 14   Bradley was discharged from making all future payments to the Royal Society. Bradley elected to the Council of the Royal Society for the fourth time. [*Memoirs* p. lxix].

December 15   Bradley observed Jupiter's 1st satellite (Io) disappear in the 15ft tube. [p. 364]. Bradley also observed β, γ Draconis. [p. 264].

December 18   Bradley observed β Draconis. [p. 264].

December 24   Bradley observed γ, β Draconis, λ, α, θ Cassiopeii, τ, γ, α Persei, *Capella*. Bradley revealed that the wall had fallen to the east. Adjusted sector moving the tube by one revolution. [p. 264].

December 27   Bradley observed *Capella*. [p. 264].

December 28   Bradley observed β Draconis, ε, ζ Ursæ Majoris, *Alcor*. [p. 264].

December 30   Bradley observed Jupiter's 3rd satellite (Ganymede) disappear in the 15ft tube. Then got the 6ft reflector ready. The 2nd satellite (Europa) just nearer than the 3rd. observed the 2nd satellite disappear using the 15ft tube. [p. 364].

1731   **January 1**   Bradley observed γ Draconis. [p. 264].

January 3   Bradley observed τ, γ, α, δ Persei, *Capella*, ε, ζ, η Ursæ Majoris, *Alcor*. [p. 264].

January 4   Bradley observed γ Draconis. [p. 265].

January 5   Bradley observed Jupiter's 1st satellite (Io) disappear behind the planet using the 15ft tube. Jupiter was very bright. [p. 364].

January 18   Bradley observed τ, γ, α Persei, 9 Aurigæ, *Capella,* [p. 265].

January 24   Bradley observed *Capella.* [p. 265].

February 3   Bradley observed γ Draconis, *Capella.* [p. 265].

February 5   Bradley observed *Capella,* 9 Aurigæ, 35 Camelopardalis. [p. 265].

February 6   Bradley observed β, γ Draconis, *Capella.* [p. 265].

Bradley observed Jupiter's 1st satellite (Io) disappear behind the planet using the 6ft reflector. It was near Jupiter's limb and the reflector was used because Jupiter was bright. [p. 364].

February 7   Bradley observed γ Draconis. [p. 265].

April 11   Bradley observed Jupiter's 1st satellite (Io) disappear behind the planet using the 15ft tube. Good observation. [p. 364].

April 27   Bradley observed *Capella.* [p. 265].

May   Bradley wrote a memorandum concerning the house in Wanstead. [p. 265]. [*Memoirs* p. xxiv].

Bradley lost the election to Jos. Andrews, Fellow of Magdalen College to become the Keeper of the Ashmolean Museum, the location of his lectures on experimental philosophy. His one supporter was Hearne who had originally regarded Bradley in very negative terms. [Hearne No.129, p. 138]. [*Memoirs* p. xxxviii].

Hadley published his account of his octant for measuring angular distances at sea. [*Phil. Trans.* Vol. 37, p. 147] [*Memoirs* p. xl]. Benjamin Franklin supported fellow Philadelphian Thomas Godfrey's claim. Both Godfrey and Hadley had equal claims but the Admiralty tested Hadley's instrument. [*Memoirs* p. xl]. It was Newton who had outlined the principles behind the instrument some years before. [*Phil. Trans.* Vol. 42, p.155].

May 3   Bradley went to Oxford. He returned to Wanstead on August 3. It must have been around this time that Bradley began to recognize the impracticality of living in Wanstead when his Oxford duties had expanded to such an extent, now presenting three courses of twenty lectures a year in experimental philosophy in addition to his duties as the Savilian Professor of Astronomy. He also had commitments at and with the Royal Society.

July   Maupertuis's paper on rotating fluid bodies was read to the Royal Society. It was a major paper written by an Academician of the Royal Academy of Sciences supportive of Newtonian methods. This was just before he published his ideas at the Royal Academy in Paris.

August 4   Bradley observed ε, ζ, η Ursæ Majoris, β Draconis. [p. 265]. Bradley examined his instrument and recognized that there was some degree of settling, moving the instrument to the north.

August 7   Bradley observed η Ursæ Majoris. [p. 266].

August 9   Bradley observed *Capella.* [p. 266].

August 10   Bradley observed *Capella.* [p. 266].

August 14   Bradley observed β, γ Draconis. [p. 266].

August 17   Bradley observed *Capella,* η Ursæ Majoris, β, γ Draconis. [p. 266].

August 18   Bradley observed β, γ Draconis. [p. 266].

August 19   Bradley observed *Capella.* [p. 266].

August 20   Bradley observed *Capella,* ε, η Ursæ Majoris, β, γ Draconis. [p. 266].

August 21   Bradley observed *Capella,* γ Draconis. [p. 266].

August 22   Bradley observed β, γ Draconis. [p. 266].

August 23 Bradley observed β, γ Draconis. [p. 266].

September 3 Bradley observed β, γ Draconis. [p. 266].

September 7 Bradley observed β, γ Draconis. [p. 266].

September 9 Bradley observed *Capella*. [p. 266].

September 10 Bradley observed ε, ζ, η Ursæ Majoris, β, γ Draconis. [p. 266].

September 11 Bradley observed η Ursæ Majoris. [p. 266].

September 12 Bradley observed *Capella*, η Ursæ Majoris, β, γ Draconis. [p. 266].

September 13 Bradley observed *Capella*, ε, ζ, η Ursæ Majoris, β, γ Draconis. [p. 267].

September 15 Bradley observed ε, ζ, Ursæ Majoris, β, γ Draconis. [p. 267].

September 18 Bradley observed ζ, η Ursæ Majoris. [p. 267].

September 19 Bradley observed β, γ Draconis. [p. 267].

September 20 Bradley observed ζ Ursæ Majoris, β, γ Draconis. [p. 267].

September 22 Bradley observed *Capella*, η Ursæ Majoris, β, γ Draconis. [p. 267].

September 25 Bradley observed β, γ Draconis. [p. 267].

October 1 Bradley observed ε, ζ, η Ursæ Majoris. [p. 267].

October 2 Bradley observed γ Draconis. [p. 267].

October 4 Bradley observed η Ursæ Majoris, β, γ Draconis. [p. 267].

December 18 Bradley observed β, γ Draconis. [p. 267].

December 26 Bradley observed τ, γ, α Persei. [p. 267].

December 31 Bradley observed γ Draconis. [p. 267].

Winter The Lords of the Admiralty ordered Hadley's instrument to be tried at sea. [*Memoirs* p. xl]. [*Phil. Trans.*Vol. 37, p. 341. Bradley took part in these trials aboard the Chatham yacht.

1732   **January 1** Bradley observed ε, ζ, η Ursæ Majoris, *Alcor*. [p. 267].

January 4 Bradley observed *Capella*. [p. 268].

January 8 Bradley observed β, λ, α Cassiopeii, τ, γ, α Persei, *Capella*. [p. 268].

January 9 Bradley observed ε, ζ, η Ursæ Majoris, *Alcor*, γ Draconis, τ Persei. [p. 268].

January 10 Bradley observed Jupiter's 1st satellite (Io) disappear behind the planet in the 15ft tube. [p. 365].

January 11 Bradley observed ε, ζ, η Ursæ Majoris, *Alcor*. [p. 268]. Bradley left for Oxford, returning to Wanstead shortly before 7 April.

April 7 Bradley observed *Capella*. [p. 268].

April 13 Bradley observed Jupiter's 1st satellite (Io) disappear behind the planet in the 15ft tube. [p. 365]. Bradley also observed η Ursæ Majoris. [p. 268].

April 17 Bradley observed α Cassiopeii, η Ursæ Majoris, *Capella*. [p. 268].

April 22 Bradley found wire broke at the top notch. He then observed *Capella*. [p. 268]. After this Bradley moved his main abode from Wanstead to Oxford.

April 28 Death of Thomas Parker. George Parker succeeds to the title of the 2nd Earl of Macclesfield. When Bradley moved to Oxford he was only 17 miles from his friend and patron's residence at Shirburne Castle.

April Maupertuis complains that his paper on rotating fluid bodies has not been published in the *Philosophical Transactions*. It was in the possession of John Machin who saw some confusion in the work although conforming to Newtonian methods. He argued that the paper took the law of gravity as a given. This was amended in the final paper. [British Library, Additional MS 4285, fol. 212].

May Bradley's now extensive duties and engagements in Oxford, now teaching three courses a year in experimental philosophy led to his enforced move from Wanstead to Oxford. [*Memoirs* p. xxxix]. At the beginning of May he moved to the house in New College Lane that came with the Savilian professorship. [p. 22]. Mrs Pound's house no doubt under the care of one of Matthew Wymondesold's servants.

Halley also had a residence close by Bradley in Oxford as the Savilian Professor of Geometry. There is no evidence that Bradley ever gave lectures in geometry. [*Memoirs* p. xxxix].

Bradley's move to Oxford did not separate him from Elizabeth Pound. She moved to Oxford to live with him until 1737 when illness forced her to move back to Wanstead and the care of her brother Matthew Wymondesold.

Bradley wrote to Hadley in support of the octant. [p. 505]. [*Memoirs* p. xl].

The motion of the Moon's rising node had now carried it back to Capricorn. [p. 24] [*Memoirs* p. lxiv].

June 13 Bradley observed η Virginis preceding Jupiter by 9sec in time, 2¼min in RA. Also 4′ 7″ north of Jupiter's centre. Used 15ft tube. [p. 365].

June 14 Bradley observed η Virginis preceding Jupiter by 27sec in time 6′ 46″ RA, also 6′ 18″ north of Jupiter's centre. Used 15ft tube. [p. 365].

Jun 30 Bradley observed Jupiter's 1st satellite (Io) emerge from the planet in the 15ft tube. [p. 365].

July 15 George Graham in London corresponds to Bradley in Oxford. Letter concerned the experiment in Jamaica. [p. 395].

July 18 Bradley replies to Graham about the experiment in Jamaica. [p. 395].

July 22 Graham wrote to Bradley again about the experiment in Jamaica. [p. 396].

July 25 Graham again wrote to Bradley concerning the Jamaican experiment. [p. 396].

July 31 Joseph Harris wrote to Bradley about the experiment in Jamaica. Harris had returned to London due to a decline in his health. [p. 398].

September 2 Bradley observed β, γ Draconis in Wanstead. [p. 268].

September 3 Bradley observed β, γ, ε, ζ Ursæ Majoris, β, γ Draconis. [p. 268].

September 4 Bradley observed γ, ε Ursæ Majoris, β, γ Draconis. [p. 269].

September 5 Bradley observed β, γ Draconis. [p. 269].

September 6 Bradley observed β, ε, ζ Ursæ Majoris, β, γ Draconis. [p. 269].

November 30 Bradley observed the Sun at Oxford. [p. 365].

December Since observing at Wanstead with his zenith sector Bradley was now certain that the greater part of the residuals in his observations were due to a lunar-induced nutation. Bradley wrote to Hadley about his octant. [p. 505].

December 1 Bradley observed the Sun at Oxford. [p. 365].

December 2 Bradley observed the Sun at Oxford. [p. 365].

December 6 Bradley observed an eclipse of the Sun at Oxford. Clouds prevented him from seeing the eclipse starting. Observed 21min 30sec to the end. Timed by watch 4min 50sec slower than the Ashmolean Museum clock. [p. 365].

December 22 Bradley arrived in Wanstead. He made no observations until December 30. [*Memoirs* p. xxxix]. Bradley rectified the spot. The instrument had altered 12″ between September and December. [p. 269].

December 24 Matthew Wymondesold Jnr. Died. He was later buried at Lockinge near Wantage close to their estate in Dorset. Bradley attended the funeral. [p. 401].

December 25 Bradley and Elizabeth Pound (with Matthew Wymondesold's other son) spent Christmas in Wanstead at her brother's house. [p. 401, 403] [*Memoirs* p. xxxix].

December 30 Bradley began making observations. [p. 269]. [*Memoirs* p. xxxix]. In future when Bradley visited Elizabeth's house to make observations with his zenith sector he did so during hours that were not inconvenient to any residents of the house. [*Memoirs* p. xl]. Bradley observed β, λ Cassiopeii. [p. 269].

1733   **January**   Bradley wrote from Oxford to Dr Robert Smith in Cambridge concerning Hadley's telescope.

January 1   Bradley observed α, β Cassiopeii. [p. 269].

January 21   Bradley observed γ Draconis, β, α Cassiopeii, τ, γ Persei. [p. 269].

January 22   Bradley observed β, γ Draconis. [p. 269].

January 31   Bradley observed *Capella*, 9, δ Aurigæ, 18, 35 Camelopardalis. [p. 269].

May 2   Marked 1732. Bradley observed an eclipse of the Sun. Measured the diameter of the Sun. Observed the eclipse for 1hr 44min. [pp. 366–367].

August 2   Bradley observed Jupiter's 3rd satellite Ganymede disappear behind the planet using the 15ft tube at Oxford. Jupiter was low in the sky. [p. 367].

August 25   Bradley found that the index of the sector had shifted by 8.5″ since January. Bradley observed β, γ Draconis. [p. 269].

August 26   Bradley observed β, γ Draconis. [p. 269].

August 29   Bradley observed β, γ Draconis. [p. 269].

November   Bradley published his paper, 'An Account of Observations in London and Black-River in Jamaica to Determine the Lengths of Isochronal Pendulums'. [*Phil. Trans.* No. 432, Vol. 38, p. 302]. Observations made by George Graham in London from 20 to 30 August 1731 and Colin Campbell in Jamaica 23 January and 18th February 1732. [p. 62].

November 24   J. Stirling corresponded with Bradley from London about the experiment in Jamaica. [p. 398].

1734   **February 21**   Bradley shortened his pendulum 124 divisions to make it vibrate sidereal seconds. Further observations of Mr. Campbell were useful in determining the effects of temperature. [p. 397].

March   Bradley continued making observations of vibrations of the pendulum as a continuation of those made by George Graham in London and Colin Campbell in Jamaica. Referring to George Graham, Bradley called it 'our experiment'. [p. 395]. Bradley obviously perceived that the pendulum experiments were an important support to his observations of what he now fully recognized to be a lunar-induced nutation of the Earth's axis, a consequence of the Earth being an oblate spheroid.

June 11   Bradley observed *Capella*, ε, ζ, η Ursæ Majoris, [p. 270].

June 12   Bradley observed β, α Cassiopeii, α Persei, *Capella*, ε, ζ, η Ursæ Majoris. [p. 270].

June 13   Bradley observed β, α Cassiopeii. [p. 270].

June 16   Bradley observed ε, η Ursæ Majoris. [p. 270].

June 18   Bradley observed ζ, η Ursæ Majoris. [p. 270].

June 19   Bradley observed β, γ Draconis. [p. 270].

June 21   Bradley observed β, γ Draconis. [p. 270].

June 28   Bradley examined Halley's 8ft quadrant at Greenwich with George Graham. [*Memoirs* liii]. He discovered that the instrument was badly out of true.

July 3   Bradley observed *Capella*. [p. 270].

July 5   Bradley observed β, γ Draconis. [p. 270].

July 9   Bradley observed ε, ζ, η Ursæ Majoris. [p. 270].

July 10   Bradley observed ε, ζ, η Ursæ Majoris, β, γ Draconis. [p. 270].

July 11   Bradley observed α Persei, *Capella*, ζ, η Ursæ Majoris. [p. 270].

July 16   Bradley observed *Capella*, η Ursæ Majoris, β, γ Draconis. [p. 270].

July 17   Bradley observed ε, η Ursæ Majoris. [p. 271].

July 29   Bradley observed β, γ Draconis. [p. 271].

July 31   Bradley observed ε Ursæ Majoris, β, γ Draconis. [p. 271].

August 3   Bradley observed *Capella*, ζ, η Ursæ Majoris, β, γ Draconis. [p. 271].

August 6 Bradley observed *Capella*, β, γ Draconis. [p. 271].

August 7 Bradley observed *Capella*, ζ, η Ursæ Majoris, β, γ Draconis. [p. 271].

August 8 Bradley observed β, γ Draconis. [p. 271].

August 11 Bradley observed γ Draconis. [p. 271].

August 12 Mrs Elizabeth Jenkins took up residence in Elizabeth Pound's townhouse in Wanstead. Bradley continued to observe with his zenith sector which remained in position until 1747. Bradley agreed only to use the instrument when it didn't inconvenience Mrs Jenkins' household. Only twice in 13 years was the instrument possibly interfered with. [*Memoirs* p. xxxix].

December 11 From August 11 to December 11 the index altered by only 7″. Bradley observed β, γ Draconis, β, λ, α Cassiopeii. [p. 271].

December 22 Bradley observed β, γ Draconis, β, λ, α Cassiopeii, *Capella*. [p. 271].

December 23 Bradley observed β, γ Draconis. [p. 271].

December 25 Bradley observed β, λ, α Cassiopeii, *Capella*. [p. 272].

December 27 Bradley observed β, γ Draconis. [p. 272].

1735 James Stirling of Balliol College published a paper on the figure of the Earth. [*Phil. Trans.* Vol. 34, p. 98]. [*Memoirs* p. xlii]. Stirling wrote to Bradley about this inquiry. [p. 398]. [*Memoirs* p. xlii]. Bradley replied to Stirling. [p. 400].

James Stirling impugned Campbell's experiment. [p. 399], [*Phil. Trans.* Vol. 34, p. 104]. John Machin wrote on the subject. [p. 397].

July 26 Bradley, Machin, and Graham examined Halley's 8ft quadrant. Bradley determined the angle of error in the vertical adjustment amounted to 34½″, estimating that it was shifting 2″ a year. Again he used plumblines to test the quadrant. [*Memoirs* p. liii].

September 9 Bradley observed η Ursæ Majoris, γ Draconis. [p. 272].

September 10 Bradley observed η Ursæ Majoris, β, γ Draconis. [p. 272].

September 13 Bradley observed β, γ, ε, ζ Ursæ Majoris, β, γ Draconis. [p. 272].

1736 **August 31** Bradley observed β, γ Draconis. [p. 272].

September 2 Bradley noted that between August 31 and September 2 someone meddled with the string. Mrs Jenkins was in residence in Elizabeth Pound's townhouse where the zenith sector was housed. Whether this was accidental or due to a malignant act, he was unable to determine. It must have been serious because Bradley normally dealt with various adjustments but now he was forced to call in George Graham to rectify various parts of the sector. Bradley was unable to observe for several days.

September 8 Bradley observed ζ, η Ursæ Majoris, β, γ Draconis. [p. 272].

September 9 Bradley observed β, γ Draconis. [p. 273].

September 10 Bradley observed ε Ursæ Majoris. [p. 273].

September 12 Bradley observed ε, ζ, η Ursæ Majoris. [p. 273].

1737 **January 3** Bradley observed γ Draconis. [p. 274].

January 4 Bradley observed β, γ Draconis. [p. 274].

January 9 Bradley observed *Capella*. [p. 274].

January 18 Bradley observed *Capella*. [p. 274].

January 19 Bradley observed β, γ Draconis, *Capella*. [p. 274]. Bradley adjusted the sector.

February Bradley observed the major comet of this year from January to March while at Oxford. [*Phil. Trans.* No. 446, Vol. 40, p. 111].

February 17 Bradley observed the comet of 1737. [Table p. 52].

February 18   Bradley observed the comet of 1737. [Table p. 52]. Bradley also observed an eclipse of the Sun. Measured the Sun's diameter. Observed for 2hrs 40min. [pp. 367–368].

February 21   Bradley observed the comet of 1737. [Table p. 52].

February 22   Bradley observed the comet of 1737. [Table p. 52].

February 25   Bradley observed the comet of 1737. [Table p. 52].

February 27   Bradley observed the comet of 1737. [Table p. 52].

March 4   Bradley observed the comet of 1737. [Table p. 52].

March 12   Bradley observed the comet of 1737. [Table p. 52].

March 14   Bradley observed the comet of 1737. [Table p. 52].

March 17   Bradley observed the comet of 1737. [Table p. 52].

March 19   Bradley observed the comet of 1737. [Table p. 52].

March 20   Bradley observed the comet of 1737. [Table p. 52].

March 22   Bradley observed the comet of 1737 for the final time. [Table p. 52].

June 22   Bradley observed ζ, η Ursæ Majoris. [p. 274].

July 1   Bradley observed η Ursæ Majoris. [p. 274].

July 3   Bradley observed η Ursæ Majoris. [p. 275].

July 6   Bradley observed ζ, η Ursæ Majoris. [p. 275].

Summer   Elizabeth Pound falls ill. She was forced by circumstance to return to Wanstead and to the full time care of her brother's household.

September 1   Bradley observed β, γ Draconis. [p. 275].

September 2   Bradley observed β, γ Draconis. [p. 275].

September 4   Bradley observed β Draconis. [p. 275].

September 6   Bradley observed ϵ, ζ, η Ursæ Majoris, β, γ Draconis. [p. 275].

September 7   Bradley observed β, γ Draconis. [p. 275].

September 8   Bradley observed β, γ Draconis. [p. 275].

September 10   Bradley observed γ Draconis. [p. 275].

September 12   Bradley observed γ Draconis. [p. 275].

September 27   Maupertuis corresponds with Bradley requesting him to correct or check the results of the Lapland expedition, using the zenith sector constructed by Graham. He particularly wanted his observations of δ and α Draconis to be verified. Bradley translated the letter into English for the benefit of George Graham in London. [pp. 404, 406]. [*Memoirs* p. xlii]. Bradley shares his concern with Graham that the French astronomers had not reversed their sector to ensure the veracity of their results. [p. 408].

October 16   Bradley observed Jupiter's 2nd satellite (Europa) emerge from Jupiter. He observed the 1st satellite (Io) emerge from the planet. Moon at full and near Jupiter.

October 19 +   Bradley calculated the elements of the 1737 comet from his own observations. The calculations took up 60 pages of foolscap.

October 19+   Bradley was the first astronomer to adopt Halley's approach to the calculation of cometary pathways. He was highly regarded by Newton. [See also *Principia* Bk. III, Prop. 42.]

October 27   Bradley replied to Maupertuis confirming that the calculations to reduce their observations including aberration were correct. He now revealed that for the past ten years he had been observing another phenomenon which he now identified as a lunar-induced nutation of the Earth's axis. This he had observed through half a complete retrogression of the Moon's nodes. [*Memoirs* p. lxv]. It was this that led to de Fouchy to speak inaccurately of Bradley publishing his theory in 1737.

December Edmond Halley, in spite of his advanced years maintains his observations to the best of his declining abilities, until illness and the paralysis of his right hand made any observations increasingly impossible. It was about this time that Halley wanted to resign from his post as the Astronomer Royal in favour of his protégé James Bradley. His request was consistently refused.

1738   **June 18** Bradley observed ζ, η Ursæ Majoris. [p. 275].

June 19 Bradley observed *Capella*. [p. 275].

June 27 Bradley observed ζ, η Ursæ Majoris. [p. 275].

June 29 Bradley observed ζ, η Ursæ Majoris. [p. 275].

June 30 Bradley observed *Capella*. [p. 276].

July 2 Bradley observed ζ, η Ursæ Majoris. [p. 276].

July 15 Bradley observed ζ, η Ursæ Majoris. [p. 276].

July 25 Bradley observed β, γ Draconis. [p. 276].

July 28 Bradley observed β, γ Draconis. [p. 276].

July 29 Bradley observed β, γ Draconis. [p. 276].

August 5 John Bevis wrote to Bradley concerning Delisle's observations of Jupiter's satellites at St Petersburg from 21 May 1735 to 3 June 1738.

September 9 Bradley determined that the index differed by only 3″ from July 29. [p. 276]. Observed γ Draconis. [p. 276].

September 12 Bradley observed γ Draconis. [p. 276].

September 13 Bradley observed β, γ Draconis. [p. 276].

September 14 Bradley observed γ Draconis. [p. 276].

September 15 Bradley observed γ Draconis. [p. 276].

September 16 Bradley observed γ Draconis. [p. 276].

September 23 Bradley observed η Ursæ Majoris, β Draconis. [p. 276].

October 16 Bradley observed Jupiter's 1st and 2nd satellite (Io and Europa) emerge from the planet. Jupiter's following limb shared the RA of o Pisce. [p. 368].

November 8 Bradley observed Jupiter's 1st satellite (Io) emerge from Jupiter with 15ft tube. [p. 369].

November 10 Bradley observed Jupiter's 1st satellite (Io) emerge from Jupiter with 15ft tube. [p. 369].

November 12 Bradley observed Jupiter's 2nd satellite (Europa) emerge from Jupiter with the 15ft tube. The 3rd satellite (Ganymede) came into view but observation ended with clouds. [p. 369].

December 10 Bradley observed Jupiter's 1st satellite (Io) emerge from the planet. Cloudy at intervals. [p. 369].

December 12 Bradley observed at Shirburne, Aldebaran disappear behind Jupiter 5hr 0min 33sec by the clock then emerge at 6hr 3min 23sec. Determined the latitude of Shirburne. Notes. [p. 369].

December 22 Bradley returns to Wanstead from Oxford and adjusts sector, etc. 6½″. Bradley observed β, λ Cassiopeii, θ, τ, γ Persei. [p. 276].

December 23 Bradley observed β, α Cassiopeii, τ, γ, α Persei. [p. 276].

December 24 Bradley observed τ, γ α Persei. [p. 277].

December 28 Bradley observed β, λ, α Cassiopeii, τ, γ Persei. [p. 277].

December 30 Bradley observed γ Draconis, *Capella*, δ Aurigæ, 18, 35 Camelopardalis. [p. 277]. Sector left until January 14.

1739   **January 14** Bradley observed γ Draconis, *Capella*, δ Aurigæ, 18, 35 Camelopardalis. [p. 277].

January 15 Bradley observed γ Draconis. [p. 277].

January 24 Bradley observed β, γ Draconis. [p. 277].

February 2 Bradley observed *Capella*, δ Aurigæ, 35 Camelopardalis. [p. 277].

February 3 Bradley observed *Capella*. [p. 277].

February 4 Bradley observed *Capella*, δ Aurigæ, 18, 35 Camelopardalis. [p. 277].

February 6 Bradley observed *Capella*, δ Aurigæ, 18, 35 Camelopardalis. [p. 277].

April 19 Bradley returned to Wanstead and noted that index had moved 3.7″. Bradley observed γ, ε, ζ, η Ursæ Majoris, *Alcor*. [p. 277].

April 24 Bradley observed γ Ursæ Majoris. [p. 278].

April 25 Bradley observed ε, ζ, η Ursæ Majoris. [p. 278].

April 26 Bradley observed γ, ε, ζ, η Ursæ Majoris, *Alcor*. [p. 278].

April 27 John Bevis in London wrote to Bradley in Oxford concerning transits of *Polaris*. [p. 416].

May 8 John Bevis in London wrote to Bradley in Oxford stating he may send his observations of *Polaris* to Foulkes in Paris. [p. 416].

July 24 Bradley observed an eclipse of the Sun using the 15ft tube. Observed for 2hrs 20mins. [pp. 369, 370].

August 18 Bradley returned to Wanstead. Index altered 16″ from April 26. Bradley observed γ Draconis. [p. 278].

August 19 Bradley observed β, γ Draconis. [p. 278].

August 20 Bradley observed β, γ Draconis. [p. 278].

August 24 Bradley observed β, γ Draconis. [p. 278].

August 26 Bradley observed β, γ Draconis. [p. 278].

August 27 Bradley observed β, γ Draconis. [p. 278].

August 29 Bradley observed β, γ Draconis. [p. 278].

August 30 Bradley observed β, γ Draconis. [p. 278].

August 31 Bradley observed η Ursæ Majoris, β, γ Draconis. [p. 279].

September 2 Bradley observed β, γ Draconis. [p. 279].

September 3 Bradley observed η Ursæ Majoris. [p. 279]. Returned to Wanstead 20 January 20 1740.

October Bradley wrote to Bevis concerning cometary and pole star observations. [p. 417].

October 8 Bradley observed Jupiter's 1st satellite (Io) in the 15ft tube. [p. 370].

October 25+ Bradley wrote to Lord Macclesfield. Oxford to London? [p. 419]. Lord Macclesfield wrote to Bradley requesting help concerning rectifying the instrument at Shirburne. [p. 421]. Bradley wrote to Lord Macclesfield giving help rectifying the new instrument at Shirburne.

November 9 Bradley observed Jupiter's 1st satellite (Io) in the 15ft tube. [p. 370].

December The Earl of Macclesfield's new observatory was completed around this time. The design and construction of the observatory (then the best in England) by Bradley led to the design and construction of the New Observatory at Greenwich ten years later.

Bradley and Bliss were to find Lord Macclesfield's observatory very convenient as it was only 17 miles from Oxford. A transit instrument and a quadrant were both constructed by Sisson. The quadrant was divided using Graham's division of 96 as well as 90°.

Bradley made his earliest meridian observations at Shirburne Castle during this month.

1740   At or around this time Claude Langlois was appointed *ingénieur en instruments de mathématiques* by the Royal Academy of Sciences in Paris, being allocated quarters at the Louvre, where he was to provide academicians with the scientific instruments they required for their investigations.

Early   Simpson stated that Dr John Bevis was the first (as far as he could determine) who proved experimentally that phenomena observed in RA that were conformable to Bradley's hypothesis determined from observations in declination, probably during 1739. [p. *Essays 1740*, p. 10]. However, Eustachio Manfredi had published his observations in 1729.

January 20   Bradley observed β, γ Draconis, γ, α Persei, 9, δ Aurigæ, *Capella*, 18, 35 Camelopardalis. [p. 279].

January 21   Bradley observed β Cassiopeii. [p. 279].

January 22   Bradley observed τ, γ, α Persei. [p. 279].

January 26   Bradley observed β, γ Draconis. [p. 279]. Broke wire. [p. 279].

January 29   Bradley observed θ Persei, *Capella*, 18 Camelopardalis. [p. 279]. Put on a new wire. [p. 279].

May   Bradley returns to Wanstead and observed from May 30.

May 30   Bradley observed ε, ζ, η Ursæ Majoris. [p. 279].

May 31   Bradley observed ε, ζ, η Ursæ Majoris. [p. 279].

June 1   Bradley observed ε, ζ, η Ursæ Majoris. [p. 279].

June 2   Bradley observed α Cassiopeii, *Capella*, γ, ε, ζ, η Ursæ Majoris. [p. 279].

June 3   Bradley observed *Capella*, γ, ε, ζ, η Ursæ Majoris. [p. 280].

June 4   The 2nd Earl of Macclesfield began the regular series of his own observations. These were maintained to the end of his life in 1763. He and Bradley became regular partners. The observatory was sited 100 yards south of the gateway to Shirburne Castle.

June 4   Bradley observed *Capella*, ε, ζ, η Ursæ Majoris at Wanstead. [p. 280]. Lord Macclesfield begins observations at Shirburne castle from this date.

June 6   Bradley observed γ, ε, ζ Ursæ Majoris. [p. 280]. Returns to Wanstead August 9.

August 9   Bradley observed β, γ Draconis. [p. 280].

August 18   Bradley observed β, γ Draconis. [p. 280].

August 19   Bradley observed β, γ Draconis. [p. 280].

August 21   Bradley observed β, γ Draconis. [p. 280].

August 24   Bradley observed β, γ Draconis. [p. 280].

August 25   Bradley observed β, γ Draconis. [p. 280].

August 28   Bradley observed β, γ Draconis. [p. 280].

August 29   Bradley observed β, γ Draconis. [p. 280].

August 30   Bradley observed γ Draconis. [p. 280].

September 5   Bradley observed β, γ Draconis. [p. 280].

September 8   Bradley observed β, γ Draconis. [p. 280].

September 10   Elizabeth, widow of James Pound (deceased in 1724), died in Wanstead. She left half of her extensive fortune (about £20,000) to Bradley and half to Bradley's cousin Sarah, who was Elizabeth's step-daughter.

September 20   Bradley observed β, γ Draconis. [p. 281]. These were the first observations made by Bradley after the death of Elizabeth. She had been ill for about three years and Bradley returned to Wanstead about a month before her death and merely maintained a routine of basic observations. He made no more observations until ten days after her death, a sign surely of his sense of loss.

November 5   Bradley observed Jupiter's 4th satellite (Callisto) as it disappeared behind the planet in his 15ft tube at Oxford. [p. 370].

1741   **January 26**   Bradley observed β, γ Draconis, *Capella*. [p. 281].

February 6   Bradley observed Jupiter's 3rd satellite (Ganymede) emerge from the planet. [p. 370].

February 8   Bradley observed Jupiter's 2nd and 1st satellites (Europa & Io) emerge from the planet. [p. 370].

February 13   Bradley observed Jupiter's 3rd satellite (Ganymede) disappear behind the planet. [p. 370]. The Earth's Moon was close to Jupiter.

February 15   Bradley observed Jupiter's 2nd satellite (Europa) emerge from the planet. [p. 370].

February 17   Bradley observed Jupiter's 1st satellite (Io) emerge from the planet. [p. 370].

February 24   Bradley observed Jupiter's 1st satellite (Io) emerge from the planet. [p. 370].

March 12   Bradley observed Jupiter's 1st satellite (Io) emerge from the planet in 15ft tube. Moon near to Jupiter. [p. 370]. Notes about Jupiter's 2nd satellite (Europa). [p. 371].

March 26   Bradley observed Jupiter's 1st satellite (Io) emerge from the planet. [p. 371].

May 3   Bradley observed Jupiter's 3rd satellite (Ganymede) emerge from the planet. [p. 371].

August 27   Bradley observed β, γ Draconis. [p. 281].

August 28   Bradley observed β, γ Draconis. [p. 281].

August 29   Bradley observed γ Ursæ Majoris, β, γ Draconis. [p. 281].

August 31   Bradley observed γ, ε Ursæ Majoris, β, γ Draconis. [p. 281].

September 1   Bradley observed ε Ursæ Majoris, γ Draconis. [p. 281].

September 2   Bradley observed β, γ Draconis. [p. 281].

September 3   Bradley observed ε, ζ Ursæ Majoris. [p. 281].

September 4   Bradley observed ζ, η Ursæ Majoris. [p. 281].

September 6   Bradley observed β, γ Draconis. [p. 281].

September 20   Bradley observed γ Draconis. [p. 281].

September 21   Bradley observed ε, ζ, η Ursæ Majoris, β, γ Draconis. [p. 281].

September 23   Bradley observed ε, ζ, η Ursæ Majoris, γ Draconis. [p. 281]. Next observations at Wanstead 5 September 1742

1742   During this year Maupertuis was elected as the Director of the Royal Academy of Sciences in Paris. That a Newtonian was selected for this important post indicates a 'sea change' in the ethos of the Royal Academy.

January 14   Lord Macclesfield corresponds with his one-time mathematics tutor William Jones concerning his hope that James Bradley be not overlooked once Edmond Halley has died. Halley was gravely ill at this moment. [*Memoirs* p. xlvi]. He died later this day.

January 16   William Jones replies to Lord Macclesfield. He was aware of Halley's impending demise and shares his Lordship's concerns about Bradley's candidacy for the office of Astronomer Royal even though he was the outstanding candidate. [*Memoirs* p. xlvii].

February +   Bradley remained in Oxford until June, detained by his lecture courses in astronomy and experimental philosophy. [*Memoirs* p. xlix].

February 2 Sir Robert Walpole's administration was defeated in a division in the House of Commons. Walpole finally resigns from his office of First Lord of the Treasury (Prime Minister) 11 February 1742. [*Memoirs* p. xlix].

February 3 Walpole appoints Bradley as the third Astronomer Royal, succeeding Halley. It was the first action effected by Walpole after his defeat in the House. [*Memoirs* p. xlix].

February 11 Sir Robert Walpole resigned from office.

February 21+ A comet was first observed at the Cape of Good Hope, then in Europe. Bradley determined the elements of the comet's orbit. Gael Morris computed longitudes and latitudes of this comet. Only one observation has been found which can be linked to this comet between 21 February and 15 March. [p. 371].

February 22 James Bradley was created a Doctor of Divinity at Oxford University by the rare procedure of by Diploma. [*Memoirs* p. xlix].

February 24 Bradley first observed the comet by determining its distance in latitude and longitude from β Cygni and α Lyræ. [p. 371].

February 25 Bradley began observing a comet before determining the elements of its orbit. Bradley stated that the first account he had of this comet was from an observation made on Monday 20 February 1742 at Melksham in Wiltshire. [*Memoirs* xciii].

June Bradley moved to the Royal Observatory at Greenwich to take up his new duties as the third Astronomer Royal in succession to his mentor Halley. [p. 381].

When Bradley took up the office of Astronomer Royal he was entitled to no more than £100 p.a., the sum assigned to John Flamsteed in 1675. Even this modest sum was further reduced by various fees. Out of this further reduced amount Bradley had to pay for the services rendered by his nephew, a sum of £26 p.a.

James Bradley appoints his fourteen-year-old nephew, the son of his eldest brother William, as his assistant at Greenwich, employing him for £26 p.a.

Bradley discovered that Halley had allowed the transit instrument to fall badly out of alignment. Bradley noted that Halley used only two wires, making it impossible to use in conjunction with other instruments. Bradley ordered Sisson to fit two extra vertical lines.

June 15 Bradley made his earliest observations as the Astronomer Royal with Graham's 8ft quadrant. [*Memoirs* p. lii].

June 15+ Bradley sought to determine the errors of the Graham 8ft quadrant in the seven years since he and Graham had adjusted it last in 1735. It was obvious that Halley had failed to notice, or sought to remedy an error of 3/16th inch in the plane of the quadrant.

July For the fifth time James Bradley was elected a member of the Council of the Royal Society. From around this time he began work on a table of refractions, determined from observations made at Shirburne, a problem that engaged him until his demise in 1762.

July 5 Bradley discovered that Halley's clocks at Greenwich were not compensated for changes in temperature. He began to examine the 8ft mural quadrant.

July 6 Bradley completed his examination of the 8ft quadrant. It gave zenith distances too small by 34½″. This was a measure of its motion since it was set up and adjusted by George Graham and James Bradley last during 1735. [*Memoirs* p. lii]. About 800 more observations were made before the end of 1742. [*Memoirs* p. liii].

July 24 The line of collimation of the transit instrument at the Royal Observatory was adjusted. (RGO 3/31 doc. 5).

July 25   After spending several weeks repairing and adjusting many of the instruments in the Royal Observatory, he makes his first observations using Halley's transit instrument. By the end of 1742, Bradley and his nephew made some 1,500 observations with the transit instrument. [*Memoirs* p. lii].

July 31   On this day the transit instrument was set by the mark made by Edmond Halley on the park wall. [*Memoirs* p. lii].

August 1   From this date until August 14 several stars were observed with the quadrant and the transit instrument and by comparing the times of their transits the errors of the plane of mural quadrant were found to be as entered in a table drawn up before August 14.

August 18   The mural quadrant was rectified on this date.

August 20   Bradley altered the plane of the quadrant as appears by the Table of Errors then found. (RGO 3/31 doc 5, fol. 2r – 2v).

August 29   George Graham had the quadrant telescope taken from Greenwich to London where he moved the vernier further from the centre.

September 3   The clock was firmly fixed against the brick wall. [*Memoirs* p. lii]. It was from this date that John Bradley began working as his uncle's assistant. He was trained in the required skills in the three months since arriving with his uncle at the Royal Observatory in June 1742.

September 5   Bradley observes β, γ Draconis. These were his first observations at Wanstead after becoming Astronomer Royal. [p. 281].

September 6   Bradley observed γ, ε, ζ Ursæ Majoris, β Draconis at Wanstead. [p. 281].

September 11   The line of collimation of the quadrant telescope was adjusted and after several transits were observed with both the quadrant and the transit instrument a table of errors was collected on or about 19 September. On this date, the quadrant telescope had been returned from Jonathan Sisson.

October 11   George Graham fitted a gridiron pendulum as developed by John Harrison to the clock at Greenwich. [*Memoirs* p. lii].

October 16   From this date until the return of his uncle on 4 December, all of the entries are in John Bradley's hand. It reveals that James Bradley had implicit faith in the actions of his young nephew whilst he was in Oxford.

December 4   Bradley returned to Greenwich from Oxford after presenting a course in experimental philosophy.

December 15   Nathaniel Bliss in Oxford corresponded with James Bradley in Greenwich concerning Lord Macclesfield's observations of Jupiter's satellites at Shirburne. [p. 422].

December 31   Transit observations for 1742 occupy 177 folio pages. Quadrant observations take up 148 folio pages from 25 July to 31 December 1742.

1743   The salons in Paris began to abandon the Cartesian account of the world, popularized in works such as Fontenelle's *Entretiens sur la pluralité des mondes* in favour of works such as Algarotti's *Il Newtonianismo per la dame*. Another indication of the changes in Parisian natural philosophy.

January +   Busy though Bradley was as he put the instruments of the Royal Observatory through its paces he still found time to plan and initiate a course of experiments to determine the exact length of a seconds pendulum. [*Memoirs* p. lvi].

March   Lord Macclesfield began making observations at Shirburne Castle to compare with those made at Greenwich. [*Memoirs* p. liv].

August 1   Bradley wrote a memorandum about the acquisition of a shorter-focus eyeglass for the transit instrument. With a focus of 1½″ with a magnification of up to 40×.

August 8   The industriousness of Bradley and his fifteen-year old-nephew John may be judged by this single day. On this day, 255 observations were made with the transit instrument inherited from Edmond Halley. Several times during this year over 200 observations were made with the transit at Greenwich. On the same day, 181 observations were made with the 8ft quadrant also commissioned by Halley and constructed by George Graham. [*Memoirs* p. liii]. Not since Flamsteed began work with his new arc in 1689 had such productivity been witnessed at the Royal Observatory.

September 2   Bradley observed β, γ Draconis at Wanstead. [p. 282].

September 3   Bradley observed β Draconis at Wanstead. [p. 282]. The next recorded observations with the zenith sector was on 2 September 1745, two years later.

September 13   Bradley returned to a subject he formerly studied at Wanstead, determining the length of a seconds pendulum. After planning he began a series of observations until October 13 and again at the end of December.

September 13   Bradley determined the length of seconds pendulum as 39.1349 inches. [p. 384]. This was the beginning of his studies of seconds pendulums at Greenwich.

September 14   Bradley determined pendulum length at 39.1357 inches. [p. 384].

September 15   Bradley determined pendulum length at 39.1369 inches. [p. 384].

September 19   Bradley determined pendulum length at 39.1352 inches. [p. 384].

September 23   Bradley determined pendulum length at 39.1384 inches. [p. 384].

September 24   Bradley determined pendulum length at 39.1383 inches. [p. 384].

September 25   Bradley determined pendulum length at 39.1395 inches. [p. 384]. Ball altered.

September 27   Bradley determined pendulum length at 39.1395 inches. [p. 384].

September 28   Bradley determined pendulum length at 39.1396 inches. [p. 384]. Rod screwed into another hole.

September 30   Bradley determined pendulum length at 39.1376 inches. [p. 384]. Wire screwed into the opposite hole.

October 3   Bradley determined pendulum length at 39.1421 inches. [p. 384]. Rod screwed into the hole used on September 30.

October 4   Bradley determined pendulum length at 39.1390 inches. [p. 384].

October 5   Bradley determined pendulum length at 39.1412 inches. [p. 384].

October 6   Bradley determined pendulum length at 39.1410 inches. [p. 384].

October 7   Bradley determined pendulum length at 39.1415 inches. [p. 384].

October 8   Bradley determined pendulum length at 39.1410 inches. [p. 384].

October 9   Bradley determined pendulum length at 39.1410 inches. [p. 384].

October 10   Bradley determined pendulum length at 39.1405 inches. [p. 384].

October 11   Bradley determined pendulum length at 39.1422 inches. [p. 384].

October 12   Bradley determined pendulum length at 39.1417 inches. [p. 384]. Rod taken down, altered and set up again.

October 21   Bradley observed an eclipse of the Moon in his 7ft tube. Conditions were cloudy and hazy. Observed for 1hr 10mins.

October 24   Bradley observed the transit of Mercury at Greenwich. Clouds and haze did not allow clear observation of the ingress of the planet. After 2hr he could see with more certainty. Observed the phenomenon for 2½hr. [pp. 371, 372].

December   The French astronomer Le Monnier published his *Théorie des Comètes* late in 1743. He acknowledged his debt to James Bradley. [*Memoirs* p. xliv].

December 9   The object that would become the great comet of 1744 was observed by Klinkenberg at Haarlem.

December 23   The great comet was first reported in the vicinity of London by Dr John Bevis who sent an account of it to Bradley. [*Memoirs* p. lviii]. Bevis had been the discoverer of the object we now call the *Crab Nebula* (M1) in 1731. The comet was also observed by Lord Macclesfield at Shirburne Castle. [p. 425].

December 24   Lord Macclesfield corresponded with Nathaniel Bliss, Savilian Professor of Geometry at Oxford, giving his account of the comet. He says that he will send an account to Bradley and to Graham. [p. 423]. [*Memoirs* p. lviii].

December 26   Bradley observed the great comet with the 7ft tube taking its RA, declination. [p. 372].

December 27   Bradley observed the comet. Determined the RA and declination. [p. 373].

December 28   Bradley observed the comet. Determined the RA and declination. [p. 373]. Bradley also determined pendulum length at 39.1462 inches. [p. 384].

December 29   Bradley determined pendulum length at 39.1414 inches. [p. 384].

December 30   Bradley observed the comet. Determined the RA and declination. [p. 373].

December 31   James and John Bradley made nearly 18,000 observations at Greenwich during 1743. [*Memoirs* p. lv]. Bradley observed the comet. Determined the RA and declination. [p. 373].

1744   By 1744, Bradley was recording barometric pressures and thermometric observations during his researches on atmospheric refraction in order to improve the reduction of his observations at the Royal Observatory.

January 4   Bradley continued with his pendulum experiments. Using a small ball Bradley determined the pendulum length of the seconds pendulum at 39.1460 inches. [p. 385]. Small ball.

January 6   Bradley observed and determined the RA and declination of the great comet using the 15ft tube. [p. 373].

January 10   Bradley observed and determined the RA and declination of the great comet. [p. 373].

January 11   Bradley observed and determined the RA and declination of the great comet. [p. 373].

January 12   Nathaniel Bliss corresponded with Bradley about the comet as observed at Shirburne. [p. 426].

January 13   Bradley observed and determined the RA and declination of the great comet. [p. 374].

January 15   Bradley observed and determined the RA and declination of the great comet. [p. 374].

January 16   Bradley observed and determined the RA and declination of the great comet. [p. 374].

January 20   Bradley determined pendulum length at 39.1406 inches. [p. 384]. Great ball.

January 21   Bradley observed and determined the RA and declination of the great comet. [p. 374]. Bradley also determined pendulum length at 39.1490 inches. [p. 384]. This was determined again on 20 July.

January 23   Bradley observed and determined the RA and declination of the great comet. [p. 374].

January 29   Bradley observed and determined the RA and declination of the great comet. [p. 374].

January + Bradley wrote a short letter to Macclesfield concerning solar and lunar transits. [p. 429].

February + Bradley observed the finest comet of the eighteenth century. [Lalande, Section 3209] [*Memoirs* p. lvii].

February 1 Loÿs de Cheseaux wrote an account of the great comet describing the light exceeding even that of *Sirius*. [*Memoirs* p. lvii]. Bradley observed and determined the RA and declination of the great comet. [p. 374].

February 6 Bradley observed and determined the RA and declination of the great comet. [p. 374].

February 8 Bradley noted that the light of the comet was now equal to that of Jupiter in opposition. [*Memoirs* p. lvii].

February 9 Bradley twice observed and determined the RA and declination of the great comet. [p. 375].

February 10 Bradley observed and determined the RA and declination of the great comet. [p. 375].

February 12 Bliss corresponded with Lord Macclesfield indicating that Bradley hoped that his Lordship had observed the comet with its light now equalling that of Venus. [*Memoirs* p. lviii]. Bradley observed and determined the RA and declination of the great comet. [p. 375].

February 13 Bradley observed and determined the RA and declination of the great comet. [p. 375].

February 17 Bradley observed and determined the RA and declination of the great comet. [p. 375].

February 18 Bradley and many others noted that the comet was almost equal in brilliance to Venus and was larger in size. [*Memoirs* p. lvii].

February 19 Bliss wrote to Macclesfield stating that the great comet of 1744 had been observed on the meridian at Shirburne. [*Memoirs* p. lxxxv].

February 20+ The light of the comet now approached magnitude−7.

February 21 Le Monnier at Paris wrote in French to Bradley at Greenwich concerning the remarkable 1744 comet. [p. 427].

March Lacaille wrote to Bradley expressing his regret that he had found Halley's Lunar Tables unusable. [p. 430] [*Memoirs* p. xcvi].

June 25 James Bradley married Susannah Farmer née Peach. They were to have one child, a daughter, also named Susannah who was born in Greenwich in 1745.

July+ After spending the spring in Oxford he continued working on seconds pendulums. [*Memoirs* p. lvi].

July 20 Bradley determined pendulum length at 39.1443 inches. [p. 385].

July 21 Bradley determined pendulum length at 39.1355 inches. [p. 385].

July 27 Bradley determined pendulum length at 39.1332 inches. [p. 385].

July 28 Lacaille wrote to Bradley. This letter contains the French astronomer's high appraisal of the English astronomer. [p. 429]. Bradley twice determined pendulum length at 39.1244 inches and 39.1249 inches.

August 2 Bradley determined pendulum length at 39.1292 inches. [p. 385].

August 3 Bradley twice determined pendulum length at 39.1305 inches and 39.1326 inches. [p. 385].

August 4 Bradley determined pendulum length at 39.1374 inches. [p. 385].

September 15 Bradley determined pendulum length at 39.1385 inches. [p. 385].

September 16 Bradley twice determined pendulum length at 39.1339 inches and 39.1324 inches. [p. 385].

September 26   Bradley determined pendulum length at 39.1358 inches. [p. 385].

October 6   Bradley determined pendulum length at 39.1372 inches. [p. 385].

October 7   Bradley determined pendulum length at 39.1411 inches. [p. 385]. The next experiment from 3 February 1745.

November+   Bradley took the 8ft quadrant down for extensive modifications.

December+   Bradley spent time reducing the entire British Catalogue to epoch 1744. In fact, prior to 1750 Bradley examined the entire catalogue twice more. This reduced catalogue may well have escaped notice for it was described as containing 'the southern stars'. See [Catalogue of the MSS. of the Royal Observatory Greenwich p. 3]. Lord Macclesfield probably had his own share of the work involved, for Bradley's patron had become his closest partner. [*Memoirs* p. liv].

1745   During 1745 John Bird fitted a new micrometer to the telescope of the mural quadrant. A three-inch focus eyepiece was substituted with one of two-inch focus.

February 3   Bradley continued with the Greenwich pendulum experiments. Bradley determined pendulum length at 39.1475 inches. [p. 385].

February 4   Bradley determined pendulum length at 39.1472 inches. [p. 385].

February 11   Bradley determined pendulum length at 39.1409 inches. [p. 385]. Small ball.

February 12   Bradley twice determined pendulum length at 39.1400 inches and 39.1402 inches. [p. 385].

February 18   Bradley determined pendulum length at 39.1415 inches. [p. 385].

April 24   John Bevis corresponded with Bradley about the Tables of Jupiter's satellites recognizing that Bradley was the true author of Halley's Jovian Tables. [p. 431].

June   John Bird discovered that Sisson's quadrant, in use at Shirburne, was 10″ in defect. [*Memoirs* p. lxxxiii].

June +   The figure of Graham's 8ft quadrant when examined with one of Graham's levels when it was put up was enlarged by 5″. [p. 478] [*Memoirs* p. lxxxiv].

July 12   Bradley determined pendulum length at 39.1356 inches. [p. 385]. Using the great ball.

July 13   Bradley determined pendulum length at 39.1344 inches. [p. 385].

July 16   Bradley determined pendulum length at 39.1342 inches. [p. 385].

July 18   In order to improve the precision of the quadrant over and above Halley's observational practice, a new micrometer screw was applied to the 8ft quadrant. [*Memoirs* p. lv]. There were 39¼ threads in one inch. [*Memoirs* p. lv].

July 29   Bradley twice determined pendulum length at 39.1296 inches and 39.1300 inches using the small ball. [p. 385].

August+   By this time Bradley was capable of reading off angles as small as a single second of arc with the quadrant. This followed the fitting of the new micrometer screw on 18 July.

August 1   Bradley determined pendulum length at 39.1318 inches. [p. 385]. Bradley undertakes his next pendulum experiments from 28 January 1749. [p. 385].

August 15+   Bradley fitted an eyeglass with a shorter focal length in the 8ft quadrant's telescope.

Summer?   James Bradley composed the paper 'State of the Instruments at the Greenwich Observatory when Dr Bradley became Astronomer Royal'.

August?   Lady Catharine Manners, wife of Prime Minister Henry Pelham, was appointed Ranger of Greenwich Park. [*Gent. Mag.* Vol. 15, p. 109]. The Pelhams occasionally resided in Greenwich. It led to the development of a firm friendship between the Bradleys and the Pelhams. [*Memoirs* p. lxxxi]. The 4th Earl of Chesterfield also lived nearby. This was eventually to lead to the 1750/51 Parliamentary Bill to reform the calendar.

September   Bradley discerns various defects in the quadrantal arc.

September 2   Bradley makes his first visit to Wanstead since 3 September 1743. He stayed for a few days with Matthew Wymondsold, making observations at his late aunt's house with the zenith sector. Bradley observed β, γ Draconis. [p. 282].

September 3   Bradley observed γ Draconis. [p. 282].

September 5   Bradley observed ε, η Ursæ Majoris, β, γ Draconis. [p. 282]. Euler published his Lunar Tables, sending a copy to Bradley. [p. 433]. [*Memoirs* p. xcvi]. Lalande stated that Halley's Lunar Tables could not be depended on for 1ⁱ [Lalande Sect.1464] [*Memoirs* p. xcvi].

Leonhard Euler corresponded with Bradley.

September 9   A memorandum reveals that Lord Macclesfield's observation book was based on Flamsteed's catalogue, held at Shirburne Castle Observatory. [*Memoirs* p. liv].

September 10   Bradley having returned to Greenwich from Wanstead observes the zenith distances of γ and β Draconis using the 8ft mural quadrant.

September 11   Bradley observes the zenith distances of γ and β Draconis.

September 12   Bradley observes the zenith distances of γ and β Draconis.

September 13   Bradley observes the zenith distances of γ and β Draconis.

September 14   Bradley observed the zenith distances of γ and β Draconis. Bradley makes notes and calculated the distance between the two stars to be 58′ 16″. His next visit to Wanstead was during the following year on 15 September 1746.

1746   **April?**   Around this time Bradley fully recognized the futility of continuing with the instruments at his disposal at the Royal Observatory. Those he possessed largely required major works to make them usable for the survey of the stars observable from Greenwich. In addition he recognized the need for new instruments, most pressing of which was a new quadrant so that he possessed one to survey stars passing through the meridian to the north of Greenwich and another to survey those to the south. Bradley rigorously tested the scales of the Graham quadrant. By 1746, Bradley was aware of an eccentricity in the arc in which the 45° part of the scale was 15¾″ out of alignment with the ends at 0° and 90°. The likely cause of the error was its bi-metallic construction. It was a phenomenon that Bradley was aware of for he was familiar with the function of the gridiron pendulum where the differences of the coefficients of expansion of different metals were used to keep the expansion of a pendulum within narrow bounds in differing temperatures.

May 11   C. Weinstein corresponded with Bradley concerning Euler's letter of the previous 5 September.

June 1   A new semicircle was put to the transit instrument, probably of larger radius than that originally attached to the instrument. [*Memoirs* p. liv].

July   Euler wrote to the Royal Society to state that James Bradley had been received as a member of the Royal Academy of Berlin.

September 15   Bradley observed β, γ Draconis at Wanstead. [p. 282].

September 19   Bradley observed ε, ζ, η Ursæ Majoris, β Draconis. [p. 282].

September 20   Bradley observed β, γ, ζ, η Ursæ Majoris, β, γ Draconis. [p. 282].

September 23   Bradley observed γ, ε Ursæ Majoris. [p. 282]. The next observation in Wanstead was 27 February 1747. [p. 282].

1747   **February?**   Further examination of the quadrantal arc revealed that the eccentricity did not exceed 1/400th of an inch.

February 27   Bradley visited Wanstead for the first time since 23 September 1746. Bradley observed β, α Cassiopeii, α Persei, *Capella*, δ Aurigæ, 35 Camelopardalis, β, γ, ε Ursæ Majoris. [p. 282].

Summer?   John Bradley aged nineteen and Captain John Campbell aged twenty-seven developed a mutual interest in the method of lunar distances to determine the longitude at sea. Campbell observed lunar distances at Greenwich with James and John Bradley.

August 31   Bradley begins his final series of observations at Wanstead with the zenith sector at the late Elizabeth Pound's town house. Bradley observed γ Draconis. [p. 283].

September 1   Bradley observed β, γ Draconis. [p. 283].

September 2   Bradley observed ζ, η Ursæ Majoris, β, γ Draconis. [p. 284].

September 3   Bradley observed ε, η Ursæ Majoris. [p. 284]. This completed the entire sequence of observations made with the zenith sector at Wanstead begun 19 August 1727.

September 3   Bradley completed the series of observations begun in 1727 confirming his second great discovery, the nutation of the earth's axis, although some resistance was still met with from the usual suspects. Like Halley before him Bradley required an entire retrogression of the Moon's nodes. This investigation exceeded twenty years. [*Memoirs* p. lxii]. Up to this discovery the annual precession of the Earth's axis was not known with any great degree of precision. All that could be known with any degree of certainty was that it was between 50″ and 51″ p. a. [*Memoirs* p. lxii].

October 19   Sarah Pound, daughter of James Pound and his first wife Sarah, died aged 34 in Greenwich. She was a wealthy woman with at least £10,000. [*Memoirs* p. ii].

November 27   Delisle in Paris wrote to Bradley at Greenwich. [p. 433].

December 31   Bradley wrote his letter to Lord Macclesfield on his discovery of the nutation of the Earth's axis, to be read to the Royal Society.

1748   **January 7**   The first part of Bradley's paper on the nutation of the Earth's axis in the form of a letter addressed to George Parker, 2nd Earl of Macclesfield, was read at a meeting of the Royal Society. [p. 17]. The paper was a final vindication of Newtonian over Cartesian theory.

January 14   The concluding part of Bradley's paper on the nutation of the Earth's axis in the form of a letter addressed to George Parker, 2nd Earl of Macclesfield, was read at a meeting of the Royal Society. [p. 17]. The observed motion of η Ursæ Majoris led Bradley to suppose the pole to move in an ellipse rather than a circle.

January 14   After the reading of the nutation paper, Bradley was awarded the Copley Medal, the highest scientific award of the Royal Society by Martin Foulkes, the Society's President, who was instrumental in the election of Bradley as the Savilian Professor of Astronomy at Oxford in 1721. Bradley was detained by his teaching duties in Oxford. Bradley was unable to receive the award for his discovery of the aberration of light because the Copley Medal was introduced in 1731, two years after his paper on the aberration.

February   Bradley's paper on the nutation of the Earth's axis published in the *Philosophical Transactions*. [*Phil. Trans.* No.485, Vol. 45, p. 1]. The manuscript was dated 31 December 1747.

April 21   Bradley observed a comet using the zenith sector, now removed from Wanstead and situated at the Royal Observatory. Transits of this comet were inserted in Volume II of Greenwich observations. [p. 425]. There was no mention of these observations in the quadrant book. [p. 375]. Using the 17 ft tube the RA and declination recorded. [p. 376].

April 22   Bradley observed the comet via the sector. He made ten separate observations of the RA and declination. Notes. [p. 377]. Bradley used the sector because of its complete dependency and reliability.

April 23   Bradley observed the comet via the sector. He made six observations of the RA and declination. Cloudy at times. [p. 377].

April 24 Bradley observed the comet via the sector. He made three observations of the RA and declination. [p. 377].

April 27 Bradley observed the comet via the sector. He made 3 observations of RA and declination. [p. 377].

May 2 Lacaille wrote to Bradley from Paris (after France and Great Britain had been at war since 1744) complaining that no intelligence had reached Paris from London in three years, no doubt seeking to know whether Bradley's paper on the 'deviation' (Lacaille's term for the nutation) had been published. [p. 438]. [*Memoirs* p. lvii]. Lacaille continued to use the term 'deviation' rather than 'nutation'. Lacaille's term may have been used by Bradley until he was sure it was a nutation.

Bradley observed the comet via the sector. He made two observations of the RA and declination. [p. 377].

May 12 Bradley observed the comet via the sector. He made three observations of the RA and declination. [p. 377].

May 14 Bradley observed the comet via the sector. He made four observations of the RA and declination. [pp. 377–378].

May 15 Bradley observed the comet via the sector. He made five observations of the RA and declination. [p. 378].

May 16 Bradley observed the comet via the sector. He made four observations of the RA and declination. [p. 378].

May 17 Bradley observed the comet via the 7 ft tube. He made nine observations of the RA and declination. [pp. 378–379].

May 18 Bradley observed the comet via the 7 ft tube. He made three observations of the RA and declination. [p. 379].

June Bradley corresponded with Lacaille about nutation. Lacaille approached Jacques Cassini to obtain sight of his copy of Bradley's nutation paper. Cassini had long maintained a resistance to Bradley's theory of aberration because it contradicted the Cartesian assertion that the velocity of light was instantaneous.

June Lacaille may have borrowed Jacques Cassini's copy of Bradley's nutation paper. [p. 457] before translating it into French, parts of which he read to the Royal Academy of Sciences in Paris. [*Memoirs* lxix].

June Bradley calculated the elements of the comet he observed from 21 April to 18 May. [*Memoirs* p. xciii].

June 3 Delisle corresponded with Bradley. [p. 442].

June 13 Delisle again corresponded with Bradley. [p. 443].

July 14 Alex Irvine wrote to James Ferguson about a solar eclipse. [p. 447].

Summer James Bradley revealed the dire state of the instruments and facilities at the Royal Observatory to the Visitors.

August 1 James Ferguson wrote to Bradley giving a report of a solar eclipse observed by Alex Irvine. [p. 446].

August 12 Bradley received a lengthy letter from the German astronomer A. N. Grischow (1726–1760) which recalled observations made with 'Bradley's' sector (Graham was very elderly at this time entering the final years of his life). Grischow sent tables of corrections of lunar tables.

August 19 Bradley also received a letter from the mathematical prodigy Alexis Claude Clairaut (1713–1765) who had studied the infinitesimal calculus when he was 10, becoming a member of the Academy at 18.

August 22 Lacaille corresponded with Bradley concerning the nutation. [p. 454].

August 24 Bradley was created one of the Foreign Associates of the Royal Academy of Sciences in Paris. Bradley received the congratulations of Grischow and Lacaille. [p. 451]. [*Memoirs* p. lxxi]. Bradley wrote to the Comte de Maurepas acknowledging the great honour bestowed upon him.

August 29 Lord Macclesfield took loan of long-focus glasses of 210 feet and 120 feet from the Royal Society.

November Bradley's paper on the nutation, translated by Lacaille, was published in the *Journal de Trevoux*.

James Bradley prevailed on the Board of Visitors to support a petition to the Board of Admiralty to allow the purchase of new instruments for the Royal Observatory.

November 9 The President of the Royal Society (Martin Foulkes) informed the Council that he and Dr Bradley had been informed that the Lords of the Admiralty promised their assistance to re-equip the Royal Observatory. A schedule detailing the needs amounted to £1,000. This was immediately granted by King George II. [*Memoirs* p. lxxiv]. These moneys came from the sale of old navy stores. [*Memoirs* p. lxxv]. In addition to these sums, new buildings were erected to house the new and repaired instruments. Another considerable sum must have been laid out, probably by the Board of Admiralty or the Board of Longitude. [*Memoirs* p. lxxv].

November 9+ Up to the time that Bradley was able to acquire new instruments and a new observatory, he was supposed to maintain the observatory, repair instruments, and pay for assistants, all out of an annual grant of just £100.

December Bonaventura Suarez of the Jesuit Mission to Paraguay corresponds with Bradley offering his observations of the 1748 comet. [p. 444].

December 23 John Bevis wrote to Bradley referring to Lacaille. [p. 456].

1749 **January 28** Bradley continues experiments to determine the length of a seconds pendulum. Using the big ball he determines a length of 39.1488 inches. [p. 385].

January 29 Bradley determines a length of 39.1491 inches. [p. 385].

June 20 Matthew Raper corresponded with Bradley. [p. 459].

June 30 Matthew Raper wrote to Bradley thanking him for help viz. the calculation of parallaxes. [p. 460].

July The zenith sector suspended at Elizabeth's town house in Wanstead was removed, to be placed in the New Observatory at Greenwich. It had not been used since 3 September 1747, the date of his last observations leading to the discovery of the nutation of the Earth's axis.

August The Council of the Royal Society considered a petition to the Board of Admiralty prepared by Bradley, signed by Martin Foulkes, President of the Royal Society, and the Council. Not presented to the Lords of the Admiralty until October due to alterations to the petition.

Summer During a period when new instruments were being made and old instruments were being modified or repaired, Bradley reduced the Wanstead observations made with the zenith sector from 19 August 1727 to 3 September 1747. He reduced all of the observations of γ Draconis, β Draconis, α Cassiopeii, β Cassiopeii, *Capella*, α Persei, γ Persei, 35 Camelopardalis (HR2123), γ Ursæ Majoris, ε Ursæ Majoris, η Ursæ Majoris. [pp. 302–338].

A letter was written to Lord Macclesfield possibly from James Bradley. [p. 461].

'Hodgson of Christ's Hospital published tables of Jupiter's satellites, studiously avoiding all notice of mention of the work of Bradley, even though there can be no doubt he was well acquainted with Bradley's observations and tables'. [See Bailli's *Essai sur la Théorie des Satellites de Jupiter* 1766].

D'Alembert determined the ellipse of the nutation.

Bradley requested the telescope maker Short to grind object glasses for the new and repaired instruments at the Royal Observatory.

October 9 Martin Foulkes, President of the Royal Society, informed the Council of the Royal Society that he and Bradley had attended the Lords of the Admiralty abd that their Lordships had promised their assistance in forwarding the Society's petition. The sum required was estimated to be £1,000. Consequently, it was granted by George II without delay. [*Memoirs* p. lxxiv]. George II directed an order that a sum of £1,000 be paid to Bradley in order to repair the old instruments and provide new ones. These moneys came from the sale of old navy stores. [*Memoirs* p. lxxv]. The sum included £85 16s for instruments already paid for out of Bradley's own pocket. In addition to these sums, new buildings were erected to house the new and repaired instruments; costs possibly came from the Board of Ordnance. Another considerable sum must have been laid out, probably by the Board of Admiralty or the Board of Longitude. [*Memoirs* p. lxxv].

Winter Lord Macclesfield drew up a paper on the solar and lunar years for 1749–1750 after discussions with Bradley. Published in the *Philosophical Transactions*. [*Phil. Trans.* Vol.XLVI, p. 417].

The New Observatory was being constructed. Today better known as the Transit House through which the Prime Meridian passes now some 16 feet to the east of Bradley's original meridian in the transit room of the New Observatory.

1750   **February 8** At 8.23 am an earthquake shock was recorded at the Royal Observatory.

February 16 John Bird's new quadrant was suspended in the west side of the pier in the New Observatory. [*Memoirs* p. lxxvii].

June Bradley set up the new John Bird transit instrument devising conspicuous meridian marks. It located the telescope in the middle of the axis.

July Bird's transit possessed five wires, Bradley determining their exact separations. Bird's transit wires were silver 1/750th inch thick.

July 20 Schumacher corresponded with Bradley from St. Petersburg. [p. 465].

August What must not be lost sight of from this year is that Bradley and his assistants (John Bradley, Charles Mason from 1756, and Charles Green from 1760) made over 60,000 precise and accurate stellar observations amounting to 5,000 a year on average. These observations were finally reduced by Friedrich Wilhelm Bessel from 1808 to 1818 and published as a catalogue of 3,222 stars of unprecedented reliability in his *Fundamenta Astronomiæ* in 1818.

August 10 Bird's new quadrant was set to face north in order to study *Polaris* and other circumpolar stars. Bradley made his earliest observations with the new instrument from this date.

September Bradley received letters from St Petersburg inviting him to be a corresponding member of the Russian Imperial Academy, and desiring him to order on their behalf a mural quadrant by John Bird modelled on the new quadrant now in use at the Royal Observatory.

October Bradley and Delisle exchanged correspondence. Delisle was acting on behalf of Lacaille who was travelling to the Cape of Good Hope to observe the southern stars and constellations, to measure solar and lunar parallaxes and to observe Mars and Venus.

October 12 Bradley in Greenwich corresponded with Delisle in Paris about the publication of all recorded observations of Jupiter's satellites. [p. 462].

October 12 Bradley in Greenwich corresponded with Dr Mortimer in London about his correspondence with Delisle. [p. 462].

October 21 Lacaille began his travels from Paris to the Cape of Good Hope to observe the southern stars.

October 25 Bradley began his important series of observations as part of his great survey many of the stars (down to magnitude 8.5) visible from Greenwich. His survey of the stars to the north (the circumpolar stars) began using Bird's quadrant. This remained in position until 24 July 1753.

November Bradley appreciated that his pre-1750 transit and quadrant observations could not be reduced for absolute refraction. They were, therefore, of limited value. From 1750 all observations made at Greenwich were accompanied by barometric and thermometric records, the latter both internal and external.

The quadrants both proved to be highly dependable instruments. In 1818 Bessel, who was to reduce Bradley's great series, discovered that the mean error from 300 sightings of five selected stars was just 1.45″ [*Fundamenta Astronomiæ*].

Bradley was made a corresponding member of the Imperial Academy of Sciences at St Petersburg. Bradley superintended the quadrant that John Bird made for the Imperial Observatory up to 1752. [pp. 466, 476].

November 21 Lacaille set sail from L'Orient for the Cape.

November 24 Bird's quadrant was balanced. The Greenwich observations made by Bradley and his assistants were maintained for 12 years from 1750 to 1762. They were continued by Nathaniel Bliss and Charles Green to 1764.The observations were published in two volumes in 1798 and 1805, occupying 931 large folio pages, and their number cannot be less than 60,000.

December 31 By the end of this year Bradley realized that the local state of the atmosphere affected atmospheric refraction.

1751 **January 1** Bradley publishes the latitude and longitude of the Royal Greenwich Observatory. [*Phil. Trans.* Vol. 77, p. 154], [p. 75].

John Bradley wrote from Greenwich to James Bradley in Oxford about observations he had made with the Bird 8ft quadrant which he had reduced to 1 January 1751.

January 31 Delisle wrote to Bradley from Paris to inform him that Lacaille was travelling to the Cape of Good Hope (a Dutch possession) to observe the stars of the southern hemisphere. [p. 463].

February 25 Lord Chesterfield introduced a Parliamentary Bill for the Reform of the Calendar for Great Britain. This was for the introduction of the Gregorian Calendar.

March 18 The second reading of Lord Chesterfield's Bill for the Reform of the Calendar seconded by Lord Macclesfield. Lord Macclesfield's speech at the second reading sees specific mention of Bradley. [p. 20]. Bradley composed three general tables included at the end of the Bill. [p. 461]. Sent in a letter to Lord Macclesfield. [*Memoirs* p. lxxxii].

April 2 Grischow, Professor of Astronomy and Secretary to the Imperial Academy at St Petersburg corresponds with Bradley in Greenwich concerning Bird's quadrant. [p. 466].

April 20 Lacaille arrived at the Cape of Good Hope in order to survey the stars and other celestial objects of the southern hemisphere.

April From April 1751 until March 1753, Lacaille observes the southern stars in South Africa. Observed and charted some 10,000 stars in a single year as well as 42 southern nebulæ. Lacaille surveyed the height of Table Mountain. He also gave names to 14 southern constellations of his own devising.

May 22 Lord Chesterfield's Bill for the Reform of the Calendar received the Royal Assent.

May   The vicarage at Greenwich becomes vacant following the death of the incumbent, Rev. R. Skerret. [*Memoirs* p. viii]. It was offered by Prime Minister Pelham to Bradley in order to augment his income beyond the paltry £100 per annum. [*London Magazine*, 1751, p. 236]. Bradley refused the offer. Rigaud believed Bradley refused the offer on religious grounds, but he seems to have conveniently forgotten that his Savilian chair at Oxford was conditional on the relinquishment of all church preferments.

July   John Bird bisected all of the divisions he had inscribed on Lord Macclesfield's quadrant in 1745. [*Memoirs* p. lxxxiv]. A beam compass was applied to various places on the limb and was not off 'true' by more than 1″. [*Memoirs* p. lxxxiv].

July?   John Bradley observed the occultation of Venus by the Moon. [*Phil. Trans.* Vol. 47, 1752, pp. 201–202].

August 6   Lacaille began his survey of the stars and other celestial objects at the Cape of Good Hope.

October?   Dr Lowth, Professor of Poetry and Oxford scholar and later the Bishop of London gave an elegant speech at the Sheldonian Theatre in the presence and in honour of James Bradley. Although opposed by some within the university, this reveals the high regard he was generally held throughout his alma mater.

November 16   George Graham died. He was buried in Westminster Abbey close by his mentor Thomas Tompion.

1752   **January 25**   Bradley elected to the Council of the Royal Society. Bradley's name proposed on the death of Dr Mortimer. [*Memoirs* p. lxxi]. Votes cast were Dr James Bradley 95, Dr James Parsons 7, Jos. Andrews 1. In 1731 Andrews had defeated Bradley for the election of the Keeper of the Ashmolean Museum at Oxford.

February 15   Pelham made arrangements for a warrant for £250 p. a. to be paid to Bradley quarterly [in addition to his annual stipend of £100]. Later when George II died in 1760, George III renewed the Royal Warrant [which was continued to Bliss and Maskelyne]. It was paid quarterly at £62 10s. [*Memoirs* p. xcii].

April 9   John Bradley in Greenwich corresponded to James Bradley in Oxford which included polite enquiries about his aunt Mrs Bradley and his six-year-old cousin Miss Bradley. It is a rare insight into Bradley's personal life. [p. 469].

April 30   There was notice of a letter written to the Secretary of the Royal Society, Mr Davall, on his method of making large lenses by Short which can only be published at his death which was to be 1768. [*Phil. Trans.* Vol. 69, p. 507].

July   Bradley supervised the construction of the 8 ft quadrant intended for the Russian Imperial Observatory near St Petersburg. It was a copy of the quadrant constructed by John Bird for the Greenwich observatory, which was the first major work by Bird though modelled on the quadrant made for Halley in 1725 by George Graham.

July 15/26   Two months before the adoption of the Gregorian Calendar Le Monnier in Paris sent various observations to Bradley in Greenwich. [p. 471].

July 18   Lacaille completed his survey of the stars and other celestial objects on this date.

July 22   Bradley in Greenwich corresponded with Grischow in St Petersburg informing him that John Bird had finished constructing the quadrant. He had examined it and found it to be satisfactory (praise indeed from Bradley). He advised Grischow that because of the great weight and bulk of the quadrant it should not be moved more than necessary or erected anywhere before its intended location.

July   Delisle requested Bradley to observe various comparison stars as a control to be compared with several to be observed by Lacaille at the Cape.

Summer Dr John Bevis was a regular correspondent of Bradley's. [*Memoirs* p. v].
Bevis published Halley's tables of Jupiter's satellites recognizing and asserting that
most of the observations were made by Bradley. [*Memoirs* p. v].

Guillaume de Saint-Jacques de Silvabelle corresponded with the Royal Society. His
paper resolved the nutational ellipse it was translated by Dr John Bevis and published
in the *Philosophical Transactions*. [*Phil. Trans.* Vol. 48 p. 325].

August There was an exchange of correspondence with Le Monnier in Paris. Le
Monnier had been a member of the Lapland expedition in 1736 under the leadership
of Maupertuis when he was only twenty years of age. He was often hot tempered and
was disliked by Lacaille. Although he made many observations of stars and planets up
to the year of his demise in 1799, he was untidy and often used unreliable clocks and
instruments.

August 22 Bradley replied to an earlier letter from Delisle in July which had been
delayed because like many in Paris, the latter did not appreciate that Greenwich was
not in London but several miles downriver from the capital. Bradley informed Delisle
of the reasons for the delay in his reply to his inquiries.

September John Bradley tested a new transit to measure differences in RA between
pairs of bright stars.

November Lacaille derived a horizontal parallax of the Sun of 10⅓″. The generally
accepted modern value of 1967 stands at 8.794″. Bradley and Pound had derived a
value of 9″ to 12″. Bradley later improved this measurement by 1.3″ to about 10″.3.

November 30 Delisle corresponded with Bradley on observations. Letter was
received 13 January 1753. (Was it also misdirected as was the earlier letter?)

1753 **January 5** J. Howe corresponded with Bradley about Bradley's failures with the
reduction of atmospheric refraction.

January 13 Bradley received Delisle's letter sent from Paris 30 November 1752.

January Bradley carefully examined Graham's 8 ft quadrant which proved to be only
2″ less than a full quadrant. [*Memoirs* p. lxxvii].

A double eyepiece was applied to the transit instrument at the Royal Observatory.

March 8 Lacaille finally left the Cape of Good Hope on this date.

June 2 T. Melvill in Geneva corresponds with Bradley about refractive indices of
different substances. [p. 483]. He asserts that aberration must be a consequence of the
relative velocity of the Earth in its orbit to that of the velocity of light. [*Memoirs* p.
xxxii].

July 31 John Bird's quadrant had been directed since 10 August 1750 to the
observation of the northern circumpolar stars, including *Polaris*, but after this date it
was directed to the observation of stars to the south of the Observatory.

August Graham's quadrant was removed from where it had been for 28 years and
new divisions were engraved upon the limb by Bird.

Bird's quadrant was taken to the west side of the supporting pier occupied before by
Graham's quadrant. The older quadrant was utilized to observe the stars in the north
and the Bird quadrant was now reassigned to the stars of the south. [*Memoirs* p.
lxxvii].

December Melvill died at Geneva.

1754 **February 7** The latest date in a book detailing stellar observations of circumpolar
stars. [p. 418].

March or May The course of lectures presented by James Bradley in experimental
philosophy attended by Roger Heber who wrote his own notebook of the lectures and
some of his own notes concerning these lectures.

June 28 Lacaille returned to Paris after completing his survey of the stars and other
celestial objects of the southern hemisphere.

July 3 Lacaille was received like 'a conquering hero' at the Royal Academy of Sciences. He was embarrassed by his reception and maintained a low profile, preferring the company of his friends. See I. S. Glass, Oxford, 2013.

Summer Joseph Nicholas Delisle (1688–1768) was placed in charge of the Russian Imperial Observatory at St Petersburg taking possession of Bird's 8 ft quadrant continuing the relationship with Bradley forged as a go-between with Lacaille. James Bradley, already a corresponding member, was elected and admitted a full member of the Imperial Academy of Sciences at St Petersburg.

When Lord Macclesfield's eldest son contested the election for Oxfordshire vehement cries were heard against him 'Give us back the eleven days we have been robbed of'.

December 17 Thomas Barker corresponds with James Bradley concerning his table of the parabola. [*Phil. Trans.* 49 p. 347].

December Alexandre Guy Pingré wrote to Bradley from Paris in Latin. [p. 487].

1755 **January 1** The observations made by Bradley and his assistants at Greenwich from 1750 to 1762 are reduced to Epoch 1 January 1755 by Friedrich Wilhelm Bessel and published in Königsberg 1818.

January 4 New counterweights were applied to the new transit instrument and now it behaved as though it weighed just 3lb.

January James Bradley replies to Thomas Barker's communication of 17 December 17 1754. [pp. 488–489].

June 5 The Council of the Royal Society of which Bradley was a member discussed the need for a reflector at the Royal Observatory. A sum of £115 was set aside from the £1,000 Bradley received from government resources, in order for Short to construct a Newtonian telescope to Bradley's specifications. This order had still not been met when Bradley died in 1762.

November Gesnerus corresponded in Latin with Bradley from Göttingen.

December Charles Walmesley, a mathematician with a European reputation, published a paper on the precession of the equinoxes in the *Philosophical Transactions* transmitted through the offices of James Bradley. [p. 498]. [*Phil. Trans.* Vol. 44, p. 700].

December 1 Cleveland, Secretary to the Board of Admiralty corresponded with Bradley concerning Mayer's Tables. [p. 84].

December 3 Charles Walmsley corresponds with Bradley from Rome. [p. 490] [*Phil. Trans.* Vol. 44, p. 700].

December 8 Thomas Barker of Lyndon near Uppingham corresponded with Bradley on cometary orbits. Barker was an outstanding meteorologist and brother – in – law of Gilbert White the naturalist who was married to his sister Ann. Barker was the nephew of Newton's acolyte William Whiston who had assisted Bradley at Wanstead.

December 17 Bradley replied to Thomas Barker on cometary orbits. [p. 488]. [*Phil. Trans.* Vol. 44, p. 347]. The exchange of correspondence with Barker suggested an easy method of finding Halley's Comet on its return in 1758. Thomas Barker (1722–1809) authored a book on the discovery of comets and how to determine their orbits.

Winter John Bradley, eldest son of John Bradley was born. Bradley's nephew now had a growing family and needed to increase his income.

1756 **February 10** Bradley corresponds with Cleveland, Secretary to the Board of Admiralty regarding Mayer's Solar and Lunar Tables. [p. 84].

March 6 Following Bradley's affirmation of the accuracy and dependability of Mayer's Tables, the Board of Longitude determined to conduct shipboard trials to test the Lunar Distances Method. From 1747, Captain John Campbell and John Bradley shared an interest in the development of 'lunars' in the determination of the longitude at sea.

April 17   Gael Morris corresponds with Bradley about Mayer's observations of Jupiter. [p. 492].

June 29   Charles Walmesley corresponds with Bradley from Rome over his concern that his mails concerning precession and nutation may have been misdirected now that France and Great Britain were yet again at war with each other.

September   John Bradley made his last recorded observations at the Royal Observatory. After fourteen years he left his post as assistant to the Astronomer Royal. He was offered a berth with Captain John Campbell on HMS Essex in order to test Tobias Mayer's Lunar Tables.

October   Charles Mason, assistant to James Bradley makes his earliest observations at the Royal Observatory from this time.

1757   Thomas Barker published his account on the location of cometary orbits. It included a table for determining parabolic trajectories and orbits. Barker's sister Anne was married to Gilbert White, author of the *Natural History and Antiquities of Selborne*. Barker was the nephew of the Newtonian acolyte William Whiston. Barker's greatest contributions were in the science of meteorology.

March 30   John Campbell corresponds with James Bradley from Spithead about trials with a new instrument at sea. This was Mayer's reflecting circle utilizing the method of lunars. [p. 84].

April 1   Cleveland, Secretary to the Board of Admiralty corresponds with James Bradley on the directions being given to Campbell. [p. 494].

April   Mayer's Lunar Tables were tested at sea along with Mayer's newly invented device using the method of lunars. [p. 84].

Campbell and John Bradley were at sea up to 1759 on board HMS *Essex* testing the method of lunar distances and Mayer's Tables and Method.

Birth of John Bradley's second son William (1757 –1833). He later became an officer on board the ill-fated HMS *Sirius*, flagship of the first fleet to Botany Bay. The ship was wrecked off Norfolk Island in 1790. William Bradley wrote a journal documenting his time in Australia, 'A Voyage to New South Wales, Dec. 1786 to May 1792'.

April 27   J. Milnes of the Admiralty Office corresponded with Bradley on the employment of John Bird on a new instrument.

May   Bradley (though Britain was at war with France) corresponds with Lacaille apologizing for the delay in reply due to family illness. This was almost certainly his wife's terminal illness and death.

Lacaille replies to Bradley. It marks how close the two men were that they remained in close touch even in the midst of a major conflict between the two countries. [p. 495]. [p. 501]. They were in correspondence over the problems of atmospheric refraction. Both developed different methods of reducing observations in differing atmospheric conditions. These became the Greenwich equations and the French equations.

May 17   Susannah Bradley was buried on 17 May at Minchinhampton Parish Church near to Chalford, Gloucestershire close to Bradley's mother and where he would be 'laid to rest'.

June   James Bradley appointed as a member of the Institute of Bologna, Italy's leading scientific institution. That Bradley's discovery of the aberration of light, vindicating Galileo's assertion that the Earth moved now led to Papal permission for Catholics to propose heliocentric hypotheses.

Charles Walmesley (1722–1797) continued in correspondence with Bradley. Walmesley was a Benedictine monk, mathematician, and a Fellow of the Royal Society (from 1750) who published papers in the *Philosophical Transactions*. He was widely celebrated on the continent for his mathematical studies, including an investigation into the motion of the lunar apses. He was the head of the Benedictine Order in England.

September   Observations were made at the Royal Observatory from September to October of a comet. [*Phil. Trans.* Vol. 50 p. 408].

October   Bradley received a batch of correspondence from John Campbell at sea about his lunar observations.

1758   **March 10**   Corresponding from Pisa, Charles Walmesley was pleased to inform James Bradley that he had been appointed a member of the Bologna Academy. [p. 497]. This was a singular recognition by a Roman Catholic institution to a man who had vindicated Galileo Galilei.

October 21   Now domiciled in Bath, Charles Walmesley corresponds with Bradley concerning precession. [p. 498].

November 9   W. Bowyer corresponded with Bradley concerning the fact that the Moon's same face is always directed towards the Earth. [p. 495].

December 1   Bradley corresponds with Walmesley. It is a kindly letter giving a further invitation to Greenwich.

1759   **March 6**   Gael Morris corresponds with Bradley on computations of Captain John Campbell's observations of 9 September 1758. There was a reference to mutual friends in Gloucestershire. [p. 499].

April 30   Bradley begins observations of the returning comet named for his mentor Edmond Halley. [p. 379]. There were comparatively few observations of this comet by Bradley even though it had been awaited by astronomers all over the world. [*Memoirs* p. xcv].

May 1   Bradley observed Halley's Comet with the zenith sector determining RA and declination. [p. 379]. In France Lalande began his observations of Halley's Comet.

May 2   Bradley observed the Comet with the sector determining the RA and declination. [p. 379].

May 5   Bradley made six observations of the comet with the sector determining the RA and declination. [p. 379].

May 6   Bradley made four observations of the comet with the sector determining the RA and declination. [p. 379].

May 16   Bradley made six observations of the comet with the sector determining the RA and declination. [pp. 379–380].

May 18   Bradley made six observations of the comet with the tube determining the RA and declination. [p. 380].

May 19   Bradley made seven observations of the comet with the tube determining the RA and Declination. [p. 380].

May 20   Bradley made ten observations of the comet with the tube determining the RA and Declination. [p. 380].

May 25   Bradley observed the comet with the sector determining the RA and declination. It was very faint and the observations were uncertain. [p. 380]. There was also an observation recorded in Charles Mason's hand and another observation made at Shirburne. [p. 380].

May   Bradley derived the orbital elements of Halley's Comet.

June 21   A certificate was sent to Bradley from Paris even though France was at war with Great Britain. It was signed by La Condamine, Clairaut and Dolours de Mairan and read to the Academy.

July   Bradley examined Bird's and Graham's quadrants. The greatest difference in any part of the limb was just 6″ of arc.

1760   **January 17**   Bradley was appointed as an Academician of the Royal Academy of Sciences in Paris even though Great Britain and France had been at war since 1756.

February 9   Walmesley corresponded with Bradley from Bath concerning Isaac Newton's published work on the Moon's motion. [p. 500].

April 14   Bradley corresponds with Cleveland, Secretary to the Admiralty concerning Mayer's Tables. [p. 86]. He related the Tables to the observations made at sea by Captain John Campbell both in 1757 and from 9 September 1758 to 17 September 1759. Bradley further adds that with some alterations in the equations they greatly improve Mayer's Tables.

May   Bradley presented his final course of twenty lectures in experimental philosophy at Oxford. He then resigned from the position of reader in experimental philosophy due to failing health.

May 22   The Council of the Royal Society under the impetus of James Bradley resolve to support expeditions to observe the coming transit of Venus in order to determine the scale of the solar system through the determination of the mean distance between the Earth and the Sun, the length of an astronomical unit.

September 22   Costard's letter was written about this date.

November   The last dates when entries written in Charles Mason's hand appeared in an observation book at Greenwich.

Charles Green succeeded Charles Mason as the Assistant to the Astronomer Royal.

November 18   Ferner corresponds with Bradley about Lacaille and Lunar Tables. [p. 500].

Autumn   Bradley made arrangements for Nevil Maskelyne and Charles Mason to observe the transit of Venus at St Helena. Then it was arranged that Waddington would take on the role as Maskelyne's assistant. Bradley arranged for Mason to set up another observatory at Bencoolen in the Indies. Bradley wrote a memorandum for Mason's use. [pp. 389–390].

Charles Mason appoints his friend Jeremiah Dixon as his assistant for the observations to be made at Bencoolen.

Winter   Bradley wrote 'Directions for using the common Micrometer'. [*Phil. Trans.* Vol. 62, p. 46] [p. 70].

1761   **January**   Charles Mason and Jeremiah Dixon left Portsmouth late, in the frigate HMS *Sea Horse* headed to Bencoolen in Sumatra to observe the transit of Venus due on 6 June 1761.

January 12   HMS *Sea Horse* attacked by a more powerful French frigate in the English Channel. Eleven men killed and 37 wounded. Ship returned to port for repairs. The Royal Society maintained its insistence that it still head for Bencoolen in Sumatra even amidst rumours that the station had been taken by French forces.

January 31   James Bradley attended his last meeting of the Council of the Royal Society before resigning due to ill health.

February 11   Walmesley corresponded with Bradley on perturbations between Venus and the Earth. [p. 502]. An interesting topic given the impending transit of Venus. If the orbits of the planets of Earth and Venus are affected by their mutual gravity, then the observations of the transit required further reduction.

Spring   Bradley's final instructions to Mason for the coming transit of Venus was the last paper of a public nature which he drew up. [p. 388].

Mason and Dixon set sail from Plymouth for Bencoolen in the East Indies aboard the Royal Navy frigate HMS *Sea Horse*. [*Memoirs* p. c] [Minutes of the Council of the Royal Society].

April 27   HMS *Sea Horse* arrives at the Cape of Good Hope. Mason and Dixon decided to disembark and set up an observatory in order to observe the transit of Venus due on June 6. This prudent decision was taken because time was now short. Bencoolen had now been taken by the French and the sea lanes from the Cape to the Indies were being patrolled by French privateers.

June 6   When the transit of Venus was due, Bradley was so ill that he requested Bliss to take his place at Greenwich. Bliss was assisted by Green who used a telescope that had been fitted for Mason. Maskelyne and Waddington at St Helena observed the transit in very poor conditions for it was cloudy and windy. Although the phenomenon was observed in better conditions by Lord Macclesfield at Shirburne by far the best conditions of the British attempts to observe the transit were experienced by Mason and Dixon at the Cape where it was clear and fine.

June 9   Thomas Barker corresponded with Bradley about his observations of the transit of Venus. [p. 503].

September 1   John Winthrop of Cambridge, Massachusetts, corresponded with Bradley about his observations of the transit of Venus. [p. 504].

November 16   James Bradley makes his Will.

December 3   Codacil added to Bradley's Will leaving his books to his son – in – law Samuel Peach jnr.

1762   **February 20**   Tobias Mayer dies in Gôttingen aged thirty-nine. His widow was given a share of the Longitude Prize to save her from penury.

March 21   Nicolas-Louis de Lacaille died in Paris at the early age of forty-nine.

April 7   Charles Mason returned to Greenwich in the company of Nevil Maskelyne. Mason resumed work on lunar tables and a reduced catalogue of 387 stars, later published in the 1773 *Nautical Almanac*. It was also published in 1798 as an appendix to Thomas Hornsby's edition of Bradley's observations at Greenwich from 1750 to 1755.

June 30   At Chalford in Gloucestershire James Bradley 'rode out for the air' and felt ill on his return to the Peach household. [*Memoirs* p. c].

July 13   James Bradley 'exchanged this life for a better'. He was in great pain due to an abdominal inflammation. At his demise various locals in Chalford ascribed his painful death as divine punishment for robbing people of eleven days of their lives! The three leading positional astronomers of the mid-eighteenth century Bradley, Lacaille and Mayer, all perished within five months of each other.

Events following the demise of James Bradley (1692–1762) with an emphasis on his legacy

1762   **September 1**   Bradley's Will was proved before his executors.

September   In Paris de Fouchy read his Eloge for James Bradley, remarking that there was no scientific society in Europe which was not proud of attaching him to its establishment.

Nathaniel Bliss succeeded his kinsman James Bradley as the fourth Astronomer Royal. He continued with Bradley's programmes with the assistance of Bradley's final assistant Charles Green.

Autumn   In the interest of reducing the great series of observations made under Bradley's direction an analysis of Bradley's calculations was made when further investigating a 'Rule for Refraction'.

Late   Charles Mason and Jeremiah Dixon set sail for the American colonies. Between 1763 and 1767 they surveyed the boundaries between Pennsylvania, Maryland and Delaware.

1763   **Winter**   John Bradley had commissioned a small wooden observatory to be built on the south bastion of the garrison at Portsmouth.

There was a paper written in John Bradley's hand on finding the apparent refraction by the tangent. [*Memoirs* p. lxxvii].

Spring   Charles Mason and Jeremiah Dixon went to the American colonies at the request of the authorities of Maryland, Pennsylvania, and Delaware in order to survey their disputed borders. This conflict was becoming increasingly heated and was in danger of leading to conflict between these colonies. They surveyed a line later called the 'Mason–Dixon Line'. The work kept them there until 1767.

Bradley's work on refractions was published by Nevil Maskelyne in his *British Mariner's Guide* p. 120, later in [*Phil. Trans.* Vol. 54, p. 265 and Vol. 67, p. 157]. Bradley's formula which was widely used throughout the British Empire gives $57''$ tang. $(ZD - 3r)$ $b/29.6 \times 400/t + 360 = r$ The coefficient of r was not settled until after many trials. [*Memoirs* p. lxxxvii].

Summer   Sir W. Chambers published plans and drawings of Molyneux's residence on Kew Green – formerly the property of the Capel family. [*Memoirs* p. xv]. George 2nd Earl of Macclesfield, PRS died.

Summer   The Seven Years War ended.

1764   John Campbell RN was elected Fellow of the Royal Society. He had spent three years in command of HMS *Essex* (1757–1759) testing the method of lunar distances, Mayer's Lunar Tables and Mayer's method.

John Bradley was chosen by the Board of Longitude to play an active role in the trials of John Harrison's chronometer H4. To avoid any 'tampering' during a second sea trial of H4 John Bradley was one of the three shore-based key-holders.

July   Charles Green returned to Greenwich after accompanying Nevil Maskelyne on a voyage to Barbados, at the request of the Board of Longitude, in order to test John Harrison's chronometer H4.

September 2   Nathaniel Bliss, the fourth Astronomer Royal, died. The Board of Longitude purchased many of his observations from his widow Elizabeth Bliss, because they were considered useful in the resolution of the longitude problem.

1765   The contents of the 2nd Earl of Macclesfield's library at St. James' Square was sold at auction. His main library at Shirburn Castle stayed in the possession of his family. [*Memoirs* p. xlv].

John Campbell RN Fellow of the Royal Society was appointed as one of the Visitors of the Royal Observatory.

February 8   Nevil Maskelyne was appointed as the fifth Astronomer Royal. Charles Green continued as an assistant until making his final entry 15 March 1765. Maskelyne and Green had earlier tested John Harrison's H4 chronometer on the voyage to Barbados. It appears the two men were incompatible.

March 25   Charles Green officially left the employ of the new Astronomer Royal on Lady Day. Green accompanied James Cook on the voyage to Tahiti as the expedition's astronomer to observe the transit of Venus due on 3 June 1769.

June 13   A few months after the Board of Longitude paid Elizabeth Bliss for many of her late husband's observations made at the Royal Observatory, it appears that Bradley's executors were prepared to release the third Astronomer Royal's Greenwich observations. By 10 June 1767, they had changed their minds.

1766   The French astronomer Bailli, in his '*Essai sur la Théorie des Satellites de Jupiter*' gives justice to the work of Bradley. [*Memoirs* p. xxxi]. Bailli gives Bradley the credit for detecting the great part of the inequalities since recognized in the motions of the Jovian satellites. [*Memoirs* p. v].

1767   John Bradley obtained the post of second mathematical master of the Royal Naval Academy at Portsmouth under George Witchell, FRS. [*Memoirs* p. li]. Nevil Maskelyne, fifth Astronomer Royal, recommended that John Bradley, after fourteen years of work at the Royal Observatory at Greenwich and three years of sea trials of the lunar distances method of determining the longitude at sea was the ideal person for the Lizard expedition to observe the transit of Venus in 1769 for the Board of Longitude.

Charles Mason and Jeremiah Dixon had finally surveyed 244 miles of the boundary between Maryland and Pennsylvania west from the Delaware River

June 10 James Bradley's executors, the Peach family, rescinded their earlier agreement to release his observations and Greenwich registers. It may have been that the Board of Longitude's decision to pay Elizabeth Bliss for her late husband's observations may have altered their earlier agreement to release Bradley's observations. The Board initiated a legal suit.

1768   On his return from Pennsylvania, Charles Mason was requested to travel to Ireland to observe the transit of Venus due 3 June 1769.

John Bird published 'The Method of Constructing Mural Quadrants'.

James Short died 'at no great age'. He finally completed the reflector commissioned by James Bradley in 1755 between 1762 and 1768.

Short's paper was opened by the Council of the Royal Society and printed [*Phil. Trans.* Vol. 69, p. 507] under the title of 'A Method of Working Object Glasses of Refracting Telescopes Truly Spherical'.

November 12 Nevil Maskelyne recommended to the Board of Longitude the Lizard Point (the most southerly point of the British mainland) as the site to locate latitude and longitude 'once and for all'. [*RGO* 14/5, p. 175]. It was another task for John Bradley when observing the transit of Venus in 1769.

1769   **April 29** Rev. Samuel Peach, the son in law of James Bradley took his MA at Hertford College at Oxford. He was the Rector of Compton Beauchamp in Berkshire.

May 4 John Bradley was impatient that he had received no order for the ship to take him to the Lizard. His observatory at the bastion in Portsmouth had been dismantled to be transported to the Lizard. He took an equal – altitude and a transit instrument, a reflecting telescope of 2ft focus, all made by John Bird, and an astronomical clock with a gridiron pendulum.

May 12 John Bradley and his team arrived at Falmouth after a voyage from Portsmouth aboard HMS *Seaford* under Captain John McBride (having just returned from the Falklands)

May 13 Bradley ventured ashore.

May 19 John Bradley corresponded with Nevil Maskelyne informing him that the expedition had arrived safely at the Lizard. [*Memoirs* p. li].

May John Bradley and the team made observations of *Arcturus* and *Spica* as well as Jupiter's 2nd satellite (Europa).

June 3 John Bradley observed the transit of Venus. The longitude of the Lizard Point was determined on this date.

Charles Mason and Jeremiah Dixon repeat their 1761 observation of the transit of Venus in Ireland, the last until 1874. Observations were intensified this year because the 1761 results were generally disappointing. There was also the need to find the longitude at sea of the Lizard Point and the Stag Rocks.

June 4 John Bradley corresponded with Nevil Maskelyne about using the Bird quadrant. Later Maskelyne combined all the forty-four observations made for a mean latitude of Lizard Point of 49° 57' 30''.

June 4 John Bradley observes a partial solar eclipse (50% covered) it began in the evening at 6hr 14min 54sec and ended at 7hr 57min 17sec, a duration of 1hr 42min 23sec. Observed in the 2ft reflector at 120 magnifications.

June 8 As well as the transit of Venus on 3 June, two emissions of Jupiter's first satellite (Io) were observed simultaneously with observations of these emissions at Greenwich, firstly on this date and again a week later on 15 June. On 8 June, the differences in the time was 20min 53sec

June 15 Simultaneous observations of the emission of Jupiter's first satellite (Io) were made by John Bradley at the Lizard and by Maskelyne at Greenwich. The difference being 21m 52s.

June 18 John Bradley wrote to Maskelyne now obviously only too ready to return to Portsmouth. He complained that the weather was very poor and he still hasn't received any order for his party to return. He asked whether he had observed the 'black drop effect' as it was later termed. Bradley struggled to explain the phenomenon to the Astronomer Royal.

July 2 John Bradley wrote to Maskelyne complaining that he was unable to hear the ticking of his clock (a technique inculcated by his uncle James Bradley) above the noise of the persistent wind at the Lizard.

July 26 John Bradley corresponds with Commissioner Hughes. John Bradley was finally relieved via Captain Robert Keeler of the sloop HMS *Cruizer* arriving in Falmouth to take the party back to Portsmouth with the minimum of notice. They finally left after 51 days leaving the observatory building behind. John Bradley's expedition to the Lizard left its marks on him for the rest of his life, both physically and psychologically with his rheumatism and his nervous complaint.

Nevil Maskelyne made some alterations to the New Observatory at Greenwich and he also put up a second support in the quadrant room on the opposite side. [*Memoirs* p. lxxvi].

John Bradley received a total of £67 17s 0d for the Lizard expedition [£36 7s 0d to cover travel expenses there and back]. A further 20 guineas was payment for time and effort. Bradley's assistant was Nehemiah Hunt, Master of the HMS *Arrogant* who was paid 10 guineas. (A guinea is equal to £1 1s or £1.05p).

1770 Extracts from Maskelyne's observations on the latitude and longitude of the Royal Observatory. [p. 73]. A loose paper of James Bradley's with observations reduced to 1751 January. [*Memoirs* pp. 78–80] [*Phil. Trans.* Vol. LXXVII, p. 154].

1771 The Nautical Almanac contains an account of John Bradley's labours at the Lizard and his observations of the transit of Venus.

Mary Bradley the only daughter of John Bradley (he had three sons) was born in Portsmouth. She did not marry and was still alive in Southsea at the time of the 1841 census.

January 10 Susannah Bradley married her first cousin Rev. Samuel Peach MA at Rodmarton. [*Memoirs* p. ci].

January 29 Charles Green died from dysentery at Batavia during his voyage with James Cook after exploring the east coast of 'Terra Incognito Australis' and the Great Barrier Reef. HMS *Endeavour* docked at Batavia to be refitted.

1772 **September** Maskelyne began to use decimal notation. Initially he developed a notation of ⅓ of an arc-second.

November 28 Samuel Peach appeared ready to settle with the Board of Longitude to be relieved of the lawsuit by the Board. Yet he still demanded a gratuity. On 3 April 1861, George Biddell Airy expressed his strong obverse opinion of the conduct of Peach in this affair. Later in the same letter Airy evinced an equally low opinion of the actions of Thomas Hornsby. [Correspondence between Airy and Bartholomew Price of Pembroke College, Oxford, 3 April 1861].

1773	Charles Mason calculated a catalogue of 387 'fixed' stars from the Greenwich observations from 1750 to 1762. Annexed to the Nautical Almanac for 1773. Hornsby inserted these in the first volume of Bradley's observations at the Royal Observatory from 1750 to 1755, published in 1798.
1774	**July 27** Charles Mason corresponded with Thomas Hornsby, Savilian Professor of Astronomy, about his fear that the continued retention of Bradley's registers was undermining the third Astronomer Royal's reputation, once pre-eminent in Europe.
1775	The Radcliffe Observatory opened at Oxford.
1776	Bradley's registers surrendered to Lord North as Chancellor of Oxford University.
1780	Undated letter from James Bradley to Mr Nash dated 24 August but with no year concerning the axis of the Earth's motion. [*Gentleman's Magazine* 1780] [*Memoirs* p. 506].
1784	Friedrich Wilhelm Bessel born in Bremen. This German astronomer and mathematician was the first to reduce all of Bradley's Greenwich observations made from 1750 to 1762 in his *Fundamenta Astronomia* published in 1818.
1786	**October 25** Charles Mason died in Philadelphia in the newly independent United States of America.
1789	Jesse Ramsden completed the circle of five feet for Piazzi.
1793	**March** There was now a regular series of observations for the quadrant at Shirburne Castle from January 1743 to March 1793.
1794	**June 14** John Bradley (born 1728) died. He had three sons and a daughter. He was succeeded as the second mathematical master by his younger son James (1765–1820) who had worked with his father since leaving school.
1798	The first volume of Bradley's Greenwich observations edited by Thomas Hornsby was published unreduced covering the years from 1750 to 1755.
1803	A report of the position of two stars observed by James Bradley on 30 March 1719 and 1 October 1722 matching the later observations by John Herschel and J. South, revealing the accuracy of Bradley's observations. [*Memoirs* p. iv]. [*Phil. Trans.* 1824, p. 105].
	The former residence of Samuel Molyneux, the White House, latterly a royal residence of George III was taken down.
1805	The second volume of Bradley's Greenwich observations edited by Abram Robertson was published unreduced covering the years from 1756 to 1762
1806	Troughton's circle at Westbury undermined the misapprehension held by John Bird that his quadrant of 1750 would maintain its use.
1807	Wilhelm Olbers handed both volumes of Bradley's Greenwich observations to Friedrich Wilhelm Bessel suggesting he reduce Bradley's observations.
1810	**January** Friedrich Wilhelm III, King of Prussia, appointed Bessel as the Director of the newly instituted Königsberg Observatory.
1812	**September 21** Susannah Peach, née Bradley, died.
1818	Bessel published *Fundamenta Astronomiæ*. Bradley's observations revealed that the latitude of Greenwich is 51° 28′ 39″. Bessel reduced all of Bradley's stellar observations made at Greenwich made from 1750 to 1762 to epoch 1755 to produce a catalogue of 3,222 stars. [*Memoirs* p. xcii].
1820	Death of James Bradley, son of John Bradley, great nephew of James Bradley DD, FRS.
1831	A dial was erected on the location of Molyneux's house on Kew Green on the misapprehension that the aberration of light was discovered at Kew in 1725 rather than at Wanstead in 1728. It was later moved to a location in the Royal Botanical Gardens in front of Kew Palace.
1832	Stephen Peter Rigaud published the *Miscellaneous Works and Correspondence of James Bradley* together with the *Memoirs of Bradley*.

APPENDIX 2

The Aberration Papers of James Bradley

The third, final, and as yet unpublished version of Bradley's account of his discovery of the aberration of light.

Bodleian Library: The Department of Western Manuscripts: James Bradley's Manuscript Papers: MS 20*:fols. 18r to 30v.: together with an addenda of explanatory editorial notes and comparisons with the published text.

January 5th, 1728/29. Wansted

fol. 18r

You having been pleased to express your satisfaction with what I had an opportunity some time ago of telling you in conversation concerning some observations that were making by our late worthy and ingenious friend the Honourable Samuel Molyneux Esquire, and which have since been continued and repeated by myself, in order to determine the parallax of the fixed stars; I now beg leave to lay before you a more particular account of them.[1]

Before I proceed to give you the history of the observations themselves it may be proper to let you know, that they were at first begun in hopes of verifying and confirming those that Dr. Hooke formerly communicated to the public; which seemed to be attended with circumstances that promised greater exactness in them, than could be expected in any other that had been made and published on the same account. And as his attempt was what principally gave rise to this, so his method in making the observations was in some measure that which Mr. Molyneux followed;[2] for he made the choice of the

fol. 18v

same star, and his instrument was constructed upon almost the same principles. But, if it had not greatly exceeded the Doctor's in exactness,[3] we might yet have remained in great uncertainty as to the parallax of the fixed stars, as you will[4] perceive upon the comparison of the two experiments.

This indeed was chiefly owing to our curious member Mr. Graham, to whom the lovers of our science are[5] also not a little indebted for several other exact and well contrived instruments. The necessity of such will scarcely be disputed by those that have had any experience in making astronomical observations; and the inconsistency which is to be met with among different authors in their attempts to determine small angles, particularly the annual parallax of the fixed stars, may be a sufficient proof of it to others.[6] Their disagreement indeed in this article is not now, so much to be wondered at, since I doubt not, but it will appear very probable, that the instruments commonly made use of by them, were liable to a greater error than many times that[7] parallax will amount to.[8]

The success then of this experiment evidently depending so much on the accurateness and conveniency of the instrument; those were principally to be taken care of;[9] in what manner this was done[10] it is not my present purpose,[11] to tell you of[12] but from the result of the observations which I now send you; it shall be judged necessary to communicate to the curious[13] the manner of making[14] them. I may hereafter perhaps give a description, not only of Mr Molyneux's, but also of my own instrument, which has since been erected for the same purpose and upon

fol. 19r

The like principles, though it be somewhat different in its construction, for a reason you will meet with presently.

Mr Molyneux's apparatus was completed and fitted for observing about the end of November 1725; and on the third day of December following the bright star in the head of Draco (marked γ by Bayer) was first observed as it passed near the zenith, and its situation carefully taken with the instrument. The like observations were made on the 5th, 11th and 12th days of the same month, and there appearing no material difference in the star's situation a farther repetition of them at this season seemed needless, it being a part of the year wherein no sensible alteration of parallax in this star could soon be expected.[15] It was chiefly therefore curiosity[16] that tempted me (being then at Kew where the instrument was fixed) to prepare for observing the star on December 17th when having adjusted the instrument as usual I perceived that it had passed a little more southerly this day, than, when it was before observed. Not suspecting any other cause of this appearance, we first concluded that it was owing to the uncertainty[17] of the observations themselves, and that either this or the other were not so exact as we had before supposed. On this account we purposed to repeat the observation again, in order to determine from whence this difference proceeded, and upon doing it on December 20th I[18] found the star passed still more southerly than in the former observations. This sensible alteration the more surprised us in that it was the contrary way from what it would have been had it proceeded from an annual parallax of the star;[19] but being now pretty well satisfied that it could

fol. 19v

be entirely owing to the want of exactness in the observations and having no notion of anything else that could cause such an apparent motion as this in the star, we began to think that some change in the materials & etc. of the instrument itself[20] must have occasioned it. Under these apprehensions we remained some time, but being at length fully convinced by several trials of the great exactness of the instrument, and finding by the gradual increase of the difference of the star's distance from the pole, that there must be some regular cause[21] that produced it, we took care to examine nicely, at the time of each observation, how much it was; and about the beginning of March 1726 the star was found to be[22] 20″ more southerly than at the time of the first observation. It now seemed to have arrived at its utmost limit southward, because in several trials[23] made about this time no sensible difference was observed in its situation. By the middle of April it appeared to be returning back again towards the north, and about the beginning of June it passed at the same distance from the zenith as it had done in December when it was first observed.

From the quick alteration of this star's declination about this time (it increasing a second in three days) it was concluded, that it would now move northward as before it had gone southward of its present situation; which accordingly happened,[24] for the star continued to move northward till September following when it again became stationary, being then near twenty seconds more northerly than[25] in June and no less than 39″ more northerly than it was in March.

fol. 20r

From September the star returned towards the south again, till it arrived in December,[26] to the same situation, it was in that time twelve months; allowing for the difference of declination[27] on account of the precession of the equinox.[28]

This was a sufficient proof that the instruments had not been the cause[29] of this apparent motion[30] in the star; and to find one adequate to such an effect seemed a difficulty, a nutation of the Earth's axis[31] was one of the first things that offered, but this[32] soon[33] appeared to be insufficient;[34] for though it might have accounted for the change in declination in γ Draconis: yet it would not at the same time agree with the phenomena in other stars; particularly in a small one almost[35] opposite to γ Draconis in right ascension, or at almost the same distance from the north pole of the equator; for though this[36] star seemed to move the same way as a nutation of the Earth's axis would have made it, yet it changing its declination but half as much as γ Draconis in the same time (as appeared upon comparing the observations of both made upon the same days at different times of the year) this plainly proved that the apparent motion in the stars was not owing to a real nutation; since if that had been the cause the[37] alteration in both[38] the stars would have been nearly equal.

The great regularity of the observations,[39] left no room to doubt, but that there was some regular[40] cause that produced this unexpected motion, which did not depend on the uncertainty and variety of the seasons of the year.

Upon comparing the observations with each other it was discovered, that[41] in both the forementioned stars the apparent[42] difference of declination

fol. 20v

the[43] maxima,[44] was always nearly proportional to the versed-sine[45] of the Sun's distance from the equinoctial points. This was an inducement to think of the cause, whatever it was, had some relation to the Sun's situation with respect to those points.[46] But not being able to frame any hypothesis[47] at that time, sufficient to[48] solve all the phenomena, and being very desirous to search a little farther into this matter; I began to think of erecting an instrument for myself at Wanstead, that having it always at hand I might with more ease and certainty enquire into the particular laws[49] of this new motion. The consideration likewise of being able by another instrument to confirm the truth of the observations hitherto made with Mr Molyneux's was no small inducement to me; but the chief of all was, the opportunity I should thereby have of trying in what manner other stars were affected by the same cause, whatever it was. For Mr Molyneux's instrument being originally designed for observing γ Draconis[50] (in order as I said before to try whether it had any sensible annual parallax)[51] was contrived so, as to be capable of but little alteration in its direction, not about 7 or 8 minutes of a degree; and there being very few stars with half that distance from the zenith of Kew bright enough to be well observed, he could not with his instrument thoroughly examine how this cause affected stars differently situated with respect to the equinoctial and solstitial points of the ecliptic.

These considerations determined me; and by the contrivance and direction of the same ingenious person Mr Graham[52] my instrument was fixed up August 19th 1727. As I had no convenient place where I could make use of so long a telescope as Mr Molyneux's; I contented my

fol. 21r

Self with one of but little more than half the length of his, (viz: of about 12½ feet, his being 24¼) judging from the experience I had already had, that this[53] radius would be long[54] enough to adjust the instrument to a sufficient degree of exactness; neither have I[55] since had any reason to change my opinion;[56] for from all the trials I have yet made, I am very well satisfied that when it is carefully rectified, its situation may be securely depended upon to half a second or perhaps even[57] to less.[58] As the place where my instrument was to be hung in some measure determined its radius so did it also determined the length of the arch or limb on which divisions were made on it,[59] for the arch could not conveniently be extended farther than to reach about 6¼° on each side my zenith. This indeed was sufficient, since it gave me an opportunity of making choice of several stars very different[60] both in magnitude and situation; there being more than two hundred inserted in the British Catalogue that may be observed with it. I need not have extended it so far but that I was willing to take in Capella, the only star of the first magnitude that comes so near my zenith.

My instrument being fixed I began immediately to observe such stars as I judged most proper to give me light into the cause of the motion already mentioned. There was variety enough of small ones; and not less than twelve that I could observe through all seasons of the year, they being bright enough to be seen[61] in the day[62] time when nearest the Sun. I had not been long observing when I perceived that the notion we had before entertained of the stars being farthest north or south when the Sun was

fol. 21v

about the equinoxes,[63] was true only of those[64] that were near the solstitial colure. After I had continued my observations a few months,[65] I discovered, what, I then apprehended to be a general law observed by all the stars, viz: That each of them became stationary or was farthest north or south when the Sun was nearly in quadrature to it in right ascension, that is, when it passed over my zenith about six of the clock, either in the morning or evening.[66] I perceived likewise, that, whatever situation the stars were in with respect to the cardinal points of the ecliptic, the apparent motion of every one tended the same way when they passed the instrument about the same hour of the day or night; for they all moved southward while they passed in the day and northward in the night, so each was the farthest north when it came about six o'clock in the evening and farthest south about six in the morning.

Though I have since discovered that the maxima[67] in most stars do not happen exactly, when they come[68] these hours yet not being able at that time to prove the contrary and supposing that they did; I endeavoured to find out what proportion the greatest alterations of declination in different stars bore to each other; being very evident that they did[69] not[70] all change their declination[71] equally.[72] I have therefore taken notice that it appeared from Mr Molyneux's[73] observations that γ Draconis altered its declination about twice as much, as the forementioned small star almost opposite to it; but upon examining the matter more particularly I found that the greatest alteration of declination

fol. 22r

In these stars, was, as the sine of the latitude of each respectively.[74] This made me suspect that there might be the like proportion between the maxima of other stars; but finding that the observations of some of the stars would not perfectly correspond with such an hypothesis, and not knowing whether the small difference I met with, might not be owing to the uncertainty and error of the observations; I deferred the farther examination into the truth of this hypothesis till I should be furnished with a series of observations made in all parts of the year,[75] which might enable me to determine not only what errors the observations are liable to, or how far they may be safely depended upon; but also to judge whether there had been any sensible change in the parts of the instrument itself.[76]

Upon these considerations I laid aside all thoughts at that time about the cause of the forementioned phenomena, Hoping that I should the easier discover it, when I was better provided with proper means to determine more exactly what they were.[77]

When the year was completed, I began to examine and compare my observations, and having pretty well satisfied myself as to the general laws of the phenomena,[78] I then endeavoured to find out the cause of them. I was already convinced that the apparent motion of the stars was not owing to a nutation of the Earth's axis.[79] The next thing that offered itself was an alteration in the direction of the plumb-line, with which the instrument was constantly rectified,[80] but this upon trial proved insufficient. Then I considered what refraction[81] might do, here also nothing satisfactory occurred. At last

fol. 22v

I conjectured that all the phenomena hitherto mentioned proceeded from the progressive motion of light and the Earth's annual motion in its orbit about the Sun.[82] For I perceived that if light is propagated in time, the apparent place of a fixed object would not be the same as when the eye is at rest, as when it is moving in any other direction than that of the line passing through the eye and object, and that when the eye is moving in different directions, the apparent place of the object would be different.

I considered this matter in the following manner, I imagined CA to be a ray of light proceeding from an object and falling perpendicularly upon the line BD;

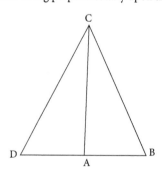

Then if the eye is at rest in A the object must appear by that ray in the direction AC, whether light is propagated in time or in an instant. But if the eye is moving from B towards A, and light is propagated in time;[83] the velocity of light will be to the velocity [of] the eye in a given ratio (suppose[84] as AC to AB) and whilst the eye moves from B to A light will move from C to A so that[85] particle of light by which the object will be discerned when the eye by its motion comes to A,[86] is at C when[87] the eye is at B. Joining the points C, B, we may suppose the line BC to be a tube (inclined to the line BD in the angle CBD) of such a diameter as to admit of but one particle of light; then it is easy to conceive that the particle of light at C (by which the object will be seen when the eye in its motion arrives at A) will pass through the tube BC if it is inclined to BD in the angle CBA, and accompanies the eye in its motion from B to A. But it cannot come to the eye placed behind such a tube, if

fol. 23r

it[88] has any other inclination to the line BD. If instead of supposing CB so small a tube we imagine it to be the axis of a larger, then the particle of light at C cannot pass through the axis unless it is inclined to BD in the angle CBD. In like manner, if the eye moved the contrary way from D towards A with the same velocity, then the tube must be inclined in the[89] angle BDC. So that, though the true or real place of an object[90] perpendicular to the line in which the eye is moving, yet the visible place thereof will not be so, since that, no doubt, must be in the direction of the tube; but the difference between the true and apparent place will be (caeteris paribus) greater or less according to the different proportion between the velocity of light and that of the eye;[91] if light is propagated in an instant, then[92] indeed there can be no difference between the real and visible place of an object although the eye is in motion; for in that case,[93] AC being infinite with respect to AB, the angle ACB vanishes. But if light is propagated in time (which I presume will readily[94] be allowed by most of the philosophers of this age) then it is[95] evident that there will be always a difference between the true and visible place of an object, unless the eye is moving directly towards or from the object; and[96] in all cases the sine of the difference between the real and visible place of the object, will be to the sine of the visible inclination of the object to the line in which the eye is moving, as the velocity of the eye to the velocity of light.[97]

If light moved but a thousand times faster than the eye,[98] and an object (supposed to be at an infinite

fol. 23v

distance) was really replaced perpendicularly over the plane in which the eye moves, it follows from what has been already said, that the apparent place of such an object, will always be inclined to that[99] plane in an angle of 89° 56′½, so that it will constantly appear 3′½ from its true plane, and seem so much less inclined to the plane that way towards which the eye tends; that is, if AC is to AB or AD as a thousand to one, the angle ABC will [be][100] 89° 56′½ and ACB = 3′½ and BCD = 2ACB = 7′. According to this supposition therefore the visible or apparent place of the object, will be changed 7′ if the direction of the eye's motion is at one time contrary to what it is at another.

If the Earth revolves around the Sun, and the velocity of light is to the velocity of the Earth's motion in its orbit (which I will at present[101] suppose to be a circle) as a thousand to one, then it is easy to conceive, that a star really situated in the very pole of the ecliptic, would to an eye carried along with the Earth seem to change its place continually, and describe a circle round that pole, every way distant therefrom 3′½ so that its longitude would be varied through all the points of the ecliptic in a year, but its latitude would always remain the same. Its right ascension and declination would also change according to the different

situation of the Sun in respect of the equinoctial points; and its apparent distance from the north pole of the equator would be $7'$ less at the Autumnal than at the Vernal Equinox. **A***
[See addenda]. Other stars that were not in the plane of the Ecliptic would seem to describe as sort of elliptical figure,[102] whose greater axes would all be equal to the diameter of the little circle, that the star in the pole of the ecliptic

fol. 24r

would seem to describe, and the lesser axis in each would be the greater, as the sine of the star's latitude to radius. Those stars that lay in the ecliptic would change their longitude, but not their latitude; but the right ascension and the declination of all would be changed more or less according to their different situations with respect to the ecliptic and their aspects with the Sun. ***A (See addenda).**

The greatest alteration[103] of the place of a star in the pole of the ecliptic (or which in effect amounts to the same, the proportion between the velocity of light and the Earth's[104] motion in its orbit) being known;[105] it will not be difficult to find what would be the difference upon this account between the true and apparent place of any star at any time; and on the contrary the difference between the true and apparent place of any star being at any time known, the proportion between the velocity of light and the Earth's motion in its orbit may be found.

As I only observed the apparent difference of declination of the stars,[106] I shall not at present take any farther notice in what manner such a cause as I have supposed would occasion a change in their apparent places in other respects; **B*** **(see addenda).** And from what has already been said, it may be easily gathered, that

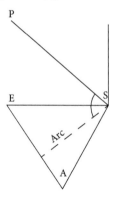

if E represent the pole of the ecliptic, P the pole of the equator, ES the complement of the star's latitude, SP the complement of its declination, SA an arc of a great circle falling perpendicularly on EA a circle of longitude passing through the Sun's place in the ecliptic

fol. 24v

then (considering small arcs as right lines) the difference between the true and apparent declination of the star, will be to the semidiameter of the little[107] circle that a star in the pole of the ecliptic would seem to describe, as the[108] rectangle under the cosine of the arc SA[109] and the cosine of the angle ASP, is to the square of the radius; when therefore the Sun is in opposite points of the ecliptic the sum of the differences[110] between the apparent and true declination, will be the diameter of the little circle in the same ratio. And the direction of the Earth's motion with respect to the[111] fixed stars being contrary, when it is in opposite points of its orbit, the change of a star's declination on account of the successive propagation of light will be also, and therefore the sum[112] of the differences between the apparent and true declination of a star at those times; this difference being known from observations, the diameter of the little circle may be determined and the velocity of light on which it depends. I consider the Earth as moving equally in a circle and not in an ellipse on purpose to avoid too perplexed a calculus, which after all the trouble of it would scarce sensibly differ[113] from that which I make use of, especially in those conclusions which I shall at present draw from the foregoing hypothesis.*B (See addenda).

This being premised, I will now proceed to determine from the observations, what the real proportion is between the velocity of light, and the velocity of the Earth in its orbit, upon the supposition

fol. 25r

That the phenomena before mentioned do depend on the causes I have here assigned. I must first let you know that in all the observations hereafter mentioned, I have made an allowance for the change of the declination of the stars on account of the precession of the equinox, upon the supposition that the alteration from this cause is proportional to the time or regular through all parts of the year. I have deduced the real annual change of declination in each star from observations themselves; and I rather choose to depend upon them on this point, because all I have yet made concur to prove, that the stars near the equinoctial colure change their declination at this time & $1''$½ or $2''$ in a year more than they should do if the precession[114] was only $50''$, as is now generally supposed.[115] I have likewise met with some small variations in the declination of other stars in different years, which do not seem to be owing to the same cause,[116] particularly in those near the solstitial colure, which on the contrary have altered their declination less than they ought if the precession is $50''$. But whether these small variations depend upon a regular cause or are occasioned by any change in the materials & etc. of my instrument, I am not yet able[117] fully to determine. However I thought it might not to be amiss just to mention to you how I have endeavoured to allow for them; though the result would have been nearly the same, if I had not considered them at all. What that is I will show, first from the observations of γ Draconis,[118] which was found to be[119] $39''$ more southerly in the beginning of March than in September.

From What has been premised, it will appear that the greatest alteration of declination in γ Draconis,[120] on account of the successive propagation of light, would

fol. 25v

be to the diameter of the little circle which a star (as before remarked) would seem to describe about the pole of the ecliptic, as $39''$ to $40.4''$. The half of this is the angle ACB (as represented in Fig. 1)[121] this therefore being $20.2''$, AC would be AB, that is, the velocity of light to the velocity of the eye (which in this case may be supposed the same as the velocity of the Earth's annual motion in its orbit) as 10,210 to one; from whence it follows that light moves or is, propagated as far, from the Sun to the Earth in $8'$ $12''$.[122]

It[123] is well known that Mr Romer who first attempted to account for an apparent inequality in the times of the eclipses of Jupiter's satellites by the hypothesis of the velocity of light from the Sun to us, but it has since been concluded by others from the same eclipses, that it is propagated as far in 7 minutes. The velocity therefore of light as deduced from the foregoing hypothesis is as it were a[124] mean betwixt what had at different times been determined from the eclipses of Jupiter's satellites.

The different methods of[125] finding the velocity of light thus agreeing in the result, we may reasonably conclude, not only that these phenomena are owing the causes to which they have been [assumed[126]] but also, that light is propagated (in the same medium) with the same velocity after it had been reflected as before;[127] for this will be the consequence, if we allow that the light of the Sun is propagated with the same velocity (before it is reflected) as the light of the fixed stars. And I imagine this will[128]

fol. 26r

Scarce be questioned if it can be made appear that the velocity of the light of all the fixed stars is equal, and that their light moves through equal spaces in equal times at all distances from them, both which points (as I apprehend) are sufficiently proved from the apparent alteration of declination of stars of different lustre, for that is not sensibly different in such stars as appear near one another, though they are of very different magnitudes. And whatever their situations are (if I proceed according to the foregoing hypothesis) I find the same velocity of light from my observations of small stars of the fifth or sixth[129] magnitude as from those of the second or third[130] which in all probability are placed at very different distances from us. The small star (for example) before spoken of that is almost opposite to[131] γ Draconis (being the 35th[132] Camelopardalis[133] in Mr Flamsteed's Catalogue) was 19″ more northerly about the beginning of March than in September. Hence I conclude according to my hypothesis that the diameter of the little circle described by a star in the pole of the ecliptic would be 40.2″.

The last star of the Great Bear's tail of the second magnitude (marked η by Bayer) was 36″ more southerly about the middle of January than in July. Hence the maximum or greatest alteration of declination of a star in the pole of the ecliptic should be 40.4″ exactly the same as was before found from the observations of[134] γ Draconis.

The star of the 5th magnitude in the head of Perseus (marked τ by Bayer), was 25″ more northerly about the end of December than on the 29th of July following

fol. 26v

hence the maximum would be 41″. This star is not bright enough to be seen[135] as it passes over my zenith about the end of June, when it should be according to the hypothesis farthest south. But because I can more certainly depend upon the greatest alteration of declination in those stars which I have frequently observed about the times when they become stationary, with regard to the motion I am now considering; I will now set down a few more instances of such, from which you may be able to judge, how near it may be possible from[136] these observations, to determine with what velocity light is propagated.

A Persei Bayero was 23′ more northerly in the beginning of January than in July. Hence the maximum would be 40.2′. α Cassiopeæ[137] as 34″ more northerly about the end of December than in June. C* (see addenda). Hence the maximum would be 41.1″.

χ Ursæ Majoris was 34″ more southerly about the middle of December than in June. *C (see addenda). Hence the maximum would be 40.8″. β Draconis[138] was 39″ more northerly in the beginning [of] September than in March. Hence the maximum would be 40.2″. Capella [α Aurigæ] was about 16″ more southerly in August than in February. Hence the maximum would be about 40″. But this star being farther from my zenith than those I have made use of, I cannot so well depend upon my observations of it, as of the others; Because I meet with some small alterations of its declination[139] that do not seem to proceed from the cause I am now[140] speaking of.

I have compared the observations of several other stars and they all conspire to prove that the maximum is about 40″ or 41″; I will therefore suppose that it is

fol. 27r

it is[141] 40″½ or (which amounts to the same) that light moves as far as from the Sun in 8′ 13″. The[142] near agreement which I meet with among my observations induces me to think that the maximum (as I have here fixed it) cannot err so much as a second from the truth; it is therefore probable that the time which light spends in passing from[143] the Sun to us may be determined by these observations within[144] 5″ or 10″; which is such a degree of certainty as we can never hope[145] to attain from the eclipses of Jupiter's satellites.

Having thus determined what the maximum of greatest change of declination would be in a star placed in the pole of the ecliptic, I will now deduce from it (according to the foregoing hypothesis) the alteration of declination in one or two stars at such times as they were actually observed; in order to see how the hypothesis will correspond with the phenomena through all parts of the year. I should be too tedious if I set down my whole series of observations, I will therefore make choice only of such as are most proper for my present purpose, and will begin with those of γ Draconis.[146]

This star appeared farthest north about the 7th day of September[147] 1727 as it ought to have done according to my hypothesis. The following table shows how much more southerly the star was found to be by observation in the several parts of the year and also how much[148] more southerly it ought to have been according to the hypothesis.

fol. 27v

	Difference of declination by observation	Difference of declination by the hypothesis		Difference of declination by observation	Difference of declination by the hypothesis
1727			1728		
October 20	4½″	4½″	March 24	37″	38″
November 17	11½″	12″	April 6	36″	36½″
December 6	17½″	18½	May 6	28½″	29½″
28	25″	26″	June 5	18½″	20″
1728			15	17½″	17″
January 24	34″	34″	July 3	11½″	11½″
February 10	38″	37″	August 2	4″	4″
March 7	39″	39″	September 6	0″	0″

From hence it appears that the hypothesis corresponds with the observations of this star through all parts of the year, for the small difference between them seems to arise from the uncertainty of the observations which is occasioned (as I imagine) chiefly by the tremulous or undulatory motion of the air and of the vapours in it, which causes the stars sometimes to dance to and fro so much that it is difficult to judge when they are exactly on the middle of the wire[149] that is in the[150] common focus of[151] the glasses of the telescope.

I must confess that the agreement of the observations with each other[152] as well as with the hypothesis[153] is much greater than I could reasonably expect to find before I[154] had compared them; and it may[155] possibly be thought to be too great by those who have been used to[156] astronomical observations[157] and know how difficult it is to make such as are[158] in all respects exact. But if it would be any satisfaction to such persons (till I have an opportunity of describing my instrument and

fol. 28r

And the manner of using it) I could assure them that in above 80 observations[159] which I made of this star in one year there is but one (and that is noted as very dubious on account of clouds) which differs from the foregoing hypothesis so much as two seconds and this does not differ 3″. **D* (see addenda).** And although this is the fact I will allow that it is not an absolute proof of the truth of my hypothesis;[160] however ***D (see addenda)** I cannot but think it very probable that the phenomena proceed from the cause I have assigned, since the foregoing observations make it sufficiently evident that the effect of the real cause (whatever it is) varies in this star in the same proportion that it ought according to the hypothesis, But least γ Draconis[161] may be thought not so proper to show the proportion in which the apparent alteration of declination is increased or diminished, as those stars which lie near the equinoctial colure, I will give you also the comparison between the hypothesis and the observations of η Ursæ Majoris[162] which was farthest south about the 17th day of January, 1728 agreeable to the hypothesis. The following table shows how much more northerly it was found[163] by observation in several parts of the year, and likewise how much it should have been according to the hypothesis.

	Difference of declination by observation	Difference of declination by the hypothesis		Difference of declination by observation	Difference of declination by the hypothesis
1727			March 21	11½″	10½″
September 14	29½″	28½″	April 16	18½″	18″
24	24½″	25½″	May 5	24½″	23½″
October 16	19½″	19½″	June 5	32″	31½″
November 11	11½″	10½″	25	35″	34½″
December 14	4″	3″	July 17	36″	36″
1728			August 2	35″	35½″
February 17	2″	3″	September 20	26½″	26½″

fol. 28v

I find upon examination that the hypothesis agrees altogether as exactly with the observations of this star as the former, for in about 50 that were made of it in a year, I do not meet with a difference of so much as 2″ except in one, which is marked as doubtful on account of the undulation of the air, and this does not differ 2½″ from the hypothesis.

The agreement between the hypothesis and the observations of this star is the more to be regarded, since it proves that the change of declination on account of the precession of the equinox is as I before supposed regular through all the parts of the year; so far at least, as not to occasion a difference great enough to be discovered with this instrument. It likewise proves the other part of my former supposition, viz: that the annual alteration of the declination of stars near the equinoctial colure is at this time greater than a precession of 50″ would occasion; for this star was at least 20″ more southerly in September 1728 than September[164] 1727, that is about 2″ more than it would have been if the precession was but 50″. But I may hereafter perhaps be better able to determine this matter from my[165] observations of such stars as lie near the equinoctial colure at about the same distance from the[166] north pole of the equator,[167] and almost opposite, as to right ascension.

I think it needless to give you the comparison between the hypothesis and the observations of any more stars, since the agreement in the foregoing is a kind of demonstration (whether it is allowed that I have discovered the real cause of the phenomenon or not)

fol. 29r

That the hypothesis gives at least the true law[168] of variation of declination in different stars with respect to their different situations and aspects with the Sun. And if this[169] is the case, it must be granted that the annual parallax of the fixed stars is much smaller than hitherto[170] has been supposed[171] by those who have pretended to deduce it from their observations.[172] I believe I may venture to say that in either of the two stars last mentioned, it does not amount to 2″. I am of the opinion that if it was 1″, I should have perceived it in the great number of observations that I made, especially of γ Draconis,[173] which agreeing with the hypothesis (without allowing anything for parallax) nearly as[174] well when the Sun was in conjunction[175] with, as when it was in opposition to this star, it seems very probable that the parallax of it is not so great as one second.[176]

There appearing[177] therefore after all no sensible parallax in the fixed stars, the Anti-Copernicans have still room on that account to object against the motion of the Earth; and they may (if they please) have a much greater objection against the hypothesis by which I have endeavoured to solve the forementioned phenomena by denying the progressive motion of light as well as that of the Earth.[178] But as I do not apprehend that either of these postulates will be denied me by the generality of the astronomers and philosophers of the present age, so I cannot[179] doubt of obtaining their assent to the consequences which I have deduced from them, if they are such as have

fol. 29 v

the approbation of so great a judge of them as your self.

E* (see addenda). If the distance of the fixed stars is so immensely great as my observations imply, it will no longer seem strange that their diameters should be so small as you have indisputably proved them (in your remarks on Mr Cassini's Essay on the parallax of Sirius) by the most proper method we can have of determining them, viz: by the occultations of the brighter stars by the Moon. For if we suppose them no bigger than the Sun, and that they are place above 400,000 times farther from us as it (as γ Draconis,[180] a star of the second magnitude seems at least to be)[181] then their diameters would not be 1/200th part of a second.[182] There is room enough therefore still left to make allowance, both for the different distances of the stars of the first and second magnitudes and[183] also for the difference of their real bigness without supposing their diameters so much as one second,[184] which the observations of Aldebaran and Spica Virginis will not allow is to do. But if we should attempt to deduce the real magnitude of the fixed stars from their apparent diameters as seen through a telescope, the aperture of whose object glass so contracted, as to take off the spurious rays and make the stars appear distinct and round, we shall find them of an enormous and incredible bigness. For even γ Draconis[185] seems in the night through the telescope of my instrument (when I leave an inch aperture) 4″ or 5″in diameter; whence (if we proceed in this manner) we should find its diameter at least 4 or 5 times bigger than the diameter of the Earth's annual orbit.

fol. 30r

But I presume no one will think this at all probable, since we[186] may from other considerations be convinced, that this apparent diameter of a fixed star is an optic fallacy. *E (see addenda).

What I have farther to add relates to the observations of Dr Hooke and Mr Flamsteed, which have been produced as arguments for the sensible annual parallax of the fixed stars. I must own to you, that before Mr Molyneux's instrument was erected, I had no small opinion of the correctness of Dr Hooke's observations;[187] the length of his telescope and the care he pretends to have taken in making his observations exact, having been strong inducements with me to think them so. And since I have been convinced both from Mr Molyneux's observations and my own, that[188] the Doctor's are really very far from being exact or agreeable to the phenomena; I am greatly at a loss how to account for it. I cannot well conceive that an instrument of the length of 36 feet, constructed in the manner he describes his; could have been liable to an error of near 20″ (which doubtless was the case)[189] if rectified with so much care as he represents.

The observations that were made by Mr Flamsteed of the different distances of the pole star from the pole in different times of the year, which were through mistake looked upon by some as a proof of the annual parallax of it, seem to have been made with much greater care than[190] those of Dr Hooke; for though they do not all exactly correspond with each other, yet[191] from the whole

fol. 30v

Mr Flamsteed concluded that the star was nearer the pole in December than in May or July by about 35″–40″ or 45″. According to my hypothesis this star ought to be 40″ nearer the pole in[192] December than in June. The agreement therefore of the observations with it, is greater than could reasonably be expected,[193] considering the radius of the instrument and the manner in which it was constructed.[194]

I am now come to the conclusion, and have no other apology to make for having troubled you with this long account than that I have sincerely endeavoured to search out the truth;

and I doubt not but that will be allowed of as a sufficient one;[195] since I am persuaded that the meanest attempt to improve arts and sciences will be always candidly looked upon by you, who have so eminently distinguished yourself by your great success therein.

I am Sir

Your most obedient
humble Servant
J. Bradley.

Addenda

Major textual variations between the published account in the *Philosophical Transactions*, No.406, Vol.35, pages 637 to 661, Re-published in *The Miscellaneous Works and Correspondence of James Bradley*, pages 1 to 16, and the later version above, designated as Appendix 2, transcribed and edited by Dr John Fisher FRAS, together with a running commentary

Introduction

There are many minor and trivial variations between the texts of the two documents. But these have largely been passed over as being the results of miscopying by the author, or as minor variations introduced by the author to express himself perhaps more clearly. Deletions, substitutions, and insertions have been listed as endnotes attached to the text of the document.

The passages here designated A to E are more worthy of comment, for they are major variations from the paper published by the Royal Society in the *Philosophical Transactions* in January 1729.

Passage A* to *A (fol. 23v, line 17 to fol. 24r, line 5)

This passage, consisting of eight lines, has been added by the author to the hurried version read by Halley to the Royal Society and subsequently published. Compared to the version published in the *Philosophical Transactions* it is interposed between the paragraph ending, '.....vernal equinox'. & the paragraph beginning with the words, 'The greatest alteration .....'.. on page 649 (or page 8 in the *Miscellaneous Papers*.)

The passage is included by Bradley as an attempt to clarify some aspects of the geometry of 'the new discovered motion'. It explicitly reveals the shape of the inferred ellipse according to the ecliptical latitude of the star. As such it describes the motion of any fixed object according to the hypothesis of the 'new discovered motion'. It bears some resemblance to the projected motion of a star according to the theory of annual parallax according to John Flamsteed in his letter to Sir Christopher Wren in November 1702 (though not published until 1750). However, the 'new discovered motion' is 90° out of phase with that expected from annual parallax, and proceeds phenomenologically in the opposite direction.

As this passage largely clarifies Bradley's concept of the 'new discovered motion' in addition to the account published by the Royal Society, it is internal evidence that this document is a later, more considered version than the published paper. It is very unlikely that Bradley would have edited such a passage out of the text in view of the way it clarifies the hypothesis.

Passage B* to *B (fol. 24r, line 15 to fol. 24v, line 15)

This passage, consisting of lines and a geometrical diagram, **replaces** a passage in the published version. The passage it replaces in the *Philosophical Transactions* begins near the end of p. 651, between the words beginning, 'but, supposing the Earth to move' to the words ending, 'from the foregoing hypothesis'. In the *Miscellaneous Papers*, the passage includes almost all of p. 9 and the first four lines of p. 10.

This replacement passage, taken in conjunction with the earlier inserted passage **A* to *A**, is easier to comprehend than the wordier passage included in the published version. There can be little doubt that the 'replacement passage' is an improvement, possessing greater clarity and a geometrical explanation. The addition of a geometrical diagram, to which the text refers, is an aid to comprehension. The replacement passage in Appendix 2 is perhaps only half the length of the passage published in the *Philosophical Transactions*. It is further internal evidence that the replacement is a later version closer to Bradley's final intentions.

Passage C* to *C (fol. 26v, line 10 to fol. 26v. line 12)

This short passage of only two lines is sure evidence of a printer's error. It has created an intriguing omission, and the acceptance of a calculable error for 293 years. The omitted passage should have been included in a passage on p. 655 of the account published in the *Philosophical Transactions*. In the *Miscellaneous Works* the omitted passage belongs to p. 12.

In order that the reader can identify the omitted section more clearly, the passage will be quoted in full, together with the two encompassing sentences. Please take notice of the simple fact that the wording of the end of the sentence immediately prior to the omitted passage resembles the wording at the end of the omitted passage itself, It is easy to comprehend how a printer, possibly working under time pressure, might inadvertently make this intriguing omission.

'α Cassiopeæ was 34″ more northerly about the end of December than in June. **Hence the maximum would be 41.1″. χ Ursæ Majoris was 34″ more southerly about the middle of December than in June.** Hence the maximum would be 40.8″.'

For the entire period from its publication, readers of Bradley's published account have read that the maximum of α Cassiopeæ was 40.8″ (whereas this is the maximum calculated for χ Ursæ Majoris). Bradley calculated the maximum for α Cassiopeæ as 41.1″.

Passage D* to *D (fol. 28r line 4 to fol 28r line 5)

This is the least significant of the five passages included in this addenda. It follows the passage contained in the *Philosophical Transactions* on p. 657 ending, 'and this does not differ 3′'. In the *Miscellaneous Works* this is located at the very foot of p. 13.

However, this passage is important in one sense at least, for it clearly reveals the conjectural status of the hypothesis as it was in Bradley's mind as he wrote his account. Thus in the version that was published by the Royal Society in the *Philosophical Transactions*, Bradley wrote, 'This therefore, being the fact', whereas in the version designated as Appendix 2, Bradley wrote instead, 'And although this is the fact I will allow that it is not an absolute proof of the truth of my hypothesis; however'. Bradley thus attempts to qualify his claim, it is conjectural.

Passage E* to *E (fol.29v line 2 to fol.30r line 2)

In many respects this is the most interesting of all the textual variations from the version published in the *Philosophical Transactions*. The entire account is addressed to Edmond Halley, and Bradley is aware that his friend is sceptical of Bradley's claim to be able to discern motions as insignificant as 0.5″ (Halley worked to tolerances of 5″). Bradley attempted to relate the consequences of the extreme precision of his observations with some recent work undertaken by Halley, on the extinction of stars when occulted by the Moon.

This passage is not included in the version included in the *Philosophical Transactions* (nor subsequently in the *Miscellaneous Works*). Bradley sought to corroborate his observations of the 'new discovered motion' and its consequences for the belief that the stars are considerably further from the Solar System than most astronomers had been willing to concede. Working to such precise limits had led Bradley to conclude that he had not found evidence for annual parallax of even 1″, with the consequence that the star γ Draconis must be at least 400,000 times the distance as that separating the Earth and the Sun. Arguing from the principle of analogy, Bradley reasoned that the visual diameter of most of the nearest stars could not much exceed 1/200th of a second. This confirmed the observations made by Halley of lunar occultations of first magnitude stars which were extinguished 'instantaneously', which suggested that such stars possessed extremely small angular diameters. This finding (now by inference corroborated by Bradley's observations) contradicted the 'appearance' of first magnitude stars in contemporary telescopes where images in excess of 4″ or 5″ were common.

Of course, Bradley was not to know of huge giant stars such as *Betelgeuse* with diameters in excess of 10 AU which would reach out as far as the planet Jupiter, but such stars are an exception rather than the rule, most stars living most of their lives on the *main sequence* as observed in HR (Hertzsprung-Russell) diagrams. This is an interesting passage for it reveals that Bradley was thinking constructively about stars as physical bodies rather than just points of light. These indeed may be amongst the first attempts to evaluate the diameters of the stars in the light of two separate sets of observational evidence.

This passage follows that found in the published version the *Philosophical Transactions* at the bottom of p. 660, or at the bottom of p. 15 in the *Miscellaneous Works*.

The 'Postscript' in the Published Account

The final argument in favour of Appendix 2 being a later and a more authoritative version of Bradley's account of the discovery of the 'new discovered motion' is the simple fact that the 'Postscript' to the published version which appears to have been added as an afterthought at the end of a document written under time pressure. In Appendix 2 this material is embedded in the text. That it immediately follows the passage E*—*E, where Bradley sought to corroborate his hypothesis by reference to work undertaken by Halley is significant, for Bradley also sought further corroboration with reference to work undertaken by John Flamsteed.

Conclusion

The evidences of the passages A to E, and the placement of the Postscript into the body of the paper suggests that Appendix 2 is a later, more inclusive paper than the account published by the Royal Society. The postscript to the published version suggests a paper hurriedly prepared and completed. A brief examination of Bradley's observation books suggests the possibility that he had spent some seven to ten days preparing his account, for very little observational work was undertaken during the last week of 1728, contrary to the normal pattern of previous years. It is certain that Bradley wrote and re-wrote and further refined his account three or four times during this period. The first document (Bradley MS20* fols. 1 & 2 and fols. 15 & 16), the second document (Bradley MS20* fols. 3 to 14 & fol. 17) and the account published in the *Philosophical Transactions*, (No.406, Vol.35, pp. 637 to 661), were all penned during this period, immediately before the version now presented as Appendix 2 in this account of the life and work of James Bradley. It was originally transcribed and edited in 1992–93 by the present author and included in an MSc dissertation as Document 3. All of the different versions of Bradley's account appear to have been written under time pressure in time to be presented to the Royal Society on 9 and 16 January 1729. The version presented here as Appendix 2 was dated the 5 January and appears to have been written with less urgency, supposedly as the account that Bradley wished to see published. It was therefore somewhat unfortunate that the 'hurried' account handed to Halley for presentation to the Royal Society was the version that was published. It is unlikely that Bradley was asked for a further account of his discovery of 'the new discovered motion' and so this account, much to be preferred to that included in the *Philosophical Transactions*, remained unpublished and largely ever since, unread. It is perhaps surprising, in view of the textual differences and the obvious lateness of this document, that Stephen Peter Rigaud did not include some references to it in his *Memoir*. There is every justification for the publication of this document as Bradley's intended full account of the 'new discovered motion' soon to be termed 'the aberration of light'. **See** Appendix 3.

Notes

1. This document is addressed to James Bradley's mentor Edmond Halley, at this time the second Astronomer Royal of England.
2. This clearly sets out that the leader of the attempt to observe annual parallax at Kew was Samuel Molyneux.
3. Bradley suggests that he had great confidence in Hooke's observations, even though he only made four observations and later expressed doubts about their possible veracity. In my discussion of the Kew experiment, I suggest that Molyneux wished to confirm Hooke's results in order to attack Newton's assertion that the stars possessed insensible parallaxes, an important requisite in Newtonian theory in support of the continuing stability of the solar system. Bradley, throughout his account sought to distance himself from any suspicion that the motives of the Kew experiment were in any way untoward. On the contrary, he sought to convince his audience that the experiment was undertaken in a spirit of disinterested enquiry. Bradley also distances himself from Hooke, almost insinuating that Hooke was nothing less than fraudulent (see fol. 30). On Bradley MS 20* fol. 68r Bradley writes, 'N.B. According to my Observations the Star would have been 5½″ more northerly Oct 21 than 'twas July ye 7th or 8th

whereas Dr Hook make it 23″ more southerly Diff = 28½″ And by my Observations twas 2″ more southerly on Oct 21 than on August 6th whereas Dr Hook makes it then 17″ more southerly, that is 15″ Difference from mine.' On fol. 30 Bradley confesses that he is 'at a loss how to account for it', but leaves the reader with little choice but to suppose Hooke was guilty of immense carelessness or even deceit.

4. 'see where I have given you' deleted / 'perceive upon' substituted.

5. 'also' inserted.

6. This emphasis on the accuracy of Graham's instruments serves a two-fold purpose. Firstly, Bradley seeks to underline the precision of his own results in order to validate the 'law' and the 'hypothesis' of the 'new discovered motion'. Secondly, it is part of Bradley's literary strategy of distancing himself from his supposed confidence in Hooke's results.

7. 'Para' deleted.

8. Bradley is here referring to parallaxes of less than 1″ (as inferred from his observations) and not the commonly accepted value of 30″ to 40″. Bradley is thus referring to inaccuracies of the order of 5″ to 10″ and not in excess of 1′ or 2′.

9. 'but' deleted.

10. 'will be too fortright to' deleted / 'it is not' substituted

11. 'will require several draughts & etc.' Deleted / 'to tell you' substituted.

12. 'the several parts thereof' deleted / 'but' inserted.

13. The word 'curious' is here used in the sense of an *objective* eagerness to learn or inquire. Thus the very first observation of the aberrant behaviour of γ Draconis was prompted by 'curiosity'. Throughout his account, Bradley emphasizes its objective character, even though it may at times have been considerably more subjective. There is in Bradley's account a strong element of self-justification.

14. 'them' inserted.

15. K. V. Hewitt, *The Astronomical Work of the Rev. James Bradley*. MSc Dissertation, London University September 1953. pp. 87–88, has concurred with Rigaud's opinion in describing Bradley's observation of the 20 December 1725 as 'decisive'. Yet, at this point in the programme this, observation could not have been anything but an anomaly. An observation such as this, taken in isolation, could never be decisive. The belief that this single observation was significant was very much a product of what I called the 'received account' I refer to in the sixth section of my MSc dissertation *James Bradley and the New Discovered Motion*.

16. The fact that Bradley made this observation on the 20 December 1725 *when no significant motion due to parallax could be expected*, indicates the thoroughness of his entire approach to observation. When he observed the aberrant motion of γ Draconis, he immediately suspected that the instrumentation itself was at fault in some measure. It is precisely these sorts of observations, made when no motion was expected in γ Draconis, which were used by Bradley to thoroughly cross check the instrumental set-up.

17. Bradley mentions seven separate conjectures or hypotheses in the course of his account, though these are not presented in their correct chronological order, this being a reflection of his carefully developed literary strategy. Throughout his paper, Bradley reveals these discarded or modified conjectures as evidence of the

thoroughness of his investigation. This *first conjecture,* proposes that he first considered *observational error,* due to the difficulty of 'reading off' small angles of $1''$ or less. Halley was both amused and irritated by Bradley's insistence that he could measure such small angles in spite of the obvious vagaries of instrumental and observational error. Bradley's confidence that he could measure such small angles was supported by two important factors. Firstly, George Graham's instruments were, arguably, the most accurate ever constructed. Secondly, his observational skills, which were a development of techniques acquired from his uncle, James Pound, one of the finest observers in England, enabled him to perceive motions as small as $\frac{1}{2}''$. Even so, his caution led him first to suspect *observational error.* One of Bradley's most significant attributes as an observer and as a theoretician was his cautious scepticism.

18. It would appear from the personal pronoun that Bradley was the leading observer of the group (due undoubtedly to his unequalled skills in the observation of small angles), Molyneux being the main recorder. This appears to be evidenced by the fact that the original Kew Observation Book was written in Molyneux's hand (from which Bradley later made his own copy). Molyneux was the leading actor. The instrument was commissioned and purchased by Molyneux from George Graham and suspended at his own mansion in Kew. George Graham was also an observer during the programme, his knowledge of the limitations of the instrument being invaluable.

19. Bradley's assertion that the alteration in the position of γ Draconis was *the contrary way from what it would have been had it proceeded from an annual parallax of the star* is evidence that from this period, after the end of December 1725, Bradley recognized that the motion had nothing to do with annual parallax. First, endnote 16 reveals that the motion of γ Draconis was not expected at that time of the year. It was *out of phase* (by three months or 90°) with the motion to be expected from annual parallax. Second, Bradley noted that the motion of γ Draconis was opposite to the motion to be expected from annual parallax. Bradley clearly recognized at a very early stage in the investigation that he and his partners were observing a novel motion.

20. This is the *second conjecture* mentioned by Bradley. It is that of *instrumental error.* Although Bradley had great confidence in the accuracy and efficacy of Graham's instruments, the apparent perversity of the phenomenon they were observing was so great that it was necessary to determine first, whether the instrumentation was set up correctly, and whether it really was as accurate as he believed it to be.

21. The shift away from an explanation based on a combination of observational and instrumental errors was probably a gradual process, never entirely abandoned until Bradley was able to make *multiple confirmations* of the phenomenon over a wide range of ecliptical longitudes and latitudes. By March 1726, Molyneux, Graham, and Bradley recognized they were almost certainly observing a new phenomenon.

22. 'near' deleted.

23. 'made' inserted.

24. By March 1726, Molyneux, Graham, and Bradley were feeling secure enough to apprehend that *a regular and predictable phenomenon* was being observed. James Bradley's own basic presuppositions would have expected some rational or predictable explanation.

25. 'it was' deleted.

26. 'Precisely' deleted.

27. At the correction referred to at note 26, between the words 'December' and 'to', Bradley deleted the word 'precisely'. It is a significant deletion. *When Bradley was writing his account* he presumed that Halley, his correspondent, recognized that he made allowance for the regular precession of the equinoxes. Yet Bradley also recognized (by early or mid 1728) that another motion was certainly to be discerned in the residuals between theory and observation. The stars (making allowance for precession) did not return *precisely* to their original positions. Even as he wrote his account, Bradley was already well advanced in the programme of determining the causes of this anomaly, recognizing already that it could not be accounted for by annual parallax.

28. 'Equinoctial points' deleted. / 'equinox' substituted.

29. Bradley was sure it was a natural motion as he had been observing for nine months. For one thing, the motion was too regular and predictable. Nevertheless, he never entirely abandoned the suspicion until he obtained independent corroboration with his own instrument, suspended at Wanstead on 19 August 1727.

30. Although Bradley reveals, in different drafts of his paper on the 'new discovered motion' (this is the fourth), he felt he had great difficulty in expressing himself both clearly and convincingly, and he is at all times careful in his use of words. Thus, when Bradley uses the word 'apparent', he wishes to express the difficulty he still had at this stage of the inquiry, of determining the 'reality' of the motion observed. Bradley's cautious scepticism is at all times an integral aspect of his approach to observation.

31. The *third conjecture* of Bradley's account was that of *the annual nutation of the Earth's axis*. Bradley's account is not a simple narrative of his investigation but a self-justification of his entire programme. At this point in his account, he completely ignores Molyneux's hypothesis apparently presented in an untraced paper of the 18 June 1726 mentioned by Rigaud, that of refraction due to the alteration in the figure of the atmosphere due to the Earth moving in a resisting medium. This embarrassing conjecture was not one he believed he could discuss in a paper in which he was seeking to convince his contemporaries who large supported a post-Newtonian physics. He may have concurred with this conjecture out of respect for Molyneux, his social superior. Equally we can be sure he was uncomfortable with it, for it was rapidly abandoned. Even so, to account for such a large motion in terms of an annual nutation was equally a radical challenge to the validity of Newtonian theory. It was this 'nutation' that couldn't have been maintained beyond the beginning of March 1726, that Molyneux so eagerly informed Conduitt and Newton of, claiming it destroyed 'completely the Newtonian system'. Even though I am sure neither Bradley nor Graham were apprised of Molyneux's premature approach to Newton, I am sure Bradley's caution would have avoided the use of the term 'annual'. In referring to a 'nutation', Bradley was able to assuage any possible criticism, for according to Newtonian theory, there were other possible nutations, due to the differing combined effects of the Sun and the Moon on the equatorial regions of the Earth.

32. 'was' deleted.

33. 'found' deleted (with 'a') / 'appeared' substituted.

34. Bradley claims that this nutation soon proved to be insufficient. Certainly, entries in the Kew Observation Book suggest that the hypothesis had to be abandoned by the

beginning of March 1726 at the latest. This may underline just how quickly Molyneux made his premature approach to Conduitt with his claim that he and his partners had discovered a nutation which '....he thought destroyed entirely the Newtonian System'. Bradley would not be party to such an approach for he would have advised caution above all, particularly in an approach to Sir Isaac Newton of all people.

35. 'opposite' inserted.
36. 'small' deleted.
37. 'appearance' deleted.
38. 'the' inserted.
39. 'of [illegible word]' deleted. / 'of defi ... [illegible word]' deleted.
40. 'and constant' deleted.
41. 'in both the forementioned stars' inserted.
42. 'change' deleted. / 'difference' substituted.
43. 'from' deleted. / 'other' substituted, then deleted.
44. 'maximam' transposed to 'maxima'.
45. 'sines' transposed to 'sine'
46. Bradley's conjecture was that the motion of γ Draconis was not unique, but was a manifestation of a phenomenon that must be displayed by all of the stars; the whole of the future Wanstead experiment was predicated on this presumption. Bradley sought a result that was common to all of the objective stars of his experiment, and designed it accordingly.
47. Bradley is no longer seeking to frame hypotheses; he is seeking a phenomenological law that describes the motion of all the stars. Only then will he attempt to hypothesize the possible causes of that law. His approach contrasts markedly from the procedures exercised at Kew.
48. 'explain' deleted.
49. This underlines the change in the methodology. His whole approach to the problem has shifted and the focus of his investigation has been diverted away from a search for causes to that of determining the parameters of the phenomenon he is sure will be displayed by all of the stars.
50. 'only' deleted.
51. At this point in his account, Bradley applied his literary strategy of convincing his audience that he and his partners were motivated by disinterested curiosity, that of determining whether γ Draconis had *any sensible parallax*. Bradley, however, introduces what may have been an unwitting inconsistency in his account which quite clearly contradicts his assertion on fol. 18r that the experiment at Kew was set up 'in hopes of verifying and confirming those that Dr Hooke formerly communicated to the public'. The significance of this apparently trivial inconsistency is, however, revealed once it is recognized that the experiment may have been perceived by Molyneux, at least, as an attempt to discredit the veracity of Newtonian natural philosophy in that an annual parallax of some 30″ or more would place the nearer stars sufficiently close to the solar system to disrupt it. It was, nevertheless, under the guise of an attempt to verify Copernican heliocentric theory. Taken in conjunction with the memorandum penned by John Conduitt it can be argued that this was precisely Molyneux's real purpose.

52. 'Mr Graham' inserted.
53. 'radius' inserted.
54. '[illegible letter]' deleted.
55. 'since' inserted.
56. 'since' deleted.
57. 'to' inserted.
58. 'than that' deleted.
59. 'with' deleted. / ',' omitted.
60. 'both' deleted.
61. 'very plain' deleted.
62. 'time' inserted.
63. See also note 46. By early 1728, Bradley was sure that the phenomenon of the 'new discovered motion' was connected in some way with *the relative motion of the Earth, Sun and stars*. The thoroughness of his methodology, combined with his skilful observational techniques, were beginning to reveal the underlying patterns which were slowly pointing to some common law of motion.
64. 'that were' inserted.
65. By March 1728. That is after about six months of observation, when the stars being observed had passed through their projected annual motions and it was possible to make conjectures about their completed annual motions.
66. The pivotal turning point of Bradley's investigation which vindicates his entire programme, both in its chosen methodology and in its instrumentation, in the awareness that each of the stars observed followed exactly the same pattern of motion. Bradley was now convinced he *was* observing a universal phenomenon. Yet his 'discovery' of the pattern of events described was as much a product of his preconceptions as anything that happened in the 'real' world. It was a direct consequence of a programme designed specifically to isolate such patterns from out of a 'background' of anomalous and unaccountable motions. He has yet only a vague conception of what this universal law of motion is and even less comprehension of its possible causes, but Bradley is now capable of creating the necessary theoretical framework that will enable him to interpret his observations.
67. 'in most stars' inserted.
68. 'about' deleted.
69. 'all' inserted and deleted.
70. 'all' inserted.
71. 'alike' deleted.
72. That the stars under observation did not alter their declinations equally led Bradley to determine the pattern of these variations and he was soon led to the conclusion that it was a function of ecliptic latitude, being greatest at the pole of the ecliptic, and least at the plane of the ecliptic.
73. 'observations' inserted.
74. This is confirmed in his *fourth conjecture, the law of the sine of the latitude* or *the law of the new discovered motion*. Bradley, however, is concerned even early in 1728, that the observations do not *exactly* fit this projected law of motion, and it is from this period that Bradley begins his programme to determine the causes of

this anomaly; a programme which led to the confirmation of the nutation of the Earth's axis.

75. Bradley's caution is indicative of his entire approach to observation. The systematic thoroughness is only one aspect of its gathering maturity. This is why it is difficult to reconcile Bradley's claimed acceptance of Hooke's observations, unless he was acting out of character, perhaps in response to the socially superior Molyneux and his request to join him at Kew to make observations to verify Hooke's claim to have observed the annual parallax of γ Draconis. By Bradley's standards, Hooke's approach to observation was unbelievably incompetent. It is little wonder that Bradley seeks to distance himself from a supposed attachment to Hooke's results.

76. Throughout his account Bradley is attempting, through his carefully constructed literary strategy, to convince his audience, including any possible detractors, of the thoroughness of his approach. Here he is implying that he is aware of a possible 'confirmatory bias' in his approach, being evidently aware that Hooke was guilty of such a bias. He is claiming that he suspended judgement until he had completed his observations, in order to determine whether there was any possible systemic or instrumental error.

77. Bradley was sure he had determined the 'law of the new discovered motion' but it is a part of his literary strategy to suggest that he would only seek the causes of this law when he had completed his programme.

78. Again we see Bradley's carefully argued strategy, suggesting that the 'law' was solely the product of empirical research rather than a validation of his presuppositions. The discovery of the 'causes' would necessarily follow (according to this public account) from just such an empirical approach.

79. See note 31. This had been the third conjecture mentioned in his account.

80. This, *the fifth conjecture* of his account, that of *collimation error*, at once reveals that it is in fact a 'reconstructed account', with the objective of convincing possible detractors of the thoroughness of his approach. It is inconceivable that such a careful observer, on the verge of presenting an account which confirmed his deepest presuppositions, would have seriously entertained the notion that such a regular and well-observed phenomenon could have been accounted for by collimation error. It is merely another aspect of his literary strategy.

81. The *sixth conjecture of his account*, that of *the possible effects of refraction*. This is proof that Bradley's account is not a simple narrative of his programme. See also note 3. This conjecture was not entertained late in 1728, as a naive reading of this account might infer, but in March 1726, well over a year before he suspended the zenith sector in Wanstead. A paper that examined the possible effects of refraction was dated 18 June 1726, supposedly written by Molyneux. We can be convinced that this conjecture possibly pre-dated the supposition that the phenomenon might be explained as a nutation. Explained as the Earth moving through a dense medium it was nothing less than a 'Cartesian' hypothesis which conjectured the possible figure of the Earth's atmosphere as it moved through this medium. Bradley does not refer to the hypothesis in any detail, merely presenting it as a rejected conjecture, as part of his wider strategy, designed to convince his audience of the completeness of his entire programme. He does not in any way defer to the notion that such a contention might validate a model of the universe that was quite opposed to the basic presumptions of

Newtonian theory. On the contrary, he distances himself from such a possibility by claiming that he has entertained the notion and has rejected it.

82. *The seventh and final conjecture; the hypothesis of the new discovered motion*, now termed the 'aberration of light'. In Bradley's account the hypothesis seems to appear as though pulled like a rabbit out of a conjuror's hat. It is this which probably led Thomson to write his account of Bradley's discovery of aberration. Yet the basis of Bradley's hypothesis is well grounded in his own astronomical practice. The four integral components of the hypothesis were (1) The belief that the Earth was in motion around the Sun. (2) The belief that the velocity of light was finite and not instantaneous. (3) The acceptance that light progressed in discrete particles. (4) That the 'constant of aberration' could be obtained through a simple process of trigonometry; As the base of the triangle was determined by the velocity of the Earth in its orbit, and the constant of aberration was obtained by observation. Bradley's hypothesis provided natural philosophers with an accurate determination of the velocity of light. The first two components were grounded in Bradley's consistent and precise observations of Jupiter and the Galilean satellites and in the production of accurate ephemerides of the Jovian system on behalf of Halley, who was the correspondent to whom this letter was addressed. The third component was acquired reluctantly when he adopted the generally accepted model espoused by Isaac Newton, which perceived light as a stream of corpuscles, abandoning the more neutral term 'ray' during the earlier versions of this paper. The final component was a simple application of vectors. Even if Thomson's account was based on some oral tradition now lost, it fails to explain the origins of the hypothesis. Only if the hypothesis is seen in the context of Bradley's astronomical practice can such an account make sense; and as such it is largely superfluous. The hypothesis of the new discovered motion was firmly grounded in the theory and practice of a skilled and experienced astronomer.

83. 'then' deleted.
84. 'as' inserted.
85. 'that' was written twice.
86. 'was' deleted.
87. 'they' transposed to 'the'.
88. 'the tube' deleted.
89. 'angle' inserted.
90. 'is at' deleted.
91. 'and' deleted.
92. 'indeed' inserted.
93. 'D' deleted.
94. 'be' inserted.
95. 'sufficiently' deleted.
96. 'supposing the velocity of light shall always the same perpendicular to the velocity of the eye, then' deleted / 'in all cases' inserted.
97. 'that is' deleted / ':'. Inserted.
98. Bradley seeks to demonstrate that the movement of a star due to the mechanism of the 'new discovered motion' is directly a consequence of the ratio of the velocity of

the Earth around the Sun relative to the velocity of light. The smaller the angle, the greater is the velocity of light relative to the velocity of the Earth. Bradley claims a hypothetical method of determining the velocity of light, provided that the motion of the Earth is accepted.

99. 'plane' inserted.
100. Bradley inadvertently wrote 'by'/ edited to 'be'
101. 'suppose' inserted.
102. 'figures' amended to 'figure'.
103. 'in' deleted. / 'of' overwritten.
104. 'motion' inserted.
105. '[illegible word]' deleted.
106. Coming directly after the passage A* to *A, which clarifies the theory of the law of the new discovered motion (this passage is not found in the earlier published version), Bradley infers that the 'aberrational ellipse' is obtained from the varying degrees of motion of each of the eleven or twelve stars consistently observed in declination. In order to have confirmed the ellipse of the new discovered motion, he should have timed the exact moment of passage of each star through the meridian, in order to determine any possible changes in right ascension; repeating the task undertaken (unknown to Bradley) by Manfredi. Bradley's own presumptions allowed him to project his conceptual framework onto the raw data, by which he was then able to interpret it.
107. 'circle' inserted.
108. 'rectangle under the' inserted.
109. '[illegible word]' deleted.
110. 'of declinat' deleted.
111. 'Earth's motion' deleted.
112. 'summ' amended to 'sum'
113. 'wha' deleted.
114. 'pecession' edited to 'precession'.
115. Bradley's expressed confidence in his data is a consequence of: (1) *The accuracy of his instrument.* Such was the accuracy of Graham's 12½' zenith sector, that when it was later transferred to Greenwich, it was used to cross-check the vertical collimation of the meridian quadrants. (See Allan Chapman, 'Pure Research and Practical Teaching: The Astronomical Career of James Bradley. 1693–1762. (Pre-published typescript provided by the author. 1993. p. 6.)) (2) *The corroboration of the earlier Kew results by the Wanstead observations.* (3) *The systematic approach deployed by Bradley.* He was evidently the first observer who recognized the importance of constant monitoring of an instrument for possible errors. (See Chapman, 1993, (above) for this important point). This is why the fifth conjecture, that of possible collimation error was obviously included as a rhetorical flourish; Bradley's whole approach makes such a possibility at such a late stage of the investigation quite nonsensical. (4)The fact that all twelve stars under constant observation behaved (approximately) in accordance with the hypothesis. Bradley obviously identified an anomaly as early as February or March 1728 and this led to the continuation of his observations until 1747 when he published his paper on the nutation of the Earth's axis. (5) The

fact that he was, due to his systematic approach, able to determine the upper limits of his errors. Bradley may have been the first observer to identify the importance of error.

116. These observations of such variations were the earliest intimations of the phenomenon of the nutation of the Earth's nutation, the subject of a paper in 1748 after observing the phenomenon for twenty years.

117. 'fully' inserted.

118. 'γ Drac:' amended to 'γ Draconis'.

119. '39‴' inserted.

120. 'γ Drac:' amended to 'γ Draconis'.

121. 'which in this case is abo' deleted. / 'this therefore being' inserted.

122. Modern measures put this at close to 8′ 20″. This is a measure of the remarkable accuracy of Bradley's observations.

123. 'It is well known that' inserted.

124. 'Maxim' deleted. / 'mean' substituted.

125. 'determining' deleted. / 'finding' substituted.

126. The text is difficult to decifer. It may be 'assumed' / 'assumed' included by the editor to fit the text.

127. Bradley's presumption that the velocity of light was a constant was a reflection of his fundamental belief in the order and uniformity of nature.

128. 'scarce' deleted. / 'not' inserted and then deleted. / 'be' omitted.

129. 'magnitude' deleted and restored.

130. 'which' deleted and restored.

131. 'γ Drac:' amended to 'γ Draconis'

132. '35th Camelopard' amended to '35th Camelopardalis'

133. 'Hevel' omitted.

134. 'γ Drac' amended to γ Draconis.

135. 'when' deleted. / 'as' inserted.

136. 'such' deleted. / 'these' inserted.

137. 'α Cassiop.' amended to 'α Cassiopeæ'

138. 'β Drac' amended to 'β Draconis'.

139. 'of Capella' deleted.

140. '[illegible word]' deleted.

141. 'it is' is written twice in succession when beginning a fresh sheet of paper.

142. 'great' deleted. / 'near' inserted.

143. 'its' deleted.

144. 'with' amended to 'within'.

145. 'for' deleted. / 'to attain' inserted.

146. 'γ Drac:' amended to γ Draconis.

147. '1727' inserted.

148. 'to the' deleted. / 'more' inserted.

149. 'that is' inserted.

150. 'common' inserted.

151. 'the glasses of' inserted.

152. 'as well as with the hypothesis' inserted.

153. Here Bradley confesses and reveals the internal conflict which arose from his axiomatic belief in the rationality of the world and the incapacity of empirical observation and measurement to clearly demonstrate this order beyond the limits of observational and systemic error. The tension between his fundamental acceptance of the order of the universe, a prime requisite to any approach to nature based on extreme precision, and his deeply ingrained caution and scepticism concerning the capacity of observation to determine that order is one of the fundamental imperatives that drove Bradley's work towards its consummation in the 1748 paper on the nutation, and the unprecedented series of observations made at the Royal Observatory from 1750 to 1762

154. 'began a week' deleted. / 'had compared' inserted.

155. 'possibly' deleted. / 'possibly' rewritten.

156. 'make' deleted.

157. Bradley is mindful of the possibility of 'confirmatory bias' in the unexpected 'accuracy' of his results. He is determined to avoid the trap of seeing only that which he wishes to see (Hooke's major error). He is also mistrustful of the ability of empirical observation to come so close to prediction without there being some form of bias at play. It is an essential aspect of Bradley's strategy to acknowledge the possibility that such a bias is possible and that he has considered this. At the same time he wishes to show that his hypothesis is supported, so far as it is possible to be supported, by his observations.

158. '[illegible word]' deleted.

159. Compare Bradley's eighty observations of this one single star with the four made by Hooke over a period of only 3½ months.

160. Here Bradley further reveals the psychological tensions between the axiomatic belief and epistemic doubt in the unexpectedly close correlation between theory and observation. Even though the 'law' appears to be confirmed, he recognizes that the 'hypothesis' is not. The hypothesis is grounded on several unconfirmed conjectures based very firmly in his past astronomical practice, such as his belief in the finite velocity of light and the motion of the Earth in an orbit around the Sun, neither of which had been established independently of the other.

161. 'γ Drac' amended to 'γ Draconis'.

162. 'η Urs: Maj' amended to 'η Ursæ Majoris'

163. 'found' inserted.

164. 'Septemb.' amended to 'September'.

165. 'future' deleted.

166. 'No' inserted. / amended to 'north'.

167. 'but' deleted. / 'and' inserted.

168. The distinction between the 'law' of the new discovered motion and the 'hypothesis' is here made explicit. Bradley is sure he has determined 'the true law of the variation of declination', whether or not he has successfully discovered 'the real cause of the phenomenon or not'.

169. 'be' overwritten by 'is'.

170. 'has' inserted.

171. This may have been a backwards glance at his late partner's stated credentials and his belief in the validity of Hooke's claims to have discovered the annual parallax of

a star. The word 'supposed' used in conjunction with the word 'pretended' reveals his withering contempt for Hooke's claims and his concerns that his late partner had allowed himself to be fooled by such specious claims.

172. The vehemence (by Bradley's standards) of this sentence is indicative of the shift that has taken place in Bradley's evaluation of the worth of Hooke's observations. This may also be a part of Bradley's literary strategy. Whether or not Bradley was ever confident of Hooke's observation s is questionable, and his supposed confidence may have been expressed merely as a subterfuge to hide the original purposes of the Kew observations, which Bradley may later have recognized as an attempt to discredit the validity of Newton's writings. Given this, it may have been perceived necessary by Bradley to claim confidence, only later to disavow this with what may be described as feigned anger.

173. 'Drac:' amended to 'γ Draconis'.

174. 'exactly' deleted. / 'well' inserted.

175. 'with' inserted.

176. Bradley has now re-defined the problem of determining annual parallax, placing it effectively beyond the technological possibilities of his own epoch. In doing this, he has now reconciled Newton's long-held insistence of a larger scale of stellar distances with empirical evidences of a distance scale several orders greater than many astronomers had usually been prepared to concede. The very fact that the discovery of annual parallax had now become such a remote possibility, led to the acceptance of the new discovered motion as empirical evidence of the Earth's motion, even though the motion of the Earth was one of the unconfirmed conjectures upon which the hypothesis was constructed.

177. 'therefore' inserted.

178. There was not the slightest doubt in Bradley's mind, of the veracity of the Copernican–Keplerian system. It was an integral part of Bradley's astronomical background and practice, cemented by years of precise observations of Jupiter and Saturn and their respective satellite systems. With respect to his studies of Jupiter's three innermost Galilean satellites he was the first to identify that their motions in their orbits were in resonance with each other.

179. '[Illegible word]' deleted.

180. 'γ Drac:' amended to 'γ Draconis'.

181. The distance of 400,000 AU was a consequence of Bradley's assertion that the upper limit of any expected annual parallax was $1''$ of arc. In modern practice, which utilizes the mean radius of the Earth's orbit as the baseline, an Astronomical Unit (AU) and not the mean diameter (the orbis magna), this corresponds to an annual parallax of $0.500''$. The nearest star to the solar system is Proxima Centauri which has an annual parallax of about $0.750''$, vinidicating Bradley's confident assertions.

182. From the inception of the telescopic observation of the stars, it was initially a puzzle why it was that planetary images increased in diameter with increased powers of magnification, whilst stellar images decreased their apparent diameter. It was recognized that stellar images were largely a consequence of telescopic artifacts which exaggerated their apparent sizes. Yet, it was believed their diameters were sensible. Bradley's reasoning, however, when combined with the results of his observations,

suggests that it was not possible to 'see' the stars, for an estimated diameter of 1/200th″ was much too small to be observed by the human eye, even with instrumental aid.

183. 'their' deleted.

184. With an upper parallactic displacement of 1″, a stellar diameter of even 1″ would infer a diameter equal to the diameter of the Earth's orbit.

185. 'γ Drac' amended to 'γ Draconis'

186. 'are' deleted. / 'may' inserted.

187. I have come to the conclusion that Bradley is being disingenuous here. Remember that after observing Jupiter and its satellites, many stars, and many other astronomical objects, he had developed a reiterative methodology that was completely irreconcilable with Hooke's four observations over 3½ months. I think he is seeking to underline his literary strategy in his insistence that this was a disinterested enquiry.

188. 'Dr' deleted.

189. See Bradley MS 20* fol. 68. (Appendix 3). Bradley records Hooke's four observations.

6th July 1669	2′ 12″ North.
9th	2′ 12″ North.
August 6th	2′ 6″
October 21st	1′ 48″ or 1′ 50″.

190. 'Dr' overwritten by 'those'.

191. 'upon' deleted. / 'from' inserted.

192. 'June than' deleted.

193. Bradley is here using Flamsteed's observations of *Polaris* as a corroboration of his hypothesis deduced from his observations of γ Draconis.

194. This was the 140° arc constructed and divided by Abraham Sharp in 1689. The instrument was removed from the Royal Observatory by Margaret Flamsteed as a part of her late husband's estate.

195. 'by you' deleted.

Two Transitional Aberration Documents: Transcribed and edited by John Fisher

Document A

Bodleian Library: The Department of Western Manuscripts: James Bradley's Manuscript papers: MS20*

fol. 60v.

The Parallax of ye Stars[1] on account of the motion of light[2] causing them[3] to appear always[4] removed from their true places in lines parallel to ye Tangents in the respective points of ye Earths Orbite;[5] tis thence evident that if the earths orbite were perfectly circular and ye Sun in its centre[6] the stars would describe exactly the same Ellipses by their apparent motion, as if they were subject to an Equal[7] annual parallax upon account of Distance;[8] only the longer Axis in the Parallax arising from Light[9] would[10] lie in the same Direction as the Shorter Axis in the Parallax arising from Distance;[11] the visible place of the star in the former case[12] being 3 signs farther back in the Ecliptick,[13] than in the Latter.[14]

Since therefore in the Parallax on Acct of Distance[15] the latitude of a star is least when it appears in conjunction with the Sun and greatest when in opposition, it follows that the Parallax on Account of Light[16] will occasion the Star to appear near the Ecliptick of that Quadrature which precedes its Conjunction or when ye Sun 9S[17] from ye star, and farthest from ye Ecliptick when ye sun is three signs from ye star,[18] while therefore the Sun is passing from ye conjunction to the opposition with ye star, that is whilst ye longitude of [the][19] Sun from ye Star is less than 6 signs,[20] the visible latitude is greatest;[21] but if ye longitude[22] of ye Sun from ye star be greater than[23] 6s[24] then its visible latitude will be less than its true.

Document B

Bodleian Library: The Department of Western Manuscripts: James Bradley's Manuscript papers: MS20*

fol. 56r

Rules for computing the Aberration[25] or Parallax of the Stars arising from the Motion of Light.

To find the Parallax of Declination.

Let A be the Angle of Position (or the Angle of the Star made by two great circles drawn from it thro the poles of the Ecliptick & Equator), then say as the sine of the Stars latitude, is to the Radius, so is the Tangent[26] of A, to the Tangent of B. Then if the Star's[27] Longitude be[28] in [the] Ascending Semicircle[29] subtract B from the Stars Longitude, but if the descending Semicircle add B and you will have a Point, in the Ecliptick[30] to which the Sun arriving the star will have no Parallax of Declination being then the same with its true. And while the Sun is passing from[31] its conjunction with this point to its opposition with it, the apparent Declination[32] will be less[33] than the true if the stars Latitude & Declination are of the same Denomination (but Greater if of contrary Denominations).[34] The Quantity of the Parallax of Declination increases in the proportion of the Sine of the Sun's Longitude from the Point found as above, and the Greatest Parallax[35] will be as the Sine of the Angle A to ye sine of Angle B before determined.[36]

Notes

1. 'The Parallax of the Stars on account of the motion of light' is Bradley's transitional term to describe the hypothesis and the law of the new discovered motion prior to describing it as 'the Aberration or Parallax of the Stars arising from the Motion of Light', after becoming aware of Manfredi's terminology in Document B. An attentive reading of even this short document reveals that if Bradley had persisted in the use of the terminology used in this paper, general acceptance may have been inhibited. Certainly, the term 'parallax of the stars on account of the motion of light' is confusing and inaccurate.
2. 'is always being' deleted.
3. 'the' is amended to 'them' and 'Stars' is deleted.
4. 'nearer to ye Ecliptick' deleted.
5. 'then' deleted / Above the line are the words, 'paccd Parallax proceeding from distance being always in ae placement', omitted by editor.
6. 'The' deleted. / 'then' deleted.
7. 'annual' inserted.
8. Annual parallax on account of distance is annual parallax.
9. Annual aberration.
10. 'be at right Angle to' deleted.
11. Annual parallax
12. Annual aberration
13. This refers to three of the zodiacal constellations. It points to aberration being a full 90° out of phase with annual parallax.
14. Annual parallax.
15. Annual parallax.
16. Annual aberration.
17. Nine signs or 270°
18. 90° from the star. That is when the Sun is in quadrature.
19. 'the' added by the editor.
20. That is 180°.
21. 'than its true' is written above the line after 'greatest'. Bradlley did not attempt to incorporate it into the text.
22. 'long-' is rendered as 'longitude' by the editor.
23. '[illegible word]' omitted.

24. signs or 180°.
25. This is arguably the first time Bradley ever used the term 'Aberration' in conjunction with the new discovered motion. It dates possibly from late 1729 or early 1730.
26. 'Tang' edited to 'Tangent'.
27. 'Latitude be North' inserted but omitted by the editor to retain the sense of the text.
28. 'being' amended by Bradley to 'be'.
29. 'add' deleted
30. '(But the contrary if ye Latitude be south)' was written over the final line of the sentence but omitted by the editor.
31. 'the' deleted.
32. 'distance of ye star from ye pole' inserted then deleted.
33. 'greater' placed over 'less', then deleted.
34. 'and will be to its greatest parallax of Declinat'. Deleted.
35. Bradley overwrites 'Parallax' with 'Aberration', still unsure about whether he should use it to describe the phenomenon of the new discovered motion
36. There is no full stop. It may therefore have been abandoned.

Bibliography of the Primary and Secondary Sources

Primary sources

The primary sources located at the Department of Western Manuscripts at the Bodleian Library, Oxford

Bradley MS 1. Bradley's notebook for lectures at Oxford. 46 folios. Written 1730 to 1750. In English. Formerly designated 16404.

Bradley MS 2. An account of James Bradley taken from the *Eloge*. 32 folios. Written after 1762. Translated from French into English. Formerly designated 16405.

Bradley MS 3. Lectures on 'The more known properties of the Air established...' Plus notes on members attending his lectures. 198 folios. Written 1745 to 1760. In English. Formerly designated 16406.

Bradley MS 4. Lectures on Specific Gravity, Optics, etc. 64 folios. Written 1745 to 1760. In English. Formerly designated 16407.

Bradley MS 5. Lectures on Pneumatics, Tides, Optics and Astronomy. 95 folios. Written 1745 to 1760. In English. Formerly designated 16408.

Bradley MS 6. Lectures on Optics. 30 folios. Written 1745 to 1760. In English. Formerly designated 16409.

Bradley MS 7. Lectures on Mechanics. 38 folios. Written 1745 to 1760. In English. Formerly designated 16410.

Bradley MS 8. Lectures on Motion. 39 folios. Written 1745 to 1760. In English. Formerly designated 16411.

Bradley MS 9. Lectures on the Density of Fluids, and on Imparted Motion. 37 Folios. Written 1745 to 1760. In English. Formerly designated 16412.

Bradley MS 10. On Hydrostatics, partly 'by Mr. Saunderson' and on the Magnet. 100 folios. Written 1745 to 1760. In English. Formerly designated 16413.

Bradley MS 11. On Natural Phenomena. 85 folios. Written 1745 to 1760. In English. Formerly designated 16414.

Bradley MS 12. An abridgement of Lectures 1 to 5 (part) on the Properties of Bodies and Mechanics, delivered in Oxford in 1747. (Not in James Bradley's hand). 36 folios. Written 1745 to 1747. In English. Formerly designated 16415.

Bradley MS 13. An abridgement of Lectures 10 to 19 on Optics and Hydrostatics and for Lecture 5 (in part). (For Lectures 6 to 9 see Bradley MS 17). 36 folios. Written 1745 to 1760. In English. Formerly designated 16416.

Bradley MS 14. The Kew Observation Book. K14. 141 folios. Written 1720 to 1727. In English and Latin. Formerly designated 16417.

Bradley MS 15. Observations on the fixed stars made at Wanstead from 1727 to 1747. 111 folios. Written 1727 to 1747. In English. Formerly designated 16418.

Bradley MS 16. Tables of astronomical data – Sun, Moon, Jupiter and its satellites. 184 folios. Written 1720 to 1721. In English and Latin. Formerly designated 16419.

Bradley MS 17. Papers relating to James Bradley's lectures on natural philosophy at Oxford. From 1740 to 1760 (not complete). 166 folios. Written 1740 to 1760. In English. Formerly designated 16420.

Bradley MS 18. Notes by James Bradley about annuities, interest and income from professorship in 1724. Lectures etc. Miscellaneous astronomical observations. 45 folios. Written 1717 to 1762. In English. Formerly designated 16421.

Bradley MS 19. Latin lectures delivered at Oxford by James Bradley. It includes his 'Oratio inaugurals' 26th April, 1722. 191 folios. Written 1722 to 1733. In Latin. Formerly designated 16422.

Bradley MS 20. Tables of aberration and angles of position. 29 folios. Written 1730 to 1755. In English. Formerly designated 16423.

Bradley MS 20* Drafts of papers by James Bradley on Aberration. 104 folios. Written 1729 to 1750. In English and Latin. Formerly designated 16424.

Bradley MS 21. A book of observations made at Wanstead by James Pound and James Bradley. It includes a rough 'Oxford Observation Book'. 190 folios. Written 1718 to 1740. Formerly designated 16425.

Bradley MS 22. Dr. Bradley's papers including those belonging to Dr Pound. Astronomy, Natural Philosophy and Mathematics. Includes James Bradley's early work at Wanstead. 108 folios. Written 1715 to 1724. In English and Latin. Formerly designated 16426.

Bradley MS 23. James Pound's account book 1715 to 1724 with some receipts and bills. 110 folios. Written 1715 to 1724. In English. Formerly designated 16427.

Bradley MS 24. James Pound's reports of his voyages in the Indies, etc. 181 folios. Written 1705. In English. Formerly designated 16428.

Bradley MS 25. Calculations and determinations of the orbital elements etc., of the Comets of 1723, 1742, 1748, and 1757. 323 folios. Written 1723 to 1759. In English. Formerly designated 16429.

Bradley MS 26. Computations relating to the Sun 1743 to 1758. 197 folios. Written 1743 to 1758. In English. Formerly designated 16430.

Bradley MS 27. Computations relating to the Moon 1722 to 1759. 653 folios. Written 1722 to 1759. In English. Formerly designated 16431.

Bradley MS 28 Determination of longitude by observations at sea by Captain Campbell, 1757 to 1759. 73 folios. Written 1756 to 1760. In English. Formerly designated 16432.

Bradley MS 28* Calculations concerning the precession of the equinoxes. absolute and relative motion of certain fixed stars compared with Flamsteed's Observations. 24 folios. Written 1746 to 1760. In English. Formerly designated 16433.

Bradley MS 29. Observations and calculations by James Bradley of the fixed stars. 441 folios. Written 1746 to 1760. In English. Formerly designated 16434.

Bradley MS 30. Catalogus Stellarum. Not in James Bradley's hand. 34 folios. Written 1720. In Latin. Formerly designated 16435.

Bradley MS 31. James Bradley's observations at Greenwich with two transit instruments from 25 July 1742 to January 1742/43. Lord Macclesfield's similar observations at Shirburne, September to December 1742 and March 1743. 44 folios. Written 1742 to 1743. In English. Formerly designated 16436.

Bradley MS 32. Miscellaneous astronomical observations by James Bradley, including copies of some by Edmond Halley, and observations by Charles Green of an eclipse, May 1762, and of an Eclipse in Oxford in 1732. 73 folios. Written 1740 to 1760. In English. Formerly designated 16437.

Bradley MS 33. Miscellaneous observations and computations. 91 folios. Written 1740 to 1760. In English. Formerly designated 16438.

Bradley MS 34. Reduction of the Wanstead observations. 22 folios. Written 1740 to 1747. In English. Formerly designated 16439.

Bradley MS 35. Tables of Jupiter's satellites. Copies of some foreign observations. 102 folios. Written 1735 to 1760. In English. Formerly designated 16440.

Bradley MS 36 Early observations and calculations 1719 to 1721. The parallax of the Sun and Mars. 33 folios. Written 1719 to 1721. In English. Formerly designated 16441.

Bradley MS 37. Complete observations of pendulums. In Wanstead in 1719. At Jamaica by Mr. Campbell and Mr. Harris in 1732 and at Greenwich from 1743 to 1744. Drafts of James Bradley's paper to the Royal Society in 1733. 113 folios. Written 1719 to 1750. In English. Formerly designated 16442.

Bradley MS 38. A few observations and calculations by James Bradley. Aurora Borealis. Variations of the compass 1729 to 1731. 43 folios. Written 1727 to 1731. In English. Formerly designated 16443.

Bradley MS 38* A description of an instrument set up at Kew. Samuel Molyneux's original draft. 43 folios. Written 1725 to 1727. In English. Formerly designated 16444.

Bradley MS 39. A description of an instrument set up at Kew. James Bradley's copy of Samuel Molyneux's draft. 93 folios. Written 1725 to 1727. In English. Formerly designated 16445.

Bradley MS 40. Papers related to Nutation. Observations, calculations and drafts of the letter to Lord Macclesfield. 110 folios. Written 1725 to 1750. In English. Formerly designated 16446.

Bradley MS 41. The printed Bill and Act (24 Geo.2, cap 23: AD1751) for regulating the commencement of the year. Notes and tables on the subject by James Bradley. 35 folios. Printed and written 1750 to 1751. In English. Formerly designated 16447.

Bradley MS 42. Observations and calculations on refractions. Observed refractions. 56 folios. Written 1750 to 1754. In English. Formerly designated 16448.

Bradley MS 43. Miscellaneous astronomical tables by James Bradley. Tables of the Sun's declination, mean precession in longitude, etc. 93 folios. Written 1730 to 1760. In English and Latin. Formerly designated 16449.

Bradley MS 44. Correspondence to James Bradley and his assistants. 151 folios. Written 1732 to 1761. In English, Latin and French. Formerly designated 16450.

Bradley MS 45. Correspondence to James Bradley and his assistants. 72 folios. Written 1732 to 1761. In English, Latin and French. Formerly designated 16451.

Bradley MS 46. Papers in print prior to 1829. Not written by James Bradley. 56 folios. Written 1740 to 1760. In English. Formerly designated 16452.

Bradley MS 47. Miscellaneous papers. Not written by James Bradley. 155 folios. Written 1732 to 1766. In English, Latin and French. Formerly designated 16453.

Bradley MS 48. Notes of John Whiteside's Opticks. The 17th, 18th and 19th of a course of mathematical lectures and experiments. 64 folios. Written 1720 to 1725. In English. Formerly designated 16454.

Heber, Roger, *Memorandums of Doctor Bradley's Course of Experimental Philosophy,* by R. Heber, MS. Eng. Misc. e. 15.

The Archives of the Royal Greenwich Observatory

The Papers of James Bradley, Astronomer Royal 1742 to 1762 located at the Department of Manuscripts at Cambridge University Library

RGO 3 MS 1   Observations: Transits of major stars including adjustments to the telescope and observational method. Written 1743. Working copy.

RGO 3 MS 2   Observations: Transits of major stars including adjustments to the telescope and observational method. Written 1744 to 1746. Working copy.

RGO 3 MS 3   Observations: Transits of major stars including adjustments to the telescope and observational method. Written 1746 to 1749. Working copy.

RGO 3 MS 4   Observations: Transits of major stars including adjustments to the telescope and observational method. Written 1749 to 1750. Working copy.

RGO 3 MS 5   Observations: Transits of major stars including adjustments to the telescope and observational method. Written 1750 to 1755. Working copy.

RGO 3 MS 6   Observations: Transits of major stars including adjustments to the telescope and observational method. Written 1755 to 1758. Working copy.

RGO 3 MS 7   Observations: Transits of major stars including adjustments to the telescope and observational method. Written 1758 to 1762. Working copy.

RGO 3 MS 8   Observations, using a Quadrant, of major stars, including adjustments to the telescope and observational method. Written 1743 to 1744. Working copy.

RGO 3 MS 9   Observations, using a Quadrant, of major stars, including adjustments to the telescope and observational method. Written 1744 to 1747. Working copy.

RGO 3 MS 10 Observations, using a Quadrant, of major stars, including adjustments to the telescope and observational method. Written 1747 to 1753. Working copy.

RGO 3 MS 11 Observations using a Quadrant, of major stars, including adjustments to the telescope and observational method. Written 1750 to 1757. Working copy.

RGO 3 MS 12   Observations using a Quadrant, of major stars, including adjustments to the telescope and observational method. Written 1753 to 1757. Working copy.

RGO 3 MS 13   Observations using a Quadrant, of major stars, including adjustments to the telescope and observational method. Written 1758 to 1762. Working copy.

RGO 3 MS 14   Observations, using the New Quadrant, of transits of major stars, 1762 to 1765, also transits of major stars, 1764 to 1765, as observed by Charles Green, assistant to Nathaniel Bliss. Included are notes on adjustments to the telescope and observational method and compiled tables for the following items.

   1. Daily rate of the transit clock, deduced from the transits of the fixed stars, 1762 to 1764.
   2. Calculated mean times of the moon's passing the meridian, 1762 to 1765 (added to the volume in 1861).
   3. Apparent right ascensions of the planets, 1762 to 1765.
   4. The lunar eclipse of 17 March, 1764.
   5. Quadrant observations of the major stars by Charles Green, 1764 to 1765.

   Written 1762 to 1765. Working copy.

RGO 3 MS 15   Observations: Transits of major stars, including adjustments to the telescope and observational method. Written 1743. Fair copy of RGO 3.

RGO 3 MS 16   Observations; Transits of major stars, including adjustments to the telescope and observational method. Written 1744 to 1746. Fair copy of RGO 3 MS 2.

RGO 3 MS 17   Observations: Transits of major stars, including adjustments to the telescope and observational method. Written 1746 to 1749. Fair copy of RGO 3 MS 3.

RGO 3 MS 18   Observations: Transits of major stars, including adjustments to the telescope and observational method. Written 1749 to 1750. Fair copy of RGO 3 MS 4.

RGO 3 MS 19   Observations: Transits of major stars, including adjustments to the telescope and observational method. Written 1750 to 1755. Fair copy of RGO 3 MS 5.

RGO 3 MS 20   Observations; Transits of major stars, including adjustments to the telescopes and observational method. Written 1755 to 1758. Fair copy of RGO 3 MS 6.

RGO 3 MS 21   Observations: Transits of major stars, including adjustments to the telescopes and observational method. Written 1758 to 1762. Fair copy of RGO 3 MS 7.

RGO 3 MS 22   Observations, using a Quadrant, of major stars, including adjustments to the telescope and observational method. Written 1743 to 1744. Fair copy of RGO 3 MS 8.

RGO 3 MS 23   Observations, using a Quadrant, of major stars, including adjustments to the telescope and observational method. Written 1744 to 1747. Fair copy of RGO 3 MS 9.

RGO 3 MS 24   Observations, using a Quadrant, of major stars, including adjustments to the telescope and observational method. Written 1747 to 1753. Fair copy of RGO 3 MS 10.

RGO 3 MS 25 Observations, using the Quadrants, of major stars, including adjustments to the telescope and observational method. Written 1750 to 1757. Fair copy of RGO 3 MS 11.

RGO 3 MS 26 Observations, using the New Quadrant, of major stars, including adjustments to the telescope and observational method. Written 1753 to 1757. Fair copy of RGO 3 MS 12.

RGO 3 MS 27 Observations, using the New Quadrant, of major stars, including adjustments to the telescope and observational method. Written 1758 to 1762. Fair copy of RGO 3 MS 13.

RGO 3 MS 28 Observations: Transits of major stars 1762 to 1765, including adjustments to the telescope and observational method and also transits of major stars observed by Charles Green, 1764 to 1765. Written 1762 to 1765. Fair copy of RGO 3 MS 14.

RGO 3 MS 29 Observations, using the New Quadrant, of major stars, 1762 to 1765, including adjustments to the telescope and observational method, and also Quadrant observations of major stars, 1764 to 1765 taken by Charles Green. Written 1762 to 1765. Fair copy of RGO 3 MS 14.

RGO 3 MS 30 Observations made with the Zenith Sector, received by Nevil Maskelyne, 11 March 1770 from William Wales. It includes a Latin piece on the orbits of Jupiter's satellites and contains adjustments to the telescope and observational method. Written 1749 to 1760. Working copy.

RGO 3 MS 31 A miscellaneous collection of handwritten notes; they include:
1. A note dated 28 June 1734 on the errors of Halley's Quadrant from experiments performed by Bradley.
2. A note dated 26 July 1735 correcting the errors of the Quadrant and the previous year's over-estimated calculations.
3. A note dated 6 July 1742 on correcting errors in the Quadrant and repairs to the telescope.
4. A note dated 8 July 1742 comparing the Quadrant's present position with its adjustments since 1726.
5. An undated note probably 1742. Describes Halley's observational technique. Explains improvements made to aid observations in difficult weather and refine the results taken.
6. Notes on the alterations to the line of collimation, 1744 to 1745.
7. Notes on the alterations to the line of collimation, 1744 to 1745. A Bound Manuscript. It contains:
   a. Observations of *Polaris* and other stars.
   b. Tests on the New Quadrant.
   c. Observations made with the New Transit Telescope and description of its size and location.
   d. A law of refraction and its calculations.
   e. A note on the replacement of instruments from a £1,000 grant.
   f. Quadrant and Zenith Sector calculations.
8. Errors of solar tables. All written 1743 to 1756.

RGO 3 MS 32. Various loose papers in a file; they include:
1. Bradley's table of aberration of certain fixed stars, 1746 to 1753.
2. Adjustments made to the telescope, 18 July 1745.
3. Observations made of Wanstead Tower and St. Paul's, 1746.

4. Various unprovenanced calculations, some for the period 1689 to 1691.

5. Calculations written on a Royal Society invitation to Bradley, 26 June, 1757.

6. A table of observations of the moon in Gemini made by the new Quadrant, 1745 to 1746.

7. Tables and observations relating to the Quadrant, 1745 to 1746.

8. Tables relating to the Transit Telescope, 1743.

9. Papers relating to the occultations of Jupiter, 6 June 1744. Observed at Oxford and Shirburne. Observations of Jupiter and Mercury from 1743 to 1749.

10. Notes establishing the index from the Quadrant, 1745 to 1753.

11. Calculations and tables to ascertain the zenith from the Old and New Quadrants, 1746 to 1753.

12. Collimation error of the Quadrant, 1742 to 1752.

13. Unprovenanced chart giving the positions of indeterminable stars.

14. Errors of the planes of the Quadrant of various times from 1742 to 1745. A bound copy of 'Comparisons between the arc of the sector and Quadrants', with diagrams and equations, 1756 to 1758.

15. Errors and rates of clocks, 1743 to 1748. There is part of a letter concerning the building of an observatory.

16. Various data relating to adjustments taken on the Quadrant.

17. Observations with the Zenith Sector, 1749 to 1758.

RGO 3 MS 33   Calculations relating to observations of the Moon. Written 1752 to 1762.

RGO 3 MS 34   Ecliptic tables of Jupiter's satellites from 1700 to 1739 compiled by Bradley. Written 1719 to 1739.

RGO 3 MS 35   Sun's mean right ascension and longitude compared with tables, 1743 to 1758. A table of mean right ascension of stars in 1758. Written 1743 to 1758.

RGO 3 MS 36   A catalogue of stars as observed by John Flamsteed reduced to 1744. A table of lunar parallaxes. Written 1744. Fine copy.

RGO 3 MS 37   'Index to the days on which every observation of each star, with the Quadrants was made by Bradley' or Working Catalogue marked off for the Quadrant observation 0hrs to 12hrs. Written 1753 to 1757.

RGO 3 MS 38   'Index to the days on which every observation of each star, with the Quadrants was made by Bradley' or Working Catalogue marked off for the Quadrant observation 12hrs to 24 hrs. Written 1753 to 1757.

RGO 3 MS 39   'Index to the days on which every observation of each star, with the Transit was made by Bradley' or Working Catalogue marked off for the Transit observation 0hrs to 12hrs. Written 1750 to 1756.

RGO 3 MS 40   'Index to the days on which every observation of each star, with the Transit was made by Bradley' or Working Catalogue marked off for the Transit observation 12hrs to 24hrs. Written 1750 to 1756.

RGO 3 MS 41   Computations to ascertain the Moon's position by observations and to find the apparent diameter of the Moon. The rules to compute the aberration of any star. Written 1742 to 1744.

RGO 3 MS 42   A collection of tables.

1. Refraction tables; Lord Macclesfield's observations.

2. Reduction of micrometer.
3. Tables of Jupiter's satellites (in Latin).
4. Lunar tables (in Latin).
5. Solar tables.
6. Lunar tables.
7. Tables of the nonagesimal (on the Quadrant).
Written 1745.

RGO 3 MS 43 Miscellaneous Observations: this is a bound volume of calculations, manuscripts, transcripts and letters from 1723 to 1804, divided up into eight sections.
1. Notes on the handwriting of Bradley's handwriting.
   a. Explanation of observation code for 1743. A – other observers. B – Bradley.
   b. A list of observations.
2. Quadrant observations for 1750.
3. Deductions from Bliss's observations and the Board of Longitude.
   a. Mean time of the Moon's passage with latitude and longitude from 30 August 1762 to 2 March 1765.
   b. Apparent right ascension of the planets from 29th August 1762 to 3 March 1765.
   c. Daily Rate of the Transit Clock from 27 August 1762 to 2 March 1765.
   d. Minutes of the Board of Longitude, 1 March 1804 concerning the acquisition of the above data, which was observed by Nathaniel Bliss and Charles Green.
4. Observations of eclipses, 1748 to 1751.
   a. Observations of a lunar eclipse on 9 March, 1751.
   b. Observations of an eclipse on 14 July, 1748.
   c. Observations of a lunar eclipse on 12 December, 1749.
   d. Observations of a lunar eclipse on 8 June, 1750.
5. Miscellaneous observations and calculations 1742 to 1755.
   a. Mathematical equation to find the relative position of the Sun and the Moon with sketches and calculations.
   b. Rule and equation concerning the Sun's declination.
   c. Observations of an eclipse, 14 July 1748, and calculations relating to that observation.
   d. Adjustments to the line of collimation, with a detailed explanation of the procedure, 1745.
   e. Improvements in illuminating the wires of the Quadrant and Transit telescopes by Bradley as well as other general alterations. Undated but probably 1742. See RGO 3 MS 31.5.
   f. Invitation to attend the Royal Society, 20 June 1757, with unprovenanced calculations on the reverse.
   g. Establishing the distances of meridians at Shirburne and Oxford from Greenwich 6 June, 1744.
   h. Observations of Jupiter and Mercury, 1744 to 1745.
   i. Observations of *Polaris*, 1745 to 1746.
   j. Observations of Jupiter and Mars at Oxford, 21 May 1743. See RGO 3 MS 8.

k. Unknown column of monthly figures in 'Time for 13⅓ in Azimuth.

l. Column of figures for 'The Differences in Time for 13⅓ in Azimuth'.

m. Tables for reducing 96/90 of Quadrant to degrees, 1750 to 1755.

n. Transit and zenith observations, southwards 1751 to 1752.

o. Unknown rough calculations.

p. Computation of Sun's error of right ascension, September 1749.

q. Transit observations, 1754.

r. Transit observations, 1752.

s. Computations of the Sun's passage for transit observations.

t. Part of a letter from Francis Rowden concerning the building of an observatory. See RGO 3 MS 32. MS15.

u. Transit observations, 1743 to 1746, to establish the zenith.

v. A bill for food, 1743 to 1746.

w. A table of stars passing through the threads of the Transit Telescope.

x. Quadrant observations, 1743.

y. Measurement of the Sun by clocks, 1743 to 1744.

z. Quadrant observations, 1743 to 1745.

aa. Tables relating to the zenith, 1743.

bb. Observations of the stars with regular daily intervals between recording observations, 1744 to 1748.

cc. Table of zenith distances and errors, 1742.

dd. Mean distances in right ascension between Orion and Aquila, 1 January 1743.

ee. Positions of major stars, 1743.

ff. Observations of stars with regular daily intervals between recording observations, 1743 to 1748.

6. Papers principally on comets, 1723 to 1759.

a. 'Observations upon the Comets that appeared in the months of September and October 1757' with reference to unnamed stars and observational technique including an acknowledgement to Dr. Smith's three volumes on optics.

b. Observations to ascertain the 1757 comet's right ascension, its position relative to the Earth and other stars and its declination with attended calculations.

c. Table of observations of the 1759 comet.

d. Table of observations of the 1759 to 1760 comet, fixing its relative position with reference to the 1744 comet.

e. Notes on the position of the 1744 comet within the constellation of Pegasus.

f. Notes on the position of the 1748 comet.

g. Preface for a paper on the comets of 1743 to 1744. It mentions the observations at Shirburne.

h. Two identical diagrams of planes.

i. Large drawing of arcs and planes.

j. Observations of the 1743 comet establishing its right ascension and position relative to the constellation Pegasus.

k. Chart of the position of the comets of 1743 and 1744.

l. Comments on the position of the comets of 1743 and 1744.

m. Computations on the comet by Mr. Gael Morris, 1743.

n. A letter from Mr. John Bradley on his computations of the comet at Greenwich, 1744.

o. Observations of the comet by Lord Macclesfield, 1744.

p. Observations of the comet at Shirburne with associated tables, 1744.

q. Comparative table of observations between Gael Morris and John Bradley.

r. Notes on the orbit of the comet of 1744.

s. Hypothesis in the handwriting of Nevil Maskelyne on the orbit of the 1759 comet, with calculations.

t. Problem to find the position of a ship, with diagram.

u. Position of the 1719 comet from appearance to disappearance.

v. Short note on the position of the 1723 comet with reference to the position of the comet of 1707–1708, as observed at Bologna.

w. A theory in Latin with an equation on the orbit of the planets, including calculations.

x. Observations of the 1757 comet.

y. Calculations and equations to find the trajectory of a comet and find its presumed positions in 1742.

z. Observations of a comet 1727.

aa. Observations of a comet in 1742 by Gael Morris.

bb. Observations of a comet in 1742 by Mr. Wright.

cc. Observations of a comet 1738 to 1739.

dd. A letter from Gael Morris amending his computations of the comet.

ee. Kepler's problem resolved by John Machin and probably addressed to 'Mr. Jones'. Copied by James Bradley. See *Philosophical Transactions,* Vol. 40, p. 205.

ff. Letter from an astronomer, Julius Ascetto of Turin, written in English, enquiring into his observations of the Moon, 1750.

gg. Letter from Bradley to an unknown person stating he will be in Oxford, 1742.

hh. Calculations on a letter to Bradley.

ii. Memoranda on alterations and repairs to the telescope, 1742 to 1743.

jj. The plotted eclipse of the Sun, 1737.

kk. Observations of a comet, 1748.

ll. Various unprovenanced calculations.

mm. Explanation of the generation of the parabola, with calculations.

7. Early computations, mainly cometary.

a. Chart of cometary observations, 1723.

b. Several pages of calculations.

    c. 'The Places of the Telescopic Stars from which the Comet was Observed'; a chart, 1723.

    d. An equation to solve a problem on parabolas.

    e. An account of the observations of a comet by Maraldi at Paris, 1723. It mentions the 1707 cometary sighting.

    f. Several pages of calculations.

    g. Equation and method to solve a problem involving parabolas.

8. Letters from various persons to Bradley.

    a. Letter to Bradley on atmospheric pressure with two examples of evidence provided by the author for his argument. Also a piece on the origin of springs and rainfall by an unknown admirer 'Academicus'.

    b. Equation on a letter to Mrs. Paine of Wanstead, Essex.

    c. Letter to Mrs. Woodam from Matthew Wymondesolde on travelling arrangements, 1729. On the reverse is a note on gravity.

    d. Letter from Matthew Wymondesolde on travel arrangements to Dorchester, Dorset from Wanstead, Essex, 1741.

    e. Letter from Gael Morris on the position of the Moon by observing two stars, 1742.

    f. Letter from Thomas Corbett requesting the pleasure of Dr. Bradley's company.

    g. Letter from John Bradley explaining his lack of success in observing the comet, 1743.

    h. Various unprovenanced calculations.

    i. Letter from John Bradley on general conditions for his observations, 1743.

    j. Letter from Mr. Bevis wishing to compare observations of Mercury's late transit, 1742. It includes his observational technique and his findings.

    k. Letter from John Periam of Somerset. It makes several notes on comets deduced from his observations, and their religious connotations.

    l. Letter from Richard Bates of Cardiff asking for information to improve his line of collimation, 1743.

    m. Another letter from Richard Bates of Cardiff on his observations of Mercury and asking for Bradley's advice on the construction of a grid iron pendulum, 1746.

    n. Request for figures of the Moon's distance from the Greenwich Meridian by Joseph Pollack on behalf of the 2nd Earl of Macclesfield, 1749. There is an equation on light rays on the reverse.

    o. Letter in Italian from Julius Ascetto of Turin on the recent lunar eclipse, 1750.

    p. Letter covering the enclosed papers on astronomy researched by Mr. Rogers of Bolton, Lancashire, 1753.

    q. A long letter from Richard Heaton on the position of the poles and the implications of his computations on lines of latitude and longitude for the benefit of navigators, 1752.

r. Another lengthy letter from Richard Heaton concerning his invention of the planisphere with its implications for the safety of sailors and a direct comparison with the state of science in the mid-18th century with as it was in classical Greece, 1752.

s. Covering letter from Emanuel Mendes de Costa, on meteorological observations and lunar and solar eclipses, 1753.

t. Letter from Gael Morris concerning zenith distances, 1754.

u. Letter from Thomas Cooke who analyses the differences between physical laws and the natural order as laid down by God, 1754.

v. Piece on the exact time for the Earth to orbit the Sun, 1755.

w. Correspondence from Thomas Cooke of Oxford on the conflict between astronomy and theology, 1756.

x. Letter from John Kennedy requesting information on the exact time the Earth orbits the Sun, 1755.

y. Letter from Mr. R. Roper requesting observations before 1743 and highlighting the errors in his own observations at that time, 1755.

z. Equation and method to find the Earth's orbit around the Sun.

aa. Correspondence from James Stewart Mackenzie, of Ham, Surrey disputing Bradley's calculations for measuring the altitude of the Sun, 1757.

bb. Letter from Contianus Liber on observations and comments on the Milky Way, 1759.

cc. Letter from Thomas Stevenson on the observations of Halley's Comet from Barbados in 1759. It provides details of its orbit and predicts its reappearance in 1833. There is a table of returns of Halley's Comet, 1305–1833.

dd. Further observations of Halley's Comet by Thomas Stevenson from Barbados in 1759 with details of its position and the technique employed to trace the comet's path. It mentions the comet in relation to an 'eclipse', really an occultation, of Jupiter by the Sun.

ee. A request from John Fletcher of Halling, Gloucestershire, to discuss with Bradley the calorific efficiency of fires, 1760.

ff. Letter from Christopher Irwin concerning the Sun's orbit in conjunction with an eclipse, 1760.

gg. Letter from Stephen Hales of Teddington requesting that Bradley act as an intermediary between him and a rich heiress, Mrs. Mathews of Petersham, Richmond.

hh. Letter from Samuel Smedhurst of Manchester on the transit of Venus and the difference in longitude of Manchester and London, 1762.

ii. Comparative study of observations of an eclipse of Jupiter taken at London and Manchester, 1761.

RGO 3 MS 44. Various works and correspondence for the period 1764–1774. These are as follows:

1. Eclipse of the Moon, 1 March 1764.

2. Eclipse of the Sun, 1 April 1764.

3. Observations on the transits of Venus 1769, including and eclipse of the Sun observed at Gibraltar 1769, and a bill for crockery.
4. An eclipse of the Sun, 4 June 1764.
5. Eclipses of Jupiter's satellites, 1772–1774.
6. Eclipse of the Moon, 12 December, (Undated, List of eclipses gives 13 December 1769).
7. Drawing of a solar eclipse.
8. Table of observations of the comet of Mr. Napier, 1766. See *Philosopical Transactions,* Vol. 56, p. 62, with computations and sketches.
9. Eclipses of the Sun and the Moon 1769–1770, as observed at Hawkhill, Scotland.
10. Latitudes of Hawkhill from the meridian altitudes of the Sun, 1770.
11. Transit observations at Hawkhill.
12. Meridian altitude of Cassiopeia for finding the error of the Quadrant.
13. Observations and calculations on the orbits of comets, 1769–1770.
14. Observations of an eclipse of Jupiter.
15. Calculations to find the distance of Ben Lomond from Hawkhill.
16. Short note on reflecting telescopes.
17. Reference on behalf of Thomas Roy.

RGO 3 MS 45* Miscellaneous Observations, 1755 to 1763.

1. Observations made with the movable brass Quadrant, 5 December, 1755 to 27 December, 1760. Notes on the observational technique and the clocks employed.
2. Observations with the Transit Instrument from the 31 July 1760 to 4 December 1763. Notes on the observational technique and the clocks employed.
3. Rough calculations.

*These papers were placed in RGO 3 MS 45 after being presented to the Royal Greenwich Observatory by the Radcliffe Trustees in 1934.

Miscellaneous manuscript sources

King's College Library, Cambridge

Keynes MS 130.5 John Conduitt, *Memorabilia.* (Cambridge University Library, Microfilm, MS 659).

Gloucestershire Records Office, Gloucester

Sherborne Parish Records.
Rendcomb Parish Records.

Royal Society Library

FISh Sharp Letters.

Christ Church, Oxford

William Wake Letters.

Pembroke College, Oxford

Letters: Correspondence from George Biddell Airy, 7th AR to Bartholomew Price of Pembroke College, 3 April 1861. Concerning the transfer of James Bradley's Greenwich Registers from the Bodleian Library to the Royal Observatory

Southampton Records Office

The William Molyneux and Samuel Molyneux Papers.

British Library

Add. MSS. 4478 D Dr. Birch's Common Place Book.

Secondary Sources

'Espinasse, Margaret, Robert Hooke, William Heinemann, London, 1956.

Aitken, R. G, *Edmund Halley and Stellar Proper Motions, Astronomical Society of the Pacific Leaflets*, 1942, Vol. 4, No. 164.

Aiton, Eric, *The Cartesian Theories of the Planetary Motions*, PhD Thesis, University of London, 1958.

Aiton, Eric, *The Vortex Theory of Planetary Motions*, Macdonald, London, 1972.

Aiton, Eric, 'French Engineers Become Professionals; or How Meritocracy Made Knowledge Objective', in W. Clark, J. Golinski, and S. Schaffer (eds), *The Sciences in Enlightened Europe*, University of Chicago Press, London, 1999, pp. 94–125.

Alexander, A. F. O'D. *The Planet Saturn: A History of Observation, Theory and Discovery*, Dover Publications, New York, 1962.

Alexander, H. G. (ed). *The Leibniz–Clarke Correspondence: Together with Extracts from Newton's Principia and Opticks*, edited with an introduction and notes by H. G. Alexander, Manchester University Press, Manchester, 1956.

Allen, Richard H. *Star Names: Their Lore and Meaning*, London, 1899, Reprinted, Dover Publications, New York, 1963.

Aristotle *Physics*, with an English translation by P. H. Wicksteed and F. S. Cornford, Loeb Edition, William Heinemann, London, 1980.

Ashworth, William J. 'The Calculating Eye: Baily, Herschel, Babbage and the Business of Astronomy', *British Journal for the History of Science*, 1994, Vol. 27, No. 4, pp. 409–441.

Bacon, Francis. *Novum Organum; With Other parts of the Great Instauration*, translated and edited by P. Orbach and J. Gibson, Open Court, Chicago, 1994.

Baily, Francis. *An Account of the Rev. John Flamsteed, the First Astronomer Royal, Compiled from his Manuscripts and other Authentic Documents, Never Before Published*, London, 1835.

Barker-Benfield, G. J. *The Culture of Sensibility: Sex and Society in Eighteenth-Century Britain*, University of Chicago Press, London, 1992.

Beeson, David, *Maupertuis: An Intellectual Biography*, Voltaire Foundation at the Taylor Institution, Oxford, 1992.

Bennett, J. A. 'Hooke's Instruments' in J. A. Bennett, M. Cooper, M. Hunter and L. Jardine, *London's Leonardo: The Life and Work of Robert Hooke*, Oxford University Press, Oxford, 2003, pp. 63–104.

Bennett, J. A. 'Hooke's Instruments for Astronomy and Navigation' in M. Hunter and S. Schaffer (eds), *Robert Hooke: New Studies*, The Boydell Press, Woodbridge, 1989, pp. 21–32.

Bennett, J. A., Cooper, M., Hunter, M., and Jardine, L. *London's Leonardo: The Life and Work of Robert Hooke*, Oxford University Press, Oxford, 2003.

Bennett, J. A. *The Mathematical Science of Christopher Wren*, Cambridge University Press, Cambridge, 1982.

Bennett, J. A. 'The Mechanic's Philosophy and the Mechanical Philosophy' in *History of Science*, Vol. 24, No. 1, 1986, pp. 1–28.

Berkeley, George, *A Treatise Concerning the Principles of Human Knowledge*, J. M. Dent, London, 1980.

Berlinski, David, *Newton's Gift: How Sir Isaac Newton Unlocked the System of the World*, Duckworth, London, 2000.

Bertoloni Meli D. 'The Relativization of Centrifugal Force' in *Isis*, Vol. 81, No. 306, 1990, pp. 23–43.

Bessel, F. W. *Fundamenta Astronomiæ pro annoMDCCLV deducta ex observationibus viri incomparabilis James Bradley in specula Grenovicensi per annos 1750–1762 institutis.* Königsberg, 1818.

Biagioli, Mario. *Galileo Courtier: The Practice of Science in the Culture of Absolutism*, University of Chicago Press, London, 1993.

Biagioli, M. and Galison, P. *Scientific Authorship: Credit and Intellectual Property in Science.* Routledge, London, 2000.

Birks, John L. *John Flamsteed: The First Astronomer Royal at Greenwich*, Avon Books, London, 1999.

Black J. and Porter R. *A Dictionary of Eighteenth-Century World History*, Blackwell Publishers, Oxford, 1994.

Blay, Michel *Reasoning with the Infinite: From the Closed World to the Mathematical Universe.* Translated by M. B. De Bevoise, University of Chicago Press, London, 1998.

Boss, Valentin, *Newton and Russia: The Early Influence, 1698–1796*, London, 1972.

Boyer, Carl B. *The History of the Calculus and its Conceptual Development*, Dover Publications, New York, 1959.

Brackenridge, J. Bruce. *The Key to Newton's Dynamics: The Kepler Problem and the Principia*, University of California Press, London, 1995.

Bradley, James. *Miscellaneous Works and Correspondence of the Rev. James Bradley, DD, FRS*, edited by Stephen Peter Rigaud, Clarendon Press, Oxford, 1832.

Bradley, James, *Supplement to the Miscellaneous Works and Correspondence of the Rev. James Bradley, DD, FRS*, edited by Stephen Peter Rigaud, Clarendon Press, Oxford, 1833.

Bradley, James, 'An account of some observations made in London, by Mr. George Graham, FRS; and at Black-River in Jamaica, by Colin Campbell, Esq; FRS concerning the going of a Clock, in order to determine the difference between the lengths of isochronal pendulums in those places'. In *Miscellaneous Works and Correspondence of the Rev. James Bradley, DD, FRS*, edited by Stephen Peter Rigaud, Clarendon Press, Oxford, 1832, pp. 62–69, and *Philosophical Transactions*, No 432, Vol. 38, 1734, pp. 302–313.

Bradley, James, *Astronomical Observations made at Greenwich from the year MDCCL to the year MDCCLXII*. Vol. I, MDCCL to MDCCLV, *Observations with Transit and Mural Quadrant from 1750 to 1755*, edited by Thomas Hornsby, Clarendon Press, Oxford, 1798.

Bradley, James, *Astronomical Observations made at Greenwich from the year MDCCL to the year MDCCLXII*. Vol. II, MDCCLVI to MDCCLXII, *Observations with Transit and Mural Quadrant from 1756 to 1762*, edited by Abram Robertson, Clarendon Press, Oxford, 1805.

Bradley, James, 'Correspondence of James Bradley' in *Miscellaneous Works and Correspondence of the Rev. James Bradley, DD, FRS*, edited by Stephen Peter Rigaud, Clarendon Press, Oxford, 1832, pp. 391–510.

Bradley, James, 'Demonstration of the Rules Relating to the Apparent Motion of the Fixed Stars upon account of the Motion of Light', in *Miscellaneous Works and Correspondence of the Rev. James Bradley, DD. FRS*, edited by Stephen Peter Rigaud, Clarendon Press, Oxford, 1832, pp. 287–301.

Bradley, James, 'Directions for Using the Common Micrometer', in *Miscellaneous Works and Correspondence of the Rev. James Bradley, DD, FRS*, edited by Stephen Peter Rigaud, Clarendon Press, Oxford, 1832, pp. 70–74, and *Philosophical Transactions*, Vol. 62, pp. 46–52.

Bradley James, 'Experiments to Determine the Length of the Pendulum Vibrating Seconds at Greenwich'. In *Miscellaneous Works and Correspondence of the Rev. James Bradley, DD, FRS*, edited by Stephen Peter Rigaud, Clarendon Press, Oxford, 1832, pp. 384–387.

Bradley, James, 'A Letter to Dr. Edmund Halley, Astronomer Royal and etc., Giving an Account of a New-Discovered Motion of the Fixed Stars', in *Miscellaneous Works and Correspondence of the Rev. James Bradley, DD, FRS*, edited by Stephen Peter Rigaud, Clarendon Press, Oxford, 1832, pp. 1–16, and *Philosophical Transactions*, No. 406, Vol. 35, 1729, pp. 637–661.

Bradley, James, 'A Letter to the Rt. Hon. George Earl of Macclesfield, Concerning an Apparent Motion Observed in Some of the Fixed Stars', in *Miscellaneous Works and Correspondence of the Rev. James Bradley, DD, FRS*, edited by Stephen Peter Rigaud, Clarendon Press, Oxford, 1832, pp. 17–41, and *Philosophical Transactions*, No. 485, Vol. 45, 1748, pp. 1–46.

Bradley, James, 'The Longitude of Lisbon and the Fort of New York from Wanstead and London, Determined by Eclipses of the First Satellite of Jupiter', in *Miscellaneous Works and Correspondence of the Rev. James Bradley, DD, FRS*, edited by Stephen Peter Rigaud, Clarendon Press, Oxford, 1832, pp. 58–61, and *Philosophical Transactions*, No. 394, Vol. 34, 1724, pp. 85–89.

Bradley, James, 'Memoranda Respecting the Instructions that Dr. Bradley Drew Up at the Desire of the Council of the Royal Society, Relating to Charles Mason Observing the Transit of Venus in the East Indies', in *Miscellaneous Works and Correspondence of the Rev. James Bradley, DD, FRS*, edited by Stephen Peter Rigaud, Clarendon Press, Oxford, 1832, pp. 388–390.

Bradley, James, 'Memoranda Respecting the Instrument at Wansted' in *Miscellaneous Works and Correspondence of the Rev. James Bradley, DD, FRS*, edited by Stephen Peter Rigaud, Clarendon Press, Oxford, 1832, pp. 194–200.

Bradley, James, 'Miscellaneous Astronomical Observations', in *Miscellaneous Works and Correspondence of the Rev. James Bradley, DD, FRS*, edited by Stephen Peter Rigaud, Clarendon Press, Oxford, 1832, pp. 339–380.

Bradley, James, 'Observations Made at Kew', in *Miscellaneous Works and Correspondence of the Rev. James Bradley, DD, FRS*, edited by Stephen Peter Rigaud, Clarendon Press, Oxford, 1832, pp. 116–193.

Bradley, James, Observations on the Fixed Stars Made at Wansted in Essex by J. Bradley', in *Miscellaneous Works and Correspondence of the Rev. James Bradley, DD, FRS*, edited by Stephen Peter Rigaud, Clarendon Press, Oxford, 1832, pp. 201–286.

Bradley, James, 'Observations upon the Comet that Appeared in the Months of October, November and December, 1723', in *Miscellaneous Works and Correspondence of the Rev. James Bradley, DD, FRS*, edited by Stephen Peter Rigaud, Clarendon Press, Oxford, 1832, pp. 42–48, and *Philosophical Transactions* No. 382, Vol. 33, 1724, pp. 41–50.

Bradley, James, 'Observations upon the Comet that Appeared in the Months of January, February and March, 1737' in *Miscellaneous Works and Correspondence of the Rev. James Bradley, DD, FRS*, edited by Stephen Peter Rigaud, Clarendon Press, Oxford, 1832, pp. 49–52, and *Philosophical Transactions*, No. 446, Vol. 40, 1737, pp. 111–117.

Bradley, James, 'Observations upon the Comet that Appeared in the Months of September and October, 1757, Made at the Royal Observatory' in *Miscellaneous Works and Correspondence of the Rev. James Bradley, DD, FRS*, edited by Stephen Peter Rigaud, Clarendon Press, Oxford, 1832, pp. 53–57, and *Philosophical Transactions*, Vol. 50, 1758, pp. 408–413.

Bradley, James, 'Reduction of the Wansted Observations' in *Miscellaneous Works and Correspondence of the Rev. James Bradley, DD, FRS*, edited by Stephen Peter Rigaud, Clarendon Press, Oxford, 1832, pp. 302–338.

Bradley, James, 'The Reverend Mr. James Bradley's Observations on his Tables of Jupiter's Satellites', in *Miscellaneous Works and Correspondence of the Rev. James Bradley, DD, FRS*, edited by Stephen Peter Rigaud, Clarendon Press, Oxford, 1832, pp. 81–83.

Bradley, James, 'State of the Instruments at the Greenwich Observatory when Dr. Bradley became Astronomer Royal', in *Miscellaneous Works and Correspondence of the Rev. James Bradley, DD, FRS*, edited by Stephen Peter Rigaud, Clarendon Press, Oxford, 1832, pp. 381–383.

Bradley, James, 'Zenith Observations at Wansted' in *Miscellaneous Works and Correspondence of James Bradley*, edited by Stephen Peter Rigaud, Clarendon Press, Oxford, 1832, pp. 201–284.

Brewer, John, 'Commercialization and Politics' in N. McKendrick, J. Brewer, and J. H. Plumb, *The Birth of a Consumer Society: The Commercialization of Eighteenth-Century England*, Indiana University Press, London, 1982, pp. 197–262.

Brewer, John, *The Pleasures of the Imagination: English Culture in the Eighteenth-Century*, Harper Collins, London, 1997.

Buchdahl, Gerd, 'Explanation and Gravity' in M. Teich and R. Young (eds), *Changing Perspectives in the History of Science: Essays in Honour of Joseph Needham*, Heinemann, London, 1973, pp. 167–203.

Buchwald, Jed Z. (ed), *Scientific Practice: Theories and Stories of Doing Physics*, University of Chicago Press, London, 1995.

Burke, John G. (ed), *The Uses of Science in the Age of Newton*, University of California Press, London, 1983.

Cannon, John, *Aristocratic Century: The Peerage of Eighteenth-Century England*, Cambridge University Press, 1984.

Cassini, Jacques, 'Réflexions sur une letter de M. Flamsteed à M. Wallis touchant la parallaxe annuelle de l'étoile polaire', in *Mémoires de l'Académie royale des sciences pour 1699*, Paris, 1702, pp. 177–183.

Cassirer, Ernst, *The Philosophy of the Enlightenment*, Princeton University Press, Princeton, 1957.

Cescinsky, Herbert and Webster, Malcolm R. *English Domestic Clocks*, New York, 1968.

Chapman, Allan, *Dividing the Circle: The Development of Critical Angular Measurement in Astronomy, 1500–1850*, second edition, John Wiley and Sons, Chichester, 1995.

Chapman, Allan, '*Pure Research and Practical Teaching:The Astronomical Career of James Bradley, 1693–1762*', *Notes and Records of the Royal Society*, 47 (2), 1993, pp. 205–212.

Chenakai, V. L. 'The Astronomical Instruments of the Seventeenth and Eighteenth Centuries in the USSR', in *Vistas in Astronomy*, Vol. 9, (1968), pp. 53–77.

Clark, George, *Science and Social Welfare in the Age of Newton*, 2nd edn, Clarendon Press, Oxford, 1970.

Clark, W. et al (eds), *The Sciences in Enlightened Europe*, in W. Clark, J. Golinski, and S. Schaffer (eds), University of Chicago Press, London, 1999.

Clarke, Samuel, *The Works of Samuel Clarke, DD*, edited by Benjamin Hoadly, 1738. Reprinted in four volumes by Thoemmes Press, October, 2002.

Clerke, Agnes Mary, 'James Bradley' in Leslie Stephens (ed.) *Dictionary of National Biography*, Vol. 2, London, 1882, pp. 1074–1079.

Clodfelter, Micheal, *Warfare and Armed Conflicts: A Statistical Encyclopedia of Casualty and Other Figures, 1492–2015*, 4th edn., McFarland & Company, Jefferson, NC, 2017.

Cohen, I. Bernard, *Introduction to Newton's 'Principia'*, Harvard University Press, Cambridge, MA, 1978.

Collins, H. M., *Changing Order: Replication and Induction in Scientific Practice*, Sage Publications, London, 1985.

Cook, Alan, *Edmond Halley: Charting the Heavens and the Seas*, Clarendon Press, Oxford, 1998.

Cooper, M. A. R, '*A More Beautiful City':Robert Hooke and the Rebuilding of London after the Great Fire*, Alan Sutton, Stroud, 2003.

Daston, Lorraine, 'The Ethos of Enlightenment' in W. Clark, J. Golinski, and S. Schaffer (eds), in *The Sciences in Enlightened Europe*, University of Chicago Press, London, 1999, pp. 495–504.

Dear, Peter, *Discipline and Experience: The Mathematical Way in the Scientific Revolution*, University of Chicago Press, London, 1995.

Dear, Peter, *Revolutionizing the Sciences: European Knowledge and its Ambitions, 1500 – 1700*, Palgrave, Basingstoke, 2001.

Delambre, J-B-J, *Histoire de l'Astronomie au XVIII Siecle*, Paris, 1819.

Descartes, Rene, *Discours de la method*. Paris, 1637.

Dobbs, B. J. T., *The Foundations of Newton's Alchemy, or 'the Hunting of the Greene Lyon'*, Cambridge University Press, Cambridge, 1975.

Dobbs, B. J. T. and Jacob, M. C., *Newton and the Culture of Newtonianism*, Humanity Books, New York, 1998.

Dobbs, B. J. T., 'Newton as Final Cause and First Mover', in M. J. Osler (ed.), *Rethinking the Scientific Revolution*, Cambridge University Press, Cambridge, 2000, pp. 25–39.

Dupré, John, *The Disorder of Things: Metaphysical Foundations of the Disunity of Science*, Harvard University Press, Cambridge, MA, 1993.

Evelyn, John, *The Diary of John Evelyn*, edited by Guy de la Bédoyère, Boydell and Brewer, Woodbridge, 1995.

Farber, Daniel, *Descartes' Metaphysical Physics*, University of Chicago Press, London, 1992.

Feyerabend, Paul, *Against Method: Outline of an Anarchistic Theory of Knowledge*, Verso, London, 1975.

Fisher, John Roger, *James Bradley and the New Discovered Motion: The Origins, Development and Reification of James Bradley's Hypothesis of the New Discovered Motion of the Fixed Stars*, MSc Dissertation, University of London, 1994.

Fisher, John Roger, *Astronomy and Patronage in Hanoverian England: The Work of James Bradley, Third Astronomer Royal of England*, PhD Thesis, University of London, 2004.

Fisher, John Roger, Conjectures and reputations: The composition and reception of James Bradley's paper on the aberration of light with reference to a third unpublished version, *British Journal for the History of Science* 43(1), pp. 19–48.

Fisher, John Roger, The discovery of the aberration of light by James Bradley, *Antiquarian Astronomer*, June, 2022, pp. 57–69.

Flamsteed, John, *The Preface to John Flamsteed's Historia Coelestis Britannica*, edited and introduced by Allan Chapman, based on a translation by Alison Dione Johnson, National Maritime Museum, Maritime Monographs and Reports, No. 52, Greenwich, 1982.

Fleck, Ludwik, *Genesis and Development of a Scientific Fact*, edited by Thaddeus J. Trenn and Robert K. Merton, translated from *Ensterhung unt Entwicklung einer wissenschaftlichen Tatsache: Einfuhrung in die Lehre vom Denkstil und Denkkollectiv*, by Fred Bradley and Thaddeus J. Trenn, Chicago University Press, London, 1979.

Forbes, Eric, *Greenwich Observatory: The Royal Observatory at Greenwich and Herstmonceux, 1675–1975*. Vol. 1: *Origins and Early History (1675–1835)*, Taylor and Francis Ltd, London, 1975.

Force, James E., *William Whiston: Honest Newtonian*, Cambridge University Press, Cambridge, 1985.

Fox, Robert, Gooday, Graeme, and Simcock, Tony, 'Physics in Oxford: Problems and Perspectives' R. Fox and G. Gooday (eds), *Physics in Oxford, 1839–1939: Laboratories, Learning and College Life*, Oxford University Press, Oxford, 2005, pp. 1–23.

Gabbey, Alan, 'Newton and Natural Philosophy' in R. C. Olby, G. N. Cantor, J. R. R. Christie and M. J. S. Hodge (eds), *Companion to the History of Modern Science*, Routledge, London, 1990, pp. 243–263

Garber, Daniel, *Descartes' Metaphysical Physics*, University of Chicago Press, Chicago and London, 1992.

Gillispie, Charles C. *Pierre-Simon Laplace, 1749 – 1837: A Life in Exact Science*, with the collaboration of Robert Fox and Ivor Grattan Guiness, Princeton University Press, Princeton, 2000,

Glass, I. S. *Nicolas-Louis de La Caille: Astronomer and Geodesist*, Oxford University Press, Oxford, 2013.

Gooding D., Pinch, T., and Schaffer, S. (eds) *The Uses of Experiment: Studies in the Natural Sciences*, Cambridge University Press, Cambridge, 1989.

Grant, Robert, *History of Physical Astronomy: From the Earliest Ages to the Middle of the Nineteenth Century*, Henry G. Bohn, London, 1852.

Grasshoff, Gerd, *The History of Ptolemy's Star Catalogue*, Springer Verlag, London, 1990.

Greenberg, John, L. *The Problem of the Earth's Shape from Newton to Clairaut:The rise of mathematical science in eighteenth-century Paris and the fall of 'normal' science*, Cambridge University Press, Cambridge, 1995.

Gregory, James, *Geometriæ pars universalis*, Padua, 1668. *1666*

Gregory, J. and Stevenson, J., *The Longman Companion to Britain in the Eighteenth-Century, 1688–1820*, London, 2000.

Grosholz, Emily R., *Cartesian Method and the Problem of Reduction*, Clarendon Press, Oxford, 1991.

Gross, Alan G., *The Rhetoric of Science*, Harvard University Press, London, 1990.

Hahn, Roger, *The Anatomy of a Scientific Institution: The Paris Academy of Sciences, 1666–1803*. University of California Press, London, 1971.

Hakfoort, Casper, *Optics in the Age of Euler: Conceptions of the Nature of Light: 1700–1795*, Cambridge University Press, Cambridge, 1995.

Hall A. Rupert, *All was Light: An Introduction to Newton's Optics*, Clarendon Press, Oxford, 1993.

Hall A. Rupert, *From Galileo to Newton*, Dover Publications, New York, 1981.

Hall A. Rupert, *Isaac Newton: Adventurer in Thought*, Blackwell Publications, Oxford, 1992.

Hall A. Rupert, *Isaac Newton: Eighteenth-century Perspectives*, Oxford University Press, Oxford, 1999.

Hankins, Thomas L, *Science and the Enlightenment*, Cambridge University Press, Cambridge, 1985.

Harvey, A. D., *Sex in Georgian England: Attitudes and Prejudices from the 1720s to the 1820s*, Duckworth, London, 1994.

Harwit, Martin, *Cosmic Discovery: The Search, Scope and Heritage of Astronomy*, Harvester Press, Brighton, 1981.

Hatton, Ragnhild *George I*, Yale University Press, London, 2001.

Heaps, Leo, *Log of the Centurion*, Book Club Associates, London, 1974

Heilbron, J. L., *Elements of Early Modern Physics*, University of California Press, London, 1982.

Henry, John, *The Scientific Revolution and the Origins of Modern Science*, 2nd edn, Palgrave, Basingstoke, 2002.

Hesse, Mary B., *Forces and Fields: A Study of Action at a Distance in the History of Physics*, Thomas Nelson, London, 1961.

Hewitt, K. V., *The Astronomical Work of the Reverend James Bradley, Third Astronomer Royal*, MSc Dissertation, University of London, 1953.

Hirshfeld, Alan W., *Parallax: The Race to Measure the Cosmos*, W. H. Freeman and Sons, New York, 2001.

Hooke, Robert, 'An Attempt to Prove the Motion of the Earth from Observations', in R. T. Gunther, *Early Science in Oxford*, Vol.8, 1931, pp. 17–25.

Hooke, Robert, *Lectiones Cutlerianæ, or A Collection of Lectures: Physical, Mechanical, Geographical and Astronomical, etc*, London, 1679.

Hoskin, Michael (ed), *The Cambridge Concise History of Astronomy*, Cambridge University Press, Cambridge, 1999.

Howse, Derek, *Greenwich Observatory: Vol.3, The Buildings and Instruments*, National Maritime Museum, Greenwich, 1975.

Hunter, M. and Schaffer, S., *Robert Hooke: New Studies*, The Boydell Press, Woodbridge, 1989.

Hunter, Michael, *Robert Hooke: Tercentennial Studies*, Routledge, London, 2003.

Hunter, Michael, *Science and Society in Restoration England*, Cambridge University Press, Cambridge, 1981.

Iliffe, Rob, 'Aplatisseur du Monde et de Cassini' Maupertuis, Precision Measurement, and the Shape of the Earth in the 1730s', in *History of Science*, vol.xxxi (1993), pp. 335–375.

Iliffe, Rob, 'Butter for Parsnips: Authorship, Audience and the Incomprehensibility of the Principia' in M. Biagioli and P. Galison (eds), *Scientific Authorship: Credit and Intellectual Property in Science*, Routledge, London, 2000, pp. 33–65.

Iliffe, Rob, 'Material Doubts: Hooke, Artisan Culture and the Exchange of Information in 1670s London', *British Journal for the History of Science*, 1995, Vol.28, pp. 283–318.

Iliffe, Rob, 'Philosophy of Science' in R. Porter (ed.) *The Cambridge History of Science*, Vol.4 : *Eighteenth-Century Science*, Cambridge University Press, Cambridge, 2003, pp. 267–284.

Jardine, Lisa, *On a Grander Scale: The Outstanding Career of Sir Christopher Wren*, Harper-Collins, London, 2002.

Jacob, Margaret C., *The Newtonians and the English Revolution, 1689–1720*, Cornell University Press, Ithaca, 1976.

Jacob, Margaret C., 'The Truth of Newton's Science and the Truth of Science's History: Heroic Science at its Eighteenth-Century Formulation' in M. J. Osler (ed.), *Rethinking the Scientific Revolution*, Cambridge University Press, Cambridge, 2000, pp. 315–332.

Johns, Adrian, 'The Ambivalence of Authorship in Early Modern Natural Philosophy Principia' in *Scientific Authorship: Credit and Intellectual Property in Science*, edited by Mario Biagioli and Peter Galison, Routledge, London, 2000, pp.67–90.

King, Henry C., *The History of the Telescope*, Dover Publications, New York, 1979.

Knorr-Cetina, Karin D., *The Manufacture of Knowledge; An Essay on the Constructivist and Contextual Nature of Science*, Pergamon Press, Oxford, 1981.

Koyré, Alexandre, *Metaphysics and Measurement: Essays in the Scientific Revolution*, Chapman and Hall, London, 1968.

Koyré, Alexandre, *Newtonian Studies*, University of Chicago Press, London, 1965.

Kuhn, Thomas S, *The Copernican Revolution: Planetary Astronomy in the Development of Western Thought*, Harvard University Press, London, 1957.

Langford, Paul, *A Polite and Commercial People: England 1727–1783*, Oxford University Press, 1992.

Latour, B. and Woolgar, S., *Laboratory Life:The Construction of Scientific Facts*, Princeton University Press, Princeton, 1986.

Latour, Bruno, *Science in Action*, Harvard University Press, London, 1988.

Leighton, C. D. A., 'Hutchinsonianism: A Counter-Enlightenment Reform Movement' in *Journal of Religious History*, Vol.23, Part 2, 1999, pp. 168–184.

Longhurst, Sybil, Tufnell, W., and Tufnell. A., *Sherborne: A Cotswold Village*, Allan Sutton, Stroud, 1992.

McCrea, W. H., 'The Significance of the Discovery of Aberration', *Quarterly Journal of the Royal Astronomical Society*, Vol. 4, March 1963, pp.41–43.

McGuire, J. E., 'The Fate of the Date: The Theology of Newton's Principia Revisited', in in M. J. Osler (ed.), *Rethinking the Scientific Revolution*, Cambridge University Press, Cambridge, 2000, pp. 271–295.

Mackay, Charles, *Extraordinary Popular Delusions and the Madness of Crowds*, London, 1852, new edn. New York, 1980.

Manfredi, Eustachio, *De Annuis Inerrantium Stellarum Aberrationibus*, Bologna, 1729.

Martin, Julian, *Francis Bacon, the State, and the reform of Natural Philosophy*, Cambridge University Press, Cambridge, 1992.

Maskelyne, Nevil, 'Analysis of Calculations Made by Bradley when Investigating a Rule for Refraction'. In *Supplement to the Miscellaneous Works and Correspondence of James Bradley*, edited by Stephen Peter Rigaud, Clarendon Press, Oxford, 1832, pp. 3–16.

Maskelyne, Nevil, 'Extracts from Dr Maskelyne's Observations on the Latitude and Longitude of the Royal Observatory at Greenwich', in *Miscellaneous Works and Correspondence of James Bradley*, edited by Stephen Peter Rigaud, Clarendon Press, Oxford, 1832, pp. 75–80, and *Philosophical Transactions*, Vol. 77, pp. 154–161.

Maunder, E. Walter, *The Royal Greenwich Observatory: A Glance at its History and Work*, London, 1900, Reprinted, Cambridge University Press, Cambridge, 2013.

Midgley, Graham, *University Life in Eighteenth-Century Oxford*, Yale University Press, New Haven, CT, and London, 1996.

Molyneux, Samuel, 'A Description of an Instrument set up at Kew, in Surrey, for Investigating the Annual Parallax of the Fixed Stars, with an Account of the Observations Made therewith', in *Miscellaneous Works and Correspondence of James Bradley*, edited by Stephen Peter Rigaud, Clarendon Press, Oxford, 1832, pp. 93–115.

Montgomery, Scott I., *The Scientific Voice*, The Guilford Press, London, 1996.

Murdin, Lesley, *Under Newton's Shadow: Astronomical Practices in the Seventeenth Century*, Bristol, 1985.

Newton, Isaac, *Isaac Newton's Papers and Letters on Natural Philosophy and Related Documents*, edited by I. Bernard Cohen assisted by Robert E. Schofield, Cambridge University Press, Cambridge, 1958.

Newton, Isaac, *Opticks, or a Treatise of the Reflections, Refractions, Inflections and Colours of Light*, with a Preface by I Bernard Cohen, Dover Publications, New York, 1979.

Newton, Isaac, *Philosophiæ Naturalis Principia Mathematica*, London, 1687, facsimile of the first edition published by Culture et Civilisation, Brussels, 1965.

Newton, Isaac, *The Principia: Mathematical Principles of Natural Philosophy*, a new translation of *Philosophiæ Naturalis Principia Mathematica*, by I. Bernard Cohen and Anne Whitman, University of Chicago Press, London, 1999.

Newton, Isaac, *Unpublished Scientific Papers of Isaac Newton: A Selection from the Portsmouth Collection in the University Library, Cambridge*, chosen, edited and translated by A. Rupert Hall and Marie Boas Hall, Cambridge University Press, Cambridge, 1962.

Newton-Smith, W. H., (ed.), *A Companion to the Philosophy of Science*, Blackwell Publishers, Oxford, 2000.

North, John, 'The Satellites of Jupiter: from Galileo to Bradley', in Alwyn Van der Merwe (ed.), *Old and New Questions in Physics: Cosmology, Philosophy and Theoretical Biology*, Plenum Press, New York and London, 2012, pp. 689–717.

Olby, R. C., Cantor, G. N., Christie, J. R. R., and Hodge, M. J. S. (eds), *Companion to the History of Modern Science*, Routledge, London, 1990.

Olson, Richard, 'Tory High Church Opposition to Science and Scientism in the Eighteenth-Century: The Works of John Arbuthnot, Jonathan Swift and Samuel Johnson', in John G. Burke (ed.), in *The Uses of Science in the Age of Newton*, University of California Press, London, 1983, pp. 171–204.

Osler, Margaret (ed.), *Rethinking the Scientific Revolution*, University of Cambridge Press, Cambridge, 2000.

Pannekoek, Anton, *A History of Astronomy*, George Allen and Unwin, London, 1961. Reprinted by Dover Books, New York, 1989.

Pickering, Andrew, *The Mangle of Practice: Time, Agency and Science*, University of Chicago Press, London, 1995.

Pickering, Andrew (ed.), *Science as Practice and Culture*, University of Chicago Press, London, 1992.

Plummer, H. C. 'Horrocks and his Opera Posthuma', *Notes and Records of the Royal Society*, Vol. 3, 1940–41, pp. 39–52.

Porter, Roy (ed.), *The Cambridge History of Science, Vol.4: Eighteenth-Century Science*, Cambridge University Press, Cambridge, 2003.

Porter, Roy, *English Society in the Eighteenth-Century*, revised edn, Penguin Books, London, 1990.

Porter, Roy, *Enlightenment: Britain and the Creation of the Modern World*, Allen Lane, London, 2000.

Porter, R. and Teich, M, *The Scientific Revolution in National Context*, Cambridge University Press, Cambridge, 1992.

Pyle, Andrew, *Atomism and its Critics: From Democritus to Newton*, Thoemmes Press, Bristol, 1997.

Redondi, Pietro, *Galileo: Heretic*, translated by Raymond Rosenthal, Penguin Books, London, 1989.

Rigaud, Stephen Peter, *Correspondence of Scientific Men of the Seventeenth Century*, Clarendon Press, Oxford, 1841.

Rigaud, Stephen Peter, *Memoirs of Bradley*, bound with James Bradley, *Miscellaneous Works and Correspondence of James Bradley*, Clarendon Press, Oxford, 1832.

Robinson, H. W. and Adams, W., *The Diary of Robert Hooke 1672–1680*, London, 1935.

Rossi, Paolo, *The Birth of Modern Science*, Blackwell Publishers, Oxford, 2001.

Rusnock, Andrea, 'Correspondence Networks and the Royal Society 1700–1750', *British Journal for the History of Science*, Vol. 32, No. 2, pp. 155–169, 1999.

Sabra A. L, *Theories of Light from Descartes to Newton*, Cambridge University Press, Cambridge, 1981.

Schaffer, Simon, 'Accurate Measurement is an English Science', in M. Norton Wise (ed.), *The Values of Precision*, Princeton University Press, Princeton, 1995, pp. 135–172.

Schaffer, Simon, 'Astronomers Mark Time: Discipline and the Personal Equation' *Science in Context*, Vol. 2, No. 1, 1988, pp. 115–145.

Schaffer, Simon, 'Glass Works: Newton's Prisms and the Uses of Experiment' in D. Goodman, T. Pinch, and S. Schaffer (eds), in *The Uses of Experiment: Studies in the Natural Sciences*, Cambridge University Press, Cambridge, 1989, pp. 67–104.

Schaffer, Simon, 'Natural Philosophy' in G. Rousseau and R. Porter (eds), *The Ferment of Knowledge*, Cambridge University Press, Cambridge, 1980, pp. 58–71.

Schaffer, Simon, 'Newtonianism' in R. C. Olby, G. N. Cantor, J. R. R. Christie, and M. J. S. Hodge (eds), *Companion to the History of Modern Science*, Routledge, London, 1990, pp. 610–626.

Schechner, Sara J. *Comets: Popular Culture and the Birth of Modern Cosmology*, Princeton University Press, Princeton, 1997.

Shapin, Steven, 'The Image of the Man of Science', in R. Porter (ed.), *The Cambridge History of Science*, Vol. 4: *Eighteenth-Century Science*, Cambridge University Press, Cambridge, 2003, pp. 159–183.

Shapin, S and Schaffer, S, *Leviathan and the Air Pump: Hobbes, Boyle and the Experimental Life*, Princeton University Press, Princeton, 1985.

Shapin, Steven, *A Social History of Truth: Civility and Science in Seventeenth-Century England*, University of Chicago Press, London, 1994.

Simcock, A. V. *The Ashmolean Museum and Oxford Science, 1683–1983*, Museum of the History of Science, Oxford, 1984.

Simpson, A. D. C., 'Robert Hooke and Practical Optics: Technical Support at a Scientific Frontier', in M. Hunter and S. Schaffer (eds), *Robert Hooke: New Studies*, The Boydell Press, Woodbridge, 1989, pp. 33–61.

Stewart, Larry, *The Rise of Public Science: Rhetoric, Technology and Natural Philosophy in Newtonian Britain, 1660–1750*, Cambridge University Press, Cambridge, 1992.

Stewart, Larry, 'Global Pillage: Science, Commerce and Empire' in R. Porter (ed.), *The Cambridge History of Science*, Vol. 4: Eighteenth-Century Science, Cambridge University Press, Cambridge, 2008, pp. 825–844.

Struik, D. J. (ed), *A Source Book in Mathematics, 1200–1800*, Princeton University Press, Princeton, 1986.

Sutton, Geoffrey V, *Science for a Polite Society; Gender, Culture and the Demonstration of Enlightenment*, Westview Press, Oxford, 1995.

Stedman Jones, Joseph, *Moses Principia or Principia Mathematica: Hutchinsonianism and Natural Philosophy in Mid-eighteenth-century Oxford*, MSc Dissertation, Oxford University, Oxford, 2017.

Taton, Rene, 'Sur la diffusion des theories Newtoniennes en France: Clairaut et le problem de la figure de la terre', *Vistas in Astronomy*, Vol. 22, No. 2, 1978, pp. 485–509.

Teich, M. and Young, R, (eds), *Changing Perspectives in the History of Science: Essays in Honour of Joseph Needham*, Heinemann, London, 1973.

Terrall, Mary, *The Man Who Flattened the Earth: Maupertuis and the Sciences in the Enlightenment*, University of Chicago Press, London, 2002.

Terrall, Mary, 'Representing the Earth's Shape: The Polemics Surrounding Maupertuis's Expedition to Lapland', *Isis*, Vol. 83, No. 2, 1992, pp. 218–237.

Thomson, Thomas, *A History of the Royal Society*, London, 1812.

Toomer, G. J, *Ptolemy's Almagest*, Duckworth, London, 1984.

Trinder, Barrie, *Britain's Industrial Revolution: The Making of a Manufacturing People, 1700–1870*, Carnegie Publishing, Lancaster, 2013.

Urbach, Peter, *Francis Bacon's Philosophy of Science: An Account and a Reappraisal*, Open Court, Chicago, 1987.

Van Helden, Albert, *Measuring the Universe: Cosmic Dimensions from Aristarchus to Halley*, Chicago University Press, London, 1985.

Vickers, Brian (ed.), *English Science, Bacon to Newton*, Cambridge University Press, Cambridge, 1987.

Vickery, Amanda, *The Gentleman's Daughter: Women's Lives in Georgian England*, Yale University Press, London, 1998.

Voltaire (Arouet, F-M), *Lettres Philosophique*, Paris, 1733.

Voltaire (Arouet, F-M), *Letters from England*, London, 1734.

Westfall, Richard S., *Never at Rest: A Biography of Isaac Newton*, Cambridge University Press, Cambridge, 1981.

Williams, M. E. W, *Attempts to Measure Annual Stellar Parallax: Hooke to Bessel*, PhD Thesis, University of London, 1981.

Williams, M. E. W, 'Flamsteed's alleged measurement of annual parallax for the Pole Star', *Journal for the History of Astronomy*, Vol. 10, No. 2, 1979, pp. 102–116.

Willmoth, Frances (ed), *Flamsteed's Stars: New Perspectives on the Life and Work of the First Astronomer Royal*, London, 1997.

Wilson, Margaret D. *Ideas and Mechanism: Essays on Early Modern Philosophy*, Princeton University Press, Princeton, 1999.

Wise, M. Norton, *The Values of Precision*, Princeton University Press, Princeton, 1995.

Woolf, Harry, *The Transits of Venus: A Study of Eighteenth-Century Science*, Princeton University Press, Princeton, 1959.

Wordsworth, Christopher, *Scholæ Academicæ: Some Account of the Studies at the English Universities in the Eighteenth Century*, Cambridge University Press, Cambridge, 1877.

Yeomans, Donald K. *Comets: A Chronological History of Observation, Science, Myth and Folklore*, John Wiley, Chichester, 1991.

Picture Credits

James Bradley. Oil on canvas c.1742–1747, by Thomas Hudson
© The Royal Society

John Flamsteed. Oil painting 1712, by Thomas Gibson
© The Royal Society

Edmond Halley. Line engraving c.1720s, by George Vertue, after Richard Phillips
© The Royal Society

Sir Isaac Newton. Mezzotint 1712, by John Smith after Godfrey Kneller
© The Royal Society

Thomas Parker, 1st Early of Macclesfield. Line engraving 1722, by George Vertue after
    Godfrey Kneller
© National Portrait Gallery, London

Benjamin Hoadly. Line engraving 1709, by George Vertue
© National Portrait Gallery, London

George Graham. Mezzotint c.1740, by John Faber Jr after Thomas Hudson
©The Royal Society

George Parker 2nd Earl of Macclesfield. Mezzotint 1754, by John Faber Jr after Thomas
    Hudson
© National Portrait Gallery, London

Henry Pelham. Oil on canvas c.1752, by John Shackleton
© National Portrait Gallery, London

Tubeless telescope described by Huygens Illustration 1684, published in *Astroscopia Com-
    pendiaria*
ETH-Bibliothek Zürich/Science Photo Library

Hooke's chimney telescope. Diagram 1674, from Hooke's 'An Attempt to Prove the Motion
    of the Earth'
Royal Astronomical Society/Science Photo Library

Bradley's boxwood model.c. 1729, used by Bradley to demonstrate the aberration of light.
Science Museums Group

Bradley's zenith sector telescope. 12½-foot zenith sector constructed by George Graham in
    1727, Royal Observatory Greenwich
© National Maritime Museum, Greenwich, London

Bradley's quadrant. Eight-foot Quadrant constructed by John Bird in 1750, Royal Obser-
    vatory Greenwich
© National Maritime Museum, Greenwich, London

Bradley's transit instrument. Commissioned by Bradley in 1749 and constructed by John
    Sisson, Royal Observatory Greenwich
© National Maritime Museum, Greenwich, London

A watercolour, c. 1770, of the Royal Observatory, Greenwich and its meridian buildings from the southeast.
© National Maritime Museum, Greenwich, London

The Meridian Building, the Royal Observatory, Greenwich, originally Bradley's New Observatory.
© National Maritime Museum, Greenwich, London

Bradley's tomb. Holy Trinity, Minchinhampton, Stroud, Gloucestershire
Courtesy of Graham and Margaret Garnett

Inscription from the tomb (now displayed inside the church). Holy Trinity, Minchinhampton
Courtesy of Graham and Margaret Garnett

Translation of the inscription from Bradley's tomb. Translated by Rachel Chapman
Courtesy of Graham and Margaret Garnett

Bradley memorial stone. St Mary the Virgin, Wanstead, London
Courtesy of Pat Elliott

Bradley commemorative plaque. The Corner House, Grove Park, Wanstead, London
Courtesy of Barbara Elliott

Index